HANDBOOK OF GEOPHYSICAL EXPLORATION : SEISMIC EXPLORATION

Volume 40

Active Geophysical Monitoring

HANDBOOK OF GEOPHYSICAL EXPLORATION : SEISMIC EXPLORATION

Volume 40

Active Geophysical Monitoring

Edited by

J. Kasahara
V. Korneev
M.S. Zhdanov

ELSEVIER

Amsterdam • Boston • Heidelberg • London • New York • Oxford
Paris • San Diego • San Francisco • Singapore • Sydney • Tokyo

Elsevier
Radarweg 29, PO Box 211, 1000 AE Amsterdam, The Netherlands
Linacre House, Jordan Hill, Oxford OX2 8DP, UK

First Edition 2010

British Library Cataloguing in Publication Data
A catalogue record for this book is available from the British Library

Library of Congress Cataloging-in-Publication Data
A catalog record for this book is available from the Library of Congress

ISBN: 978-0-08-097484-2

For information on all Elsevier publications
visit our web site at elsevierdirect.com

Typeset by River Valley Technologies, UK

Contents

Part II. Theory and Technology of Active Monitoring

Technology of Active Monitoring

Signal Processing and Accuracy Control in Active Monitoring

PREFACE

This is the first book devoted to the variety of topics related to active geophysical monitoring, which summarizes up-to-date information on this subject. Active geophysical monitoring involves the detection and interpretation of subsurface changes in rock states and properties induced by acoustic and electromagnetic waves periodically excited by repeated controlled sources. There is a little doubt that active monitoring of time-evolving structures and states in the tectonically active lithosphere of the Earth is very important. However, no systematic analysis of continuous geophysical monitoring has been done until now because no reliable technological systems have been developed other than those generating impulsive signals (e.g., chemical explosives, air guns, and seismic vibrators). Extensive technical and theoretical efforts have been expanded over the last few decades to develop continuously running sources (e.g., using electromagnetic controlled sources), though those efforts have not generally received much international attention.

However, recent technological advances and growing social concern about public safety and potential environmental hazards have triggered the systematic strategic development and application of new active monitoring methods. This new interest in active monitoring has been noted in extensive work done by Russian scientists since 1976, by Japanese scientists since 1994, and by the U.S. scientists since 1987. Five key ingredients in active monitoring are: (1) the use of accurately controlled sinusoidal signals for longer periods of time, enabling the effective stacking of data, which in turn leads to higher signal-to-noise ratios; (2) the development of the powerful sources of electromagnetic fields (e.g., magnetohydrodynamic – MHD - generators); (3) more accurate Global Positioning System (GPS) and time-recording technology; (4) stable, more accurate signal transmission and data acquisition hardware, and (5) greater computational power for data management and signal processing. The critical component of effective active monitoring is an automated and user-friendly data acquisition system with remote access and data transmission. The potential of this new technology is considerable and could result in active monitoring methods contributing qualitatively new and important information to solid-earth geophysics. This technology could be used for monitoring and predicting seismic activity, and (eventually) for disaster mitigation and prevention. Such goals require extensive international and interdisciplinary cooperation.

At the meeting of the International Union of Geodesy and Geophysics (IUGG) held June 30-July 11, 2003, in Sapporo, Japan, the participants in the workshop entitled Tectonophysics and Seismic Imaging of the

Lithosphere (under International Association of the Seismology and Physics of the Earth's Interior [IASPEI]) organized a new task group named Active Monitoring to promote cooperative actions regarding active monitoring science and technology among IASPEI, the International Association of Geomagnetism and Aeronomy (IAGA), and the International Association of Volcanology and Chemistry of the Earth's Interior (IAVCEI). In October 2003, responding to this decision, Hiromichi Higashihara of the Earthquake Research Institute, University of Tokyo (now the Earthquake Disaster Mitigation Research Center, National Institute of Earth Science Disaster Prevention) organized the Japanese Consortium of Earth's Active Monitoring (JCEAM). As a kick-off meeting for Active Geophysics the first international workshop on Active Monitoring in solid earth Geophysics was held June 30–July 2, 2004, in Mizunami, Japan, as the ISAPEI workshop. More than 100 scientists and engineers attended from Japan, Russia, USA, China, India, Uzbekistan, Turkey, and Argentina. Special sessions on active monitoring in geophysics have become a regular part of the Fall AGU meetings in San Francisco since 2004.

In Russia several artificial seismic and electromagnetic sources have been developed since the late 1970s. Large seismic vibrators have been used there to study the static and dynamic structures of the earth. These vibrators have a twin eccentric-source rotary-transmitter design, which generates a sweeping sinusoidal force of ~1000 kN. Radiation sessions with this transmitter have usually lasted for about an hour and could be detected at a distance of ~500 km. Field experiments using seismic vibrators have been extensively conducted in Russia. The experiments in the Baikal, Altai-Sayan, and Okhotsk-Chukotski regions have shown that reflected and refracted arrivals from the deep continental crust can be recorded at a distance of 500 km from the source. Also a number of field experiments with the powerful MHD generators have been conducted in different regions of Russia to study the earth's crust and lithosphere since the 1970s.

In Japan a so-called Accurately Controlled and Routinely Operated Signal System (ACROSS) has been in existence since 1995, with the objective of detecting subtle changes in the earth's structure. With the understanding that the application of a large force to the ground may destroy seismic source signatures, ACROSS has been designed to minimize damage to the ground. The seismic ACROSS single-eccentric source transmitter generates a continuous repetition of FM and/or AM signals up to ~200 kN. This system provides a transfer function (in tensor form) that includes information on polarization anisotropy. A variety of monitoring methods/technologies are linked with these transmitters. They included a low-friction bearing for rotary machines, synchronization of signals to universal time, optimum design of the signal mode, a more stable method of operation, an array design for both transmitters and sensors, and various modes of target-oriented data processing.

In the United States active monitoring of the San Andreas Fault occurred from 1987 to 1996, using the Vibroseis source. The observed time variations correlated strongly with seismicity. In Japan, a similar kind of monitoring is taking place in the Awaji and Tono regions. The acquired data reveal temporal variations that correlate with other geophysical measurements. Another ongoing experiment in the United States monitors stress-induced changes in seismic velocity.

We present the results of strategic systematic development and application of new technologies for active geophysical monitoring in this collective monograph. The authors demonstrate that active monitoring may drastically change solid-earth geophysics through the acquisition of substantially new information based on high accuracy and real-time observations. Active monitoring could also provide new means for disaster mitigation, in conjunction with substantial international and interdisciplinary cooperation.

Sections in this book outline the general concepts of active geophysical monitoring with powerful seismic vibrators and MHD generators, and provide the historical background for previous studies of seismically active zones in Russia, USA, and Japan. A significant part of the book is dedicated to the exposition of the theory and technology of active monitoring, including signal processing, data analysis, novel approaches to numerical modeling, and interpretation. The different authors—who constitute the world leaders in active geophysical monitoring—discuss case histories and present the results of regional active monitoring experiments in Russia, Japan, and the United States.

This book was partially sponsored by the Earth Science Division at Ernest Orlando Lawrence National Laboratory, USA. Dan Hawkes was the English editor of eight articles.

Junzo Kasahara, Valeri Korneev and Michael S. Zhdanov
Editors
Japan, USA
June 30, 2009

HANDBOOK OF GEOPHYSICAL EXPLORATION

SECTION I. SEISMIC EXPLORATION

Editors: Klaus Helbig and Sven Treitel

Volume

1. Basic Theory in Reflection Seismology
2. Seismic Instrumentation, 2nd Edition
3. Seismic Field Techniques

4A. Seismic Inversion and Deconvolution: Classical Methods
4B. Seismic Inversion and Deconvolution: Dual-Sensor Technology

5. Seismic Migration (Theory and Practice)
6. Seismic Velocity Analysis
7. Seismic Noise Attenuation
8. Structural Interpretation
9. Seismic Stratigraphy
10. Production Seismology
11. 3-D Seismic Exploration
12. Seismic Resolution
13. Refraction Seismics
14. Vertical Seismic Profiling: Principles
3rd Updated and Revised Edition

15A. Seismic Shear Waves: Theory
15B. Seismic Shear Waves: Applications
16A. Seismic Coal Exploration: Surface Methods
16B. Seismic Coal Exploration: In-Seam Seismics

17. Mathematical Aspects of Seismology
18. Physical Properties of Rocks
19. Shallow High-Resolution Reflection Seismics
20. Pattern Recognition and Image Processing
21. Supercomputers in Seismic Exploration
22. Foundations of Anisotropy for Exploration Seismics
23. Seismic Tomography
24. Borehole Acoustics
25. High Frequency Crosswell Seismic Profiling
26. Applications of Anisotropy in Vertical Seismic Profiling
27. Seismic Multiple Elimination Techniques
28. Wavelet Transforms and Their Applications to Seismic Data
Acquisition, Compression, Processing and Interpretation
29. Seismic Signatures and Analysis of Reflection Data in Anisotropic Media

30. Computational Neural Networks for Geophysical Data Processing
31. Wave Fields in Real Media: Wave Propagation in Anitsotropic, Anelastic and Porous Media
32. Nuclear Magnetic Resonance Petrophysical and Logging Applications
33. Seismic Amplitude Inversion in Reflection Tomography
34. Seismic Waves and Rays in Elastic Wave Media
35. Seismic While Drilling: Fundamentals of Drill-Bit Seismic for Exploration
36. Information-based Inversion and Processing with Applications
37. Seismic Stratigraphy, Basin Analysis and Reservoir Characterisation
38. Wave Fields in Real Media: Wave Propagation in Anisotropic, Anelastic Porous and Electromagnetic Media
(Second Edition, Revised and Extended)
39. Coding and Decoding: Seismic Data
The Concept of Multishooting
40. Active Geophysical Monitoring

PART I

GENERAL CONCEPT AND HISTORICAL REVIEW

GENERAL CONCEPT OF ACTIVE GEOPHYSICAL MONITORING

ACTIVE MONITORING TARGETS

General Concept of Active Geophysical Monitoring

CHAPTER 1

Seismicity Precursors for Active Monitoring of Earthquakes

Valeri A. Korneev[a]

Contents

1. Introduction

The town of Parkfield, located on the San Andreas Fault (SAF) in central California, has been the site of intensive, multidisciplinary earthquake studies since the 1970s. Moderate-sized earthquakes of about magnitude 6 (M6.0) have occurred on the Parkfield section of the SAF at fairly regular intervals—in 1857, 1881, 1901, 1922, 1934, and 1966 (Bakun and McEvilly, 1979). The 1857 event was a foreshock of the great Fort Tejon M7.9 earthquake, which produced a rupture along the fault at least 290 km in length from Parkfield to the southeast (Meltzner and Wald, 1999)—and the probability that another moderate-sized Parkfield earthquake might occur as a foreshock to another Fort Tejon-type event remains high.

The goal of research in the Parkfield area has been to observe the fault and surrounding crust, at close range and at high resolution before, during,

[a] Lawrence Berkeley National Laboratory, Berkeley, California, United States

E-mail address: VAKorneev@lbl.gov.

Handbook of Geophysical Exploration: Seismic Exploration, Volume 40 2010 Published by Elsevier Ltd.
ISSN 0950-1401, DOI: 10.1016/S0950-1401(10)04006-1

and after a characteristic M6 earthquake, so as to better understand the earthquake process and to provide a scientific basis for earthquake prediction and hazard assessment. Recognizing this hazard, and the regular periodicity of recurring events near Parkfield, the U.S. Geological Survey (USGS) and the State of California began a comprehensive, long-term Parkfield Earthquake Prediction Project in 1985 (Bakun and Lindh, 1985). More than 10,000 earthquakes have been recorded since 1970 in the magnitude range $0 < M < 5$. The long-anticipated M6.0 event finally occurred on September 28, 2004. Langbein et al. (2005) issued a preliminary report indicating that no immediate precursory phenomena were observed, which was confirmed by Bakun et al. (2005).

There is currently little optimism in the scientific community about the possibility of earthquake prediction (Geller, 1997; http://earthquake.usgs.govhazards/prediction.html; Geller et al., 1996). Recent discussions in *Nature* (http://www.nature.com/nature/debates/earthquake/index.html) include such statements as: "W*e do not have a method for making short-term predictions*"; "*There is a bleak future for individual earthquake prediction*"; and "T*here is no prospect of deterministic earthquake prediction in the foreseeable future*". While it is not the intention of this paper to give an overview of current earthquake prediction methods, the author notes that most methods (e.g., Bowman and Sammis (2004)) seek changes in coefficients of the Gutenberg-Richter relationship (GRR) $\log N = a - bM$, which relates the number of earthquakes (N) greater than magnitude M in some region to the magnitude itself. The GRR reflects the behavior of seismicity over periods of time sufficiently long enough to collect reliable statistics for a wide range of magnitudes.

However, using the GRR as a basis for prediction methods has several disadvantages. First, the relatively rare occurrence of large magnitude events means that there is great uncertainty in the predicted probability of a large event. Second, applications of this relationship provide no information concerning the location of an event within a cataloged region. Third, and finally, current earthquake-generating models show little or no direct relationship between changes in the GRR coefficients (a and b) and characteristic earthquake occurrences. The catastrophic events similar to M7.9 Fort Tejon (Langbein et al., 2005) and M7.7 1906 San Francisco have an average slip of about 4 m (Wald et al., 1993), which translates to an average recurrence time interval of every 100-200 years. This evaluation assumes approximately 2-3 cm/yr of average tectonic plate displacement by the SAF and also accounts for some incomplete release of accumulated strain for those events. With catastrophic events occurring so rarely, even moderate uncertainty in prediction makes it unrealistic to use GRR-derived statistics for disaster-related warnings. Moreover, predictions expressed in terms of probabilities are inappropriate for rare earthquake occurrences, since definitions of probability are based on statistical limits of multiply occurring events. Practically applicable prediction methods need to be based on causal approaches.

In this paper, a selective seismicity analysis is used in which only events having a direct relationship with strain-buildup processes are included. The idea of selectiveness is partially based on the results of a Vibroseis monitoring experiment, in which seismic waves repeatedly illuminated the epicentral region of the expected M6 event at Parkfield from June 1987 to November 1996. Data collected by the borehole network were examined for evidence of changes associated with the nucleation process of the anticipated M6 earthquake at Parkfield (Karageorgi et al., 1992, 1997; Korneev et al., 2000; Korneev and Nadeau, 2004). These investigations reported significant travel-time changes for paths crossing the fault zone in the locked southeast part of the SAF while in the northwest (creeping) part of the SAF, no changes were observed. This result suggests that little or no information about stress accumulation in the SAF can be gathered from the seismicity of the SAF's creeping part, where a weak fault steadily releases small stress changes and the seismicity mostly represents a stationary random process. Indeed, the weak creeping faults can be modeled as large-scale fractures having very low friction and the capability to immediately discharge any applied shear stresses. The stress-strain conditions on both sides of such fractures generally stay unchanged, with just small fluctuations and no dependence on the regional stress buildup. At the same time, the seismicity associated with weak faults makes a dominant contribution to regional event statistics, overshadowing the seismicity directly related to regional stress accumulation. Therefore, all events with hypocenters within the active fault zone are excluded from the results shown in this paper. The transition zone between the locked and creeping parts of the SAF is a northwesterly dipping structure, oriented at approximately 45° and extending for about 5 km along the fault (Korneev et al., 2003, Figure 8).

Not all earthquakes can be recorded by a seismic recording network. Typically, if most stations within the network detect an event within the same interval of several seconds, the network is triggered and the event is recorded, located, and made available in catalogs. For the Parkfield area, all events above magnitude 1.5 are likely recorded, as reflected in the statistics of regional seismicity, providing a good fit to the GRR (Figure 1). However, not all smaller-magnitude events are detected, because they have low amplitudes relative to seismic noise. Although this suggests that recorded events of magnitude <1.5 cannot be used for GRR statistics, such events nonetheless can give rise to a strong precursory signature (Korneev, 2005; Artamonova and Korneev, 2005), as demonstrated below.

2. Seismicity Precursors

2.1. Parkfield M6, 2004

The U.S. Geological Survey catalog was initially used to analyze the spatial and temporal distribution of Parkfield-area events from 1968 to the main

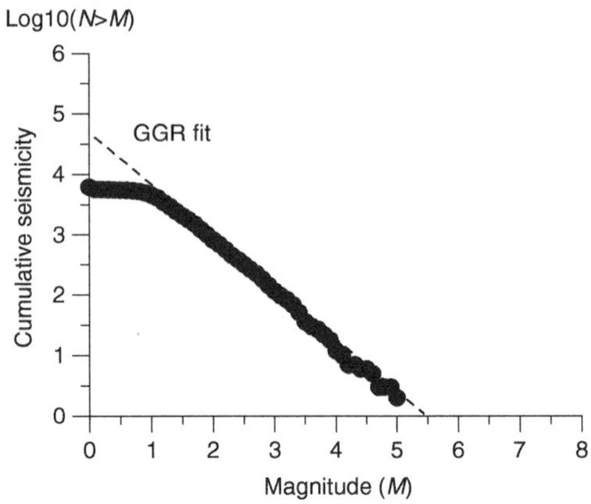

Figure 1 Cumulative seismicity vs. magnitude for Parkfield area using all events from 1968 till M6 2004 main shock on September 28. Straight dashed line shows GRR fit. Deviation from GRR is visible for magnitudes less than 1.5.

shock of the M6 2004, which occurred on September 28 (Figure 2). In March 2006, the data were reprocessed after incorporating events relocated by a double-differences (DD) method (Thurber et al., 2006), bringing minor corrections to the results. Nineteen out of 56,304 events were discarded as having rms values exceeding 1 second, ensuring that the uncertainty of event locations would not be larger than 3 km. The total number of events occurring per month was computed and analyzed. During this 36-year period of observation, four distinct peaks of seismicity are visible in the data (Figure 3). The first three peaks correspond to the aftershock series following the M5.5 1975 Parkfield, M6.5 1983 Coalinga, and a series of four M4 1992-1994 Parkfield events. The final rise in seismicity begins in 2000, attains its peak in December 2003, and then falls to below the average level before the M6 2004 event. To eliminate the influence of aftershock and creeping seismicity, all seismic events within a 6 km and outside of 15 km corridors around the central SAF zone were excluded from the data observed in the 35 km × 50 km area around the epicenter (as shown in Figure 4). The resulting seismicity is shown in Figure 5a and b. Except for the two sharp peaks in 1970 (M3.9 Parkfield aftershocks) and 1983 (M6.5 Coalinga aftershocks), the only other increase in seismicity begins in the middle of 2002 and reaches its maximum in May 2004. There are visible cyclic bursts of seismicity that occur at decreasing intervals. Expansion of the analysis area beyond the chosen size produces the same effect, but the results become increasingly contaminated by aftershock events of the M6.5 1983 Coalinga and M6.5 2003 San Simeon earthquakes. Computations show that the observed pre-event peaks are not very sensitive to the elimination corridor width until it decreases to about 3 km, at which time the creeping SAF events become statistically dominant.

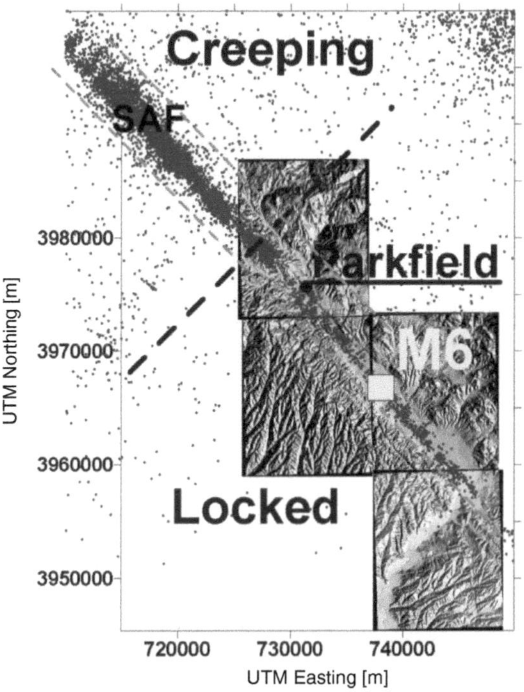

Figure 2 Seismicity of the SAF in the Parkfield area during the 36 years before the M6.0 September 28, 2004 earthquake. The topographic map of the Parkfield area is partially shown. Dots are the event epicenters. Dashed thin lines are the bounds of an excluded corridor around the SAF. The dashed thick line is a delineation zone (Korneev et al., 2003) between the locked and creeping parts of the SAF. The square marks the epicenter of the M6.0 event.

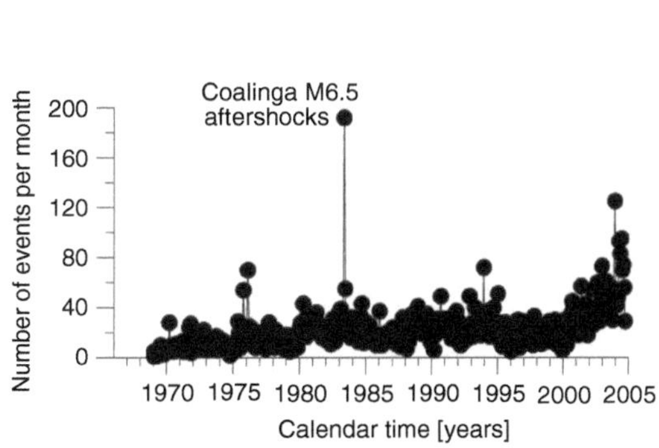

Figure 3 Seismicity of the SAF in the Parkfield area (shown on Figure 2) during the 36 years before the M6.0 September 28, 2004 earthquake. Note the steady seismicity rise starting in 2000. Visible seismicity peaks correspond to the aftershock series following the M5.5 1975 Parkfield, M6.5 1983 Coalinga, and four M4 1992-1994 Parkfield events.

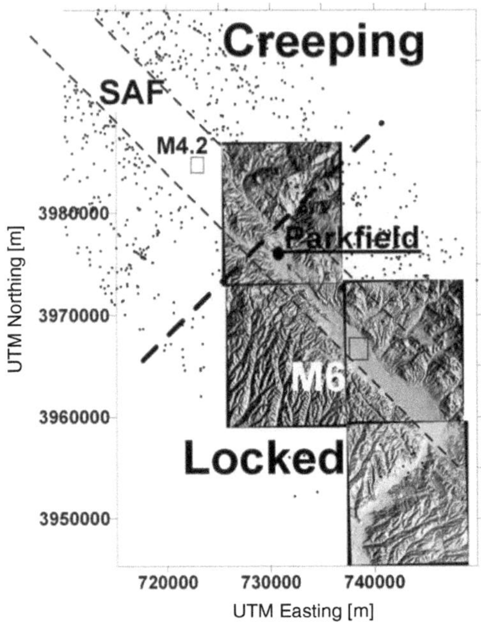

Figure 4 Seismicity of the SAF in the Parkfield area from 1968 until the M6.0 September 28, 2004 earthquake which was used in the analysis. Fault zone events within 6 km and outside of 15 km corridor around the SAF trace are excluded. Partially shown is the topographic map of the Parkfield area. Dots are the event epicenters. Dashed thin lines are the bounds of the excluded corridor around the SAF. Squares mark the epicenters of the M6.0 and M4.2 events.

Further elimination of all events with magnitudes greater than 1 produces even more distinctive results (Figure 5c, d). The peak in such "microseimicity" occurs about 6 months prior to the main M6 2004 event and is 8 times greater than the background level of roughly two recorded events per 10 days. Following the peak, there is a steady decrease in activity up to the time of the main event. Prior to the peak, no distinctive features can be seen in this microseismicity. In the year preceding the main event, the epicenters of the microseismicity are mostly concentrated along the delineation zone between the creeping and locked parts of the SAF (Figure 6). Also during this period, an area approximately 30 km in diameter surrounding the future M6 2004 epicenter contains no events. This no-activity area lies mostly on the southwest side of the SAF.

To better understand the spatial and temporal characteristics of the observed peaks, we conducted a seismicity count, using a scan-stripe oriented in the southwest-northeast direction (Figure 7) and crossing the SAF along the creeping-locked delineation line shown in Figure 2.

This scan-stripe extends up to 80 km offset from the SAF in the southwest direction, where it crosses the epicentral region of the M6.5 San

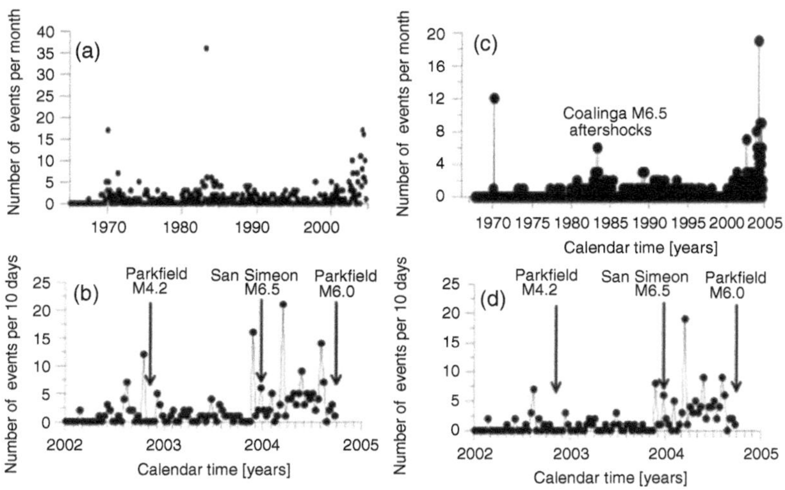

Figure 5 Average number of events per month in the Parkfield area before the M6.0 September 28, 2004 earthquake. (a) All events starting from 1968. The 1970 and 1983 spikes correspond to postseismic aftershocks of the M4 SAF and M6.5 Coalinga events. (b) Average number of events per 10 days starting from 2002. Visible are cyclic bursts of seismicity with decreasing intervals between peaks as the time approaches the earthquake. (c) Small ($M < 1$) magnitude events of the same series as on a). (d) Average number of $M < 1$ events per 10 days starting from 2002. The seismicity peak is reached 6 months before the earthquake followed by steady decrease.

Simeon 2003 event; on the other side, it has a 60 km offset in the southwest direction, where it crosses the epicentral region of the M6.5 Coalinga 1983 event. The seismicity history starting from 1968 to 2005 for the 4 km by 20 km rectangle was computed for a 4 km interval in the southeast-northwest direction, ensuring that no event is counted more than once. The results are shown in Figure 8. The distinctive strong burst of out-of-fault seismicity precedes the M6.0 Parkfield earthquake by several months (upper circled areas).

2.2. Loma Prieta M7.0 1989

A similar analysis was applied to the events preceding the M7.0 1989 Loma Prieta earthquake, which caused substantial damage in the San Francisco Bay Area region. The area to the west of the epicenter was chosen for the seismicity study because it does not contain as many active faults as other areas adjacent to the epicenter. Figure 9a shows the seismicity history for the 25 years of observation before the event. Up to two months prior to the event (Figure 9c and d), seismicity increased to approximately eight times the base level of about 6 events per month, and then decreased over the following two months. Analysis of the low magnitude (<1) seismicity yielded the same trend, although there were not enough events for statistically significant

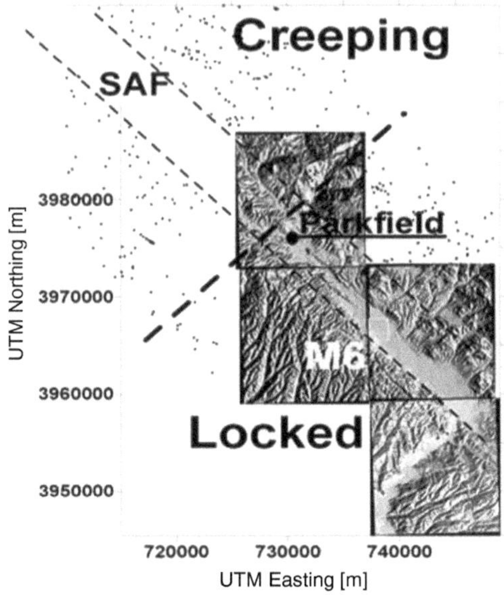

Figure 6 Seismicity of the SAF in the Parkfield area during 1 year before M6.0 September 28, 2004 earthquake. Fault zone events within 6 km and outside of 15 km corridors around the SAF trace are excluded. Note the "quiet" zone around the locked part. Most of the pre-event seismicity takes place in the vicinity of the creeping- locked delineation zone.

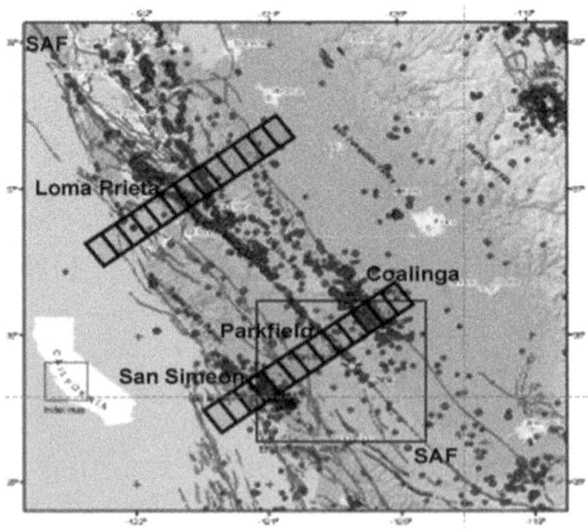

Figure 7 Central California topographic map (www.usgs.gov) and its seismicity. Two scan-stripes crossing the SAF in the Parkfield and Loma Prieta areas were used for computations of seismicity history shown correspondingly on Figures 8 and 10. The Parkfield scan-stripe crosses the epicentral regions of M6.5 San Simeon 2003 on the southwest flank and of M6.5 Coalinga 1983 on the northeast flank. The center of this scan stripe crosses the Parkfield area along the delineation zone shown on Figure 2 by a dashed blue line. The Loma Prieta scan-stripe crosses the Calaveras fault on the northeast flank.

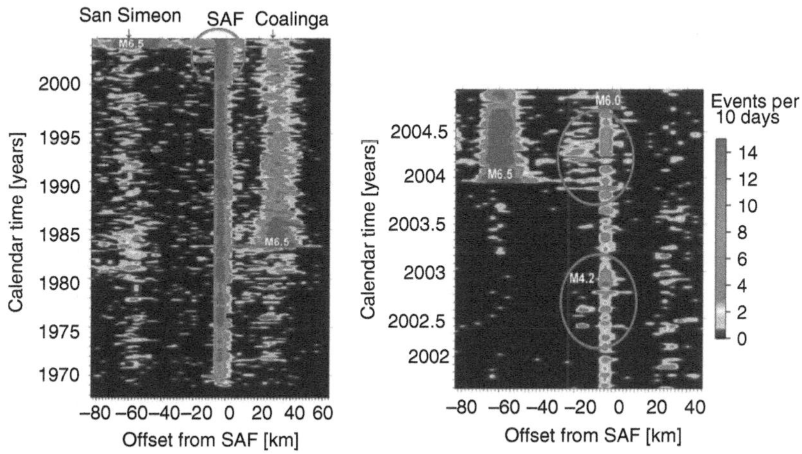

Figure 8 Seismicity history for 4 km by 20 km rectangular scanning along the scan-stripe from Figure 7 in southeast-northwest direction. The longest side of the rectangular is parallel to SAF. The offset has values in −80 km-60 km range and measured from the SAF with 4 km interval, ensuring that no event is counted more than once. (a) Data for the 1968-2005 interval. Visible are bursts of seismicity correspondent to M6.5 2003 San Simeon (−60 km offset) and 1983 M6.5 Coalinga (37 km offset). The circles area shows the only in 38 years of observation burst of out-of- fault seismicity preceding M6.0 Parkfield earthquake. (b) Blow-up of the last three years from (a). Upper circle contains seismicity preceding M6 Parkfield event. The lower circle shows a similar seismicity pattern for M4.2 Parkfield 2002 event. In both cases the rise of seismicity starts 5-7 months before the main event at about 10-15 km offsets from SAF, and gradually moving closer (5 km offset) to the SAF couple months before the earthquake. A color version of this figure can be found in Korneev (2006).

results. Similarly to the Parkfield case, in the last year before the earthquake, a low-seismicity area appeared around the future rupture (Figure 7b). Existence of this area was shown in Reasenberg and Simpson (1992) after comparing long periods of seismicity before and after the earthquake.

Space-temporal seismicity analysis (Figure 10) for the M7 Loma Prieta area was done in the same manner as for the Parkfield M6 event (Figure 8), with the geometry of the scan-stripe shown in the upper part of Figure 7. The offset of scanning for the 4 km by 20 km rectangle has values in the -45-80 km range, measured from the SAF at 4 km intervals. Seismicity growth preceding the M7 Loma Prieta event is shown in the circle. The M7 event was preceded by two (M5.3 1988 and M5.4 1989) Lake Elsman events. Note that while M5.3 1988 showed only a slight rise in seismicity (Figure 7d); M5.4 1989 was preceded by distinctive seismicity outbursts contributing to the pre- M7 peak. This pre-event seismicity pattern is similar to the pattern observed for the M6 Parkfield earthquake in Figure 8. Note that the Lake Elsman events do not belong to the creeping sections of the faults, and therefore they are incorporated in the seismicity count.

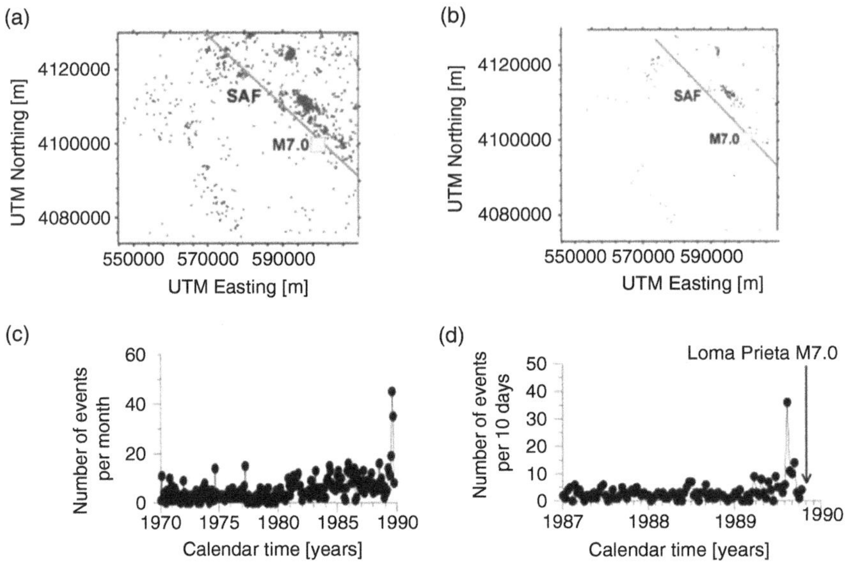

Figure 9 Seismicity of the SAF in the westerly Loma Prieta area before the M7.0 October 17 earthquake. (a) Dots are the event epicenters. The solid straight line indicates the SAF trace. (b) Same as on previous panel when just the events of the last year are plotted. Note the "quiet" zone southwest of the SAF, which is similar to that on Figure 6. (c) Average number of earthquakes per month starting from 1968. (d) Average number of earthquakes per 10 days starting from 1987. The seismicity peak is reached 2 months before the earthquake, followed by a steady decrease.

2.3. Seismic noise

According to the GRR, seismicity should exponentially increase in lower magnitudes. However, current instrumentation capabilities do not allow robust detection and location of all events, usually limiting the lowest detectable magnitude to 0. Numerous events with negative magnitudes therefore stay below the seismic-station-network resolution. The typical seismic station network has an average spacing of about 10 km and operates in a trigger fashion: when a certain threshold number of stations record an event, it counts as a triggered event and gets stored in a database. While all large-amplitude events trigger the network, events with small amplitude have less chance to be recorded because of the lower signal-to-noise ratio. Also, small-magnitude events have higher cyclic frequency content, and therefore their waves are more attenuated. As a result, the number of detected "very small" magnitude events is lower compared to the seismicity of high-magnitude events, which looks like a violation of the GRR and cannot be used for a- and b- constants evaluation. For example, Figure 1 shows the cumulative seismicity recorded before the M6 2004 Parkfield event for the area from Figure 2. Violation of the GRR is

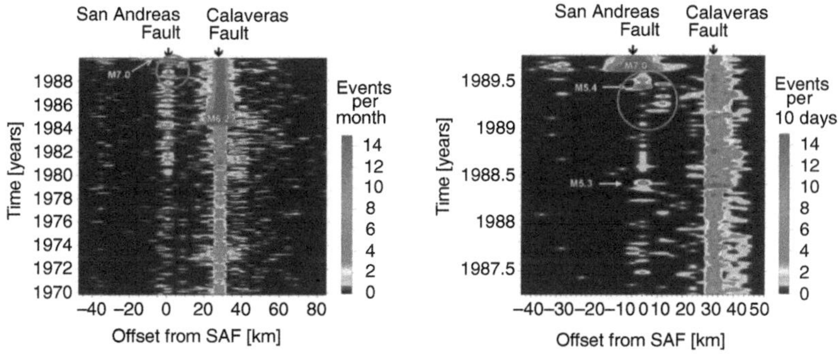

Figure 10 Seismicity history for 4 km by 20 km rectangular scan in the southeast-northwest direction across the M7 Loma Prieta epicenter: (a) 1970-1990 time interval; (b) Blow-up of the last three years from (a). The geometry of the scan-stripe is shown on Figure 7. The longest side of the rectangular is parallel to the SAF. The offset has values in −45 km-80 km range and measured from the SAF with 4 km intervals, ensuring that no event is counted more then once. The red circle contains seismicity growth preceding the M7 Loma Prieta event. Note that the M5.3 1988 event gave just a slight rise of seismicity (Figure 9d), while the M5.4 1989 event was preceded by distinctive seismicity outbursts, which together with this event gave the pre- M7 peak. This pre-event seismicity pattern is similar to the pattern observed for the M6 Parkfield earthquake from Figure 8. A color version of this figure can be found in Korneev (2006).

visible for the magnitudes below 1.5. Thus, the direct application of the GRR and monitoring of its constants is restricted by poor statistics for the rarely occurring large-magnitude events, and by resolution limitations in the detection of small magnitude events. To increase the resolution, we need much denser networks, with stations located in boreholes (as exist now for HRSN). But this is currently rather expensive.

As a different approach, seismicity monitoring can be based on the statistical connection between events of different magnitudes (given by the GRR), which leads to a hypothesis about the direct correspondence between seismic noise level and seismicity for magnitudes falling in the detectable region ($M > -1$). Seismicity changes for detectable (rare) events are likely to be accompanied by similar changes for undetectable micro-events that comprise background seismic noise. Small magnitude seismicity ($M < -1$) has higher frequencies and therefore a local character, due to the high attenuation for these frequencies. This hypothesis was tested using Parkfield data for the MMNB borehole station of the HRSN, recorded from the same micro-earthquake cluster over a 10-year interval. This station was chosen because it is located in the vicinity of the SAF locked-creeping transition zone, where in 1993-1994 a series of M4 events occurred and thus likely produced local stress changes. Noise records were taken from the initial 1.5-second intervals of traces that precede the first arrival, and the average noise amplitude was computed in the 80-100 Hz range. Assuming a 3D

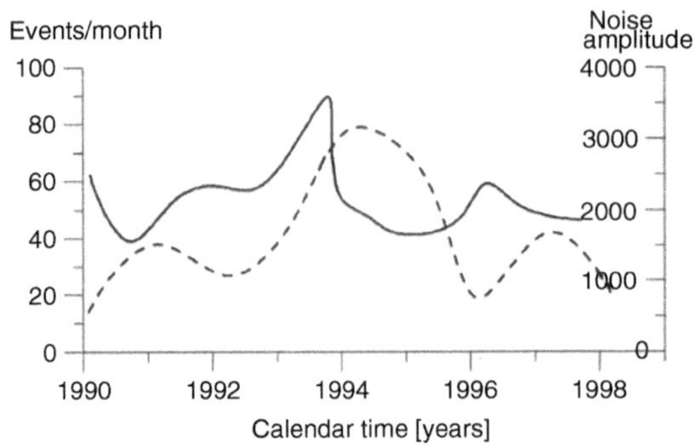

Figure 11 Seismic noise at 80 Hz (solid line) and released seismic energy (dashed line) for the MMNB station at the SAF. Energy is computed within 5 km radius from the MMNB location using a 1 year averaging window. The seismic noise peak occurs about 6 months before the energy peak, which is related to the 1993-1994 M4 series at Parkfield. Note the repeat of this pattern with peak of noise at the middle of 1996 and the subsequent peak in energy in the early 1997.

distribution of seismic noise sources, the maximum contribution distance r can be evaluated from the formula

$$r = \frac{Qv}{2\pi f}, \tag{1}$$

where Q is a quality factor, v is velocity, and f is frequency. For $f = 100$ Hz, $Q = 200$, $v = 3$ km/s, the radius is approximately equal to 1 km. Thus, at high frequencies, noise measurements cover volumes of just a few kilometers in extent and have a local character. Figure 11 shows a comparison of noise amplitudes with seismicity in the 5 km vicinity of the MMNB station at Parkfield. As seen from Figure 11, the noise energy rises by about two times several months before the seismicity peak.

3. Discussion

3.1. Similarity of precursor patterns

The out-of-fault-zone microseismicity patterns of the two events studied here are quite similar, consisting of a sharp seismicity increase that reaches its maximum several months prior to the main event, and then decreases to background seismicity levels by the time the main event occurs. The pre-Parkfield M6 seismicity peaks form a unique pattern for 38 years of observation time, and they occur 2-6 months before the earthquake. This microseismicity clusters at a 15-20 km offset from the fault and then migrates closer to the SAF, peaking two months before the main event at a 5-10 km offset. Note that an off-fault seismicity increase is usually followed by an in-fault seismicity rise, which likely indicates the accelerated fault creep, which reduces stress in the surrounding crust. Both events were preceded by very similar precursory seismicity patterns. There is an indication of some crosstalk between the 2003 San Simeon event and the pre-M6 Parkfield seismicity peak, although they are separated by 3-4 months. The shape of the San Simeon aftershock series nicely reveals the Omori law decay. There is also a visible increase in seismic activity within the SAF 1-2 months before the San Simeon earthquake. This implies that the pre-event seismicity rise in Parkfield is unlikely to be a part of the San Simeon aftershock series, and rather represents a different process, although some weak stress interaction between the Parkfield and San Simeon regions might be possible. Note that the pre-M4.2 2002 seismicity peak in the SAF area occurs 4 months before the event and has a pattern similar to that of the M6 event (Figure 5b). This indicates that the M4.2 2002 Parkfield earthquake is likely a foreshock of the M6 2004 event. For the Loma Prieta earthquake, a similar pattern is observed: Seismic activity starts at a 15 km offset from the SAF 5 months before the event, and then moves closer to the SAF, peaking 2 months before at a 5 km offset from the SAF. This similarity suggests the scalability of the observed precursory seismicity phenomena to events of different magnitudes.

Seismicity peaks occurring in 1970 and in 1983 observed for the Parkfield area have distinctly different spatial and temporal characteristics in comparison with the pre- M6 event seismicity. The 1983 peak is caused by aftershocks of the M6.5 Coalinga earthquake and composed of the events located on the northeast side of the SAF. The 1970 swarm occurred within a four-hour interval on February 23, 1970, in the compact 4 sq miles area at Cholame Hills on the southwest side of the SAF. Development of the observed precursors was at least several months long, had a cyclic character, and was observed on both sides of the SAF (Figure 8).

3.2. Precursor mechanism

The author proposes an explanation for the observed seismicity precursors—with the understanding that such an explanation requires further, intensive study.

Increased seismicity in rock under increasing stress is a well-known laboratory observation[1] (e.g., Lei et al., 2000a,b; Lockner et al., 1992). Under increasing shear strain, the seismicity of an initially intact rock sample grows throughout the so-called strain-hardening regime. This seismicity

[1] Event count in laboratory rock-testing experiments is usually called *acoustic emission*. In seismology, the term *seismicity* is traditionally used for the correspondent phenomena. As follows from the results of this paper, both terms are equally suitable in describing of the observed precursory increase of micro-event frequency.

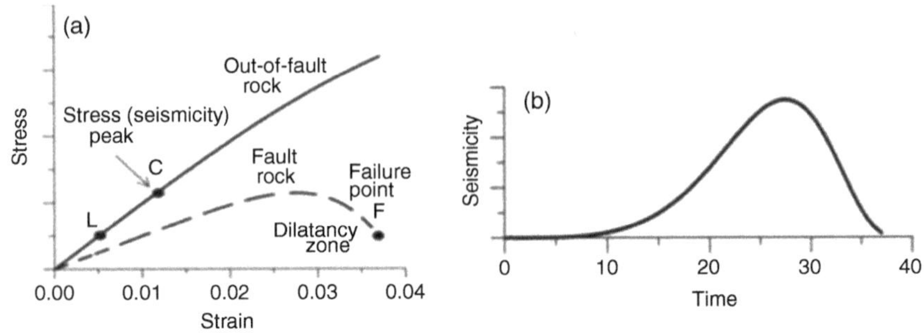

Figure 12 (a) Stress-strain curves for weak fault rock (dashed curve) and stronger out of fault rock (solid curve). Stress changes in time for both rocks are the same. For an increasing strain, the stress in the fault material goes nonlinear and reaches a maximum, entering a dilatancy zone which is characterized by development of multiple fractures and a reduction of rock stiffness, eventually progressing to the failure point F. At the same time the stronger out-of-fault rock experiences linear changes reaching a maximum strain at point C and decreasing to the point L at the failure time. (b) Seismicity history computed from stress history (shown on upper panel) using a fourth power law (Dunegan, 1996) gives a qualitative explanation for the observed precursory peaks.

reaches a peak, after which the seismicity rate drops as the rock sample enters a "softening" stage associated with strain localization and failure along a band of accumulated damage (Lei et al., 2004; Sobolev et al., 1996). Note that the constant stress load experiments do not produce a visible decrease in seismicity before rock failure, whereas constant strain-rate-load experiments produce such peaks (as supported by laboratory data (Lei et al., 2005) and numerical modeling (Tham et al., 2001)). In the presence of pre-existing faults in nature (like the SAF), a slightly different scenario is plausible. It is well known that most earthquakes occur on existing faults and that fault-zone rocks have less strength than the surrounding crust. For an increasing strain load, the stress-strain relationships start behaving nonlinearly (Sholz, 1990) and, after reaching a maximum stress value, enter a softening (or dilatancy) regime, characterized by the development of multiple fractures and reduction of rock stiffness. This process eventually progresses to rock failure (earthquake). During this process, the stronger out-of-fault rock experiences the same stress load, but does not reach a nonlinear regime (Figure 12).

The fault's stress-strain curve maximum provides a corresponding maximum for the stress in the surrounding crust, which in turn creates a peak in seismicity; according to the empirical fourth-power relationship between stress and seismicity (Dunegan, 1996). The fault-zone rock occupies relatively small volumes, because it is aligned to fault planes having widths of several hundred meters or less. At the same time, the out-of-fault rocks have much larger volumes, by virtue of being essentially 3D structures that provide better statistics for event count. This simplistic but illustrative mechanism can be further quantified by applying more complex models, such as the following.

As strain builds through tectonic loading, it appears that at sufficiently large stress levels, the crust enters a strain-hardening regime in which the crust surrounding the fault begins to microcrack, as manifested through the increased level of seismicity observed (for example in Figure 5). However, as seismicity intensifies, and strain continues to build, slip velocities on portions of the fault may begin to increase. The increased seismicity may act on the fault by (for example) changing rate-and-state parameters (Melosh, 1979) controlling the slip rate on the fault (Ruina, 1983; Dieterich, 1994). Acceleration of the SAF prior to the 2004 M4 event manifests itself in an average fault seismicity increase initially visible about 4 years before the event (Figure 3).

The spatial distribution of seismicity at Parkfield can thus be qualitatively described. Stress must concentrate in the creeping-locked transition zone, where the stably sliding (or "creeping") portion of the SAF to the north of Parkfield meets the unstable "locked" portion to the south (although in the rate-and-state description, this "locked" portion is considered to be slipping to a very small degree). This enhanced stress in the crust surrounding the creeping/locked intersection is likely to be the reason why the seismic activity is generally concentrated there.

However, the 30 km diameter zone surrounding the epicenter that exhibits low seismic activity in the year prior to the main event requires a more subtle explanation. The rock to the southwest of the SAF, where this zone lies, contains granite and is generally stronger than the sedimentary rock to the northeast. Such heterogeneity, together with the intersection of the locked and creeping sections of the SAF, result in complex local stress patterns. Only numerical simulations can indicate whether the modeled stress concentrations are consistent with the observed spatial distribution of seismicity. Usually, the fault zones are subjected to a constant strain-change rate and have a long history of periodic slip, with the same maximum stress value and along the same orientation. As a result of such a cyclic stress load, most of the fractures that were overcritically stressed are now discharged, and application of the same loading conditions triggers few new events. It is common to assume that rock masses contain numerous critically stressed fractures with random locations and orientations. At the same time, the weak fault itself can be regarded as a large-scale fracture, with the fracture tip in the transition zone. Fracture-tip vicinities are known for generation of highest stresses under loading (Sholz, 1990). While the locked and creeping sections of the fault zones show stress changes occurring primarily in the same direction, stress gradients are high in the transition zones, and stress changes lead to changes in the orientations of principal stresses. This creates new dimensions along which triggering of critically stressed fractures can occur and explains the previously described observation of increased seismicity in the transition zones.

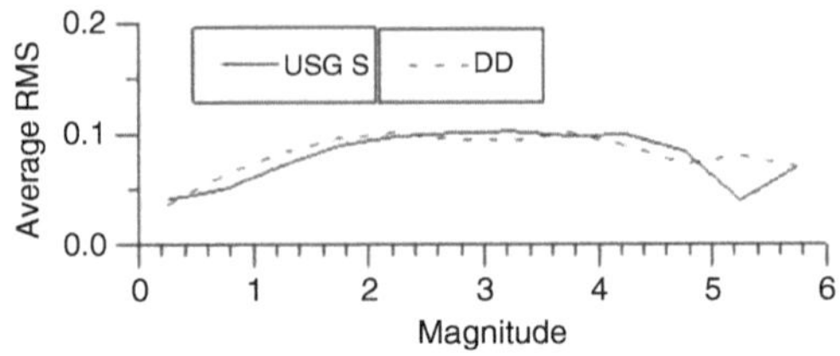

Figure 13 Average rms value vs. magnitude for USGS and DD (Thurber et al., 2006) relocated events in the Parkfield area. Note that the largest location errors belong to 2-4 magnitude intervals.

The increased sharpness of seismicity peaks for small-magnitude events can be explained by nonuniform station coverage in the region. Most of the stations (especially the High Resolution Seismic Network [HRSN]) are concentrated along the southeast-creeping part of the SAF, where pre-event seismicity is concentrated. In other words, the spatial inhomogeneity of station coverage in this case correlates with the pattern of spatial inhomogeneity of the micro-events. Obviously, small-magnitude seismicity in this region has a higher chance of being detected, providing the observed intensity of the precursor.

Thurber et al. (2006) claim that "...differential times help sharpen the Vp image along the SAF and also resolve the seismicity streaks and multiplets reported previously in the double-difference location study of Waldhauser et al. (2004)." Such impressive accuracy for relatively large (>80 km^2) seismicity count areas (Figures 4, 8 and 10), along with the confinement of errors for absolute locations of the events within a 1 km distance (Zhang and Thurber, 2003), justifies the robustness of the observed pre-event peaks. Somewhat surprising are the comparisons of the average event location accuracy by USGS and DD. The same events from USGS and DD catalogs (19,963 events in total) were used, where 19 events with rms > 1s were discarded as outliers. For the rest of the 1,994 events, the average rms value was computed as a function of the event magnitude. DD relocations gave an improvement to 65% of events, but for 35% of events rms values have increased. However, the average rms value for USGS locations is 0.0673s, while for DD locations the average rms = 0.0773 s, which is less accurate by 15%. The average rms values as functions of magnitude are plotted in Figure 13.

USGS locations look slightly better than DD locations, suggesting that while the DD method seems to work well for correcting the relative locations, the absolute locations are not improved by application of this method. Notably, the location accuracy of small-magnitude events is better than for mid-range magnitudes in both cases, which justifies the use of small-magnitude events in the seismicity analysis. Better location accuracy for small-magnitude events might be explained by the increase in rupture

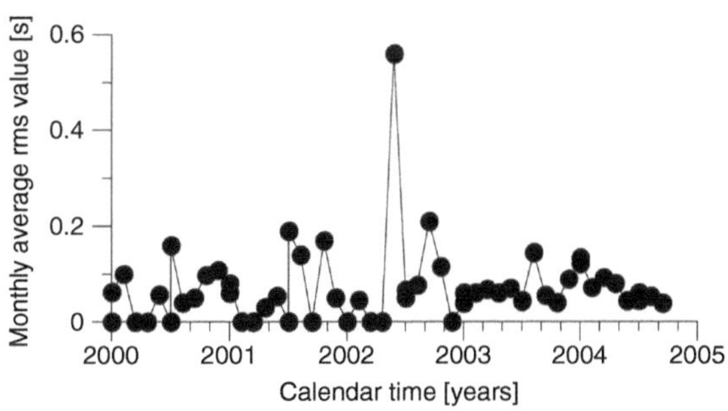

Figure 14 Monthly average rms values for events in the selection area from Figure 4. During the last two years preceding M6 Parkfield earthquake there is no anomalous rise of event location error, which eliminates the hypothesis about the artificial nature of precursory seismicity peaks.

length for events with larger magnitudes. To ensure data quality in the out-of-fault seismicity analysis, we computed the monthly average rms values for the selection area from Figure 4. During the last two years preceding the M6 Parkfield earthquake no anomalous rise of event location error is evident (Figure 14), which eliminates the hypothesis about the possible artificial nature of precursory seismicity peaks as a result of the migrated San Simeon M6.5 aftershock series.

Fault zone seismicity can be attributed to one of four proposed seismicity types, which differ by their relationships to strain changes in and around the fault. First, there is the weak (creeping) fault seismicity, which results from steady strain release and represents the dominant fraction of all regional events. Seismicity of this type is not directly related to stress buildup in the locked portion of the fault. The second type of seismic behavior is related to areas around nucleation zones of the future earthquakes, where a decrease of seismicity is observed in the pre-event stage. Such zones resemble Mogi doughnuts (Mogi, 1985), can have very elongated shapes, and are approximately equal in size to the future event rupture zone (Reasenberg and Simpson, 1992). The third type of seismicity corresponds to an aftershock series of moderate and large events, which may significantly contribute to the overall statistics but do not provide obvious clues for large earthquake prediction. The fourth type is the out-off-fault-zone seismicity occurring in the relatively intact rocks surrounding the fault zones. This type of seismicity is directly related to stress buildup in the crust and provides only a moderate number of detectable events, because these events are small. Results from this paper were mainly based on this fourth type of seismicity. The shape and size of the fourth-seismicity-type regions are likely to have a direct relationship to the magnitude of the future main event. The greater the portion of the fault bathed in seismicity, the larger the area experiencing accelerated slip, with the subsequent possibility of a larger induced earthquake.

3.3. Probability and earthquakes

All current earthquake prediction approaches use the term "probability" when describing expected events. Indeed, probability can be of use when one deals with representative statistics, in which large trial numbers allow reliable evaluation of average values. On the other hand, the probability of a rare single event (like a large earthquake that occurs every 100-150 years) carries no or little information for civilians, because probability variations are not followed by decision making and/or actions. Earthquake predictions formulated using probability language generally cannot be verified or invalidated either before or after a single event occurrence. For government agencies and insurance companies, the probability estimates can be useful for urban and financial strategy planning when they deal with multiple areas and customers. However, for any particular individual the meaning of the term "earthquake probability" is usually misleading and obscure. For the general population, probability-based predictions or forecasts do not lead to any decision making and taking of preventive actions.

Earthquake mechanisms obey laws of mechanics, applied to complex rather than random models. This means that the seeming randomness of earthquake occurrences happens because we have insufficient data about rock strength, stress values, and their evolution at depth. A close analog to this situation can be found in a coin toss. By saying that there is a 50% probability of a coin landing on either side, we recognize the fact that we have no information about this phenomenon other than that both sides of the coin have about the same physical properties, and neither one has a preference. However, if after a throw we managed to make instantaneous measurements of the coin's position and velocity—which can be done with the help of optical scanning (i.e., active monitoring)—then the future evolution of a coin becomes the subject of relatively simple mechanical laws and can be quickly modeled numerically before coin's landing. Randomness of coin behavior lies in a lack of information that can be changed after a proper investment. It also lies not in the properties of the coin itself, but in unrepeatable initial conditions. Therefore, measurements of coin parameters are useless before the coin is airborne. But after the coin is thrown, this event (precursor) should trigger the specialized measurements to precisely predict when and how the coin will land. This author proposes that a similar approach can be applied to earthquake prediction with the help of active monitoring of potential rupture zones.

3.4. Active monitoring and earthquake prediction

Most known earthquakes occur on existing faults that are generally weaker than the host rock. Dangerous faults are usually instrumented with passive listening devices that collect and transmit data by monitoring fault activity and physical parameters. Data processing aims to determine if data contains information, where data are coded signals and measurements, and information is an ability to cause change resulting in a correspondent action. In earthquake seismology, such a change should consist of precursor detection, and the corresponding action consists of resource mobilization and monitoring of target zones using active sources.

Comprehensive active monitoring of a large earthquake under preparation requires substantial investment, which by far exceeds the ability of modern technologies and economy. However, the existence of the mid-term precursors (like the one described in this paper) potentially yield information about the location and timing for such an event, and can serve as a trigger for detailed targeted observations aiming for short-term earthquake predictions having an accuracy of within several days (or even hours).

While some researchers (Alekseev et al., Chapter 3, this issue) propose the existence of precursory effects, observable on the surface and caused by development of dilatancy zones in epicentral regions of developing earthquakes, various geophysical instruments did not reveal any pre-event peculiarities before the M6 2004 Parkfield earthquake (Bakun et al., 2005). This is likely to be a result of sensor locations outside the seismogenic depths that start at approximately 2 km and extend down to about 15 km. Direct waves from sources placed on the surface usually do not penetrate down to seismogenic depths, because of the strong subsurface gradient. Some visible changes occurring at 1-2 km depths were observed in Parkfield using Vibroseis sources (Karageorgi et al., 1997; Korneev et al., 2000). However, penetration at seismogenic depths can be possible using reflections from sharp impedance discontinuities such as the Moho boundary. At the same time, laboratory studies (Sobolev et al., 1996) reveal that at pre-failure stages, softening rocks exhibit dramatic changes in both velocities and attenuation of seismic wave propagation. These changes can potentially be accurately monitored by properly placed sources and seismic networks, when waves sample fault zones at depths where most pre-event changes occur.

Recognizing the importance of the GRR, we must remember that it represents a statistical limit of seismicity distribution that requires a large observation volume and time, and its applications compromise the spatial and temporal resolution of any prediction methods that utilize this law as a cornerstone. From the results of this paper, it would seem natural that at the first (triggering) stage, we have to look at changes in natural seismicity within an area of a few hundred square kilometers (passive monitoring regime), waiting for an increase and development of a distinctive peak within a several-month period, which triggers a countdown for the actions aimed to collect detailed information about the fault zone changes and state. After reaching its maximum, natural seismicity fades and use of active controlled sources become critically important. In the low-seismicity regions, a high-frequency seismic noise can be a likely source of information about large-scale stress changes. The observed peak of seismic background noise before the 1993-1994 M4 series (Figure 6) repeats the seismicity patterns of the M6 2004 Parkfield earthquake (Figure 2). If shown for large data volumes, this direct correspondence between seismicity and seismic noise level may provide a key method for making instantaneous measurements of seismicity (stress) and its changes.

3.5. Short-term predictions and earthquake triggering

The short-term monitoring stage requires:

- Special design of observation systems that depend on local geology and fault system configurations, ensuring that seismic waves penetrate down to seismogenic depths.
- High sensitivity, in order to detect details of fault zone changes.
- Real-time data flow and analysis.
- Mobile deployment.

At the short-term monitoring stage, we should expect the development of large dilatancy zones as they approach a point of massive rock failure. These zones will likely show rapid changes in their extension, seismic velocities, anisotropy and attenuation. Recent results from continuous seismic monitoring showed the ability to detect subtle changes in rock properties. It is also natural to expect similar changes in electrical fields and processes associated with changes in fluid flow in fracture networks, and gravitational fields associated with dilatancy zones. Subsidence and surface temperature maps obtained from remote sensing data should be incorporated into combined data analysis to ensure prediction accuracy.

This author is convinced that active monitoring is the proper way for accurate earthquake forecasting. An inspiring example comes from meteorology, where such formerly thought-to-be "random" natural disasters as hurricanes are now being closely watched and tracked using satellite images. Instead of talking in probability language, the authorities here are now capable of giving clear warnings about coming disasters and responsibly declare evacuations.

It is yet unknown whether the short-term prediction of earthquakes can possibly be accurate enough to become a routine part of human life. A potential rupture zone of a catastrophic earthquake extends over hundreds of kilometers, and possible earthquake scenarios might include a wide variety of initial rupture locations. Many external and uncontrollable factors (such as lunar and solar tides, transient seismic waves, precipitation, human activity, etc.) can trigger an earthquake when the rock in the fault zone is in a critically-stressed state and just a small impact is enough to produce a massive rock failure—an earthquake. In such uncertain circumstances, any preventive action (mobilization of resources and supplies, evacuation, shutting down nuclear power plants, etc) to avoid damage and loss of life would be ineffective, costly, and unrealistic. However, if a section of a tectonic fault is diagnosed as being in a critically stressed state, and a catastrophic earthquake is considered inevitable, then human and economic damage can be substantially reduced if the earthquake is artificially triggered in a controllable manner. Such controllable triggering would allow implementing a set of damage-minimizing actions before the expected earthquake—similar to the well established practice of avalanches being triggered in populated mountain regions after all necessary warnings and precautions are made.

Triggering of earthquakes is a yet barely explored science. Such triggering can possibly be achieved by explosions (Aki et al., 1969), other earthquakes (Miyazawa and Brodsky, 2006; Helmstetter, 2003; Gomberg and Scott, 1995), application of powerful seismic vibrators (Alekseev et al., 2005), fuel injections (Ferrazzini et al., 1990; Brodsky and Prejean, 2004; Ross et al., 1999; Eberhart-Phillips and Oppenheimer, 1984), application of high-voltage electric sources (Avagimov et al., 2004) or, most likely, by combined use of these and other means.

4. Conclusions

Similar peaks in seismicity occurring several months prior to two recent large SAF events suggest that they are good candidates for earthquake prediction studies. Aftershocks and creeping fault seismic activity mask the effect and should be excluded from the data, as having little or no relation to the stress buildup in the locked fault zones. The size and location of earthquakes correspond to both the size and location of the "quiet" zones and the zones of increased seismicity occurring before the main shocks. Because of their frequent occurrence, small-magnitude events may be ideal for routine daily monitoring of stress changes. The development of active seismic monitoring techniques is necessary for investigating changes in the pre-seismic nucleation zone. The observations reported here leave open the possibility that successful earthquake prediction may yet be possible.

Acknowledgments

Data processing was done at the Center for Computational Seismology (CCS) at Lawrence Berkeley National Laboratory, which is operated by the University of California for the U.S. Department of Energy (DOE) under Contract No. DE-AC03-76SF00098. The author thanks Federica Marone for supplying earthquake data, and also Robert Nadeau, who made several helpful comments. This paper significantly benefited from the help and suggestions of Steven Pride. Ruth Harris suggested recomputing the results using relocated events in the Parkfield area, which were provided by Cliff Thurber.

REFERENCES

Aki, K., Reasenberg, P., Defazio, T., Tsai, V., 1969. Near-field and far-field seismic evidences for triggering of an earthquake by Benham explosion. Bulletin of the Seismological Society of America 93 (n6), 2197–2207.

Alekseev, A.S., Chichinin, I.S., Korneev, V.A., 2005. Powerful low-frequency vibrators for active seismology. Bulletin of the Seismological Society of America 95, 1–17.

Artamonova, M.S., Korneev, V.A., 2005. Active seismic monitoring for earthquake forecasting. Eos Trans. AGU 86 (52), Fall Meet. Suppl., Abstract NG32A-06.

Avagimov, A., Bogomolov, L., Cheidze, T., Ponomarev, A., Sobolev, G., Tarasov, N., Zeigarnik, V., 2004. Induced seismicity by trigger stimulation from laboratory and field tests: Active monitoring in geophysics, paper S2-06, Mizunami.

Bakun, W.H., McEvilly, T.V., 1979. Earthquakes near Parkfield, California: Comparing the 1934 and 1966 Sequences. Science 205, 1375–1377.

Bakun, W.H., Lindh, A.G., 1985. The Parkfield, California, prediction experiment. Earthquake Prediction Research 3, 285–304.

Bakun, W.H., Aagaard, B., Dost, B., Ellsworth, W.L., Hardebeck, J.L., Harris, R.A., Ji, C., Johnston, M.J.S., Langbein, J., Lienkaemper, J.J., Michael, A.J., Murray, J.R., Nadeau, R.M., Reasenberg, P.A., Reichle, M.S., Roeloffs, E.A., Shakal, A., Simpson, R.W., Waldhauser, F., 2005. Implications for prediction and hazard assessment from the 2004 Parkfield earthquake. Nature. doi:10.1038/nature04067.

Bowman, D.D., Sammis, C.G., 2004. Intermittent criticality and the Gutenberg–Richter distribution. Pure and Applied Geophysics 104, 1945–1956. doi:10.1007/s00024-004-2541-z.

Brodsky, E.E., Prejean, S.G., 2004. The evidence that long-range earthquake triggering involves fluids. American Geophysical Union, Fall Meeting 2004, abstract #S23C-01.

Dieterich, J.A., 1994. A constitutive law for rate of earthquake production and its application to earthquake clustering. Journal of Geophysical Research 99, 2601–2618.

Dunegan, H.L., 1996. Prediction of earthquakes with AE/MS? Why not. In: 6th Conference on AE/MS Activity in Geologic Structures and Materials, Pennsylvania State University.

Eberhart-Phillips, D., Oppenheimer, D.H., 1984. Induced seismicity in The Geysers geothermal area, California. Journal of Geophysysical Research 89, 1191.

Ferrazzini, V., Chouet, B., Fehler, M., Aki, K., 1990. Quantitative analysis of long-period events recorded during hydrofracture experiments at Fenton Hill, New Mexico. Journal of Geophysical Research 95 (B13), 21,871–21,884.

Geller, R.J., 1997. Earthquake prediction: A critical review. Geophysical Journal International 131, 425–450.

Geller, R.G., Jackson, D.D., Kagan, Y.Y., Mulargia, F., 1996. Earthquakes cannot be predicted. Science 275 (5306), 1616.

Gomberg, J., Scott, D., 1995. Stress/strain changes and triggered seismicity at The Geysers, California. Journal of Geophysical Research 101 (B1), 733–750. doi:10.1029/95JB03250.

Helmstetter, A., 2003. Is Earthquake Triggering Driven by Small Earthquakes? Physical Review Letters 91. doi:91.058501.

Karageorgi, E., Clymer, R., McEvilly, T.V., 1992. Seismological studies at Parkfield: II. Search for temporal variations in wave propagation using Vibroseis. Bulletin of the Seismological Society of America 82, 1388–1415.

Karageorgi, E.D., McEvilly, T.V., Clymer, R.W., 1997. Seismological studies at Parkfield IV: Variations incontrolled-source waveform parameters and their correlation with seismic activity 1987–1994. Bulletin of the Seismological Society of America.

Korneev, V., 2006. Seismicity precursors of the M6.0 2004 Parkfield and M7.0 1989 Loma Prieta earthquakes, Lawrence Berkeley National Laboratory, Scientific report – 60012, esd.lbl.gov/FILES/about/staff/valerikorneev/LBNLrep60012.pdf.

Korneev, V.A., 2005. Seismicity precursors of M6.0 2004 Parkfield and M7.0 1989 Loma Prieta earthquakes. Eos Trans. AGU 86 (52), Fall Meet. Suppl., Abstract S53B-1097.
Korneev, V.A., Nadeau, R., 2004. Vibroseis monitoring of San Andreas fault in California. In: Proceedings of the International Workshop in Active Monitoring, Mizinami, 88–95.
Korneev, V.A., McEvilly, T.V., Karageorgi, E.D., 2000. Seismological studies at Parkfield VIII: Modeling the observed travel-time changes. Bulletin of the Seismological Society of America 90 (3), 702–708.
Korneev, V.A., Nadeau, R.M., McEvilly, T.V., 2003. Seismological studies at Parkfield IX: Fault zone imaging using guided wave attenuation. Bulletin of the Seismological Society of America 93 (4), 1415–1426.
Langbein, J., Borherdt, R., Dreger, D., Fletcher, J., Hardebeck, J.L., Hellweg, M., Ji, C., Johnston, M., Murray, J.R., Nadeau, R.N., Rymer, M.J., Treiman, J.A., 2005. Preliminary report on the 29 September 2004, M 6.0 Parkfield, California earthquake. Seismological Research Letters 76 (1), 10–26.
Lei, X.-L., Kusunose, K., Nishizawa, O., Cho, A., Satoh, T., 2000a. On the spatio-temporal distribution of acoustic emissions in two granitic rocks under triaxial compression: The role of pre-existing cracks. Geophysical Research Letters 27 (13).
Lei, X.-L., Kusunose, K., Rao, M.V.M.S., Nishizawa, O., Satoh, T., 2000b. Quasi-static fault growth and cracking in homogeneous brittle rocks under triaxial compression using acoustic emission monitoring. Journal of Geophysical Research 105 (B3), 6127–6139.
Lei, X.-L., Masuda, K., Nishizawa, O., Jouniaux, L., Liu, L., Ma, W., Satoh, T., Kusunose, K., 2004. Detailed analysis of acoustic emission activity during catastrophic fracture of faults in rocks. Journal of Structural Geology 26, 247–258.
Lei, X.-L., Satoh, T., Nishizawa, O., Kusunose, K., Rao, M.V.M.S., 2005. Modeling damage creation in stressed brittle rocks by means of acoustic emission. In: Proceedings of the 6th International Symposium on Rockbursts and Seismicity in Mines (RsSim6), Australia, 327–334.
Lockner, D.A., Byerlee, J., Kusenko, V., Ponomarev, A., Sidorin, A., 1992. Observations of quasistatic fault growth from acoustic emissions. In: Evans, B., Wong, T.-F. (Eds.), Fault Mechanics and Transport Properties of Rocks. Academic Press, London, pp. 3–31.
Melosh, H.J., 1979. Acoustic fluidization: A new geologic process? JGR 84, 7513–7520.
Meltzner, A.J., Wald, D.J., 1999. Foreshocks and aftershocks of the great 1857 California earthquake. Bulletin of the Seismological Society of America 89, 1109–1120.
Miyazawa, M., Brodsky, E.E., 2006. Deep low-frequency tremor that correlates with the passing surface waves. Journal of Geophysical Research 113 (B01307), doi:10.1029/2006JB004890.
Mogi, K., 1985. Earthquake Prediction. Academic Press, Tokyo. Japan.
Reasenberg, P.A., Simpson, R.W., 1992. Response of regional seismicity to the static stress change by the Loma Prieta earthquake. Science 255, 1687–1690.
Ross, A., Foulger, G.R., Julian, B.R., 1999. Source processes of industrially-induced earthquakes at the Geysers Geothermal Area, California. Geophysics 64, 1877.
Ruina, A.L., 1983. Slip instability and state variable-friction laws. Journal of Geophysical Research 88, 10359–10370.
Sholz, C.H., 1990. The Mechanics of Earthquakes and Faulting. Cambridge University Press.
Sobolev, G.A., Babichev, O.V., Los, V.F., Kol'tsov, A.V., Ponomaryov, A.V., Ponyatovskaya, V.I., Rozanov, A.S., Stanchits, S.A., Khromov, A.A., Frolov, D.I., Yangquan, L., Jialiu, Z., Jiadong, Q., Shiyu, L., 1996. Precursors of the destruction of water containing blocks of rock. Journal of Earthquake Prediction Research 39.
Tham, L.G., Cheung, Y.K., Tang, C.A., 2001. Numerical simulation of the failure process of rocks. Tamkang Journal of Science and Engineering 4 (4), 239–252.
Thurber, C., Zhang, H., Waldhauser, F., Hardebeck, J., Michael, A.J., Eberhart-Phillips, D., 2006. Threedimensional compressional wavespeed model, earthquake relocations, and focal mechanisms for the Parkfield, California, region. Bulletin of the Seismological Society of America 96, S38–S49. doi:10.1785/0120050825.

Wald, D.J., Kanamori, H.V., Helmberger, D.V., Heaton, T.H., 1993. Source study of the 1906 San Francisco Earthquake. Bulletin of the Seismological Society of America 83, 981–1019.

Waldhauser, F., Ellsworth, W.L., Schaff, D.P., Cole, A., 2004. Streaks, multiplets, and holes: High-resolution spatio-temporal behavior of Parkfield seismicity. Geophysical Research Letters 31 (L18608), doi:10.1029/2004GL020649.

Zhang, H., Thurber, C.H., 2003. Double-difference tomography: The method and its application to the Hayward fault, California. Bulletin of the Seismological Society of America 93 (5), 1875–1889. doi:10.1785/0120020190.

CHAPTER 2

Large-scale Geophysical Surveys of the Earth's Crust Using High-power Electromagnetic Pulses

E.P. Velikhov[a] *and* V.P. Panchenko[a]

Contents

1. Introduction

Geoelectrical methods are used for solving a large variety of different exploration problems, from groundwater prospecting to mining and oil and gas exploration (Zhdanov and Keller, 1994). There are many different

[a] Russian Research Center Kurchatov Institute

Handbook of Geophysical Exploration: Seismic Exploration, Volume 40

ISSN 0950-1401, DOI: 10.1016/S0950-1401(10)04007-3

Table 1 Requirements of the source for EM soundings by the single-pulse method

Geophysical tasks	**Typical sounding depth** (km)	**Specific electric resistance** (Ohm m)	**Current impulse duration** (s)	**Moment electric/magnetic dipole** J_E (A m)/M_H (A m^2)
Study the deep structure of the earth's crust and upper mantle	30–100	10^3–10^4	0.1–10	$3 \cdot 10^6$–$3 \cdot 10^8$ $3 \cdot 10^{10}$–10^{12}
Prediction and "unloading" of earthquakes	15–30	10^2–10^3	0.1–10	$3 \cdot 10^6$–$3 \cdot 10^7$ $3 \cdot 10^9$–$3 \cdot 10^{10}$
On-shore oil and gas exploration	3–8	3–30	1–20	$3 \cdot 10^6$–$3 \cdot 10^7$ 10^8–$5 \cdot 10^9$
Surveying for ore deposits	1–3	10–10^3	0.1–1	10^3–10^4 10^6–$5 \cdot 10^7$
Offshore oil and gas exploration	1–10	1–10	1–10^2	10^6–10^7 $5 \cdot 10^7$–10^9

modifications of electromagnetic (EM) geophysical methods, which can be based on the application of direct current (DC), frequency domain, or time domain fields, as well as their combinations (Zhdanov, 2009). The depth of investigation of controlled-source electromagnetic (CSEM) methods in studying the deep layers of the earth's crust is limited, however, because of the limited power of the appropriate sources of electric energy available (Velikhov et al., 2000; Zeigarnik, 2006). Note that, a range of power of the controlled sources required for different geoelectrical methods differs by three to four orders of magnitude, which prevents us from building a multipurpose source suitable for a wide range of applications.

There are two alternative approaches to increasing the depth of EM sounding. The first one is based on the application of a high-power single pulse, the second one uses a series of low-power (≤100 kW) impulses provided by an EM field generator. In the latter case a long time of repeated EM soundings may be required, and the recorded signals are stacked to increase the signal-to-noise ratio. Table 1 shows a list of basic geoelectrical tasks, which are of the most importance in geological applications, and the corresponding typical requirements of the sources of a single pulse (Velikhov et al., 1975; Velikhov and Volkov, 1981).

During the last 30 years, pulsed solid propellant magneto-hydrodynamic (MHD) generators of different scales and durations of operation have been developed in Russia. These generators have no analogs in world engineering

(Velikhov et al., 1975; Velikhov and Volkov, 1981). They have been applied as multimegawatt power sources for EM sounding. The self-sufficiency of MHD generators is provided by their self-excitation from the initial (low) magnetic field/current generated by the initial excitation system based on a battery storage or a capacitor bank. The gain factor for the current is equal to 10-50 times, and for the electric power it is equal to 100-2500 times. The duration of the self-excitation phase is in a range of 1 to 2 s, and that of the propellant burning phase is 3 to 12 s. The pulsed MHD facilities have the following unique features: high unit power (5-500 MWe), high current in the load (10-200 kA); self-sufficiency, compactness, transportability, and high reliability.

The idea for the application of pulsed MHD generators in electromagnetic sounding systems was originally introduced as early as 1971 (Velikhov et al., 1975, 2000). Since then, the corresponding parameters of EM sounding devices have been determined, and the geophysical MHD facilities have been created (Velikhov, 1989).

Table 2 presents a list of MHD facilities used in geophysical surveys, which include Pamir-1, Pamir-2, Ural, Prycaspiy, Khibiny, and Prognoz-1.

In this paper we present an overview of the results obtained in geophysical applications of high-power MHD facilities. The large-scale experimental surveys were conducted using MHD generators for oil and gas exploration both onshore and offshore (e.g., Caspian Lowland, Eastern Siberia, Barents Sea). The method was used for electrical prospecting of ore deposits in the Kola Peninsula. The deep EM sounding was used to study the geoelectrical structures of the earth's crust and upper mantle in the Ural Mountains (up to 40 km) and on the Baltic Shield (up to 150 km). The MHD facilities were also used in experiments for earthquake prediction and monitoring of seismic activities (e.g., Tajikistan, Kirghizia). We also discuss in this paper the capabilities and future directions of the research and development of EM geophysical technology based on the use of the powerful electrical source.

2. General Concept of the Application of High-power Electromagnetic Pulses in Geophysical Surveys

It is well known that in conductive layers of the earth's crust, the EM field attenuates exponentially with the depth. In order to increase the depth and resolution of EM sounding of the earth's crust, one has to use more powerful sources of EM field. Even in a simple case of DC sounding, when an approximate sounding depth (h) is proportional to the half the separation (r) between the source and receiver, $h \sim 0.5r$, in order to keep the uniformity of the measurements, the required power (N_g) of the field

Table 2 MHD facilities used in geophysical applications

Parameter	Pamir-1	Pamir-2 Sever-1	Prycaspiy	Ural	Prognoz-1	Khybiny
Maximum power, MW	16	16	16	30	30	40
Voltage, V	1350	1350	800	600	600	2000
Load resistance, Ohm	1.7	0.3	0.13	0.03	0.4	0.1
Current in the load, kA	≤ 3	≤ 4.2	≤ 7	≤ 40	≤ 2.2	≤ 20
Moment of						
electric dipole, A m	$5 \cdot 10^6$	$5 \cdot 10^7$	10^8		$6, 6 \cdot 10^6$	
magnetic dipole, A m^2			$5 \cdot 10^{10}$	$4 \cdot 10^{10}$		$\sim 10^{14}$
Current pulse duration, s	2 and 7	2 and 7-9	2 and 7-8	1.5	3 and 8	3 and 7-9
Mass of facility, ton	15	25	25	20	20	40
Mass of the load (dipole), ton	10	40	100	40	40	120

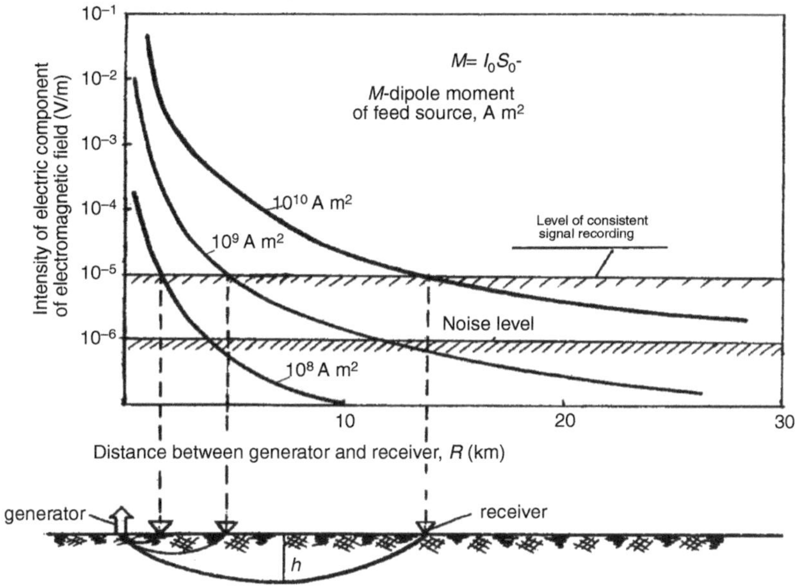

Figure 1 Relationship between the intensity of the electric field and the depth of penetration of the electromagnetic field, $h \approx (0.3\text{-}0.5) \cdot R$.

source has to be proportional to the 6th order of the depth: $N_g \sim h^6$ (Velikhov et al., 2000).

Figure 1 can be used to estimate the moment of the magnetic dipole required for sounding the earth's crust at different depths. One can see that for obtaining information on the geoelectrical structure of the earth's crust to depths up to 15-20 km, the dipole moment of the source should be $M \sim 10^{10}$ A m^2. The geophysical MHD facilities are capable of providing the required level of transmitter dipole moments of the source in a single pulse (Table 1) and obtaining information from depths of the earth's crust inaccessible with the use of conventional low-power sources of the electromagnetic field. Figure 2 schematically shows the survey configuration for sounding with one fixed high-power source over a large area covering up to 10^5 km^2.

If the depth of the earth's crust thickness under study is specified, the required frequency range of an electromagnetic signal is determined by the corresponding skin depth of the EM field. It is usually within 0.01-50 Hz. In this case the most important factor for EM sounding is an increase of the signal-to-noise ratio in the receiver. It was demonstrated in Velikhov et al. (2000) and Zeigarnik (2006) that with the same power expenditures, the method based on a single high-power pulse provides a signal-to-noise ratio 10-100 times greater than those achieved by sounding with a conventional source and the data stacking mode.

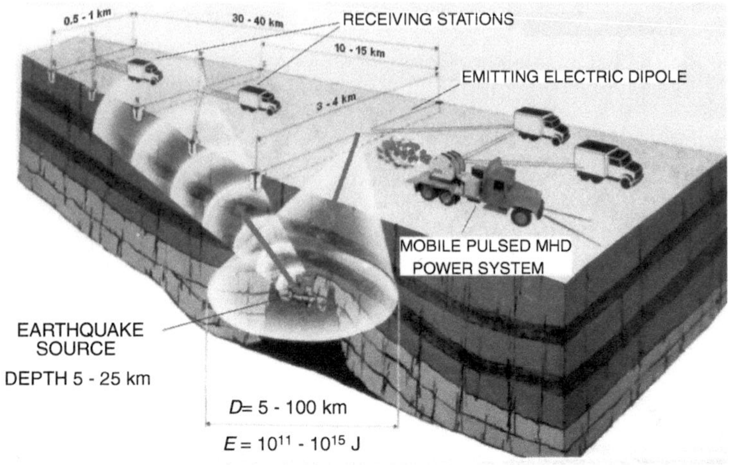

Figure 2 Cartoon of electromagnetic sounding of the earth with the use of an MHD facility.

This effect is achieved due to the following three factors:

- the capability of obtaining the maximum transmitter dipole moment under the condition of optimal matching of the MHD generator parameters with the emitting circuits;
- the non-stationarity of natural noise with large correlation distances;
- the capability of generating a sounding pulse within the predicted time lags with respect to the noise minimum.

We should note, however, that an approach based on a combination of two modes of electromagnetic sounding, one based on a single powerful pulse and another using a series of repeated low-power pulses with stacking of the observed signals, may be optimal from both technological and economical points of view. Typically, for reaching greater depths, the high-power single pulse should be used, while for detailed investigations of relatively shallow areas a series of low-power pulses produced by a standard electric generator can be carried out. This concept was realized in the "Khibiny" experiment, described below (Velikhov, 1989).

3. Analysis of the Geoelectrical Structure of the Earth's Crust and Upper Mantle

3.1. Study of the geoelectrical structure of the earth's crust in the Ural Mountains

The first investigation of a deep geoelectrical structure of the earth's crust was initiated in 1975 in the Ural mountains area located about 100 km south

Figure 3 MHD facility "Ural".

from Ekaterinburg city. This experiment was conducted with the application of the geophysical MHD facility "Ural" (Figure 3).

The major results of this experiment can be summarized as follows (Velikhov et al., 1975):

- geoelectrical information about the entire depth of the earth's crust up to 40 km under the Ural mountains was obtained;
- the extension of deep faults was determined;
- the magnitude of generated magnetic field at a distance of 5-8 km from the source (electric dipole) was found to be sufficient for detection of magnetite deposits at a depth of 1.5 km.

3.2. Analysis of a deep geoelectrical structure of the eastern part of the Baltic shield

The complex study of a geoelectrical structure of the eastern part of the Baltic shield was carried out with the application of the unique geophysical MHD facility "Khibiny" installed in 1976 on the coast of the Barents Sea at the Rybachiy Peninsula (Kola Peninsula) (see Figure 4) (Velikhov, 1989; Fortov, 2000).

The MHD facilities were installed on the neck of the Sredniy Peninsula, and the generated current was supplied through a connecting aluminum cable of 8 km length and 160 ton mass to the groundings located at opposite sides of the neck (Figure 5). Thus, the coastal line around the Sredniy and Rybachiy Peninsulas was used as a large quasi-circular transmitting loop.

The impedance in the load was 90 MOhm, the generated current reached 22 kA, the duration of the pulse was 5-6 s. Due to the application of the coastal line as a transmitting loop (Figure 5), it was possible to obtain

Figure 4 General view of MHD facility "Khybiny" consisting of two MHD generators.

a magnetic dipole moment as high as $M_H \sim 10^{14}$ A m^2, which was a record for geoelectrical engineering. The generated signals considerably exceeded the natural field noise level at distances from the MHD generator to the sea side up to 150 km and to the land side up to 500 km. The average accuracy of signal measurements was 5-7% for radiuses up to 200 km and 10-15% for distances of 250-400 km. Two measurements were performed at ultra long distances of 640 km and 750 km. The strength of the magnetic field was about 1γ ($\gamma = 10^{-9}$ T) at distances of not less than 200 km and about 10 mγ at a distance of 750 km.

The deep EM soundings with the use of the MHD facility "Khibiny" were performed along three basic lines: I, II and III (Figure 5).

The major results of the geophysical experiment "Khibiny" can be summarized as follows (Velikhov, 1989; Fortov, 2000; Zhamaletdinov, 1990):

- ten large geoelectrical blocks and zones in the upper section of the earth's crust up to depth of 10-15 km with dimensions about 50 km × 50 km have been detected (Figure 6);
- the resistivity of the blocks varies from ~10 Ohm m to 10^5 Ohm m;
- the blocks with high resistivity usually correspond to the outcrops of the most ancient granitoid rocks;
- the blocks with low resistivity, as a rule, are connected with the outcrops of the younger rocks, and they are considered to be prospective zones for exploration for ore deposits and other mineral resources.

The interpretation of the experimental data resulted in the construction of a map of the total longitudinal electrical conductivity of the upper 10 km thickness of the earth's crust shown in Figure 6. We present in Figure 7 a

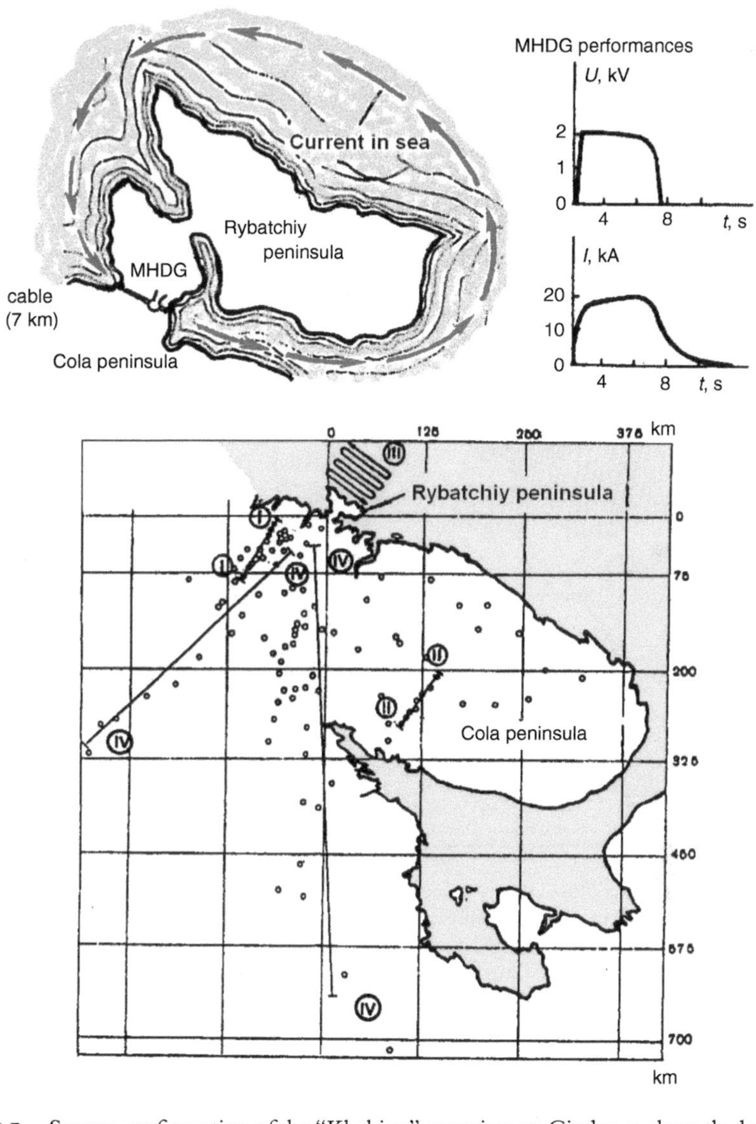

Figure 5 Survey configuration of the "Khybiny" experiment. Circles, o, show the location of the observation points. The black lines show the profiles of detailed EM soundings: I-I—Pechenga ore-bearing structure, II-II—Imandra-Varzug ore-bearing area, III-III—Barents Sea shelf, IV-IV—profile of the deep sounding of the Baltic shield.

deep geoelectrical section of the lithosphere of the Baltic shield constructed based on the results of the "Khybiny" experiment.

The resistivity of the rocks increases gradually from $\sim 5 \cdot 10^3$-10^4 Ohm m on the surface up to $\sim 5 \cdot 10^6$ Ohm m at a depth of 35-40 km, and then

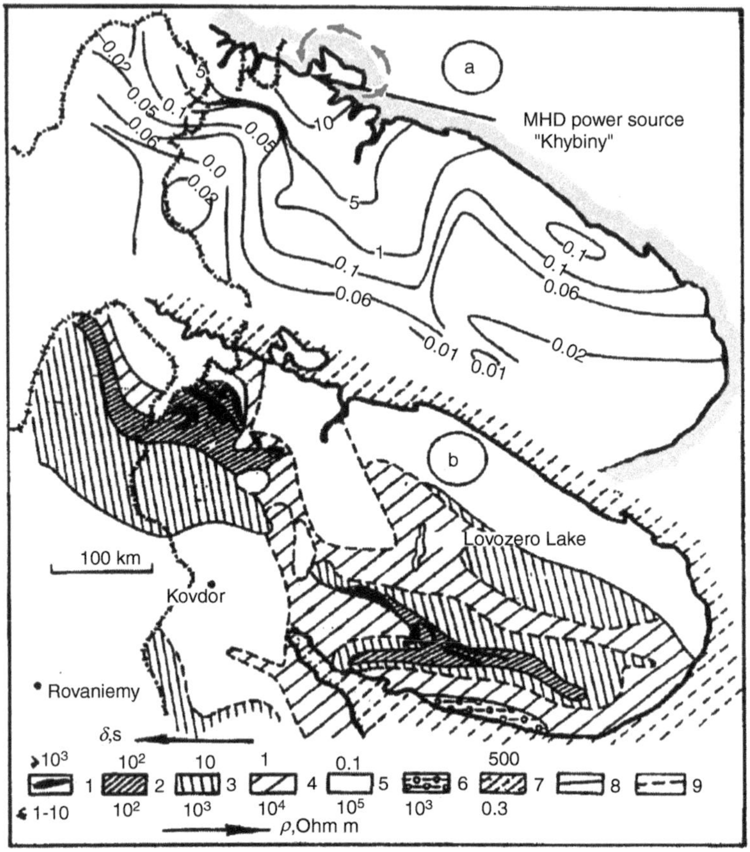

Figure 6 Schematic illustration of the results of EM soundings with the MHD power source "Khybiny": (a) isolines of absolute values of the intensity of a horizontal electric field component (mV/km) adjusted to a current of 20 kA; (b) diagram of the longitudinal conductivity of the upper 10 km thickness of the earth's crust. Notation conventions: "1-7"—blocks with different electrical conductivities; 8—boundaries of the blocks ("1-7"—determined, 9—suggested).

decreases smoothly (Figure 7). The well-conducting strata in this depth range are not revealed. The temperature of the earth's crust at a depth of 40 km does not exceed 400 °C, which corresponds well to the mean geothermal gradient of 10 °C/km. The estimated lateral resistivity of a poorly conductive section of the crystallized basement is within the range of $5 \cdot 10^9$-10^{10} Ohm m.

One of the new results of the "Khybiny" experiment was a discovery of conducting channels in the earth's crust. The electric current, induced by the MHD generator, propagates in a horizontal direction through these channels to a distance of hundreds of kilometers, providing electrical coupling

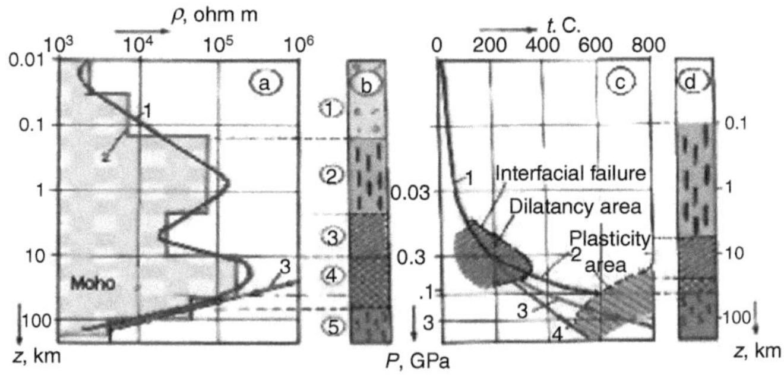

Figure 7 Vertical resistivity profile of the lithosphere in the north-eastern part of the Baltic shield and its geodynamic interpretation.

between a number of electronic-conductive graphitic structures typical for this area. For example, the quantitative interpretation of the observed distribution of magnetic field components (H_x, H_y, H_z) in the Imandra-Varguza structure (Figure 5, profile II-II), based on the EM migration method (Zhdanov et al., 1988), shows that the conducting area extends to a depth of about 10 km. This estimation is in good accord with the results of a magnetovariational survey at the western flank of the structure (Zhamaletdinov, 1990).

Figure 8 shows geological and geophysical sections for profile "G-H" "White Sea-Barents Sea" of about 300 km length. The geoelectrical section produced by the results of MHD soundings in a combination with other methods is shown in Figure 8(c). The results of the large-scale "Khibiny" experiment cardinally changed a view on the structure and nature of electrical conductivity of the Baltic shield. Based on the obtained data, the overall lateral resistance of the Baltic Shield lithosphere was defined, including the earth's crust and upper mantle ($T = \rho \cdot h = 10^{10}$ Ohm m; ρ is the resistivity of "cold" poorly conducting part of the lithosphere, h is its total thickness).

3.3. Deep geoelectrical structure of the Northern Tien Shan region

EM observations based on the MHD facilities "Prognoz-1" (a prototype of the "Ural" MHD facility, Figure 3) and "Pamir-2" have been carried out since 1982 at the Bishkek geophysical test site, Kirghizia (Velikhov et al., 2000; Zeigarnik, 2006; Fortov, 2000). The test site is located at a distance of 30 km to the south of Bishkek city. Active geophysical monitoring with the use of MHD generators was performed on the test site area and in adjoining territories. The analysis of geoelectrical structures was conducted under both static and dynamic conditions using a controlled-source EM method (for the

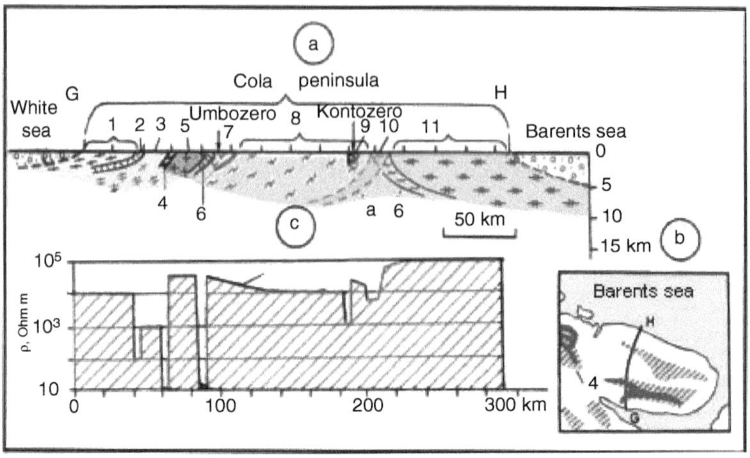

Figure 8 Geological and geophysical sections of profile "G-H" "White Sea-Barents Sea": (a) geological section; (b) location of the profile on the map; (c) geoelectrical section along the profile.

upper and intermediate part of the crust up to 25 km) and magnetotelluric (MT) sounding (for the lower part of the crust up to the Moho boundary).

The results of EM soundings have revealed the major features of the deep structures of the Northern Tien Shan seismogenic area. In particular, we have discovered two strata with increased electrical conductivity in a section of a Proterozoic bed, which was an unexpected result. One of these strata is located in the upper part of the earth's section at a depth of about 10 km in the southern part of the test site, pinching out in the northern direction. The second horizon is located at a depth of about 30 km. This result has been confirmed by the MT data. The study of geoelectrical structures and the observed variations of apparent resistivity have enabled us to explain specific features of seismicity of the region in general.

4. EM Soundings with a Powerful Source in Seismically Active Regions

4.1. EM soundings for earthquake prediction

The first experiments with geophysical applications of pulsed MHD facilities were conducted in 1973 in the Garm seismic test site in Tajikistan, located within 38.2°-39°N and 68.8°-71.5°E. The main purpose of those experiments was to study the possibility of using the variations in the electrical resistivity of rocks as earthquake precursors.

Figure 9 Pulsed MHD facility "Pamir-1" at the Garm geophysical test site.

It is known that the temporal variations of apparent resistivity of the earth's crust, ρ_a, detected in seismoactive regions with different geological conditions, may be used as a prognostic criterion for the preparation of a strong earthquake. Therefore, one can apply the method of electric sounding of the earth for earthquake prediction (Volykhin et al., 1993). We have demonstrated above that it is sufficient to apply a pulsed source of EM energy of a power 1-3 MW and a 1-3 s duration for study of the earth's crust up to a depth of 10-15 km, which may include earthquake source areas (Velikhov et al., 1975; Velikhov and Volkov, 1981). The geophysical MHD facility "Pamir-1," installed in 1973 at the Garm test site, completely met these requirements (Figure 9).

The electric dipoles were about 3 km long and had a total electric resistance of about 1.5 Ohm. The MHD generator produced the current in a load of 1 kA, that provided an electrical dipole moment of $3.5 \cdot 10^6$ A m and a consistent recording of EM sounding signals at distances up to 40 km from the source (Figure 10).

Figure 10 shows a diagram of the Garm test site, location of the MHD facility, receiving stations, and epicenters of several earthquakes. The figure also presents the time progress of the variations of apparent resistivity measured by two mutually orthogonal receiving dipoles in the location of station No. 2. Vertical lines mark the moments of occurred earthquakes. Index K means a power class of the earthquakes. The epicenters of indicated earthquakes are located within a depth range of 5-7 km. The variations of apparent resistivity obtained by EM soundings with the MHD facility reach up to 30% and by 2-5 times exceed similar variations observed at this test

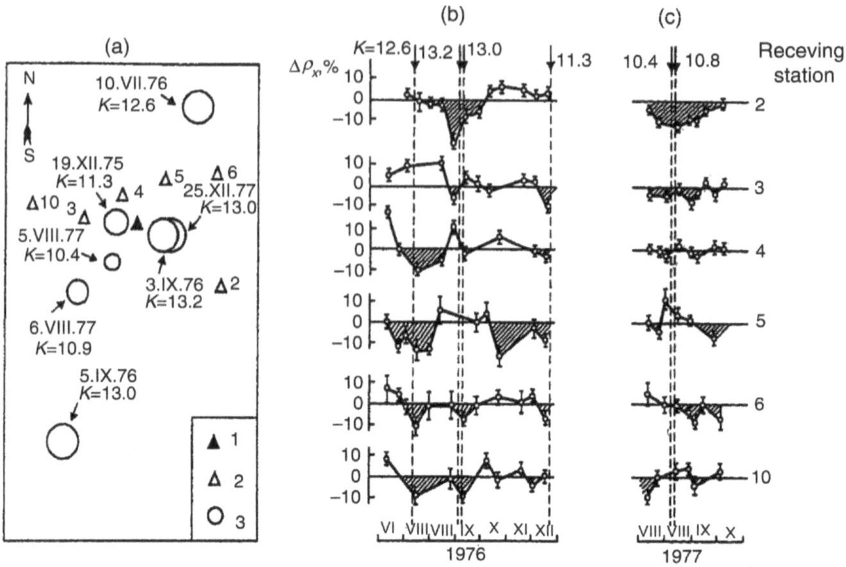

Figure 10 Scheme of the survey configuration at the Garm test site (a) and the corresponding variations of apparent resistivity data obtained with an MHD generator (b, c). 1—MHD facility "Pamir-1"; 2—receiving stations; 3—epicenters of earthquakes.

site with the use of traditional CSEM methods with a low-power generator and signal stacking. This effect can be explained by the increased depth of investigation and, respectively, the volume of rocks in the tectonically active zone covered by the EM field generated by a high-power source.

We observe in this figure a clear correlation between extreme values of apparent resistivity and the moments of strong earthquakes and an increase in amplitude of the variations in the observational points located closer to the epicenters of the earthquakes. This fact allows us to assume that the detected variations are caused by processes related to the earthquake preparation.

The data collected at the Garm test site verified the existing concept of the close relationship between the resistivity variations and the processes of preparation of the strong local earthquakes. However, the interpretation of this data was complicated by the interruption of observational series and their small total duration, as well as by an inadequate number of receiving stations.

At the Bishkek prognostic test site, located within 41°-45°N and 74°-81.4°E in Kirghizia, the systematic study of variations of apparent resistivity with the use of a high-power source began after 1983. The source was formed by two pulsed MHD generators, "Prognoz-1" (similar to the MHD facility "Ural," Figure 3) and "Pamir-2" (similar to "Pamir-1," Figure 9). This powerful source generated single pulses of a current of 2-5 kA up to 40 kA with a pulse duration of up to 10 s. In these experiments a stationary electric-pulsed installation EIS-630 was also used, which supplied periodic pulse sequences of 0.6 to 0.8 kA. The MHD generators were used for deep EM soundings in the frequency domain, while installation EIS-630 was applied for time domain sounding. An electric dipole of 4.2 km length and of 0.4 Ohm resistance was used as a transmitter. The data acquisition system was formed by a network of 35 receivers, five of which comprised a stationary array, and the others were used as a transportable array. Such a complex data acquisition system provided reliable measurements of EM sounding signals at distances of up to 60 km. The depth of investigation was 15–20 km, which provided the coverage of epicenters of a majority of local earthquakes.

The local earthquakes occurred, as a rule, after the magnitudes of the anomalous apparent resistivities reached the maximum values which were significantly different from the background level. We assume that the resistivity variations are caused by complex geodynamic processes, which are accompanied by variations of physical properties of the medium before the earthquake. The distribution in variations of apparent resistivity over the test site area is nonuniform, and, apparently, it reflects the location of the stress zones in the earth's crust. In those areas, where there is compression, the resistivity decreases, and where there is a tension, the resistivity increases. In addition, sometimes the resistivity variations concur with the remote earthquakes. Therefore, it is impossible to predict an earthquake by observation of the variations in this geophysical parameter at one point (Zeigarnik, 2006; Volykhin et al., 1993). However, using an optimal network of observation stations makes it possible to provide a probabilistic estimation of earthquake magnitude and time of occurrence.

The studies at both the Bishkek and Garm test sites proved the importance of monitoring the apparent resistivity of the earth's crust as one of the prognostic criteria of earthquake preparation. However, it seems to be problematic to establish a direct relationship between the seismic deformation process and the resistivity of the medium.

4.2. Influence of a high-power electromagnetic pulse on the spatial-temporal structure of seismicity

Recently, the influence of high-power electromagnetic pulses, generated by the MHD facilities was discovered on the variations in a spatial-temporal structure of seismicity in the region under investigation (Zeigarnik, 2006). Based on large statistical data it was shown that the firing MHD facilities at the Garm (Pamir Mountains) and Bishkek (Northern Tien Shan) test sites produced an active electromagnetic impact on a significant volume of the earth's crust, which resulted in the activation of local seismicity and in the variation in the spatial-temporal distribution that existed before the MHD pulse.

During 1976–1978 the MHD generator was fired 34 times at the Garm test site with 3 MW power, a current amplitude up to 3 kA, and energy

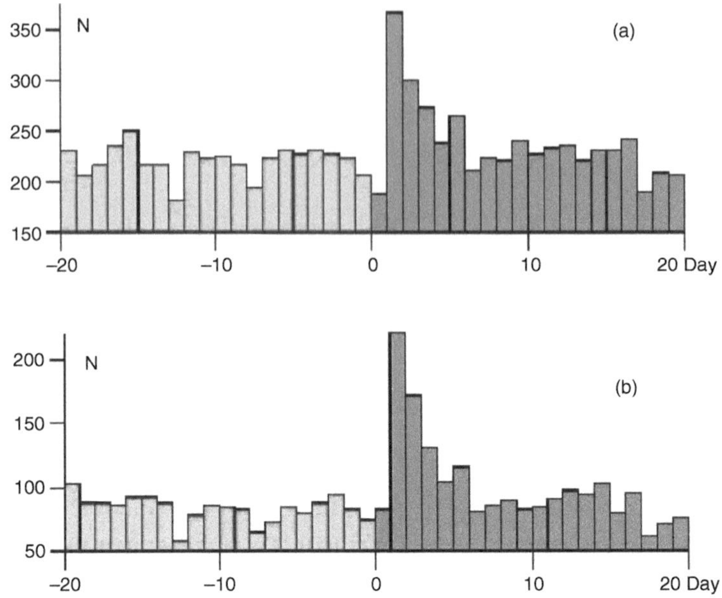

Figure 11 Time variation of daily number of earthquakes (N) in Northern Tien Shan and adjoining territories before ($t < 0$) and after ($t > 0$) runs of the MHD generator at the Bishkek test site for the entire area of observations (a) and for its upper layer of 5 km depth (b).

of 10^7 J. For the same period of observations there were 11000 earthquakes. Additional statistical data were obtained in 1983-1991 at the Bishkek test site. The observed data were processed for time intervals before and after 114 firings of the MHD generator. The earthquake catalog accepted for statistical treatment consists of 7600 events.

It was established that the ratio of the number of earthquakes which occurred within the specified time periods (±40 days at the Garm and ±20 days at the Bishkek test sites) after the MHD pulse to a corresponding number of earthquakes before the pulse is greater than one and may be two or even three (see Figure 11).

Between the firing of the MHD facility and the release of additional seismic energy there is a temporal delay, which is manifested by the growth of a number of relatively weak local earthquakes. For the Garm test site the maximum seismic activity happens on the 7th day, and for the Bishkek test site the influence of MHD pulses appears on the 2nd to 6th days and ends on the 23rd day.

The effect of the impact of EM pulses is the strongest in those geologic structures and locations, above which the MHD facility is located and in the upper 5 km layer of the earth's crust. In this case, after the EM sounding there is observed a redistribution of earthquakes by their energy class and spatial-temporal occurrence. The total energy released by earthquakes in the region of the Garm test site within 40 days after all 34 pulses of the MHD facility exceeds the energy released for the same period before the pulses by $3.8 \cdot 10^{13}$ J, i.e. by $1.1 \cdot 10^{12}$ J per MHD pulse. In the region of the Bishkek test site the total seismic energy released within 20 days after 114 pulses of the MHD generator (energy supplied to emitter is $1.1 \cdot 10^9$ J) exceeds the energy released before the MHD generator firing by $2.0 \cdot 10^{15}$ J, i.e. by $1.8 \cdot 10^{13}$ J per MHD pulse.

Normalization of the released energy values by the areas of the respective regions of the test sites provides close enough magnitudes of additionally released seismic energy equal to $6.6 \cdot 10^7$ J/km^2 per MHD pulse (Garm test site) and $6.2 \cdot 10^7$ J/km^2 per MHD pulse (Bishkek test site). This fact indicates an approximately identical reaction of the medium to high-power electromagnetic pulses in different seismically active regions. Therefore, the high-power electromagnetic pulses artificially increase the energy of seismogenic zones and initiate a release of energy accumulated in the earth's crust as relatively weak earthquakes. Therefore, the probability of an occurrence of a catastrophic event decreases (or, at least, its energy decreases). Indirectly, this assumption is confirmed by the study of the variations in the earth's global seismicity as a result of the impact of underground nuclear explosions (Zeigarnik, 2006)

This discovered effect can be considered as a basis for the development of seismic risk mitigation by the man-made release of tectonic stresses using the impact of high-power electromagnetic pulses on the zones of the earthquake source.

5. EM Exploration for Oil and Gas with the Use of Geophysical MHD Facilities

5.1. EM sounding for oil and gas on land

EM sounding in the Precaspian geological province

The EM exploration for oil and gas with the use of geophysical MHD facilities was conducted in the Astrakhan dome structure of the Near-Caspian depression in the Precaspian geological province. This is a region where operation with standard geophysical equipment was insufficiently effective. The salt-dome tectonics existing in this area make it very difficult to apply conventional seismic methods, and the large thickness (~4 km) and electrical conductivity (2-50 Ohm m) of sedimentary cover results in rapid damping of electromagnetic signals with depth. The geophysical MHD facility "Prikaspiy" (Figure 12) (similar to the "Pamir-2" MHD facility, Figure 9) was installed on two trucks, which made it possible to easily move

Figure 12 Mobile MHD generator "Prikaspiy" used for oil exploration.

the facility within the survey area. The effective output of the MHD facility was about 10 MW and the duration of the EM pulse about 7 s.

The main goal of the EM survey was detailed analysis of the geological structure of subsalt deposits, determining the petrophysical properties of hydrocarbon reservoirs, and mapping the prospective areas for oil and gas production. Two types of EM surveys with the application of the MHD facility "Prikaspiy" were used:

- type #1—EM sounding with a fixed electric dipole source of 15 km length (and total mass of 150 tons);
- type #2—EM profiling with a moving rectangular loop source of size 2 km × 4 km.

The electric dipole was grounded in the wells, the current in the load was 8 kA, and the electric impedance was 0.13 Ohm. The electrical dipole moment in this case was $1.2 \cdot 10^8$ A m, which by 4 orders exceeds the values, typical for operations with standard electric generators. The measurements were performed at 280 points in the survey area of about 1000 km^2 and along particular profiles with a distance between transmitter and receivers of up to 60 km. For comparison, the same dipole was supplied by a low current (up to 80 A) from a conventional electric generator.

The application of a high-power pulsed MHD generator increased the signal-to-noise ratio by approximately two orders of amplitude in comparison with conventional generators. As a result, the depth of the EM sounding was increased by two or three times in comparison with the standard depth of investigation in the conducting geoelectrical section typical for the Precaspian region.

The basic geophysical results of EM soundings can be summarized as follows (Velikhov et al., 2000; Fortov, 2000).

For the first time the total thickness and electrical conductivity of sedimentary cover in the Precaspian region were determined, and the geoelectrical information of subsalt limestone structure within the Astrakhan dome was obtained.

In a geological section at depths of 4000, 5500, and 7000 m, zones with improved reservoir properties were selected. In the upper horizon a zone was mapped with decreased electrical conductivity presumably related to the known gas condensate field. New information on the features of the subsalt horizons was obtained within an area of about 600 km^2.

Is was shown that for better delineation of deep structures prospective for hydrocarbons it is necessary to conduct integrated interpretation using the results of both seismic exploration and electrical sounding.

EM sounding in Eastern Siberia

Electrical sounding operations for hydrocarbons have been conducted in the Krasnoyarsk Territory of the Sobinskaya area (a region of the Podkamennaya Tunguska River) since 1986 with the application of the MHD facilities "Pamir-2" and "Sever-1" (Zeigarnik, 2006; Fortov, 2000). EM sounding with a fixed electric dipole source was used.

The MHD facility "Sever-1" was developed on the basis of the MHD facility "Pamir-2" modified for the geoelectrical conditions of Eastern Siberia (Figure 9). The transmitter was formed by an electric dipole of 12 km length with a resistance of 0.25-0.4 Ohm and induction of about 30 mH. The casing pipes of deep wells were used as grounding. The survey was performed along the profiles with a maximum distance to the source of up to 60 km. The results of the electrical survey with the MHD facility demonstrated the presence of local zones of increased conductivity within the dense carbonate-terrigenous deposits. In the northern area of the Sobinskoe uplift the thickness of Riphean deposits increases up to 2-2.5 km, and a bedding of the Sobinskoe uplift contains heterogeneous well-conducting strata.

Due to application of the high-power source for EM soundings, it was possible to separate the target horizon and Riphean deposits, and to show that the zones of decreased conductivity corresponded to hydrocarbon accumulation. It was found that the geoelectric section of the Sobinskoe uplift had block structures. Each of these blocks was characterized by a specific distribution of the conductivities along the depth. The gas condensate reservoir was outlined by areas of sharp change in electrical conductivity.

The application of the MHD facility in combination with a multichannel data acquisition system made it possible to reach a high production rate of operations in conjunction with the high spatial resolution and larger depth of investigation. The obtained results stimulated exploration for oil and gas

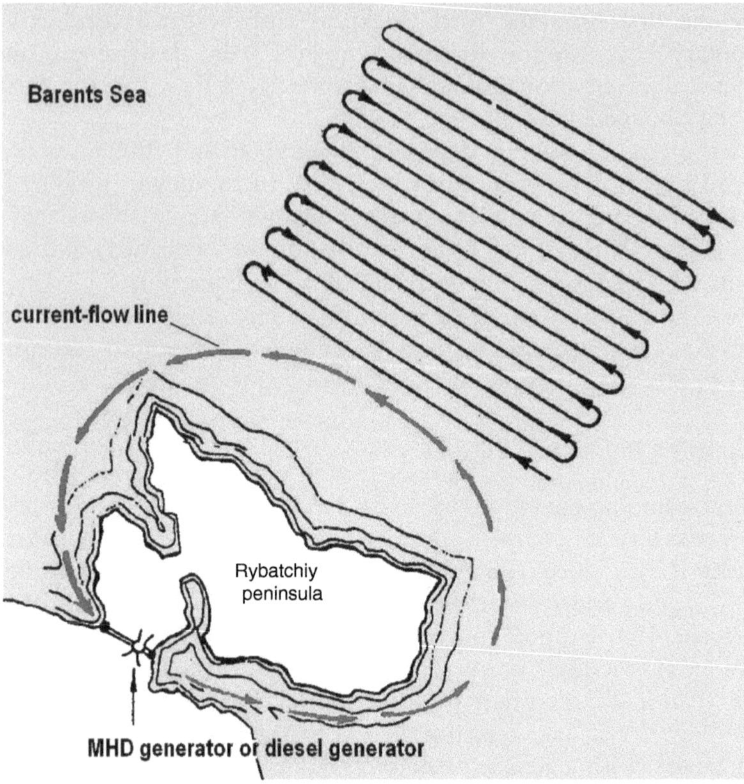

Figure 13 Diagram of electromagnetic soundings offshore, Barents Sea, with measurements using a ship-towed receiving antenna shown by the solid line.

and geoelectrical prospecting in a territory of the Tungus syneclise, where the application of seismic method was complicated.

5.2. EM exploration for oil and gas on a shelf

During the "Khibiny" experiment (see Section 2) the first ever practical electrical sounding of a shelf of the Barents Sea was performed with the high-power MHD facility "Khibiny" (Velikhov, 1989). The study was directed to determination of the total longitudinal conductivity of the sedimentary cover in an area of about 350 km^2 on the Barents Sea shelf near the Rybachiy Peninsula (Figure 13). The interpretation of the EM data was carried out by physical scale modeling. The obtained results revealed an anomaly of longitudinal conductivity in the central part of the survey area. It is possible that this anomaly was caused by hydrocarbon accumulation in this area. The obtained results were of significant value for offshore hydrocarbons exploration.

6. Deep Electromagnetic Studies in Ore-prospective Regions

In the framework of the MHD experiment "Khibiny," special attention was given to analysis of the deep structure of the highly conductive Pechenga and Imantra-Varzuga structures of the Kola Peninsula (see Section 2, Figures 5 and 6). Note that the Pechenga structure contains a widely known and well developed nickel deposit, while the Imantra-Varzuga structure is considered to contain ore-prospective zones as well.

Figure 14 schematically shows a regional map of the Pechenga structure (panel a) and a vertical section derived from seismic (panel b) and electromagnetic (panels c and d) data. The Pechenga structure is located at a distance of 100 km from the MHD facility "Khibiny." The conventional geophysical EM surveys in the Pechenga region conducted before the "Khibiny" experiment helped to determine the very low resistivity of the rocks of a productive strata of 10^{-1}-10^{-2} Ohm m, while the host rock resistivity was very high (10^{4}-10^{5} Ohm m). Thus, the resistivity contrast ratio was very high, which complicated the interpretation of the EM data. The importance of obtaining valid EM data is emphasized by the fact that the seismic method does not allow determination of whether the entire structure is productive and contains nickel or not, because seismic data cannot distinguish the pay section, clearly identifiable by high electrical conductivity.

The electromagnetic survey was performed along section "a-b" of 40 km length with a receiver step of 1 km, intersecting the Pechenga structure and going beyond its limits (see Figure 14).

Figure 14(c) shows by small circles the distribution of anomalous values of the vertical component of the magnetic field H_z^a along the profile. The solid line in the same figure represents the theoretically predicted values of the magnetic field for a conductive model of the Pechenga structure, presented in the deep section of Figure 14 (d). One can see that the synthetic predicted data fit well the observed data for the geoelectrical model shown in panel d. A similar geoelectrical structure was obtained using the method of interpretation based on electromagnetic migration developed in Zhdanov et al. (1988).

Thus, the high-power electromagnetic sounding made it possible for the first time to construct a deep geoelectrical model of the Pechenga structure. The obtained results were very important for determination of ore reserves at the given field.

The interpretation of deep EM sounding data in the ore-prospective Imandra-Varzuga zone was carried out by the method of electromagnetic migration (Zhdanov et al., 1988). The EM data observed along profile "C-D" (Figure 15(a)) were selected for interpretation. The profile intersects

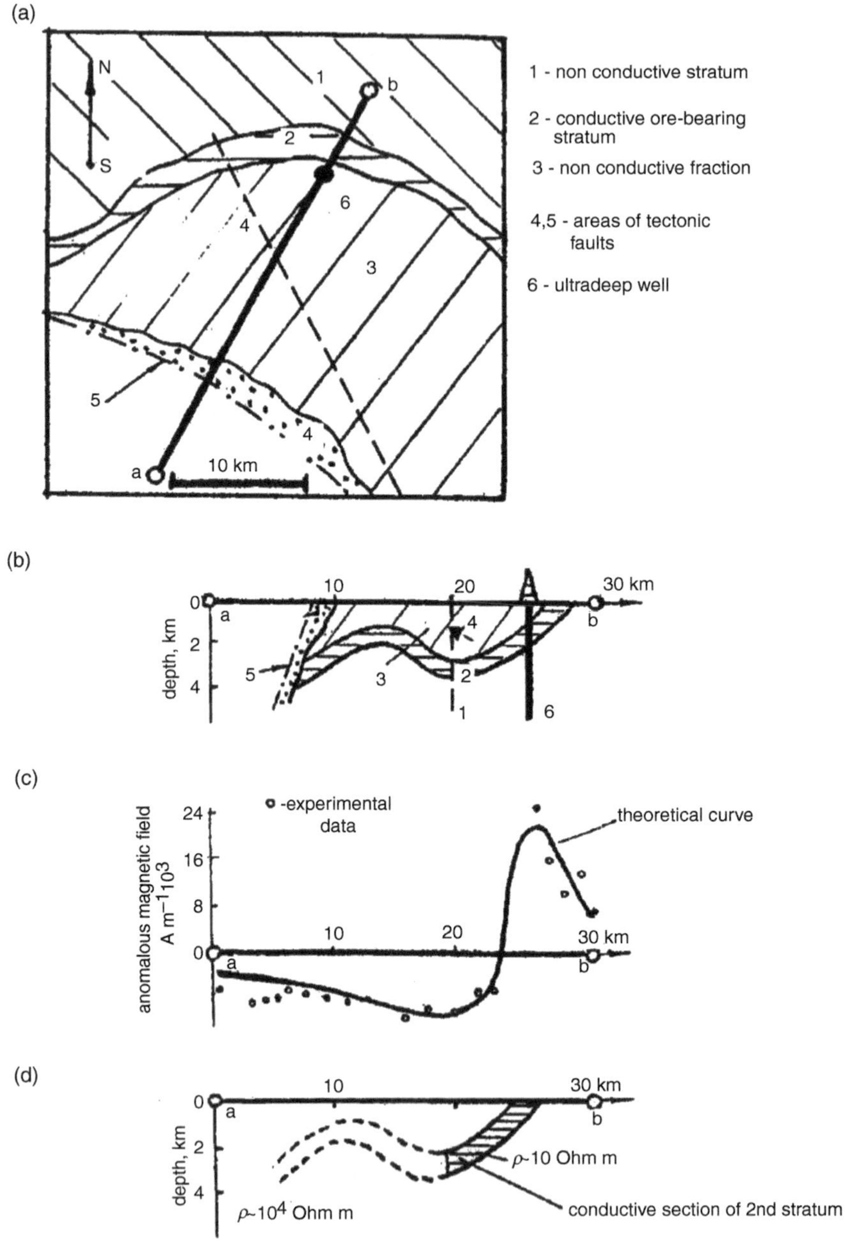

Figure 14 Deep structure of the Pechenga nickel deposit: (a) regional map; (b) model of the Pechenga structure obtained by seismic data; (c, d) models of the Pechenga structure obtained by electromagnetic sounding data.

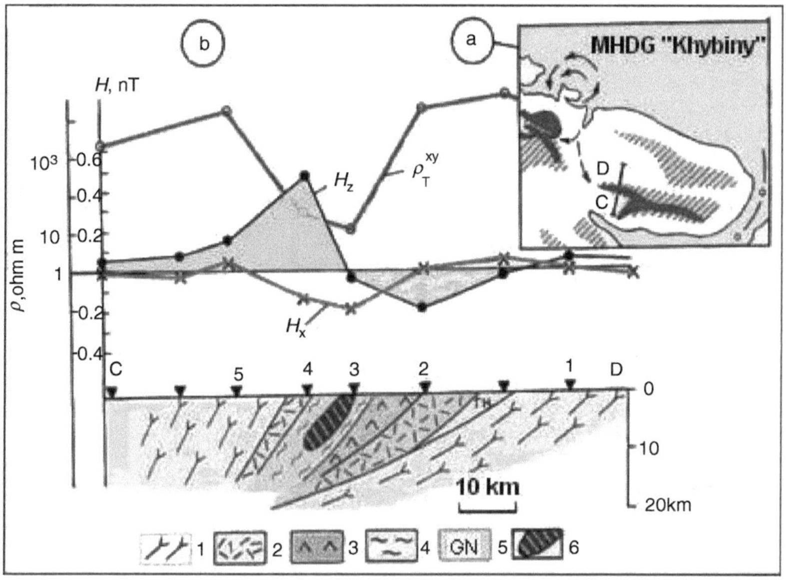

Figure 15 Deep electrical conductivity of the Imandra-Varzuga structure: (a) location of observation profile "C-D" in the Kola Peninsula; (b) profiles of recorded EM field of the MHD power source "Khybiny" over the Imandra-Varzuga structure and the geological and geoelectrical vertical resistivity section. Notation conventions: 1—granite-gneiss of the Lebyazhinsk series; 2—vulcanites of the Strel'ninsk series; 3—effusives of the Varzuga series; 4—slates and vulcanites of the Toming series; 5—gabbro-norites of the Fedorova tundra; 6—location of conductive current channels.

the western part of the structure (see Figure 5, section II-II, and Figure 8(b)). The Imandra-Varzuga zone represents a graben-syncline structure formed by volcanogenic strata extending by 350 km in the latitudinal direction. The structure has rocks with high electrical conductivity. The region under study is located at a distance of 300 km from the MHD facility, and the magnitude of galvanic current in the conductive rock is about 13 A (in the Pechenga structure about 40 A).

Figure 15(b) shows isolines of the migration field (isolines of horizontal magnetic component H_x) in the vertical plane going through the section of measurements and crossing the conductive structure. The isolines H_x were constructed for a time moment of 6.5 sec corresponding to the maximum intensity of the inductive currents. Figure 15 shows that, within a depth range of 5-10 km in a section of the earth's crust, the migration field outlines the conductive body. The obtained migration image was used for development of a starting model for determination of anomalous conductivity $\Delta\sigma$ within the conductive current channel "6" (Figure 15(b)). It was found that, within the shaded area, $\Delta\sigma$ varied from 0.1 S/m up to 0.5 S/m (Velikhov, 1989). Thus, the application of the high-power EM field generated by the MHD facility "Khibiny" in combination with the developed interpretation method allowed us to refine the ore-prospective structures and to determine quantitatively the electric conductivity of a highly conductive geologic object at a depth of 5 to 10 km which was inaccessible for soundings with conventional electric generators.

7. Conclusions

The results of this long-term experimental, numerical, and theoretical study of the applications of high-power electromagnetic soundings in the fundamental and applied geophysics can be summarized as follows:

- We have demonstrated the practical effectiveness of the application of high-power EM pulses to the solution of a wide range of geological and geophysical problems, including deep (up to ~100 km) geoelectrical sounding of the earth's crust, exploration and delineation of mineral deposits onshore and offshore, analysis of seismo-active zones, and earthquake prediction.
- The capability of analyzing the electric properties of rocks over areas up to 10–100 thousand km^2 has been shown.
- The economic efficiency of EM soundings with the application of high-power MHD facilities depends on the number of receivers used in the survey. The number of receiving stations should be more than 50–100 for simultaneous recording of the response for a single pulse.
- The effect of a high-power EM pulse, provided by an MHD facility, on the variation in the spatial-temporal distribution of seismic events has been discovered.
- The geophysical experiments with MHD generators have stimulated the development of new analytical and numerical methods of modeling and inversion of EM data, as well as the design and construction of new geophysical equipment for EM methods.
- A new concept for oil, gas, and mineral exploration onshore and offshore has been introduced and developed based on a new generation of MHD facilities.

REFERENCES

Fortov, V.E. (Ed.), 2000. Encyclopedia of Low-temperature Plasma. IV. Nauka, 507 pp. (in Russian).

Velikhov, E.P. (Ed.), 1989. Geoelectric Studies with a High-power Electrical Source in the Baltic Shield. Nauka, Moscow, 272 pp. (in Russian).

Velikhov, E.P., Volkov, Yu.M., 1981. Prospects of development of pulsed MHD power engineering and its application in geology and geophysics. Russian Scientific Center "Kurchatov Institute", Preprint IAE-3436, 28 pp. (in Russian).

Velikhov, E.P., Volkov, Yu.M., D'yakonov, B.P., 1975. Application of pulsed MHD generators for geophysical studies and earthquake prediction. In: Proceedings of VI International Conference on MHD Power Conversion, Washington DC, 5, 211-228.

Velikhov, E.P., Lisin, A.S., Matveenko, O.G., Panchenko, V.P., 2000. Application of MHD facilities for deep electromagnetic surveys of the earth's crust and exploration activities. Russian Scientific Center "Kurchatov Institute", Moscow. Preprint IAE-6177, 55 pp. (in Russian).

Volykhin, A.M., Bragin, V.D., Zubovich, A.V., 1993. Development of Geodynamic Processes in Geophysical Fields. Nauka, 158 pp. (in Russian).

Zeigarnik, V.A., 2006. Application of pulsed MHD facilities for earthquake prediction and exploration of hydrocarbon accumulations. In: Proceedings of 70 Years Academician E.P. Velikhov Anniversary Conference, Russian Scientific Center "Kurchatov Institute," 768 pp. (in Russian).

Zhamaletdinov, A.A., 1990. Model of Electrical Conductivity of Lithosphere by Results of Studies with Controlled Field Sources. Nauka, 159 pp. (in Russian).

Zhdanov, M.S., 2009. Geophysical Electromagnetic Theory and Methods. Elsevier, 856 pp.

Zhdanov, M.S., Keller, G.W., 1994. The Geoelectrical Methods in Geophysical Exploration. Elsevier, 873 pp.

Zhdanov, M.S., Matusevich, V.U., Frenkel, M.A., 1988. Seismic and Electromagnetic Migration. Nauka, Moscow, 376 pp. (in Russian).

CHAPTER 3

Elements of Active Geophysical Monitoring Theory

Anatoly S. Alekseev[a], Gennady M. Tsibulchik[a], Valery V. Kovalevsky[a], *and* Andrey S. Belonosov[a]

Contents

1. Introduction

Different kinds of geophysical precursors are used in earthquake prediction to reduce the probability of unexpected catastrophic earthquakes. Several years prior to such earthquakes, some anomalies occur within geophysical fields, such as crustal deformation, seismicity and electric conductivity. Zones that manifest such anomalies will migrate within a 200-300 km distance from the epicenter of a subsequent earthquake. The migration mechanisms for different anomalous zones, and the interrelations among them, reflect the evolution of seismotectonic stress fields, which could contain valuable information about the timing of seismic activity. Over the past 30 years, data has been collected from a number of different scientific disciplines and from several different countries, including China, Japan, USA, Greece, Turkey, and Russia.

In China, for example, information on hundreds of large earthquakes is now available from databases that are part of a substantial observational

[a] Institute of Computational Mathematics and Mathematical Geophysics SB RAS, Pr. Akad. Lavrentieva, 6, Novosibirsk, 630090, Russia

E-mail address: kovalevsky@sscc.ru (V.V. Kovalevsky).

Handbook of Geophysical Exploration: Seismic Exploration, Volume 40
ISSN 0950-1401, DOI: 10.1016/S0950-1401(10)04008-5

network. The accumulated information is used to investigate problems of earthquake prediction, using statistical analyses and various characteristics of geophysical anomalies at different pre-earthquake stages. Some previously successful earthquake predictions by Chinese geophysicists were based on synthesized information related to changes in the behavior of anomalies at medium- and short-range pre-earthquake stages (Ma et al., 1995; Mei, 1992; Zhang, G. and Zhang, Z., 1992). However, the use of multidisciplinary resources in earthquake prediction, and the results from them, raise new, important questions with regard to determining interrelationships among seismicity processes, variations in geophysical fields, and the sources of anomalies (Alekseev, 1993; Keilis-Borok and Molchan, 1968).

2. Main Properties of the Integral Precursor

Earthquakes occur through a massive rock failure, which begins in a source zone. Therefore, study of the pre-failure processes and monitoring of these processes are of major importance for earthquake prediction. An investigation of the rock failure in samples of various materials in the laboratory, as well as at large scale (in particular Earth crust blocks during earthquakes), reveal the general patterns of the rock failure process.

Step-by-step development of this process over time is the most general principle. Some kinetic laws and concepts of rock failure were established by S. N. Zhurkov and his colleagues from the Physical-Technical Institute of the Russian Academy of Sciences (RAS) in St. Petersburg (Zhurkov et al., 1977; Zhurkov, 1968). They are similar to the rock-failure-scheme concepts for large-scale objects in the Earth's crust during earthquakes that were proposed by researchers from the Institute of the Earth's Physics of the RAS (Myachkin et al., 1975, 1974). One of these researchers, G.A. Sobolev, formulated the following three principles, which he described as being of major importance in searching for earthquake precursors and predicting earthquakes (Sobolev, 1978):

1. The development of crack systems in pre-earthquake zones results from an increase in microcrack volume density, from the stage of increasing crack sizes (and decreasing numbers of cracks) to formation of large fractures;
2. The relationship describing step-by-step transition from small cracks to larger cracks, when smaller cracks reach some critical value, in accordance with the concentration criterion of S. N. Zhurkov, has the form

$$K^* = \frac{1}{\sqrt[3]{N}L}, \tag{1}$$

where N is the number of cracks of size L per unit volume, K^* is the critical average distance between cracks, measured in the units of average crack lengths. When the average distance between cracks becomes smaller than a certain critical value, there is an abrupt reorganization of the entire system of cracks, with an increase of average crack sizes (in some geometrical proportion) and a decrease of average volume concentration. Cracks tend to localize in the area of a future macrofracture. These phenomena are typical for any scale and any regime of loading.

3. Reorganization of the crack system manifests itself as a change of some characteristics of the medium in a developing earthquake source and as the formation of anomalies in some geophysical fields. In particular, concentration of the crack formation process can be evidence of a change in the seismicity regime for weak earthquakes, and in the appearance of rock anisotropy in a future earthquake source. The appearance of elastic anisotropy is most conspicuous in the formation of anomalies prior to large earthquakes (Nersesov et al., 1971).

Some geophysical fields can be affected by the opening of microcracks. In particular, gas and fluid permeability increases in those areas of the Earth's crust where this process takes place. As a consequence, the groundwater level, the intensity of gas flow, and the electrical resistance can change. The loosening of rock resulting from the increase in total crack volume must also cause local gravity anomalies.

Thus, the crack-density function as a measure of rock failure has some advantages, because this function is present in the formulation of all three principles. One more advantage is that the crack-density function can be more accurately and reliably determined from multidisciplinary data, owing to its presence in the models of various geophysical fields—the complementary principle (Alekseev, 1992).

An analysis of the pre-existing stress fields at earthquake locations (Miao, 1993; Wang and Liao, 1996) and the results from dilatancy-zone numerical modeling (presented below) suggest that cracks of some scale level can be formed at distances of 200-300 km from the source of a future earthquake.

Although the earthquake development process is "slow," lasting up to several hundreds of years, it is an energy-intensive process. Considerable rheological change in the medium takes place, and anomalous zones form within different kinds of geophysical fields. Crack openings in zones with increased shearing and tensile stresses is the most basic mechanism of medium change. Such zones are formed near sources of future earthquakes, if the spatial distribution of forces is nonuniform. Many seismologists consider that the initial stage of crack opening and the subsequent state of the medium when rock failure develops are connected to medium dilatancy (Nikolayevskii, 1982; Nur, 1971; Brace et al., 1966).

Dilatancy is the nonlinear loosening of rocks caused by crack formation from shear. This process takes place when tangential stresses exceed a certain threshold. A dilatancy zone includes points within an elastic medium, for which the following condition is satisfied:

$$D_\tau \equiv \tau - \alpha(P + \rho g z) - Y \geq 0, \tag{2}$$

where ρ is the density of rocks, g is the gravitational acceleration, α is the coefficient of internal friction, Y is the cohesion of rocks, z is the depth of the point, and P is the hydrodynamic pressure

$$P = -\frac{1}{3}(\sigma_{11} + \sigma_{22} + \sigma_{33}), \tag{3}$$

where σ_{ij} are the stresses and τ is the intensity of the tangential stresses:

$$\tau = \frac{\sqrt{3}}{2}[(\sigma_{11} - \sigma_{22})^2 + (\sigma_{22} - \sigma_{33})^2 + (\sigma_{33} - \sigma_{11})^2 \\ + 6(\sigma_{12}^2 + \sigma_{13}^2 + \sigma_{23}^2)]^{1/2}. \tag{4}$$

Condition 2 coincides with the Schleicher-Nadai's criterion of rock failure caused by shearing loads and describes the beginning of the rock-failure process. It can also be used at the "rock pre-failure" stage (when loading constitutes up to 60-90% of the critical value) for describing the shape of areas with rapid crack growth.

To demonstrate the complex character of dilatancy zones, we use the simplest model of the Earth's crust, which is taken as a uniform, isotropically elastic half-space. This complexity manifests itself even when a point force is a source of tectonic stresses. Exact solutions for elastic displacements and stresses, from a point source satisfying the conditions of zero stresses at the surface $z = 0$, were used to model the stress field in an elastic half-space (Alekseev et al., 1998b).

The domain surface $D_\tau = 0$ from Eq. (2) for the double force source at a depth of 15 km is shown in Figure 1. Here, the parameters of the elastic half-space are as follows:

$$v_p = 6000\ \text{m/s}, \quad v_s = v_p/\sqrt{3}, \quad \lambda = \mu = \rho v_s^2 = 3.48 \cdot 10^{10}\ \text{Pa}, \\ \rho = 2900\ \text{kg/m}^3, \quad g = 9.9\ \text{m/s}^2, \quad Y = 3 \cdot 10^6\ \text{Pa}, \quad \alpha = 0.5.$$

The double force is specified in the form of a moment tensor, $\boldsymbol{M} = M_0\mathbf{np}$, where M_0 is the dipole momentum (scalar), and $\mathbf{np}$ is a dyad characterizing the orientation of the force ($\mathbf{n}$) and the arm of the force ($\mathbf{p}$).

The domain surface $D_\tau = 0$ from Eq. (2) for the double-force couple source at a depth of 15 km is shown in Figure 2. The double-force couple is set in the form of a moment tensor, $\boldsymbol{M} = \frac{1}{2}M_0(\mathbf{np} + \mathbf{pn})$. In this case, the conformable matrix is symmetric.

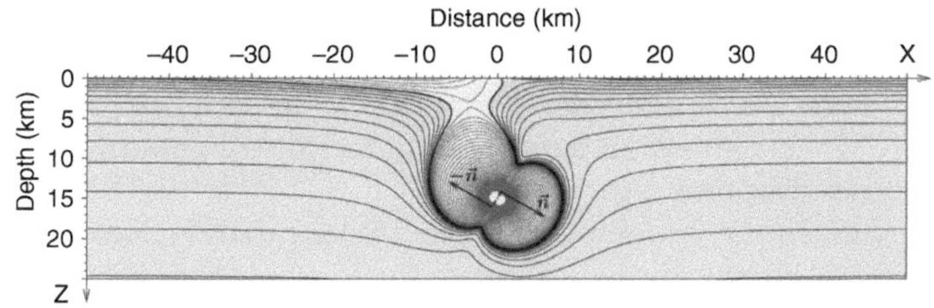

Figure 1 Exact solutions for elastic displacements and stresses from a point source satisfying the conditions of zero stresses at the free surface shows that there are two dilatancy zones—"source" zone in the vicinity of the elastic dipole application point and the "surface" zone in the layer near the free surface. Shapes of "source" and "surface" dilatancy zones in the plane $y = 0$ (15 km - source depth, double force $M_0 = 3 \times 10^{20}$ N, $\mathbf{n} = (\cos\varphi, 0, \sin\varphi)$, $\mathbf{p} = (0, 1, 0)$, $\varphi = 30°$.

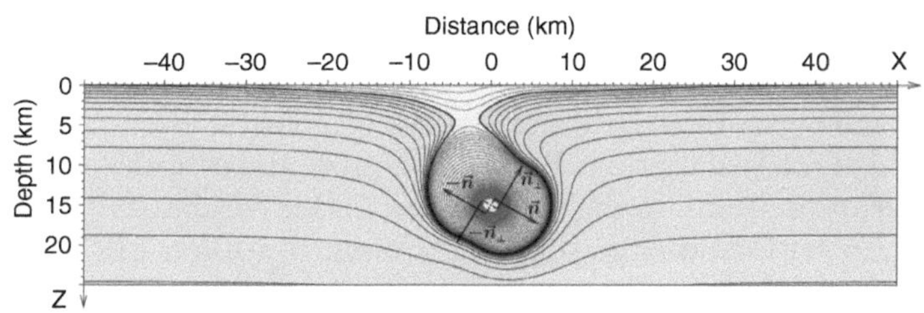

Figure 2 Same as in Figure 1 but for the double force couple $M_0 = 6.75 \times 10^{20}$ N, $\mathbf{n}_\perp = \mathbf{p}$, $\mathbf{p}_\perp = \mathbf{n}$, $\mathbf{n} = (\cos\varphi, 0, \sin\varphi)$, $p = (0, 1, 0)$, $\varphi = 30°$.

Tangential stresses inside the domain $D_\tau \geq 0$ dominate over compressional stresses. The resistance of the medium to shearing forces is overcome due to cohesion. Conditions favorable to crack increase are modeled. Note that the mechanisms of crack opening and the rheological changes of the geological medium in the zone $D_\tau \geq 0$ are not described by these solutions. The solutions are valid only for determining the transition from the elastic state to the state of nonlinear loosening.

The interesting feature in both Figures 1 and 2 is the formation of two dilatancy zones, which are the "source" zone in the vicinity of the source point and the "surface" zone in the upper part of the model. Here, the stress field from the source mostly affects tangential stresses, while the compressional stresses and the hydrostatic pressure contribute but slightly, due to the proximity of the surface.

The behavior of the surface dilatancy zone varies, depending on the following parameters: h (the source depth), M_0 (the source intensity), the angle φ (the force orientation in the source), and Y (the cohesion of the medium's elements). It can vanish with increasing source depth or merge with the source zone as the source intensity increases. In some cases, the horizontal size of the surface zone is 200 (or more) km, with a complex shape when projected onto the Earth's surface. It is easy to verify that the pattern of displacement in dilatancy zones along the surface can be complex, particularly when influenced by several sources distributed in space whose intensity varies over time.

Since anomalous geophysical fields are naturally related to surface dilatancy zones, to effectively investigate the sources of these anomalies, the location of the dilatancy zones must be determined as reliably as possible.

The condition of loosening rocks, taking into account the mechanisms of crack formation by tension of the medium, can be written in the form of a new criterion:

$$0 \le D_\sigma \equiv \begin{cases} \frac{1}{2}\sigma_1(1-\sin\varphi) - \frac{1}{2}\sigma_3(1+\sin\varphi) - Y\cos\varphi, & \text{if } -\sigma_3 \ge \sigma_p, \\ -\sigma_3 - \sigma_p, & \text{if } -\sigma_3 < \sigma_p, \end{cases} \tag{5}$$

where σ_1 and σ_3 are the largest and smallest principal stresses, respectively; φ is the angle of internal friction; Y is the cohesion, and σ_p is the rock strength. This criterion determines dilatancy zones under conditions in which the medium can withstand large shearing stresses, but offers less resistance to tensile forces.

Note that so far, the existence of surface dilatancy zones under real conditions should be considered a hypothesis. The use of this hypothesis for practical analysis of anomalies within various geophysical fields is an attempt to find reasons for the evolution of these anomalies, and to clarify the mechanisms of their interrelation.

Monitoring of the crack-density variation in the source zone is of special importance for short-range earthquake prediction. This monitoring should not be separated from observations of the surface dilatancy zone and the anomalous fields associated with it. First, the location of the future source is not known exactly, and its determination is closely related to the behavior of surface anomalous zones. Second, the reliability and accuracy of estimating the crack-density function in the source zone depends on the information regarding field anomalies in surface zones.

Before proceeding to the scheme for estimating the integral precursor in the source using multidisciplinary monitoring, it is reasonable to consider the relationship between crack-formation processes in the source and surface dilatancy areas.

Generally, a stress field is an energy-charged medium in which a relationship exists between the source and surface zones. The lines of largest tangential stress (or largest tensile stress) provide an estimate for the predominant orientation of cracks that occur in zones of the source and surface types.

When dilatancy zones of the two types are combined into one zone, there is a joint area of fracturing. This area combines the surface zones of anomalous fields with the source zone, in which the development of the rock failure process can directly influence the change in the geophysical anomalies. When dilatancy areas are separated, the source zone can retain its direct influence on the anomalies of some fields (for example, on the values of the velocities v_P and v_S) by a joint area of introduced anisotropy within the medium. The anisotropy coefficients can be expected to vary in a special way during the crack growth process, because the orientation of the axis of symmetry remains the same. This can simplify the problem of estimating the average number of cracks by a high-resolution vibroseismic method (Alekseev et al., 2001, 1999, 1998a,b).

3. Multidisciplinary Model of Integral Precursor and Combined Inverse Problems

An analysis of crack-system development at earthquake sites prior to seismic activity shows that earthquake prediction of rock failure should include a determination of the major space-time characteristics of the crack systems at these sites. Such investigations should be performed in the dilatancy zones where the crack systems are developed.

Observations of geophysical anomalies enable investigators to determine the crack-density function. It was assumed in the previous section that the crack-opening processes in dilatancy zones are related to the mechanisms forming anomalous fields. Qualitatively, the formation of anomalies in gravitational, electrical-conductivity, groundwater-level, and gas- and fluid-permeability fields can be explained by cracks. It is evident that special investigations are needed to obtain quantitative models of geophysical fields in fractured media (Brace et al., 1966).

In the process of deformation (prior to failure), loosening is characterized by the volume expansion (dilatation) $\mathbf{\Theta} = \mathbf{div U}$, where the divergence is calculated from the elastic displacement vector. It is assumed here that the vector components are sufficiently smooth (differentiable) functions. If we consider a small volume V_0, which is V_1 after deformation, then $V_1 = V_0(1 + \mathbf{\Theta})$. If the medium's density is ρ_0, after deformation it is $\rho_1 = \rho_0/(1 + \mathbf{\Theta})$. For large, deformed volumes, this loosening is considerable. It generates an anomaly in the gravitational field $V(x, y, 0) = V^0(x, y)$, which we can use to solve the inverse problem

$$\Delta V = -4\pi\rho_1\mathbf{\Theta}, \qquad V|_{z=0} = V^0(x, y), \tag{6}$$

to determine density $\rho_1(x, y, z) = \rho_0/(1 + \mathbf{\Theta})$, and loosening $\mathbf{\Theta}$ (if this inverse problem can be solved uniquely and the initial density is known).

The main difficulty—the one that leads us to consider multidisciplinary (combined) statements of inverse problems—is that the inverse problem posed by Eq. (6), does not have a unique solution. It is ill-posed, an attempt to find a three-dimensional function $\rho_1(x, y, z)$ using a known two-dimensional function $V^0(x, y)$. This is impossible without additional information. The significance of combining problem statements is in the use of additional information from the solution to state a new inverse problem for the same physical quantity.

The approach for determining the characteristics of cracking (the integral precursor) using data from geophysical anomalies can also utilize the idea of the surface dilatancy zone. Let us introduce a medium's volume expansion (loosening) function, $\theta(x, y, z, t)$. This function can be considered piecewise continuous; and it is assumed to be equal to the total relative volume of cracks in the medium's unit volume. The number of cracks in the unit volume can be determined by the formula $N = \theta(x, y, z, t)/\theta_L(x, y, z, t)$, where $\theta_L(x, y, z, t)$ is the relative average volume of a crack with length L.

Let us consider a combined inverse problem for gravitational and electric data, groundwater-level evaluation, and the seismic method for measuring the effective anisotropy coefficients of cracked rocks, on the basis of the complementary principle of geophysical methods (Alekseev et al., 1995; Alekseev, 1992)—to obtain reliable estimates of the function. Each of these methods is based on measurements at the surface $z = 0$ of a corresponding geophysical field

$$U_\nu(x, y, 0, t_k) = U_\nu^0(x, y, t_k), \tag{7}$$

here, $t_k = kT_\nu$, with T_ν the time interval between the recording times of field values during monitoring.

Methods for solving direct and inverse problems exist for all geophysical fields that are used in the problem of earthquake prediction (Alekseev, 1967; Alekseev et al., 1971, 1958; Alekseev and Mikhailenko, 1977; Alekseev and Tsibulchik, 1996; Alterman and Karal, 1968; Babich et al., 1985; Mikhailenko, 1978; Petrashen, 1978). In direct problems, the equation for the field,

$$L_\nu(U_\nu, \alpha_\nu, \beta_\nu) = f_\nu(x, y, z, t), \tag{8}$$

surface conditions,

$$l_\nu(U_\nu, \alpha_\nu, \beta_\nu)|_s = h_\nu(s, t) \tag{9}$$

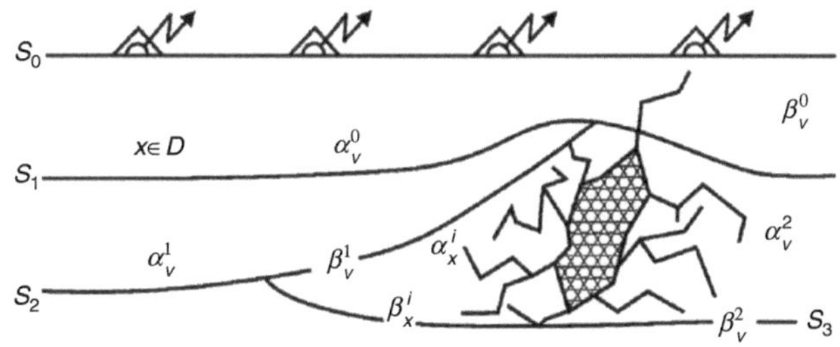

Figure 3 Statement of the combined inverse problem. The physical and geometrical characteristics of the medium $\alpha_\nu(x, y, z)$, $\beta_\nu(x, y, z)$, the field sources and the surfaces S_i.

and the initial data

$$U_\nu(x, y, z, t)|_{t=0} = U_\nu^0(x, y, z) \tag{10}$$

are assumed to be given. Here $\alpha_\nu(x, y, z)$, $\beta_\nu(x, y, z)$ are the physical and geometrical characteristics of the medium; $f_\nu(x, y, z, t)$ are the external volume sources of the field; and $h_\nu(s, t)$ are the sources at the surface S. The statement of the combined inverse problem is illustrated by Figure 3.

Numerical methods for solving direct problems exist for many of the above-mentioned fields. These methods use specified geological medium characteristics $\alpha_\nu(x, y, z)$, $\beta_\nu(x, y, z)$, the field sources, and the surface S. In inverse problems, the following information is known: the field $U_\nu^0(s_i, t)$ at a series of points s_i at the surface S and the sought-for characteristics of the medium α_ν, β_ν, or other elements of the problem (the shape of the surface S, some sources f_ν or h_ν).

The model of a multidisciplinary (combined) inverse problem determines the integral precursor $\theta(x, y, z, t)$, i.e., the relative crack-density function. In this case, all geometrical and physical parameters of the medium, with the exception of the function $\theta(x, y, z, t)$, are considered to be known, and the function $\theta(x, y, z, t)$ is considered to be independent of time during each field measurement $t_k = kT_\nu$.

An optimization method can be used to solve the combined inverse problem. Let $\beta_\nu(x, y, z, t, \theta)$ represent the operator for calculation of the field $U_\nu(x, y, z, t)$ in the direct problem for the method with the number ν. The problem lies in determining $\theta(x, y, z, t)$ which minimizes the functional:

$$I(\theta) = \underbrace{\min}_{\theta \in M_\theta} \sum_{\nu=1}^{m} \gamma_\nu [U_\nu^0(x, y) - B_\nu(x, y, 0, t_k, \theta)]^2, \tag{11}$$

where γ_ν are the weight coefficients for individual methods, M_θ is an *a priori* set of possible solutions θ; $U_\nu^0(x, y)$ is the measured field, and

$\beta_v(x, y, 0, t_k, \theta)$ is the modeled field. The functional represented in Eq. (11) assumes no statistical correlation between the measured fields.

Optimization methods traditionally involve considerable computational difficulties. They are associated with simultaneously solving a large number of direct problems for different fields. In addition, the functional being minimized often has many local minima, making searches for the global minimum difficult. To solve such problems successfully, one should use high-performance computers and good initial approximations to the sought-for functions.

The seismic method, using powerful vibroseismic sources, can yield more detailed data about the medium structure, including an evolution of fractured zones. Employing observation systems with multiple overlaps, these sources provide resolution similar to well-known results in seismic prospecting.

Here, we will not consider the capabilities of active seismology (Alekseev et al., 1997) using powerful vibrators, signals from which can be recorded at distances of up to 500-1000 km. Rather, we shall consider vibroseismic sounding of dilatancy zones as necessary instruments to increase the reliability and accuracy of the obtained information.

4. Methods for Vibroseismic Monitoring of Seismic Prone Zones

The Siberian Branch of the RAS has gathered unique experimental data from field observations (Alekseev et al., 2005, 2004, 1996, 1995). Seismograms have been obtained at distances of up to 400 km, and records of mono-frequency signals have been obtained at distances of up to 1,000 km, by using vibrators with forces of 50, 100, and 250 tons. Among the important problems for active seismology are the methods for vibroseismic monitoring of seismic-prone zones and, in particular, a method for determining the function $\theta(x, y, z, t_k)$. To determine $\theta(x, y, z, t_k)$, the Deep Seismic Sounding (DSS) scheme can be used, together with the Common Depth Point (CDP) scheme, at profiles 150-200 km in length over the source of an impending earthquake.

At the stage of long-range prediction, the period between soundings can be from 6 months to one year. At the stage of short-range prediction, soundings must be more frequent, and observation systems must be more detailed.

We assume that the medium's properties vary only slightly between measurements. These small variations can be made the major elements of variability in seismic cross sections, with the help of the "interframe correlation" method (i.e., by the subtraction of sequential images of the medium and analysis of increments).

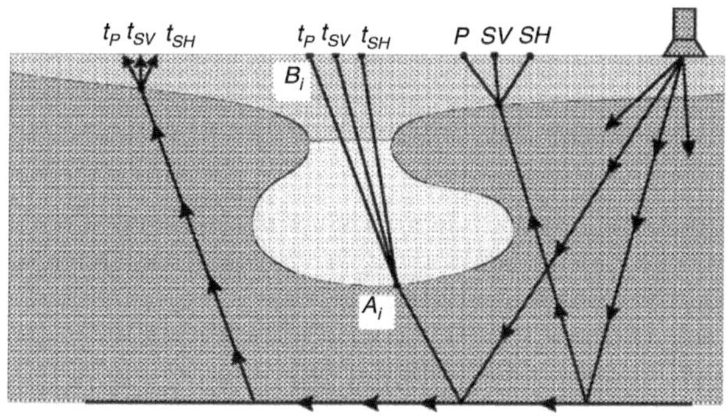

Figure 4 The general scheme of monitoring of the medium using a vibrational source. Vibroseismic observation profile of P, SV, and SH waves for the monitoring function $\theta(x, y, z, t_k)$ in the dilatancy zone. This function characterizes the development of cracks systems in the earthquake source and in anomaly zones of geophysical fields. To determine $\theta(x, y, z, t_k)$, the Deep Seismic Sounding (DSS) scheme can be used together with the Common Depth Point (CDP) scheme at profiles 150-200 km in length over the source of an incipient earthquake.

An analysis of experiments on rock failure shows that the variability in crack sizes is larger than the variability in the dominant orientation of cracks (Nur, 1971). Sometimes crack sizes vary abruptly during the transition to the next scale level of rock failure (Zhurkov, 1968). This property enables us to simplify and refine the algorithms for processing of vibroseismic observations.

A general monitoring scheme is shown in Figure 4. Automatic data processing using this scheme assumes the development of migration methods and the solution of inverse dynamic problems within the total system of equations for dynamic elasticity in an anisotropic medium:

$$\frac{\partial \sigma_{ij}}{\partial x_j} + \rho \frac{\partial \Phi}{\partial x_i} = \rho \frac{\partial^2 U_i}{\partial t^2} \tag{12}$$

with the generalized Hooke's law

$$\sigma_{ij} = C_{ik}(\nu_s, K_s, K_f, \theta)\varepsilon_{kj}, \tag{13}$$

where σ_{ij} are the stresses; ε_{kj} are the deformations; Φ is the gravitational potential; C_{ik} are the effective parameters of anisotropy for the fractured medium; ν_s is the Poisson's coefficient for the imbedding (elastic isotropic) medium; K_s is the modulus of volume deformation in the imbedding medium; K_f is the modulus of volume deformation for the liquid or gaseous phase in the porous half-space; and θ is the volume density of cracks.

Equations (12) (often lacking the gravitational potential term) and Hooke's law (Eq. (13)) are widely used in geophysics to describe seismic waves in fractured media. There are several formulations of the generalized Hooke's law, with the anisotropy coefficients approximating the wave processes in fractured media at low frequencies (Budiansky and O'Connell, 1976; Crampin, 1978, 1984). The density of cracks θ is present explicitly in Hooke's law (Eq. (13)); depending on the assumed shape of cracks. The following formula is given in Hoenig (1979):

$$e = \frac{2NA^2}{\pi P_e}. \tag{14}$$

This formula defines the density of cracks e for the densely packed N parallel-plane elliptic cracks of the area A with perimeter P_e. It is valid for any plane cracks with convex boundary shapes.

Equations (12) and (13) form the basis for seismic data processing in seismic prospecting and seismology. In practice, simplified kinematic approaches have been used so far.

The observation scheme for waves reflected and refracted from the Moho surface in the Earth's crust, using vibroseismic sounding of the source and surface dilatancy zones, is shown in Figure 4. The presence of fractures in these zones, and the changes in their volume density during the periods between monitoring sessions, can be determined from the changes in anisotropy coefficients and wave propagation velocities. The transverse wave S splits into SV and SH waves at the boundaries of dilatancy zones. The depth and shape of the boundaries, as well as the wave propagation velocities, can be determined (by well-known methods) from the lags Δt_{SV}, Δt_{SH} in the arrival times of the corresponding waves at the points B^i_{SV}, B^i_{SH}, and the source located at the point A^i.

For the fractured model of the type (Hoenig, 1979), the velocities of all three wave types v_P, v_{SV}, and v_{SH} are determined by the formula

$$v = \frac{v_0}{\sqrt{1 + ef(\gamma)}}, \tag{15}$$

where v_0 is the wave velocity in the medium before the appearance of cracks, $f(\gamma)$ is one function for all types of waves (Garbin and Knopoff, 1975; Crampin, 1978), and γ is the angle between the direction of wave propagation and the direction normal to the orientation of plane cracks. The quantity e from Eqs (13) and (14) is the sought-for function $\theta(x, y, z, t_k)$. It can be determined not only from seismic monitoring data, but also from routine seismologic observations of the velocities $v_P(t_k)$, $v_{SV}(t_k)$, and $v_{SH}(t_k)$ at seismic stations.

Estimating the sensitivity of active monitoring to changes in the elastic characteristics of the interior Earth's crust zone can be made using mathematical modeling (Kovalevsky, 2006). The model of the Earth's crust-mantle system in the form of a layer at a half-space with different velocity values of elastic waves is considered. The mathematical statement of the problem is made by approximating the acoustic wave equation. It is assumed that the vibrational source has a constant oscillation frequency and that the zone of changes in the medium has a spherical shape. The wave field in the medium is calculated using a ray approximation. Wave field variations in the medium and at the free surface are determined for the case of small velocity changes by calculating the beam pattern of a fictitious 3D source in a diffraction approach. As a result of such modeling, we estimated the sensitivity of active monitoring methods to harmonic vibrational signals.

The relationship between the relative variations in velocity within the zone of parameter variation and those of the recorded-signal amplitudes is as follows:

$$\frac{\delta c}{c} = 3 \cdot 10^{-3} \alpha \frac{\delta u}{u} \left(\frac{R_{V-Z} R_{Z-S}}{r_0 R_{V-S}} \right) \left(\frac{\lambda}{r_0} \right)^2, \tag{16}$$

where $\delta c/c$ are the relative variations in wave velocities within the zone of parameter variation, $\delta u/u$ are the relative variations in the signal amplitudes recorded on the free surface, R_{V-Z} is the distance between the vibrator and the zone of parameter variation, R_{Z-S} is the distance from the zone of parameter variation to the recording point (seismometer), r_0 is the radius of the zone, λ is the wavelength, and α is the reflection coefficient, lying within 0.15-1 for the model and the wave velocities in the core and mantle (Kovalevsky, 2006).

Experience shows that variations in the amplitudes of monofrequency signals at distances of 100-400 km from the vibrator, at the existing microseismic noise level, can be determined with an accuracy of 10^{-2}. Therefore, monitoring at the frequency $f = 6$ Hz (character wavelength $\lambda = 1$ km) and at typical source-recorder and source-anomaly distances of 50-100 km (and for the zone of parameter variation with a radius of 1-10 km) gives the following estimates of the relative variations in seismic wave velocities:

$$\begin{aligned} r_0 &= 1\ \text{km}, & \delta c/c &= 10^{-2} - 10^{-3}, \\ r_0 &= 10\ \text{km}, & \delta c/c &= 10^{-5} - 10^{-6}. \end{aligned} \tag{17}$$

These estimates show that the sensitivity of the active monitoring method is rather high for seismologic methods. This suggests its potential effectiveness for monitoring changes in the stressed-deformed state in the dilatancy zones of future earthquakes.

5. Conclusions

Although the earthquake development process lasts for a long time (up to several hundred years), it is an energy-saturated process. Substantial rheological changes in the geologic medium take place at imminent-earthquake sites, and the varying sorts of anomalous zones are formed within geophysical fields at such a site. Cracks opening in zones of increased shearing and tensile stresses is the most universal mechanism of rock changes. Most seismologists believe that the initial stage of crack opening, and the subsequent state of the medium, when rock failure processes are developing, are associated with a medium's dilatancy.

This paper describes the formation of surface dilatancy zones, which can cause variations in geophysical fields—and can thus be potential earthquake precursors. This process is illustrated by using an example of point sources for double forces and a double pair of forces. The necessary condition for the creation of dilatancy zones is an Earth surface that is free from stresses. In this case, dilatancy zones are formed from any distribution of forces, creating a nonzero component of shearing stresses. Therefore, most earthquakes are accompanied by the creation of such zones. Note that the existence of surface dilatancy zones under real conditions at this point should be taken as a hypothesis. Using this hypothesis—for practical analysis of extensive accumulated data on the monitoring of anomalies at various geophysical fields—is an attempt to find the reasons for the evolution of these anomalies and elucidate the mechanisms of their interrelation.

In this paper, a mathematical model of an integral earthquake precursor is proposed. Physically, it represents the space-time function of crack density in a zone of the highest stresses on the Earth's surface, which manifests itself in anomalous geophysical fields. Mathematically, the integral precursor is determined using the optimization method of a multidisciplinary (combined) statement of the inverse problem for the corresponding geophysical fields (the field of displacements and deformations on the Earth's surface, the electric conductivity field, anomalies of the gravitational field, the groundwater level, etc.).

A vibroseismic monitoring scheme for dilatancy zones is discussed in detail. Estimates of the accuracy obtained with the use of this scheme, which employs powerful vibrators and recording systems capable of long signal accumulation, are presented. We show that it is possible to detect relative changes in seismic wave velocity of about 10^{-5}-10^{-6} in a internal zone, with radius 10 km, using a vibromonitoring system with a 100-ton vibrator and a recording system with 50-100 km offset. These estimates show the high resolution of active vibroseismic monitoring compared to standard seismological methods for the probing of seismic-prone zones within the Earth's crust.

ACKNOWLEDGMENTS

The work has received support from the Russian Foundation for Basic Research, Grants No. 04-05-64177, 05-05-64245, 06-05-64265, 07-05-00858, 07-07-00214; Fundamental Research Program of the RAS No. 16.5 and, 16.6; Interdisciplinary Integration Projects of the SB RAS No. 16, 57 and 133.

REFERENCES

Alekseev, A.S., 1993. A multidisciplinary mathematical model of combined foreshock for earthquake prediction research. Journal of Earthquake Prediction Research 2, 137–151.

Alekseev, A.S., 1992. Complementary features of geophysical methods and the computational aspect of joint data inversion. In: Proceedings of 54th Meeting of European Association of Exploration Geophysics, Paris, pp. 750–751.

Alekseev, A.S., 1967. Inverse dynamic problems of seismics. In: Lavrentyev, M.M. (Ed.), Some Methods and Algorithms for Interpretation of Geophysical Data. Nauka Publishing House, Moscow, pp. 9–84 (in Russian).

Alekseev, A.S., Babich, V.M., Gelchinsky, B.Ya., 1958. On determination of the intensity of waves in elasticity theory by the ray method. Dokladi Akademii Nauk SSSR 1, 661–664 (in Russian).

Alekseev, A.S., Belonosov, A.S., Petrenko, V.E., 2001. On the concept of multidisciplinary earthquake prediction with the use of an integral precursor. In: Problems of the Dynamics of Lithosphere and Seismicity, Computational Seismology, vol. 32. GEOS Publishing House, Moscow, pp. 81–97 (in Russian).

Alekseev, A.S., Belonosov, A.S., Petrenko, V.E., 1999. On the multidisciplinary approach to determination of an integral earthquake precursor. Journal of Earthquake Prediction Research 10, 462–471.

Alekseev, A.S., Belonosov, A.S., Petrenko, V.E., 1998a. A mathematical model of determining the stress field and dilatancy zones by geodetic data. Bulletin of the Novosibirsk Computing Center, Series: Mathematical Modeling in Geophysics 4, 15–22.

Alekseev, A.S., Belonosov, A.S., Petrenko, V.E., 1998b. On the multidisciplinary approach to determination of an integral earthquake precursor. Bulletin of the Novosibirsk Computing Center, Series: Mathematical Modeling in Geophysics 4, 23–45.

Alekseev, A.S., Chichinin, I.S., Korneev, V.A., 2005. Powerful low-frequency vibrators for active seismology. Bulletin of the Seismological Society of America 95, 1–17.

Alekseev, A.S., Geza, N.I., Glinsky, B.M., Emanov, A.F., Kashun, V.N., Kovalevsky, V.V., Manshtein, A.K., Mikhailenko, B.G., Seleznev, V.S., Serdukov, S.V., Sobisevitch, A.L., Sobisevitch, L.E., Soloviev, V.M., Khairetdinov, M.S., Chichinin, I.S., Yushin, V.I., 2004. Active Seismology with Powerful Vibrational Sources. Novosibirsk, ICM&MG SB RAS, Filial "Geo" Publishing House of the SB RAS (in Russian).

Alekseev, A.S., Glinsky, B.M., Kovalevsky, V.V., Mikhailenko, B.G., 1997. Problems of active seismology. In: Fuchs, K. (Ed.), Upper Mantle Heterogeneities from Active and Passive Seismology. In: Nato ASI Series, vol. 17. pp. 123–130.

Alekseev, A.S., Glinsky, B.M., Emanov, A.F., Kovalevsky, V.V., Mikhailenko, B.G., Seleznev, V.S., Yushin, V.I., Khairetdinov, M.S., Kasahara, J., Sato, T., Mochizuki, K., 1996. Investigation of the structure of the vibroseismic field from powerful sources. In: Problems of Geoacoustics - Methods and Facilities. MSMU Publishing House, Moscow, pp. 155–160.

Alekseev, A.S., Glinsky, B.M., Kovalevsky, V.V., Mikhailenko, B.G., 1995. A multidisciplinary mathematical model for earthquake prediction studies and vibroseismic monitoring of seismic prone zones. In: Proceedings of 2nd International Conference of Seismology and Earthquake Engineering. Teheran: JJEES 1, 97–104.

Alekseev, A.S., Lavrentyev, M.M., Mukhometov, R.G., Nersesov, I.L., Romanov, V.G., 1971. A numerical method for determining the structure of the Earth's upper mantle. In: Mathematical Problems of Geophysics, vol. 2. CC SB USSR AS Publishing House, Novosibirsk, pp. 143–165 (in Russian).

Alekseev, A.S., Mikhailenko, B.G., 1977. Numerical modeling of seismic wave propagation in a radially inhomogeneous Earth's model. Dokladi Akademii Nauk SSSR 235, 46–49 (in Russian).

Alekseev, A.S., Tsibulchik, G.M., 1996. Inverse dynamic problems of wave diffraction in the problem of seismic monitoring. In: Vibrational sounding of the Earth vibrosounding). Nauka Publishing House, Moscow, pp. 22–25 (in Russian).

Alterman, Z.S., Karal, F.C.-jr., 1968. Propagation of elastic waves in layered media by finite-difference methods. Bulletin of the Seismological Society of America 58, 367–398.

Babich, V.M., Buldyrev, V.S., Molotkov, I.A., 1985. The Space-time Ray Method. LSU Publishing House, Leningrad (in Russian).

Brace, W.F., Paulding, B.W., Scholz, C., 1966. Dilatancy in the fracture of crystalline rocks. Journal of Geophysical Research 71, 3939–3952.

Budiansky, B., O'Connell, R.J., 1976. Elastic moduli of a cracked solid. International Journal of Solids and Structures 12, 81–97.

Crampin, S., 1984. Effective anisotropic elastic constants for wave propagation through cracked solids. Geophysical Journal of the Royal Astronomical Society 76, 135–145.

Crampin, S., 1978. Seismic wave propagation through a cracked solid: Polarization as a possible dilatancy diagnostic. Geophysical Journal of the Royal Astronomical Society 53, 467–496.

Garbin, H., Knopoff, L., 1975. Elastic moduli of a medium with liquid-filled cracks. Quarterly of Applied Mathematics 33, 301–303.

Hoenig, A., 1979. Elastic moduli of a nonrandomly cracked body. International Journal of Solids and Structures 15, 137–154.

Keilis-Borok, V.I., Molchan, G.M., 1968. Computational Seismology. GEOS Publishing House, Moscow (in Russian).

Kovalevsky, V.V., 2006. Estimation of sensitivity of the active monitoring method by harmonic signals. Bulletin of the Novosibirsk Computing Center Ser.: Mathematical Modeling in Geophysics 11, 105–114.

Ma, Li, Chen, Jianmin, Chen, Qifu, Liu, Guiping, 1995. Features of precursor fields before and after the Datong–Yanggao Earthquake swarm. Journal of Earthquake Prediction Research 4, 1–30.

Mei, Shi-rong, 1992. Progress in earthquake prediction in China during the 80ies. Journal of Earthquake Prediction Research 1, 43–57.

Miao, Liangtian, 1993. Monitoring and prediction of the Datong earthquake. Journal of Earthquake Prediction Research 2, 299–310.

Mikhailenko, B.G., 1978. Calculation of theoretical seismograms for multidimensional-inhomogeneous models of media. In: Conditionally Well-Posed Problems of Mathematical Physics in the Interpretation of Geophysical Observations. CC SB USSR AS Publishing House, Novosibirsk, pp. 75–88 (in Russian).

Myachkin, V.I., Kostrov, B.V., Sobolev, G.A., Shamina, O.G., 1975. Fundamentals of the physics of the source and earthquake precursors. In: Earthquake Source Physics. Nauka Publishing House, Moscow, pp. 6–29 (in Russian).

Myachkin, V.I., Kostrov, B.V., Sobolev, G.A., Shamina, O.G., 1974. Laboratory and theoretical investigations of the earthquake preparation processes. Izvestiya, Physics of the Solid Earth 10, 107–112 (in Russian).

Nersesov, I.L., Semenov, A.N., Simbireva, I.G., 1971. Space-time distribution of travel times of p and s waves in the garm region. In: Experimental Seismology. Nauka Publishing House, Moscow, pp. 334–345 (in Russian).
Nikolayevskii, V.N., 1982. A review: Earth's crust, dilatancy and earthquakes. In: Advances of Science and Engineering. Mir Publishing House, Moscow, pp. 133–215 (in Russian).
Nur, A., 1971. Effects of stress on velocity anisotropy in rocks with cracks. Journal of Geophysical Research 78, 2022–2034.
Petrashen, G.I., 1978. Fundamentals of the mathematical theory of elastic wave propagation. In: Issues of the Dynamic Theory of Seismic Wave Propagation, vol. 18. LSU Publishing House, Leningrad, pp. 1–248 (in Russian).
Sobolev, G.A., 1978. Investigation of the formation of shear-type fracture precursors in laboratory conditions. In: Search for Earthquake Precursors. Nauka Publishing House, Moscow, pp. 86–99 (in Russian).
Wang, Chunhua, Liao, Sugiong, 1996. Experimental study of the preparation and occurrence of strong earthquakes. Journal of Earthquake Prediction Research 5, 525 p.
Zhang, Guomin, Zhang, Zhaocheng, 1992. The study of multidisciplinary earthquake prediction in China. Journal of Earthquake Prediction Research 1, 71–85.
Zhurkov, S.N., 1968. Kinetic concept of the strength of solids. Vestnik Akademii Nauk SSSR 3, 46–52 (in Russian).
Zhurkov, S.N., Kuksenko, V.S., Petrov, V.A., 1977. On the question of prediction of rock failure. Izvestiya, Physics of the Solid Earth 8, 11–18 (in Russian).

CHAPTER 4

Seismic Active Monitoring System Concept

Rudolf Unger[a,b]

Contents

1. Introduction

Various attempts have been made to measure parametric changes in the Lithosphere, in particular for the purpose of finding potential earthquake precursors. *Passive* monitoring is performed by means of time-lapse seismic tomography or wave propagation velocity measurement using regional or teleseismic earthquakes as wave propagation sources. Problems with these methods are that the source location, onset time, and signal contents are not known exactly, and must be resolved together with the Earth parameters we want to measure. In general, this leads to residue errors too large for the space and time resolution and precision required for detecting significant time-lapse parameter changes. A partial remedy is to use controlled underground explosions, but these cannot be applied near populated areas, are expensive, and do not provide *continuous* monitoring.

[a] Unger Scientific International, Achterom 100, 3995EH Houten, The Netherlands
[b] Universidad Tecnológica Nacional, Facultad Tucumán, San Miguel de Tucumán, Argentina

E-mail address: usiruud@hotmail.com.

Handbook of Geophysical Exploration: Seismic Exploration, Volume 40
ISSN 0950-1401, DOI: 10.1016/S0950-1401(10)04009-7

In some experiments, these sources were replaced by hydraulic seismic vibrators capable of emitting precisely controlled, modulated sinusoidal ground force signals as used in seismic exploration (Clymer and McEvilly, 1981). Other efforts have been in constructing strong mechanical vibrators with rotating eccentrics (Alekseev et al., 2005). However, neither of these sources is capable of sustaining signal emissions continuously and certainly not for indefinite time. In this paper, we present the concept of a seismic transmission system capable of continuous, precise, and high-resolution active monitoring of the Lithosphere for time-lapse changes in Earth physics parameters. The system concept may be extended to application in electromagnetic and piezo-electric monitoring.

2. System Concept

The overall system objectives, concept, and design layout are described extensively in a monograph (Unger, 2002). The system concept is illustrated and summarized in Figure 1. The concept imposes the following requirements on the force signals to be emitted:

- Signature prescribed and known exactly.
- Continuously and accurately repeatable for indefinite time.
- Long-term time and phase coherent.
- High resolution, low side lobe auto-correlation function.
- Signal code duration spanning the major seismic phase arrivals (e.g., 5-20 min).
- Allow simultaneous, mutually uncorrelated emission from various source sites.
- Low-frequency (e.g., 0-10 Hz) for deep and far penetration.
- Universal Time (UT) controlled to nano-second precision.

Essential is the incorporation of two main communications technology topics:

- Pseudo-random sequence (prs) signal transmission, enabling efficient, simultaneous emission of mutually uncorrelated, pulse-like signals of sufficient duration from different source sites.
- Quadrature signal processing, resulting in complex-envelope seismograms enabling high-resolution signal detection and timing.

3. Force Signal Signatures

The system has been designed to enable the emission of a broad class of signals, and to apply digital communications technology as used in radar

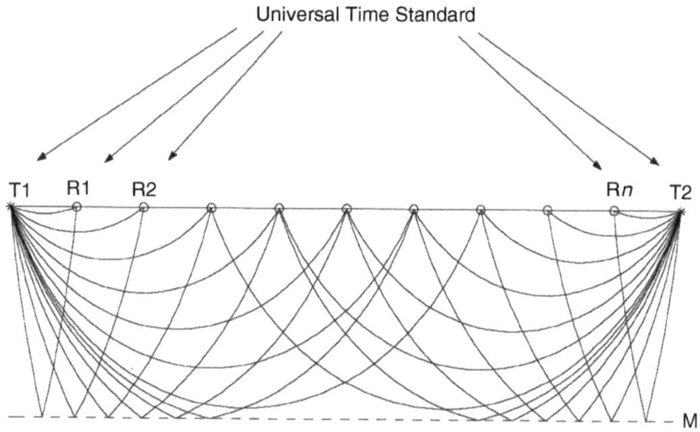

Figure 1 Two-dimensional concept of the coherent seismic transmission system. Stationary seismic vibrator sources T1, T2, ... T*m* continuously emit coded, mutually uncorrelated ground-force signals of selectable code length (e.g., 15 min sweeps). Stationary receivers R1, R2, ..., R*n* continuously record seismic motion. Universal Time standards establish precise signal time and phase coherence among transmitters and receivers. The signals are retrieved by long-term coherent stacking and matched filtering of code-length records. The rays of primary and later seismic phases provide a tomography grid. Space and time resolution are governed by receiver density and the capabilities of signal detection, timing, identifying and ray tracing. M indicates the Mohorovičić discontinuity. For clarity, only the direct and M-reflected P-wave rays are drawn.

and space communication. Signals include different forms of conventional frequency modulation sweep or "chirp" signals (Edelmann and Helbig, 1984), in general used with hydraulic vibrators in seismic exploration, and the family of maximal pseudo-random sequences (prs) modulated onto a selected monochromatic carrier wave (Steinberg and Birdsall, 1966; Golomb, 1967; Cunningham, 1979; McEliece, 1987; Unger, 2002). Both classes are a form of pulse expansion; pulse compression is achieved by means of "matched filtering", i.e., correlating the received signal with a noiseless replica of the emitted signal, the "pilot" signal. For a linear Earth transfer function, the matched filter output equals the signal auto-correlation function convolved with the Earth's impulse response or Green's function. In white noise, i.e., noise with a flat power spectrum, the coherent matched filter power processing gain equals $2WT$, i.e., the coherent matched filter improves the amplitude signal-to-noise ratio by a factor $\sqrt{2WT}$, where W is the effective signal bandwidth and T is the signal duration. In non-white noise, the processing gain is less than $2WT$. The basic properties of the chirp and prs signals are illustrated in Figure 2. For our objectives, the prs signals are the preferred emission signal class, satisfying the signal requirements mentioned as further discussed below. In our system, the prs signal transmission and processing technique has been adapted from underwater acoustics transmission technology originated by T.G. Birdsall

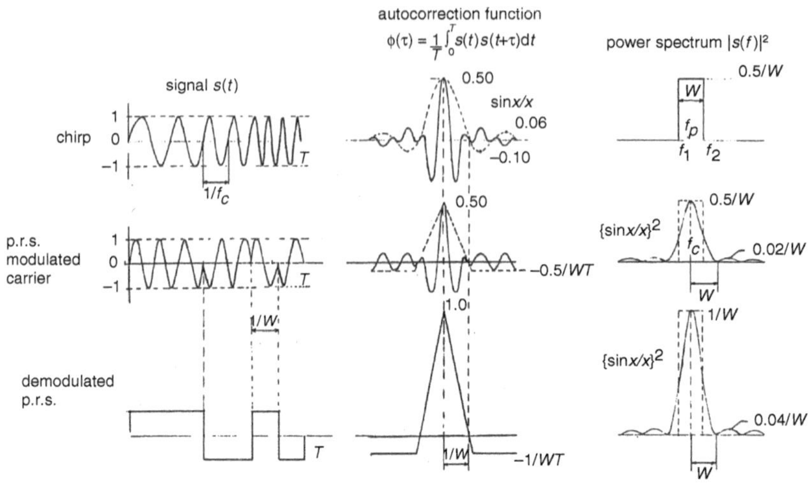

Figure 2 Characteristics of linear frequency sweep ("chirp") and prs signals (Unger et al., 1987).

(Steinberg and Birdsall, 1966), and described in detail in a technical report (Unger and Veenkant, 1967).

3.1. Pseudo-random sequence signal

The prs signals (Figure 2) have been used in a great variety of applications (radar, sonar, electromagnetic surveying, satellite communication, others), but apparently not in routine seismic surveying, and not in earthquake seismology research. For the seismic exploration application, however, some experiments have been conducted. Cunningham (1979) finds the prs signals to be superior to chirp signals, mainly because of the inherently low side-lobe of the auto-correlation function. A prs can be considered as a repeated coin-tossing experiment of which all outcomes have been pre-programmed. Thus, each prs possesses properties of randomness, but in fact is known exactly. The prs can be generated in a systematic manner in a signal generator computer simulating a series of shift-register elements with modulo-two (mod2) adder feedback (Viterbi, 1966; Golomb, 1967; McEliece, 1987), see Figure 3. A shift register with N elements results in a sequence length of $n = 2^N - 1$ digits, 2^{N-1} digits equal to a logical 1, and $2^{N-1} - 1$ digits equal to a logical 0, or vice-versa, so that $\Pr(1) \approx \Pr(0) \approx 0.5$. The logical zeros may be replaced by -1 digits as in Figure 2. Each sequence can be identified, and is completely determined, by a set of three integers $(N, r1, r2)$ representing, respectively, the number of shift register elements and the two mod2 feedback elements used. The configuration in Figure 3 generates a (5,3,5) maximal

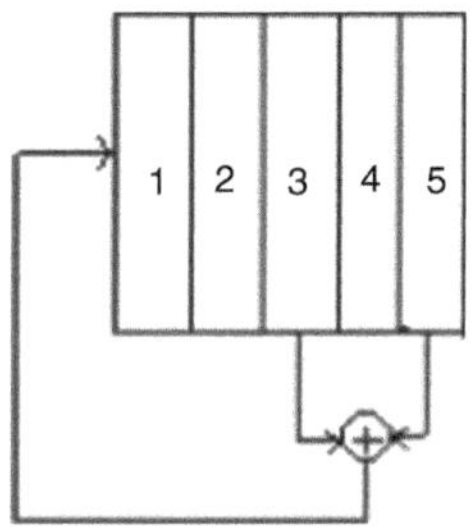

shiftregister element

1	2	3	4	5
shiftregister contents				
1	1	1	1	1
0	1	1	1	1
0	0	1	1	1
0	0	0	1	1
1	0	0	0	1
1	1	0	0	0
0	1	1	0	0
1	0	1	1	0
1	1	0	1	1
1	1	1	0	1
0	1	1	1	0
1	0	1	1	1

Figure 3 Shift register sequence generation.

pseudo-random sequence which may be tapped from any of the shift register elements as given in the corresponding columns.

Similar to the outcomes of series of coin-tossing experiments, one would expect that maximum-length prs be mutually uncorrelated. However, (1) not all shift register sequences are maximum length, i.e., certain patterns appear as repeated partial sequences so that the auto-correlation function contains more than one correlation peak within one sequence period, and (2) not all maximum-length prs are mutually uncorrelated. As an example with respect to (1) above, Figure 4 shows the auto-correlation functions for the prs generated by all ten mod2 feedback combinations with the end element of an 11-element shift register. Table 1 contains the maximum auto-correlation side-lobe value in % and in dB, relative to the zero-lag auto-correlation peak value. The prs length equals $2^{11} - 1 = 2047$ digits, so the auto-correlation noise or side-lobe amplitude of a maximum-length prs should be $1/2047 = 0.05\%$ of the zero-lag amplitude, that is, at -66 dB. We note that only the two sequences (11,2,11), (11,9,11) satisfy this maximum-length criterion. Because of their relatively low auto-correlation noise, possibly also the sequences (11,3,11) and (11,8,11) can be used, but they are not maximum-length. Observe also the symmetry in the use of feedback elements, e.g., the auto-correlation function of sequence (11, 1, 11) is identical to that of sequence (11,10,11); the same holds for the sequence pairs {(11,2,11), (11,9,11)}, {(11,3,11), (11,8,11)}, {(11,4,11), (11,7,11)}, {(11,5,11), (11,6,11)}.

Concerning problem (2), the mutual cross-correlation functions and characteristics of the "usable" sequences are given in Figure 5 and in Table 2. Apparently, not all prs are mutually uncorrelated, little is known about the conditions for this mutual non-correlation (McEliece, 1987). However, for relatively long sequences as required for our purpose, families

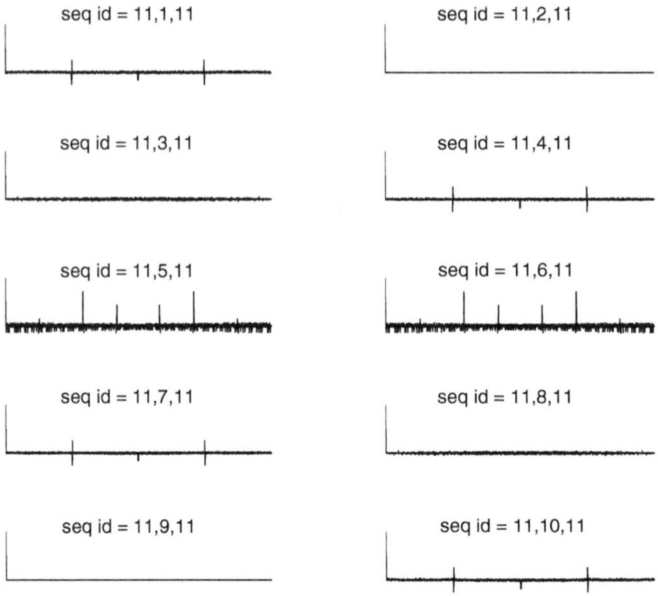

Figure 4 Prs auto-correlation functions of 11-element sequences in different feedback configurations. The zero lag is left justified.

Table 1 Prs auto-correlation noise for different 11-element shift register feedback configurations

Prs identification	Auto-correlation noise		Remarks
	%	dB	
(11,1,11)	25	−12	Max. length seq.
(11,2,11)	0.05	−66	Potent. usable seq.
(11,3,11)	5	−26	
(11,4,11)	25	−12	
(11,5,11)	71	−3	
(11,6,11)	71	−3	
(11,7,11)	25	−12	
(11,8,11)	5	−26	
(11,9,11)	0.05	−66	Potent. usable seq.
(11,10,11)	25	−12	Max. length seq.

of mutually virtually uncorrelated sequences can be selected and pre-established by computer trial, as shown above. The number of different sequences equals the number of combinations created by combining in principle, in the modulo-two adder feedback, any two of the N shift

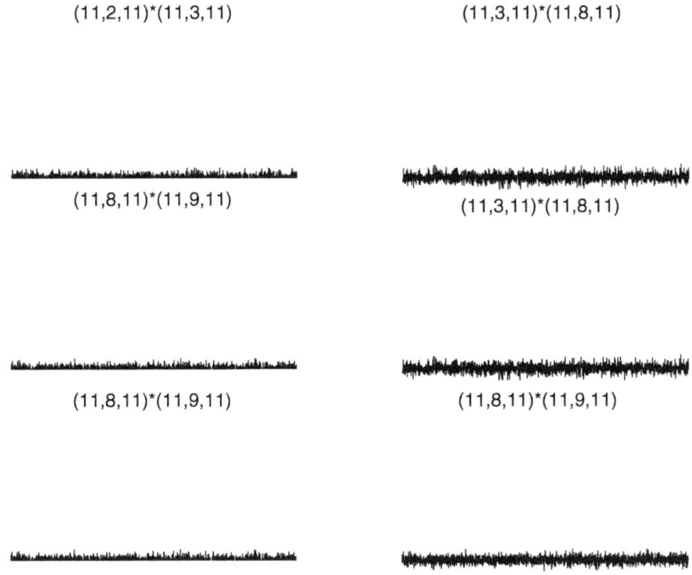

Figure 5 Mutual cross-correlation functions of the "usable" sequences identified in Table 1. The scale is the same as used in Figure 4.

Table 2 Cross-correlation noise among usable sequences

Prs1	Prs2	Cross-correlation noise	
		%	dB
11,2,11	11,3,11	8	−22
11,2,11	11,8,11	9	−21
11,2,11	11,9,11	8	−22
11,3,11	11,8,11	9	−21
11,3,11	11,9,11	8	−22
11,8,11	11,9,11	8	−22

register elements. However, computer trials indicate that one of the feedback elements has to be the end element. Thus, N shift register elements provide $N - 1$ different sequences. In our application, for a 2.5 Hz carrier wave, we suggest using 10-12 elements to obtain seismogram lengths (equal to the sequence signal periods) of 5-20 min, resulting in 9-11 possible sequences to choose from. Long sequences have the advantage of potentially providing more mutually uncorrelated sequences. On the other hand, in signal reception and processing, they are more vulnerable to noise bursts and signal emission or reception interruptions, which may render entire sequence durations useless for processing. One could try to increase the number of uncorrelated sequences by nesting multiple shift register feedback loops, but again, computer trials indicate that this is unsuccessful, suggesting the possible existence of some number theory law to that extent. Other possibilities are to use chirp signals and random number generated signals of which the mutual non-correlation is to be established by computer trial, in addition to the selected prs families. Also, in the cases where relatively few prs cross-correlation peaks occur within a correlogram, their arrival times can be anticipated exactly from computer trials as performed here; in signal processing, these spurious signals then may be removed by means of time-notch filtering. Further investigation of suitable signal codes, including signal spectrum optimization, is left for future research. 3-D transmission requires that at least two, but preferably three or more uncorrelated sequences are emitted simultaneously from different source sites. In the case of simultaneous transmissions in different regions on the same principle, their signals could interfere after long-term stacking; therefore, preferably, *all* emission signals *everywhere* should be uncorrelated. Apparently, more research is required in order to provide larger families of mutually uncorrelated signals.

3.2. Pseudo-random sequence modulation

The prs are to be modulated as a square pulse series onto a monochromatic carrier wave of the desired frequency. The modulation may be (1) binary phase, that is, using the logical $\{1,-1\}$ sequence as an amplitude modulator, resulting in phase flips of $180°$ of the carrier wave at each change from a logical 1 to -1 or vice versa, or (2) pure amplitude, using the $\{1,0\}$ sequence as amplitude modulator, resulting in digits of full amplitude signal alternated by digits of zero amplitude. Adopting the terminology from the underwater acoustics transmission technology reference (Edelmann and Helbig, 1984; Golomb, 1967), we refer to (1) as a binary phase modulated sequence (BMSEQ) and to (2) as an amplitude modulated sequence (AMSEQ). BMSEQ emits the full sinusoid signal power continuously; AMSEQ emits approximately half the full sinusoid signal power. Usually, the digit duration is chosen as an integer number of carrier wave cycles; we will do so also in our system. Figure 6 shows a 7-digit BMSEQ with one-cycle digit duration and its modulation onto a 2.5 Hz cosine carrier wave in the time- and frequency domain. Note the low side lobe ($=1/7$) in the auto-correlation function, corresponding to the low net power of the carrier wave resulting from mutual cancellation by $+1$ and -1 prs digits. In the modulated-signal spectrum the slight deviation from the $\sin x/x$ spectrum is caused by aliasing. Similarly, a 7-digit AMSEQ modulation is pictured in Figure 7.

The BMSEQ power spectrum is a $(\sin x/x)^2$ curve centered at the carrier frequency, the same (except for scaling) as that of a single pulse; its bandwidth W, as defined in Figure 2, is the inverse of the duration of a single digit

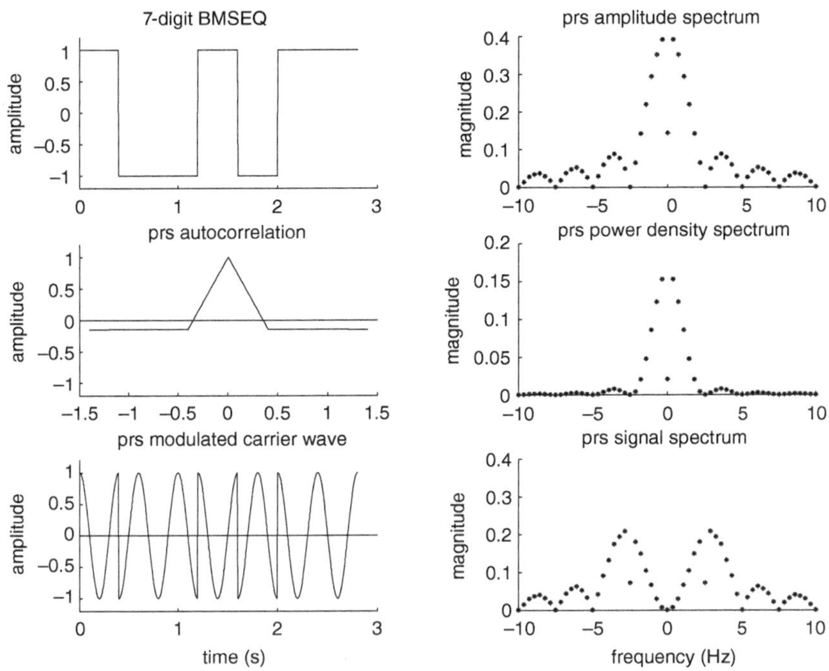

Figure 6 Characteristics of a 7-digit BMSEQ modulated onto a 2.5 Hz carrier wave: time domain (left), frequency domain (right).

or pulse width. Thus, a sequence of length T seconds contains WT pulses or digits. Due to the fact that, in BMSEQ, the numbers of positive and negative pulses differ only by one, this one pulse difference establishes the net amount of pure carrier wave power. Therefore, the carrier wave is *suppressed* by a factor WT; it appears as the side-lobe in the auto-correlation function. The spectrum of the demodulated signal has the same shape, but is centered and suppressed at dc (0 Hz). The BMSEQ auto-correlation function is given by

$$\begin{aligned} \phi(\tau) &= 1 - W|\tau|, \quad |\tau| \leq 1/W \\ &= -1/WT, \quad |\tau| \geq 1/W \end{aligned} \tag{1}$$

Thus, it has a triangular peak at zero delay ($\tau = 0$) with base width equal to $2/W$, and a constant side lobe of magnitude $1/WT$ and opposite phase. The side-lobe magnitude is thus inversely proportional to the time-bandwidth product. All BM sequences of a given pulse width have the same spectrum.

The characteristics of the AMSEQ are similar, except that in the auto-correlation function the level of the side-lobe representing the carrier wave power is at approximately one-half of the auto-correlation maximum. The spectral line at carrier frequency is *enhanced* by a factor WT, and only

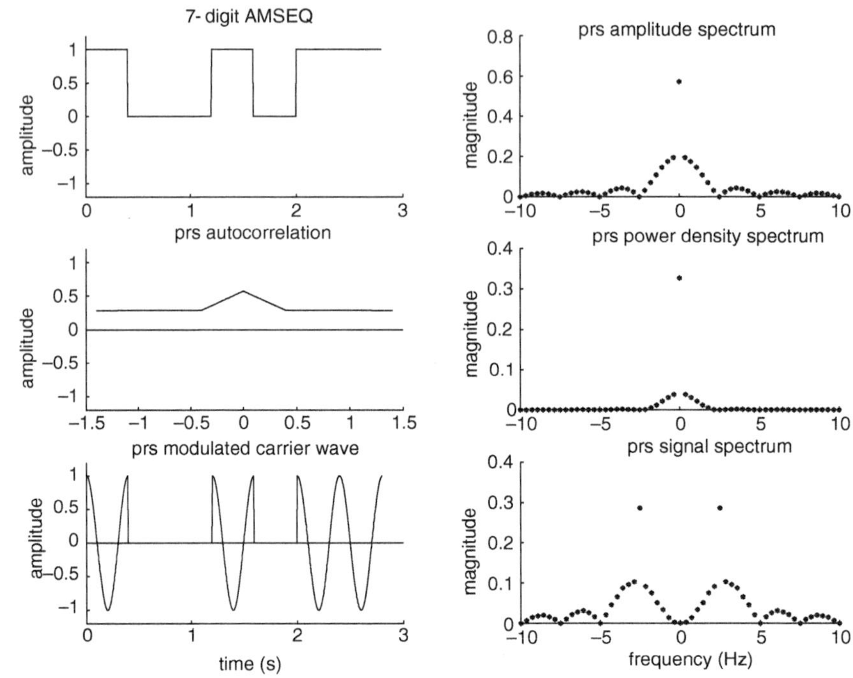

Figure 7 Characteristics of a 7-digit AMSEQ modulated onto a 2.5 Hz carrier wave: time domain (left), frequency domain (right).

approximately one-half of the energy of the BMSEQ signal is emitted. With N^1 the number of "one" digits, and N the total number of digits in an AMSEQ period, the AMSEQ auto-correlation function is expressed by

$$\begin{aligned}\phi(\tau) &= \frac{N^1}{N}(1 - W|\tau|), \quad |\tau| \leq 1/W \\ &= \frac{N^1}{2N}, \quad |\tau| \geq 1/W.\end{aligned} \tag{2}$$

For long sequences, $\frac{N^1}{N} \to 0.5$.

4. Quadrature Signal Processing

An essential component in our transmission system is the quadrature signal processing technique leading to complex-envelope seismograms consisting of the dual time series of instantaneous amplitude and instantaneous phase. Below, we define the relevant parameters and describe the signal processing methods contemplated. The text has been taken for most part from the monograph describing the total transmission system concept (Unger, 2002); parts have appeared in earlier publications (Unger and Veenkant, 1967; Unger, 1981; Unger et al., 1987). Further background material may be found in text books on communications theory (e.g., Papoulis, 1991; Schwartz, 1970). The basic method has been adapted from underwater acoustics technology developed in the 1960s, mentioned previously (Golomb, 1967; Cunningham, 1979). We incorporate a new signal detection and timing algorithm operating on coherent ensembles of the instantaneous phase seismogram.

4.1. Parameter definition

Any band limited waveform $r(t)$, not necessarily narrowband, but with its spectrum not extending down to dc, can be represented in any of the following three equivalent forms:

$$r(t) = x(t)\cos(2\pi f_0 t) - y(t)\sin(2\pi f_0 t) \tag{3}$$

$$r(t) = R(t)\cos[2\pi f_0 t + \phi(t)] \tag{4}$$

$$r(t) = R(t)\cos\left[2\pi \int_0^t f(\tau)\mathrm{d}\tau\right] \tag{5}$$

with the relationships

$$x(t) = R(t)\cos[\phi(t)] \tag{6}$$

$$y(t) = R(t)\sin[\phi(t)] \tag{7}$$

$$R(t) = [x^2(t) + y^2(t)]^{1/2} \tag{8}$$

$$\phi(t) = \arctan\left[\frac{y(t)}{x(t)}\right] \pm 2\pi k \tag{9}$$

$$f(t) = \frac{1}{2\pi}\frac{\mathrm{d}\phi(t)}{\mathrm{d}t} + f_0 \tag{10}$$

where $\{x(t), y(t)\}$ are referred to as the Cartesian or quadrature components, $R(t)$ is the instantaneous amplitude or envelope, $\phi(t)$ is the instantaneous phase, $f(t)$ is the instantaneous frequency, and $f_0(t)$ is any fixed frequency.

Equations (6)–(9) reflect the familiar Cartesian-to-polar coordinates and vice-versa conversions. Either of the function pairs $\{R(t), 2\pi f_0 t + \phi(t)\}$, $\{R(t), 2\pi \int_0^t f(\tau)\mathrm{d}\tau\}$ is known as the *complex envelope* in which $2\pi f_0 t + \phi(t)$ and $2\pi \int_0^t f(\tau)\mathrm{d}\tau$ are the angular arguments. Note that $R(t)$ is independent of the waveform's phase or frequency.

4.2. Waveform modulation/demodulation

We may view $R(t)$, $\phi(t)$, $f(t)$ as the amplitude, phase, and frequency *modulation*, respectively, of a, possibly fictitious, carrier wave $\cos(2\pi f_0 t)$ with "carrier" frequency f_0 Hz and zero phase. In active transmission, such as radio communication, active radar, etc., a carrier wave is actually present; in passive transmission, e.g., in noise or in seismograms generated by earthquakes or explosions, no actual carrier wave is present, but we may invent a fictitious one. Although the carrier frequency f_0 may be chosen arbitrarily, and even different from an actual carrier frequency if present, it is convenient to choose f_0 either as the actual carrier frequency in active transmission, or as the center of the frequency band of interest in passive transmission. This general waveform representation contains simultaneous amplitude and frequency modulation, Eq. (5), or, equivalently, simultaneous amplitude and phase modulation, Eq. (4). In line with this, we refer to $\{x(t), y(t)\}$ as the Cartesian modulation, Eq. (3).

Demodulation is achieved digitally by obtaining the Cartesian or quadrature components $x(t)$ and $y(t)$ directly from Eq. (3) by sampling the received waveform $r(t)$ at a rate of four times the carrier frequency, and substituting

$$t = t_k = \frac{k}{4f_0}, \qquad k = 0, 1, \ldots \tag{11}$$

resulting in the discrete time series

$$r(t_k) = x(t_k)\cos\left(k\frac{\pi}{2}\right) - y(t_k)\sin\left(k\frac{\pi}{2}\right). \tag{12}$$

This yields the Cartesian components time series as

$$\begin{aligned} x(t_k) &= r(t_k), \quad k = 0, 4, 8, \ldots \\ x(t_k) &= -r(t_k), \quad k = 2, 6, 10, \ldots \\ y(t_k) &= r(t_k), \quad k = 1, 5, 9, \ldots \\ y(t_k) &= -r(t_k), \quad k = 3, 7, 11, \ldots \end{aligned} \tag{13}$$

We observe that the x and y time series are shifted by 90° with respect to one another, thus constituting a Hilbert transform pair and justifying the names Cartesian or quadrature components (Schwartz, 1970; Papoulis, 1991). The dual time series, in this form, or converted to the polar components $R(t_k)$, $\phi(t_k)$, are the *complex envelopes* of the $r(t_k)$ time series. Quadrature processing is also referred to as *coherent processing*; it preserves the information contained in the instantaneous phase. The complex-envelope relations are pictured as a phasor or vector diagram in Figure 8. Rotating with the angular carrier wave frequency $\omega_0 = 2\pi f_0$, it represents the

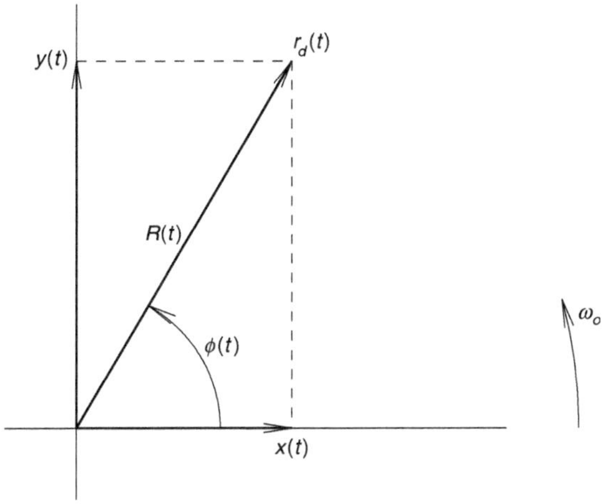

Figure 8 Waveform vector diagram representation. r_d is the demodulated waveform vector.

modulated waveform; the stationary diagram represents the demodulated waveform.

4.3. Matched filtering

In matched filtering (MF) or correlation filtering, the waveform to be processed is correlated with a noise-free replica of the emitted signal (the "pilot" signal). According to seismic wave propagation theory, the far-field particle displacement is proportional to the source force. Therefore, MF should consist of correlating the particle displacement waveform received at a seismic station with the noise-free source force signal. In quadrature processing and demodulation, each of the particle displacement quadrature components, i.e., each of the Cartesian demodulation components $x(t)$, $y(t)$ is correlated with the force modulating signal code. The two Cartesian correlograms then are converted into the polar form, resulting in the complex-envelope seismogram consisting of the correlogram dual time-series of instantaneous amplitude and instantaneous phase.

Since the pilot signal is a binary sequence consisting only of $\{1,-1\}$ or $\{1,0\}$ values, correlating the particle displacement (x, y) time series with the pilot signal time series is efficiently performed in the time domain by a set of addition and subtraction instructions in the $\{1,-1\}$ case, or additions only in the $\{1,0\}$ case, a significant additional advantage of the prs signals.

In white noise, the MF output power signal-to-noise ratio equals the ratio of signal energy over noise power density (Papoulis, 1991). As a formula, with $\boldsymbol{s}$ the signal vector, $\boldsymbol{n}$ the noise vector, $S(f)$ the signal power density spectrum received at a seismic station, T the total signal transmission and stacking time, W the effective signal frequency band, and N_0 the noise power density:

$$\left(\frac{|s|}{|n|}\right)^2 = \frac{T\int_W S(f)\mathrm{d}f}{N_0}. \tag{14}$$

In the case of colored noise, with noise power density spectrum $N(f)$ and average power density N_0, the MF output signal-to-noise ratio is arrived at by filtering the signal-plus-noise waveform such that the noise becomes white (noise "whitening"), resulting in

$$\left(\frac{|s|}{|n|}\right)^2 = T\int_W \frac{S(f)}{N(f)}\mathrm{d}f. \tag{15}$$

For a slope α in a generic log-power versus log-frequency displacement noise spectrum, we may describe the noise power density spectrum by

$$N(f) = N_0 f^{\alpha} \tag{16}$$

leading to the MF output power signal-to-noise ratio

$$\left(\frac{|s|}{|n|}\right)^2 = \frac{T\int_W f^{-\alpha}S(f)\mathrm{d}f}{N_0}. \tag{17}$$

4.4. Signal detection and timing algorithm

From signal-plus-noise vector diagram analysis it has been shown that signal phase detection is more sensitive than signal amplitude detection (Unger, 1981, 2002). In coherent ensembles (overlays) of instantaneous phase seismograms the phase will be randomly distributed if no signal is present, but will be biased, and thus clustered, in the presence of even a relatively small signal. On this basis, we construct an efficient, accurate signal detection and timing algorithm, operating on a number of MF output records forming a coherent *ensemble of complex-envelope seismograms*, as follows, see Figure 9.

- Compile a synchronized ensemble of coherent seismograms.
- Determine the ensemble phase standard deviation (PSD) at each lag in the time-series of the instantaneous phase seismogram.
- PSD minima below a set threshold reflect signal arrivals.
- Determine either the ensemble phase mean or the ensemble phase median (PMN) for the ensemble of phase values at the lag time index of minimum PSD.

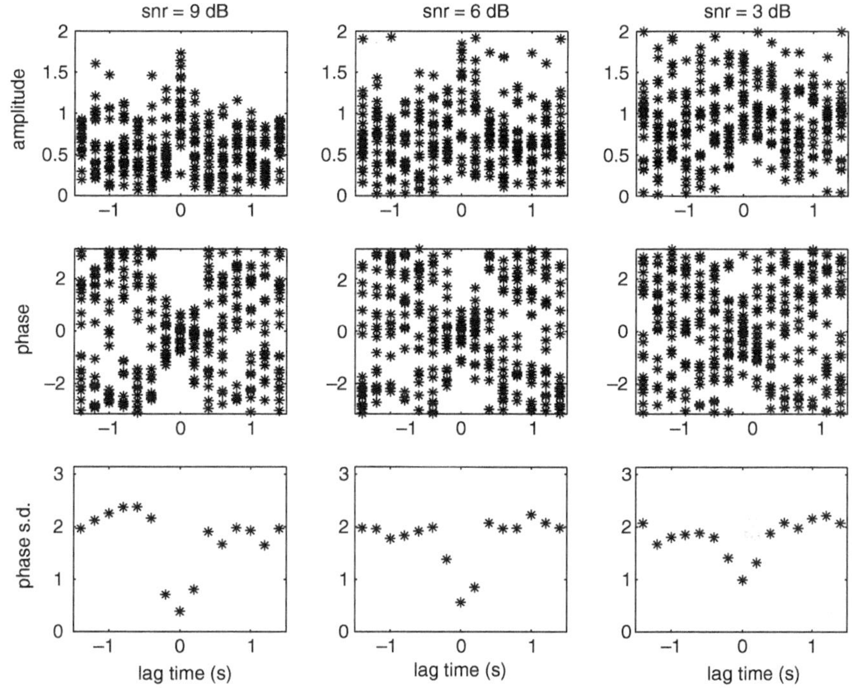

Figure 9 Ensembles of 20 complex-envelope seismogram time series for a 7-digit prs modulated onto a 2.5 Hz carrier wave at various seismogram snr levels, showing the spread of instantaneous amplitude and phase values, and the resulting ensemble PSD at each correlogram lag, for each snr.

- Relative signal arrival time = time of PSD minimum + (PMN/2π) times carrier wave period.

For a 2.5 Hz carrier wave, the signal timing error s.d. ranges from 0.3 milliseconds at 36 dB snr to 0.38 seconds at −6 dB snr, see Figure 10. Signal timing is stable down to +3 dB, with the timing error slowly increasing with snr. Below the +3 dB snr, the timing error increases rapidly, but even at −3 dB remains within one period of the carrier wave (0.4 s for this data). In the academic case that no noise is present, the timing is exact to the least significant computer bit. Down to +3 dB, these observations are in good agreement with an analytically derived PSD for a uniform instantaneous phase distribution of the noise (Unger, 2002). Below +3 dB, outliers cause deviation from the analytically determined timing error; these outliers probably can be removed by means of extreme-error filtering, or by human interaction.

In this procedure, there is a trade-off between the desired total averaging time, or the number of signal records per MF input stack, and the desired number of MF output records or seismograms per ensemble. On the

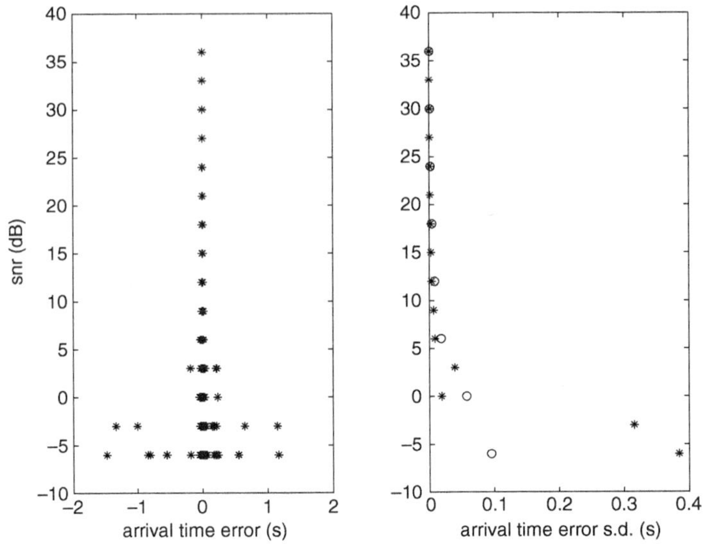

Figure 10 Signal arrival time error and error s.d. as a function of seismogram snr, obtained from 30 trials at each snr. The circles in the righthand figure mark the analytically determined time error s.d.

one hand, the snr in each seismogram increases with the square root of the number of records per MF input stack. On the other hand, a greater number of seismograms yield better statistics, in particular for the analysis of the behavior of potential earthquake precursor parameters. The maximum total raw data stacking time is determined by the minimum amount of elapsed time in which Earth parameter variation is expected to take place, whereas the minimum number of seismograms is determined by a minimum statistics criterion, e.g., at least ten seismograms per ensemble. In signal detection applications, the trade-off is a general problem of averaging procedures versus "counting" procedures. In our case, it seems prudent to put priority on satisfying the sufficient statistics criterion and relatively "short" time-lapse windows; after all, the seismogram snr always can be further increased by stacking the MF output seismograms in addition to MF input record stacking. In this way, we obtain greater flexibility in signal processing and in moving-window time-lapse parameter analysis.

Figure 11 shows the results of a signal transmission simulation (Unger, 2002) with 18 dB MF seismogram snr for a linear propagation velocity profile in the crust generating rays as in Figure 1. Note that the third arrival is not detectable in the amplitude time series whereas the phase ensemble shows a clear minimum PSD near 24 s lag time. The simulation suggests that many later signal arrivals can be detected and timed correctly so that appropriate ray tracing may yield a sufficient tomography grid.

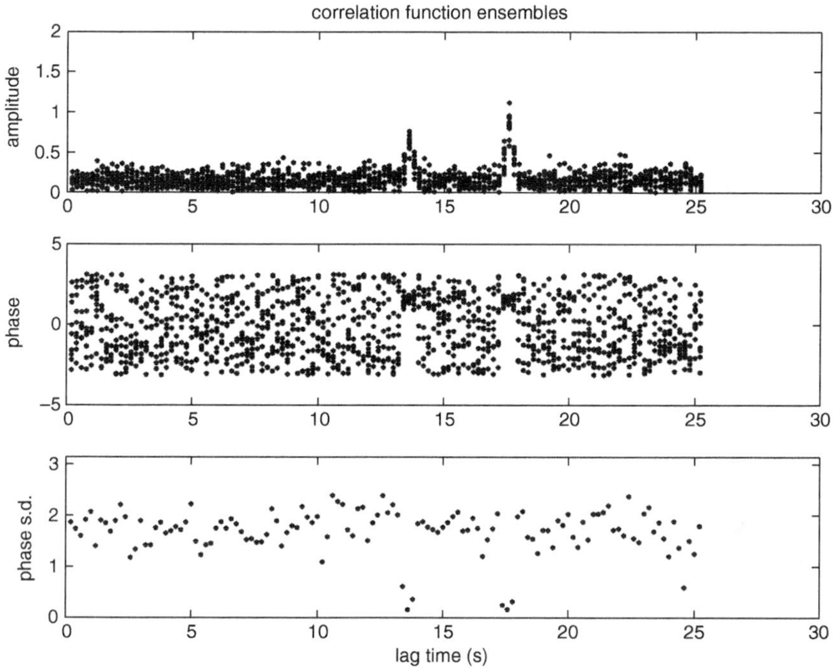

Figure 11 Coherent complex-envelope seismogram ensembles and ensemble PSD for station R3 in Figure 1.

5. Conclusions

Pseudo-random sequences modulated onto a low-frequency carrier wave form ideal force-source signals in a 4D active, continuous seismic transmission system to monitor time-lapse changes in Earth parameters. Complex-envelope seismograms obtained by coherent stacking and quadrature time domain matched-filter processing enable high-resolution signal detection and timing. For large snr's, the timing error is inversely proportional to the signal transmission duration. At 36 dB MF seismogram snr the timing error is less than 0.5 ms. In addition to application in our proposed transmission system (Unger, 2002), the method seems extremely suitable for incorporation in high-precision short baseline experiments using a piezo-electric source (Silver et al., 2004), potentially resulting in a timing precision improvement by a factor 30 or better. The method may be applicable also with the Russian large mechanical vibrators (Alekseev et al., 2005) operating with eccentrics, and producing linear sweep (chirp) force signals rather than our prs signals since they cannot accommodate the prs phase flips. It seems unlikely that existing hydraulic vibrators are capable of long-term coherent, low-frequency prs transmission. For our transmission system concept, therefore, we conceived new technology vibrators, a magnetic levitation P-wave vibrator and a linear synchronous motor S-wave vibrator (Unger, 2002, 2004).

REFERENCES

Alekseev, A.S., Chichinin, I.S., Korneev, V.A., 2005. Powerful low-frequency vibrators for active seismology. Bull. Seismol. Soc. Am. 95 (1), 1–17.

Clymer, T.W., McEvilly, T.V., 1981. Travel-time monitoring with vibroseis. Bull. Seismol. Soc. Am. 71 (6), 1903–1927.

Cunningham, A.B., 1979. Some alternative vibrator aignals. Geophysics 44, 1901–1921.

Edelmann, H.A.K., Helbig, K., 1984. Vibroseis — theory and practice. In: Continuing Education for Explorationists. EAEG, London.

Golomb, S.W., 1967. Shift Register Aequences. Holden Day, Inc., Amsterdam.

McEliece, R.J., 1987. Finite Fields for Scientists and Engineers. Kluwer, Boston.

Papoulis, A., 1991. Probability, Random Variables, and Stochastic Processes, 3rd ed. McGraw-Hill, Inc., New York.

Schwartz, M., 1970. Information, Transmission, Modulation, and Noise. McGraw-Hill, London.

Silver, P.G., Niu, F., Daley, T., Mayer, E.L., 2004. Developing a methodology for stress-induced temporal variations in seismic velocity: Successful use of barometric pressure and solid earth tides as calibration signals. SSA Presentation.

Steinberg, J.C., Birdsall, T.G., 1966. Underwater sound propagation in the Straits of Florida. J. Acoustical Soc. of Am. 39, 301–315.

Unger, R., 1981. The instantaneous amplitude, phase and frequency in seismic event detection, timing and identification. In: Husebye, E.S., Mykkeltveit, S. (Eds.), Identification of Seismic Sources — Earthquake or Underground Explosion. D. Reidel, Dordrecht, pp. 649–662.

Unger, R., 2002. A seismic transmission system for continuous monitoring of the lithosphere: A Proposition, PhD thesis monograph, Delft University Press Science, Delft, The Netherlands. http://www.library.tudelft.nl/dissertations/diss_html_2002/its_unger_20020227.html (pdf).

Unger, R., 2004. Concept and status of a controlled-signal seismic transmission system for continuous monitoring of the lithosphere II: New-technology sources, Proc. IWAM04.

Unger, R., van Kampen, W.A., Berkhout, A.J., Boiten, R.G., Ritsema, A.R., Schmidt, D.Ph., Ziolkowski, A.M., 1987. New possibilities in controlled-source seismology with a magnetic levitation vibrator 1987. Geophys. J.R. Astr. Soc. 89, 41–46.

Unger, R., Veenkant, R., 1967. Underwater sound propagation in the Straits of Florida: The MIMI experiment of 3 and 4 February 1965: Cooley Electronics Laboratory Technical Report, 183, The University of Michigan, Ann Arbor.

Viterbi, A.J., 1966. Principles of Coherent Communication. McGraw-Hill, New York.

Active Monitoring Targets

CHAPTER 5

Detection of Transient Phenomena Due to Active Scatterers

Naoyuki Fujii[a,b] *and* Mineo Kumazawa[b,c]

Contents

1. Introduction

One of the most essential targets of our concern is the constant monitoring and detection of temporal changes of the physical state of the media at and around the anticipated focal region of disastrous earthquakes in order to detect any diagnostic information useful to the prediction of the occurrence of such earthquakes. Before going further to our ultimate target, however, we need to discuss the two types of backgrounds that would enable us to trace a reasonable approach towards our target. One is the geological situation of the seismogenic zone that we are going to clarify, and the other is the physical processes leading to the time-evolution of the possible observables by remote monitoring observation.

Geologically active regions such as subduction zones are characterized by relatively short temporal variations of the crustal deformations associated with the occurrence of earthquakes and volcanic activities. The former

[a] Institute of Geosciences, Shizuoka University, Ohya 836, Suruga-ku, Shizuoka 422-8529, Japan
[b] Nagoya University, Furo-cho, Chikusa-ku, Nagoya, 464-8601, Japan; Tel.: +81-52-789-3013
[c] Tono Geoscience Center, Japan Atomic Energy Agency, Japan

Handbook of Geophysical Exploration: Seismic Exploration, Volume 40
ISSN 0950-1401, DOI: 10.1016/S0950-1401(10)04010-3

example is attributed to a sudden decrease of relative seismic velocity in the Earth crust of −0.6 percent in the source region of the $Mw = 6.6$ Mid-Niigata earthquake in 2004, that was revealed by the passive seismic image interferometry technique by correlating the seismic noise (Wegler and Sens-Schonfelder, 2007). One of the latter examples is the detection of a seismic wave velocity decrease a few weeks before the eruptions of Piton de la Fournaise by utilizing ambient seismic noise (Brenguier et al., 2008).

Typical examples of such temporal variations other than seismic and volcanic activities are the slow slip events (e.g. Hirose et al., 1999; Dragert et al., 2001) that have recently been revealed by dense GPS networks. Deep non-volcanic tremors also occurred intermittently and were distributed along the strike of the subducting slabs that were firstly recognized in southwest Japan (Obara, 2002) and in the Cascadia subduction zones (Rogers and Dragert, 2003). In addition, changes in activity of deep tremors are correlated with intermittent occurrence of slow slip events and attributed to the movement at the interface of subducting plate boundaries (Obara et al., 2004; Hirose and Obara, 2005). The intermittent occurrence of non-volcanic tremors, very low frequency earthquakes and associated slow slip events are confirmed to exist at almost all subduction zones under some particular but common conditions (Ito et al., 2007). These phenomena are likely to be associated with the movement of a super critical fluid that could be released by the dehydration processes from the subducting slab.

The main objectives of the active geophysical monitoring are to detect and clarify the temporal variations of physical properties of scattering sources that include reflectors with temporally varying reflective characteristics by using actively transmitted elastic and/or electromagnetic waves. The effort to improve the accuracy of receiving signals is also indispensable, together with estimates of the expected amount of change with time of the physical properties of the scattering sources. Their temporal change can be clearly understood if the medium were in a state that was governed by the structure sensitivity. In this paper, we discuss the possibilities to detect the transient phenomena of the scattering sources caused by the structure sensitivity, particularly associated with the movement of fluid phases (super-critical water, carbon dioxide, etc.) expected in the crust and upper mantle.

2. Structure Sensitivity of a Medium

Physical conditions of seismogenic zones have not been clarified in detail yet, we have, however, several models that may be useful to deepen our understanding. Whereas the deep-seated solid rocks had traditionally been considered to be quite stable without any significant changes in their physical properties within a relatively short time scale, our models, for example, have evolved to include the presence of variability and instability as a consequence of very small amounts of "water" contained in the textures of solid rocks as polycrystalline aggregates. The relevant physical characteristic called "structural sensitivity" in this paper, referring to solid state physics (e.g. Bullough and Simmons, 1969), will be explained later in detail. The object of direct observation by ACROSS (Accurately Controlled Routinely Operated Signal System) is addressed to "temporal variations of wave-scattering sources" or shortly "active scatterers" in this paper just for convenience (Fujii and Kumazawa, 2004). This does not mean to exclude the possible time-delay of wave propagation through a certain spatial distance.

Another important target of active monitoring is the stability of the geologic environment of underground constructions. One of the examples is the reservoir of high-level radioactive wastes for isolation. Whatever the geology of the site is, the reservoirs are surrounded by polycrystalline aggregates containing water. Therefore, it is essential to consider "structural sensitive scatterers" in this case, so it is useful to consider the relevant problems correctly.

2.1. Examples of structure sensitive property

A relatively small volume fraction of "impurity" or distributed "heterogeneity" in a medium frequently plays an important role in the whole mechanical behavior of the medium for chemical reactions, plastic deformations, and so on. Such phenomena are recognized as the mechanisms governed by the "structure sensitive" properties of the medium.

It is well known that a small amount of H_2O causes a significant decrease of the strength of minerals and rocks. This phenomenon is normally understood as the stress corrosion mechanism (Atkinson, 1984), in which the hydrogen ions easily migrate through the lattice interstitially and cut the Si–O bonds of silicate minerals in the crust and upper mantle condition. The hydrogen ions at the tip of micro cracks easily squeeze into the Si–O bonds and result in the extension of the cracks. This mechanism is a typical example of the structure sensitive characteristics of the silicate rocks.

Another example will be observed in the permeable flow of fluids (volatiles or melts) within the crust and upper mantle, in which the permeability of a small porosity medium depends strongly on the wetness between the fluids and surrounding minerals and the contiguity of pore spaces at the grain boundaries, joint structures and so on.

2.2. Active scatterers in the earth's interior

Although the term "structure sensitivity" has been used in material science, the usage is extended here to a wider range of scales, from grain sizes to a few tens of kilometers (Fujii and Kumazawa, 2004, 2005). In addition, the "structure sensitive body" can be recognized as the characteristics of the

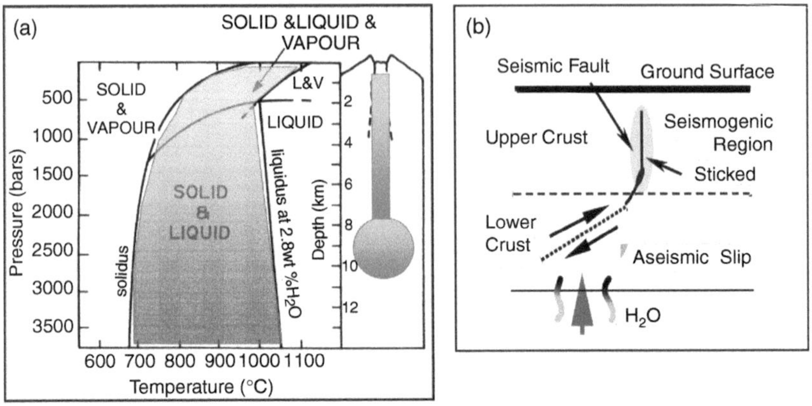

Figure 1 Examples of the distribution of structure sensitive bodies; (a) Near magma chamber; Solid-Liquid-Vapor field for an andesitic magma contained with a few wt%H_2O (Scales are approximate). (b) Aseismic slip beneath seismic faults; Slip is enhanced by invasion / infiltration of the super critical fluid to a stuck seismic fault in the seismogenic region.

active scattering sources among the distributed heterogeneities in a medium (e.g. Sato and Fehler, 1998). In the crust and upper mantle, typical candidates of the "active scatterers" are (1) abnormally localized regions such as magma chambers beneath active volcanoes, and deeper extension of active faults, (2) beneath the seismogenic regions related to brittle-ductile transitions, and (3) the deeper part of stuck regions of subducting plate boundaries (i.e. nearby regions of asperities at the subduction boundaries), etc. So, it is worth pointing out that the "active scatterers" obviously include frequently observed reflectors with changing reflective properties.

Networks of cracks and pathways of volatiles in surrounding rocks of a magma chamber can sensitively control the changes of volcanic activities related to the migration of volcanic gasses, although the actual structure of the network and its temporal changes could not be recognized. However, the scattering properties, both for transmitted elastic and electromagnetic waves, will significantly be affected by the variations of the network systems of migrating gasses. The region beneath the active faults will have similar properties in terms of the structure sensitivity due to the changes of network systems by to the localization of deformations (Figure 1(a) and (b)).

Temporal variations of scattering properties for elastic waves at a brittle-ductile transition zone will not be simply understood, because they may need additional conditions such as the criticality of stress state, magnitude and spatial scale of heterogeneity of the medium, and so on. Careful investigations of intermittent activities of deep tremors beneath the San Andreas Fault may reveal characteristic changes of physical properties from the observed variations of elastic waves (Niu et al., 2003; Nadeau and Dolenc, 2005).

Intermittent occurrence of the slow slip events associated with deep non-volcanic tremors and very low frequency earthquakes (Ito et al., 2007) are attributed to the temporal changes of the friction conditions of fault surfaces and gauges along the plate subduction boundaries in the vicinity of the asperities of the region. It seems likely that the pore pressures could change with accumulated stress during inter seismic periods (Kodaira et al., 2004). Heterogeneous distributions of pore pressures can easily affect the friction conditions due to geometrical roughness of the fault surface and changes of mechanical properties of fault gauge material, although the amount of the reflectivity (or scattering properties) for elastic waves remain uncertain.

3. Observables of Active Scatterers

Observable phenomena that cause temporal variations of the stress field related to generation of earthquakes and volcanic eruptions could mainly appear in the changes of reflected waves from "scattering sources" or "active scatterers". The heterogeneities in the lithosphere originating from both the stress state and heterogeneous distribution of fluid-bearing rocks can be the "scattering sources". Temporal variation of the properties of "scattering sources" due to the structure sensitivity of rocks could be imaged as "active scatterers" and is an essential characteristic of seismogenic zones as well as active volcanic regions.

The degree of understanding, or precision of imaging, of the "active scatterers" depend strongly on the scales and probability density of the substance relative to the wave numbers of our concern. The spatial scale of the heterogeneity can be represented by deterministic models for sufficiently small wave number, while one needs to introduce statistical models for large wave number. The number of physical parameters to represent the structure of heterogeneity could be inversely proportional to the wave number, whereas the number of anisotropy parameters would be proportional to the 3rd power of the wave number (Figure 2). In the case of monitoring by an elastic wave, our targets of tectonically interest are mostly between the two limiting cases. The scattering bodies of meters to tens of kilometers should be modeled with special care due to complex interactions both spatially and temporally among "active scatterers", especially when the driving mechanisms are presumably expected to be the pore pressure heterogeneity and movement of supercritical fluids.

Among many structure sensitive phenomena, probable changes in the transmitted seismic or electromagnetic signals are expected due to the temporal variation of the impedance and the anisotropic dispersion of the transmitted media in the subduction zones where the "scattering sources" are evolving, associated with the movement of the fluid mainly composed of super-critical water in the crust and upper mantle conditions. Active

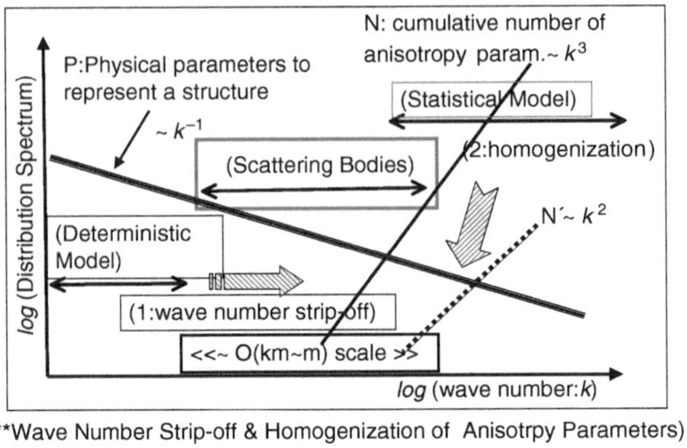

Figure 2 Logarithmic distribution spectrum of heterogeneities as a function of log (wave number) expected in the crust and upper mantle. Modeling of scattering bodies can be attributed between deterministic and statistical models.

geophysical monitoring would be the essential tool to detect and clarify such an evolving process that is governed by the "active scatterers". Recent discoveries of intermittent activities of slow slip events and non-volcanic deep tremors in the subduction zones should be the most challenging targets to clarify the characteristics of using active monitoring techniques.

3.1. Scattering characteristics of active scatterers

It is worth pointing out that the scattering sources are not always structure sensitive, because all heterogeneities in the lithosphere can scatter elastic and electromagnetic waves and where the scattering processes are closely related to the ratio of wavelength to the sizes of scatterers and the contrast of physical properties among the scatterers. As the rocks are composed of polycrystalline minerals, the macroscopic bulk material properties, such as electromagnetic (dielectric coefficients) and elastic properties under high pressures and high temperatures conditions, are normally insensitive to the small stress changes during dynamical processes in the crust and upper mantle.

A simple model of the "active scatterers" can be represented as shown in Figure 3, which is convenient for the mode analysis approach. The characteristics of the "active scatterers" can be obtained by three length scales (L, h, and g, in Figure 3) and two physical properties: velocity $S(x, \omega)$ and impedance difference $\Delta Z(\rho, S(x, \omega))$, where ω is the angular frequency of the interacting wave, ρ is the density and x is a value estimated geometrically by a combination of three length scales (L, h, and g) of the active scatterer. It will be a first step to have a realistic image of the active scatterer from the rock physics point of view by using this simple model, although a single scatterer model would be far from the reality in nature. A mode analysis approach for the detection of active scatterers, however, needs to construct mutually

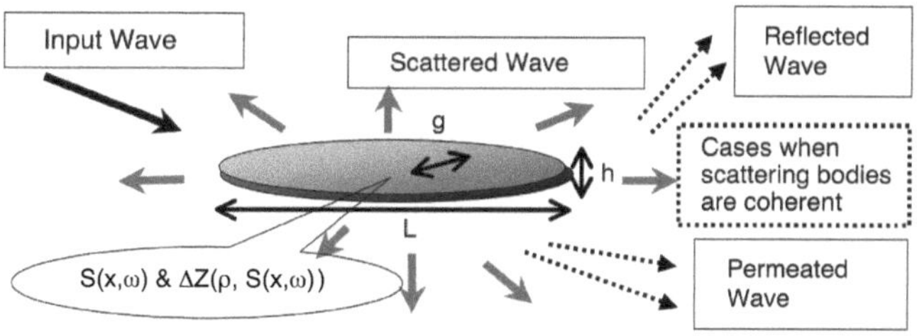

Figure 3 A simple model of active scatterer: Characteristics of the active scatterer can be made by three length scales (L, h, and g in the figure) and two physical properties: velocity $S(x, \omega)$ and impedance difference $\Delta Z(\rho, S(x, \omega))$, where ω is an angular frequency of interacting wave, ρ is the density and x is a value estimated geometrically by a combination of three length scales (L, h, and g) of the active scatterer.

independent groups of basic models systematically for further understanding of active scatterers.

Although the Moho discontinuity and the upper surface of the slab seem to be the most strong reflection boundary in the subduction zones, these heterogeneities **would not be** "structure sensitive" nor the active scatterers but material boundaries. Relatively short temporal variations of such boundaries are not expected. Once fluids were introduced into such reflection boundaries, however, the scattering properties (reflectivity or impedance, and complex dielectric coefficients) will strongly be affected due to the presence of water.

3.2. Anisotropy and dispersion of active scatterers

The main characteristic properties of the structure sensitive bodies are anisotropy due to orientation distribution of microcracks and dispersion due to the attenuation. Both properties are macroscopically observable, because the orientation distribution of microcracks and permeable structure due to network of cracks and pores can be affected by the deformation process and stress orientation in a medium. There are two types of anisotropies: one is azimuthal and the other is polarization. The former includes the anisotropic orientation distribution of microcracks due to geological evolution of tectonic stresses that can only change with long time scale (i.e. more than thousands of years?), while the latter can be strongly affected by the local deformation and stress state and possibly change with relatively short time periods. Polarization anisotropy is likely to change with the time associated with the movement of fluid through a network of pores and cracks so that the impedance of the scattering source can easily change with time. As shown by Ikuta et al. (2002), a shallow layer (<1 km) responds with ground vibrations due to earthquakes felt, and re-distribution of pore pressures in this layer made temporal changes of polarization anisotropy that disappeared within a week (Ikuta and Yamaoka, 2004).

The dispersion of transmitted waves can also change within relatively short periods associated with the re-distribution of fluids in the network of pores and cracks. In order to clarify the temporal variations of transfer functions, both in the time and frequency domains, in more detail, it is necessary to determine the seismic velocity structures precisely enough from both ray-path and mode analysis approaches. As the direct observations of the seismic ACROSS are transfer functions within frequency domain, temporal variations of dispersion relations for the medium can be easily obtained if appropriate models of dispersion are determined (Kumazawa et al., 2000, 2007; Hasada et al., 2004). Some of the preliminary results are that the apparent temporal changes in later phases of transfer functions are found (Yoshida et al., 2004) by analyzing the records of the seismic ACROSS radiated from the Tono station, where continuous operation for more than 5 years has been performed (Kunitomo and Kumazawa, 2004). As the mechanisms that can produce the variety of dispersions are so complex we need to investigate more in detail further.

3.3. Temporal variations of reflectors at subducting boundary

A seismic reflector at the subducting boundary of the Philippine Sea Plate was clearly identified by the Tokai seismic experiment (Iidaka et al., 2003), and the existence of this reflector was confirmed by using array observations for the seismic ACROSS from the Tono transmitter (Ikuta et al., 2005; Tsuruga et al., 2005, 2006). The estimated depth of the reflector is $30 \sim 50$ km depending on the model. It is suggested that the deeper extension of the Tokai Earthquake fault can be a good reflector, although the reflective property is not clearly understood. If the significant reflectivity of the plate boundary is caused by high pore pressures due to the geometrical roughness of the boundary (Kodaira et al., 2004), the time scale of the variation of the reflective property must be of essential concern, because it could vary from a week to $\sim$100 years. If this reflectivity were associated with the intermittent occurrence of deep tremors and slow slip events, it could have a duration of a few days to years, or if it were associated with the evolution of asperities at the subducting plate boundary, it could be similar to the recurrence time of the Tokai Earthquake ($\sim$100 years). Active monitoring to clarify the temporal variations of this reflector is one of the most essential targets to investigate this in more detail (Kasahara et al., 2006).

4. Discussions and Conclusions

We discuss here the possible characteristics of the temporal variations of scattering sources, particularly focus on the scattering properties due to structure sensitive bodies.

Typical candidates of "structure sensitive bodies" are discussed for three types of scattering sources: (1) abnormally localized regions such as magma chambers beneath active volcanoes, and deeper extensions of active faults, (2) beneath the seismogenic regions related to brittle-ductile transitions, and (3) the deeper part of stuck region of subducting plate boundaries (i.e. near regions of asperities at the subduction boundaries), etc.

Observable properties for the three types of scattering sources that are caused by temporal variations of the stress field related to generation of earthquakes and volcanic eruptions can be modeled through analyses of the transmitted elastic and electromagnetic waves. The heterogeneity in the lithosphere originating from both the stress state and heterogeneous distribution of fluid included in rocks can be "scattering sources" (eg. Sato and Fehler, 1998). Temporal variation of the "scattering sources" due to the structure sensitivity is an essential characteristic of seismogenic regions as well as active volcanic regions.

It is pointed out that the scattering sources are not always structure sensitive, because all heterogeneities in the lithosphere can scatter elastic and electromagnetic waves (Fujii and Kumazawa, 2005; Kumazawa et al., 2007; Kasahara et al., 2006). Polarization anisotropy and dispersion of transmitted waves are candidates for the observable properties of the temporal variations of scattering sources due to the structure sensitive bodies, however, the mechanisms governing the emergence of these phenomena have not clearly been understood.

The actual changes of the polarization anisotropy and dispersion relation depend on the processes of the movement of water vapor (or a fluid mainly composed of supercritical water) through the networks of cracks and pores, so that we need to develop models and analysis methods for the properties of scattering sources. The quantitative estimates for the possibility to detect the temporal changes of scattering properties by structure sensitive bodies await future investigation.

Acknowledgments

We thank all the members of JCEAM (Japanese Consortium of Earth's Active Monitoring) and those of the advisory board. The partial support by the Japanese Ministry of Education, Culture, Sports, Science and Technology (MEXT) (Nos. 18340132 and 18654076) and foundations from Shizuoka University are also deeply appreciated.

REFERENCES

Atkinson, B.K., 1984. Subcritical crack growth in geological materials. J. Geophys. Res. 89, 4077–4114.

Brenguier, F., Shapiro, N., Campillo, M., Ferrazzin, V., Duputel, Z., Coutant, O., Nercessian, A., 2008. Towards forecasting volcanic eruptions using seismic noise. Nature Geosci. 1, 126–130.

Bullough, R., Simmons, J.A., 1969. In: Argon, A.S. (Ed.), On the Deformation of an Imperfect Solid. In: Physics of Strength and Plasticity, MIT Press, pp. 47–64 (Chapter 5).

Dragert, H., Wang, K., James, T.S., 2001. A silent slip event on the deeper Cascadia subduction interface. Science 292, 1525–1528.

Fujii, N., Kumazawa, M., 2004. Detection of transient phenomena due to the structure sensitivity of rocks in the crust and upper mantle. In: Proc. 1st International Workshop on Active Monitoring in the Solid Earth Geophysics, S7-P05, Mizunami, Japan, pp. 353–354.

Fujii, N., Kumazawa, M., 2005. Active Scatterers: – Targets of ACROSS with Arrays of Arrays. In: Proc. 2nd International Workshop, Active Geophysical Monitoring of the Earth Lithosphere, Novosibirsk, Russia, 11–14 September.

Hasada, Y., Kumazawa, M., Tsuruga, K., Kunitomo, T., 2004. Travel time estimation from a transfer function in frequency domain: The revised Sompi event analysis. In: Proc. 1st International Workshop Active Monitoring in the Solid Earth Geophysics, in Mizunami, Japan, S5-03.

Hirose, H., Hirahara, K., Kimata, F., Fujii, N., Miyazaki, S., 1999. A slow thrust slip event following the two 1996 Hyuganada earthquakes beneath the Bungo channel, southwest Japan. Geophys. Res. Lett. 26, 3237–3240.

Hirose, H., Obara, K., 2005. Repeating short- and long-term slow slip events with deep tremor activity around the Bungo channel region, southwest Japan. Earth Planets. Space 57 (10), 961–972.

Ikuta, R., Yamaoka, K., Miyakawa, K., Kunitomo, T., Kumazawa, M., 2002. Continuous monitoring of Propagation velocity of seismic wave using ACROSS. Geophys. Res. Lett. 29. doi:10.1029/2001GL013974.

Ikuta, R., Yamaoka, K., 2004. Temporal variation in the shear wave anisotropy detected using Accurately Controlled Routinely Operated Signal System (ACROSS). J. Geophys. Res. 109 (B09305). doi:10.1029/2003JB002901.

Ikuta, R., Watanabe, T., Soma, T., Saiga, A., Yamaoka, K., Tsuruga, K., Kunitono, T., Hasada, Y., Kasahara, J., Satomura, M., Kumazawa, M., Fujii, N., 2005. Monitoring of the state of inter-plate coupling using reflection phase - Preliminary results using a short array. In: Proc. 2nd International Workshop, Active Geophysical Monitoring of the Earth Lithosphere, Novosibirsk, Russia, 11–14 Septermber.

Iidaka, T., Iwasaki, T., Takeda, T., Moriya, T., Kumakawa, I., Kurashimo, E., Kawamura, T., Yamazaki, F., Koike, K., Aoki, G., 2003. Configuration of subducting Philippine Sea plate and crustal structure in the central Japan region. Geophy. Res. Lett. 30 (5), 1219–1222.

Ito, Y., Obara, K., Shiomi, K., Sekine, S., Hirose, H., 2007. Slow earthquakes coincident with episodic tremors and slow slip events. Science 315, 503–506.

Kasahara, J., Tsuruga, K., Mikada, H., Yamaoka, K., Fujii, N., 2006. Time lapse approach in the study of earthquake generation - geophysical exploration of asperities-reflectors system (EARS) for interplate earthquake generation along a subducting oceanic plate. BUTSURI-TANSA 59 (6), 525–537.

Kodaira, S., Iidaka, T., Kato, A., Park, Jin-Oh, Iwasaki, T., Kaneda, Y., 2004. High pore fluid pressure may cause silent slip in the nankai trough. Science 304 (5675), 1295–1298. doi:10.1126/science.1096535.

Kunitomo, T., Kumazawa, M., 2004. Active monitoring of the Earth's structure by the seismic ACROSS - Transmitting and receiving technologies of the seismic ACROSS. In: Proc. 1st International Workshop Active Monitoring in the Solid Earth Geophysics, in Mizunami, Japan, S4-04.

Kumazawa, M., Kunitomo, T., Yokoyama, Y., Nakajima, T., Tsuruga, K., 2000. ACROSS: Theoretical and technical developments and prospect to future applications. Technical report, Japan Nuclear Cycle Development Institute, 9, 115-129.

Kumazawa, M., Kunitomo, T., Nakajima, T., Tsurga, K., Hasada, Y., Nagao, H., Matsumto, H., Kasahara, J., Fujii, N., Shigeta, N., 2007. Development of ACROSS (Accurately Controlled, Routinely Operated, Signal System) to realize constant monitoring the invisible Earth's interiors by means of stationary coherent elastic and electromagnetic waves. JAEA-Research 2006-33, 1–150.

Nadeau, R.M., Dolenc, D., 2005. Nonvolcanic tremors deep beneath the san andreas fault. Science 307, 389.

Niu, F., Silver, P.G., Nadeau, R.M., McEvilly, T.V., 2003. Migration of seismic scatterers associated with the 1993 Parkfield aseismic transient event. Nature 426, 544–546.

Obara, K., 2002. Nonvolcanic deep tremor associated with subduction in southwest Japan. Science 296, 1679–1681.

Obara, K., Hirose, H., Yamamizu, F., Kasahara, K., 2004. Episodic slow slip events accompanied by non-volcanic tremors in southwest Japan subduction zone. Geophys. Res. Lett. 31 (23). doi:10.1029/2004GL020848.

Rogers, G., Dragert, H., 2003. Episodic tremor and slip on the Cascadia subduction zone: The chatter of silent slip. Science 300, 1942–1943.

Sato, H., Fehler, M.C., 1998. Seismic Wave Propagation and Scattering in the Heterogeneous Earth. Springer-Verlag, New York.

Tsuruga, K., Hasada, Y., Kunitomo, T., Kasahara, J., Ikuta, R., Watanabe, T., Yamaoka, K., Fujii, N., Kumazawa, M., Nagao, H., Nakajima, T., Saiga, A., Satomura, M., 2005. Monitoring of the state of inter-plate coupling using reflection phase - Preliminary results using a long array. In: Proc. 2nd Workshop Active Geophysical Monitoring of the Earth Lithosphere in Novosibirsk, Russia, pp. 23–27.

Tsuruga, K., Kasahara, J., Mikada, H., Yamaoka, K., Fujii, N., 2006. Mapping of non-asperities and continuous, active monitoring of the interior of the earth's crust. J. Geography 115, 51–71 (in Japanese with English abstract and captions).

Wegler, U., Sens-Schonfelder, C., 2007. Fault zone monitoring with passive image interferometry. Geophys. J. Int. 168, 1029–1033. doi:10.1111/j.1365-246X.2006.03284.

Yoshida, Y., Ueno, H., Ishikawa, Y., Kunitomo, T., Kumazawa, M., 2004. The observation and analysis of ACROSS signals by seismic network - Determination of the transfer function. Paper presented at 1st International Workshop, Active Monitoring in the Solid Earth Geophysics, Mizunami, Japan, 30 June–2 July, pp. 258-261.

CHAPTER 6

Active Vibromonitoring: Experimental Systems and Fieldwork Results

Anatoly S. Alekseev[a], Boris M. Glinsky[a], Valery V. Kovalevsky[a], *and* Marat S. Khairetdinov[a]

Contents

1. Introduction

Over the past 30 years, in an effort to improve seismic hazard prediction, seismologists have turned to active geophysical monitoring methods, using powerful seismic vibrators to investigate changes in the geologic medium's stressed-deformed state within seismically active zones. This scientific direction has been vigorously pursued at institutes within the Siberian Branch of the Russian Academy of Sciences. During this period, several experimental systems for active monitoring have been created. They include powerful vibrational sources (with a vibrational force of 100 tons in the frequency range from 5 to 15 Hz), computer control systems, specialized mobile complexes VIRS-M, VIRS-K, and ROSA for precisely recording vibrational seismic signals, and data processing systems. Recently, a method for active monitoring using wideband sweep signals and narrowband monofrequency signals radiated by seismic vibrators has been developed (Alekseev et al., 2005, 2004, 1998, 1997; Andreev et al., 2002; Seleznev et al., 2004).

[a] Institute of Computational Mathematics and Mathematical Geophysics SB RAS, Pr. Akad. Lavrentieva, 6, Novosibirsk, 630090, Russia

E-mail address: kovalevsky@sscc.ru (V.V. Kovalevsky).

Handbook of Geophysical Exploration: Seismic Exploration, Volume 40

ISSN 0950-1401, DOI: 10.1016/S0950-1401(10)04011-5

Figure 1 A view of CV-100 vibrators at the Bystrovka test site (left) and Baikal test site (right): (1) ground platform; (2) heavy frame with loading mass; (3) eccentric force units; and (4) electric motors. The vibrational force on the ground is 100 tons within the frequency range of 5-15 Hz.

A powerful low-frequency vibrator (namely the CV-100) is the basis of this active monitoring system. This 100-ton vibrator consists of a ground platform, a heavy frame with a loading mass of about 120 tons, eccentric force units, and two electric motors. This vibrator (Alekseev et al., 2005) can radiate sweeps using time-varying frequency and monofrequency signals. Its computer control system provides an angular accuracy of 1-2 degrees. The duration of continuous operation ranges from 5 minutes to 1 h; usually a 20-40 minute radiation session is sufficient to record sweep signals over a distance of 300-400 km and to obtain an impulse seismogram after correlation. Detection of the monofrequency signals can be done as far as 1,400 km from the source.

At present, powerful low-frequency vibrators with eccentric excitation systems are installed at the test sites Bystrovka (near Novosibirsk), Babushkin (in the Lake Baikal region), and Goryachiy Kluch (near Krasnodar) (Alekseev et al., 2005, 2004). Figure 1 shows the CV-100 vibrators at the Bystrovka and Baikal sites.

Another type of powerful low-frequency vibrator is based on a mechanical system with variable resonance frequency. One of these, the hydro-resonance vibrator HRV-50, uses the resonant oscillation of a 60-ton water column located between two air springs in the vertical case (Figure 2). Oscillations are excited by compressed air pushed into the air springs through special valves using a computer controlled system. The HRV-50 produces a 50-ton vibrational force in the 2-10 Hz frequency range. It is a full-scale model of the 1000-ton super-powerful shaft vibrator, intended for monitoring at teleseismic distances (Kovalevsky, 2000, 1981).

In 1996-1997, the Institute of Computational Mathematics and Mathematical Geophysics of the Siberian Branch of the Russian Academy of Sciences (ICM&MG SB RAS) conducted experiments using low-frequency vibrating sources and the recording of signals at distances of 300-400 km. The goal of these experiments was to determine the active vibroseismic

Figure 2 A view of the HRV-50 hydro-resonance vibrator at the Bystrovka test site. The vibrator's case is a vertically installed railway tank. The 60-ton water column and two air springs near the top and bottom are inside the tank. Oscillations are excited by compressed air passed through special valves: (1) ground platform, (2) tank, (3) air valves. The vibrational force on the ground is 50 tons within the frequency range of 2-10 Hz.

monitoring system's sensitivity through the detection of small variations in a seismic-wave-field caused by the Earth's tidal deformation processes.

Semidiurnal and daily tides (with periods of approximately 12 and 24 h) exist on the Earth, and the deformational influence of such tides on the Earth's crust has some important ramifications for seismic monitoring. It is a global phenomenon that can be observed at any point on the Earth, exhibiting periodic changes that may be predicted with a very high degree of accuracy, and can serve as a natural standard for deformation forces in the lithosphere.

Considerable experimental effort has been devoted to investigations of how the tidal deformations of the Earth's crust influence changes in seismic wave velocities. The main question for this research has been: Do tidal deformations on the order 10^{-7}-10^{-8} result in much bigger relative variations in seismic wave velocities? The main difficulty with experiments associated with this question has been the limited accuracy of measurements. When the arrival-time variations in the P and S waves of earthquakes were used for analysis, the greatest variations in seismic waves velocities (10^{-2}) correlated with the phases of Earth's tides have been detected (Popandopulo, 1982). In investigations using explosions and earthquakes, variations in velocities have been comparable to a measurement accuracy of 10^{-3}-10^{-4} (Eisler, 1967; Gamburtsev, 1992). Experiments conducted using vibrators had the same accuracy and were carried out at an offset of 10 km (Aki et al., 1970; Fazio et al., 1973). In Scandinavia, there has been an experiment involving the recording of the monofrequency radiation of a hydropower station (by the group NORSAR) at distances of 5-14 km (Bungum et al., 1977). In this experiment, relative changes in seismic wave velocities of 10^{-3} were detected. Another experiment, carried out by the Siberian Branch of RAS, used a low-frequency vibrator with an offset of 125 km to collect vibrograms that detected variations in first-arrival times (Yushin et al., 1994). These did not show velocity variations greater than 10^{-4}.

In all these experiments, if variations in seismic velocities were observed, they had values comparable to the accuracy of the measurements. A reliable conclusion with respect to the connection between seismic velocity variations and crustal tidal deformations was not possible. However, *some* correspondence was found: if the source characteristics were not precisely known, larger variations in seismic wave velocities connected with tidal processes could be obtained in experiments. The results of these experiments have shown that for detecting variations in seismic waves caused by Earth's tides, large offsets (several hundreds of kilometers) and highly stable sources are necessary for increasing the accuracy of measurements.

2. Active Vibromonitoring Experiments

Vibroseismic interferometry was used in experiments conducted at ICM&MG SB RAS. This method is based on excitation and recording of monofrequency signals from powerful vibrators. Because of the high stability of the monofrequency signals radiated by the vibrators, their parameters are determined with high accuracy at the recording point. For the amplitude spectrum of such signals, the signal/noise ratio reaches values of 50-100 over distances of 300-400 km, for sounding sessions with durations of 20 minutes. Thus, the amplitudes of monofrequency signals are determined with an accuracy of 1-2%. Correspondingly, the accuracy of angular-phase measurements is 0.5-1 degree, which enables detection of travel-time changes within several milliseconds. Such high sensitivity allows detection of small propagation-velocity variations (Glinsky et al., 2000, 1999; Kovalevsky, 2006, 2004).

At the Bystrovka site in 1996-1997, excitation of monofrequency signals in a frequency range of 6-8 Hz (with intervals of 1 and 3 h) was carried out using the 100-ton vibrator CV-100 over a 3-4 day period. Signal recordings were made at distances of 430 km (seismic station Ust-Kan, Republic Gorny Altai, 1996) and 356 km (Savushky, Altai, 1997). In these experiments, investigators used GPS time synchronization, both at the vibrator and at the remote receiver points. At the Bystrovka test site, the recording system included a three-component seismometer installed 30 m from the vibrator.

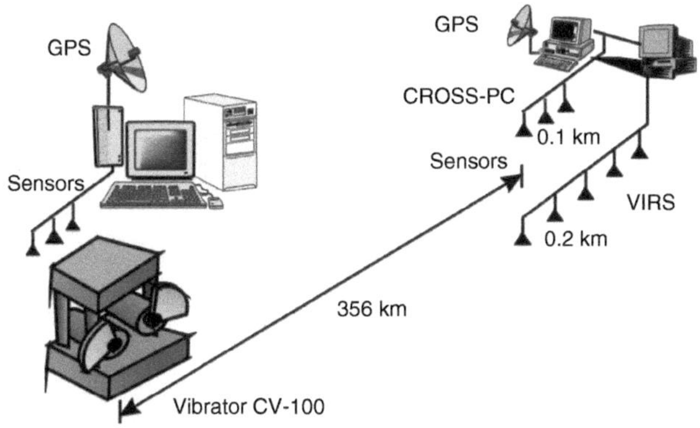

Figure 3 The scheme of the 1997 vibromonitoring experiment. The CROSS-PC and VIRS recording systems were located 356 km from the CV-100 vibrator. They have 3 and 5 three-component seismic sensors installed on the profile, with steps of 0.1 km and 0.2 km, respectively. A control recording system was placed near the vibrator. GPS was used for time synchronization. Monofrequency-radiation sessions were repeated every 3 h over 4 days.

The scheme of the experiment (at a distance of 356 km) is presented in Figure 3.

In the first experiments (in 1996), the recording of vibroseismic signals was made by the 6-channel system CROSS-PC, with three three-component seismometers (SK-1P) installed on bedrock near the Ust-Kan seismic station, Republic of Gorny Altai. GPS was used for time synchronization of the source and the recording systems. Monofrequency signals of 6.75 Hz. were used, because they had a maximal signal/noise ratio at the recording point. Periodically repeating radiation sessions were carried out to detect the daily variations in the parameters of the vibrosignals. Three series of periodic nocturnal soundings were done (on September 28-30, October 5-7, and October 14-16), in the phase of two syzygial tides and one neap tide. In each session, the duration of the sessions was 25 minutes (1500 s). Nine sessions were conducted each night. They followed one by one with a period of 1 h, from 10:30 pm until 6:30 am local time. (The choice of this schedule was determined by the low level of microseismic noise at night.)

Thus, 27 measurements of monofrequency signals were taken at a distance of 430 km from the vibrator, in three series of soundings over 3 days duration each. This measurements were carried out irregularly in time during this 3-day series: nine measurements were conducted every night, while none were conducted during the day. As a result, investigators obtained nine time series with variations in monofrequency signals (for all three X, Y, and Z components) over an irregular time scale.

In the experiments of 1997, two recording systems were used at a distance of 356 km from the vibrator, near the village of Savushky, Altai region.

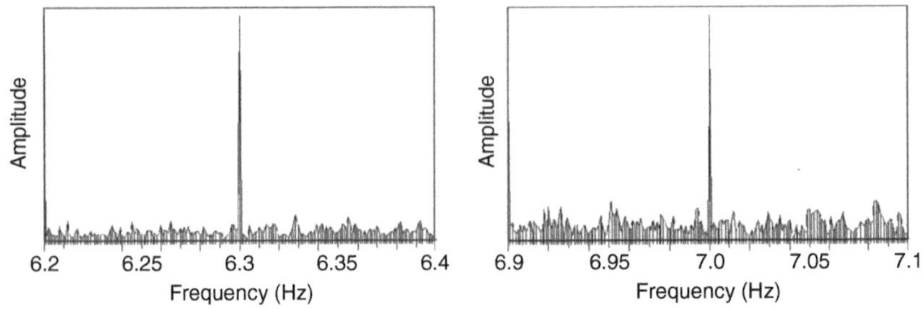

Figure 4 Spectra of monofrequency signals at frequencies of 6.3 Hz and 7.0 Hz. The distance from Vibrator CV-100 is 356 km (Savushky, Altai). A radiation signal with a duration of 1200 s provides a signal-to-noise ratio of 20-40, and an accuracy in determining signal amplitudes and phases of 2-4% and 1-2.5 degrees, respectively, at the recording point.

The VIRS recording complex had a 15-channel seismic array that included five three-component seismic sensors (SK-1P); The CROSS-PC system had three three-component seismic sensors. Monofrequency signals of two frequencies (6.3 Hz and 7.0 Hz) were radiated for durations of 25 minutes each every 3 h. The experiment lasted 4 days without interruption, resulting in a regular time series with 32 measurements for all sensor components. GPS was used for time synchronization.

3. Data Processing and Results

Processing of the recorded data was based on spectral analysis of the vibrational signal in the presence of the noise background. The 1200 s part of the record, 300 s after the beginning, was used for this analysis, which allowed us to use a stable-in-time monofrequency signal for the processing and to exclude the period of wave-field buildup connected with the arrivals of various waves. The amplitude and phase of the spectral components in a 0.0008 Hz interval around the excitation frequency were measured as parameters of the signal. For the session duration of 1000-1500 s, the amplitude of the spectral line of the signal is 20-40 times larger than the average spectrum level of noise.

Figure 4 shows the spectra of driving monofrequency signals with frequencies of 6.3 Hz and 7.0 Hz. The signal recorded at the Bystrovka test site had an amplitude that exceeded the noise level by a factor of 1000. Therefore, the parameters of the radiated signal were determined with an error of 0.1% for the amplitude and 0.1 degree for the phase.

We evaluated the accuracy of our estimates from the power spectral density (PSD) of noise within a narrow band of frequencies near the spectral line of the recorded signal. To evaluate PSD, we used a frequency interval of 0.1 Hz, near the driving monofrequency, which included 100-150 spectral lines and gave a representative sample for estimating the statistical characteristics of noise. The PSD of noise was estimated separately for each session—it improved the reliability of estimates for the nonstationary seismic noise.

We determined that the usual error in estimating the signal amplitude (for the noise level at the recording points in these experiments) was about 2-5% for the signal component with maximal amplitude, and the error of the phase was 1.5-2.5 degrees.

From these experiments, we obtained the time series of amplitudes and phases for each component of the three-component seismometers. The components with the maximal signal/noise ratio were chosen for further analysis. Thus, for a frequency of 7.0 Hz, the components X4, Y3, Y4, Y5, and Z3 were chosen; and for a frequency of 6.3 Hz, the components X3, X4, Y4, Y5, Z4, and Z5 were chosen.

Preparing the time series of vibrosignal phases for further analysis was conducted in several stages. Time synchronizations of the recording system (by GPS) were made several times during the experiment. Consequently, it was necessary to account for the precise starting times of the record files in determining the phases of signals. Similar time series were constructed for the amplitudes of signals. (The preparation of the amplitude series was somewhat easier than the preparation of the phase series, since the former was not connected with the precise recording times.) In detecting the amplitudes of spectral lines for various components, the amplification factors related to the recording complex and vibrator force were taken into account. The time series of phases and amplitudes at components X, Y, Z for frequencies of 6.3 Hz and 7.0 Hz are shown in Figures 5 and 6, respectively.

To find latent periodicities, we analyzed the time series of signal amplitudes and phases to determine the relationship between variations in signal amplitudes and phases, and Earth tides. A Fourier transform for the time series—with a non-uniform distribution based on approximation by periodic functions using the least-squares method—was used to process the results of the experiments at a distance of 430 km. The amplitude of the spectrum component was determined from functional minimization.

Spectral components for all time series of the amplitudes and phases were determined within a 6 h to 30 h interval, after which the spectrum of each time series was normalized to maximum. The normalized spectra were averaged, from which the spectrum of the time parameters averaged over all time series of the vibrosignals was obtained:

$$S(T) = \frac{1}{M}\sum_{i=1}^{M}\frac{S_{\varphi i}(T)}{\max\limits_{T}(S_{\varphi i}(T))}, \qquad (1)$$

where $S(T)$ is the spectrum of variations in parameters averaged on the time series of the vibrosignals, i is number of time series, M is the total number of

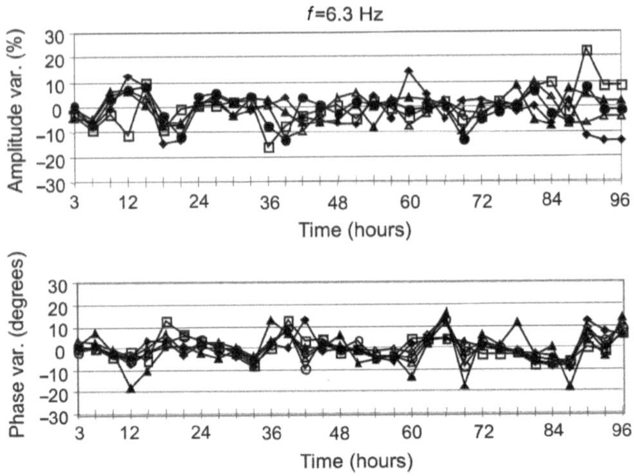

Figure 5 Time series of the amplitude and phase variations for a frequency of 6.3 Hz. They include 32 points, each corresponding to one of 32 sessions carried out every 3 h over 4 days. Distance from the CV-100 vibrator is 356 km. Five components with the maximal signal/noise ratio were chosen. Visible periodicities become apparent after spectral analysis.

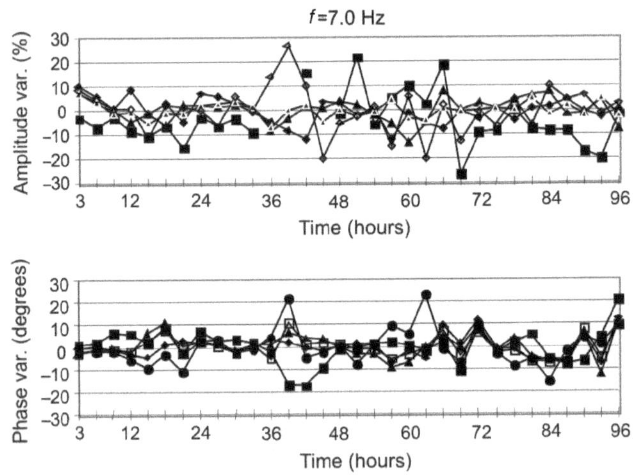

Figure 6 Same as in Figure 5, but for a frequency of 7.0 Hz.

phase and amplitude time series, T is the period, and $S_{\phi i}(T)$ is the amplitude in the spectrum of time variations.

The average spectrum of the parameter variations is presented in Figure 7. There are maxima of the spectrum with periods of 12 and 24 h; the values of the signal amplitude variations are 2-4%, and the appropriate

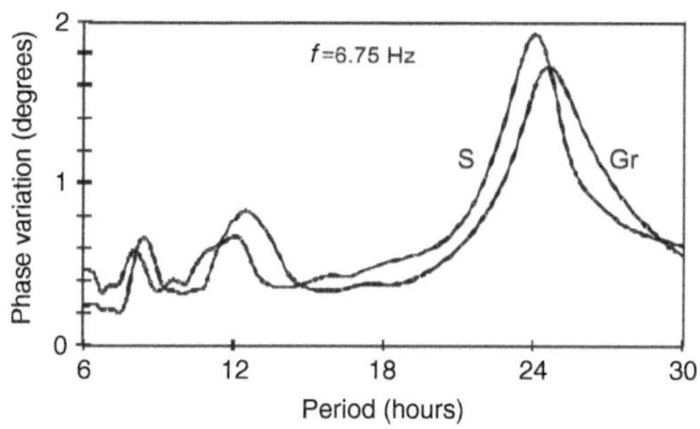

Figure 7 The average spectra of variations in the phase of monofrequency vibrosignals (S) and tidal variations of gravity (Gr). The signal frequency is 6.75 Hz. The distance from the CV-100 vibrator is 430 km. Spectra have correlated maxima with periods of 12 and 24 h for seismic and gravity data. The subordinate spectrum maximum is caused by the nonuniform time scale of the data.

value of the phase variations are 1-2 degrees. Note that the accuracy of these estimates is higher than the accuracy of an individual estimate in measuring amplitude and phase within one session. For the estimation of one value, we use the information from an ensemble of nine time series of variations, with 27 points in each series. Therefore, it can be assumed that the dispersion of the amplitude estimate decreases approximately 5-10 times. It may reduce the error in determining amplitudes by 0.5-1% and in determining phases by 0.2-0.5 degrees. Figure 7 shows also the spectrum of tidal acceleration, obtained from gravitational measurements. There are maxima with periods of 12 h and 24 h, corresponding to the semidiurnal and daily periods in the spectrum of tidal acceleration.

Despite the fact that time series with irregular points give subordinate maxima in the spectrum, as is clear from Figure 7, the 1996 experiments performed at a distance of 430 km showed for the first time the presence of daily and semidiurnal periodicity in the time variations within vibrosignal parameters, and have enabled us to assume their connection with tidal deformation processes in the Earth's crust.

Experimental techniques were considerably improved in 1997 in experiments at a distance of 356 km from a vibrator (near the village of Savushky, Altai region). In those experiments, we conducted a uniform series of radiation and recording sessions with a periodicity of 3 h over 4 days. CROSS-PC and VIRS recording systems were used, with three and five three-component seismic sensors installed on the profiles, with steps of 0.1 km and 0.2 km, respectively. This uniform series of radiation and recording sessions allowed us to construct the correct spectrum on a regular time scale without subordinate maxima, and to choose a time series of amplitudes and phases from channels with the highest signal/noise ratios.

In determining the spectrum of time variations, we took into account that the accuracy of measurements for various values in the time series was

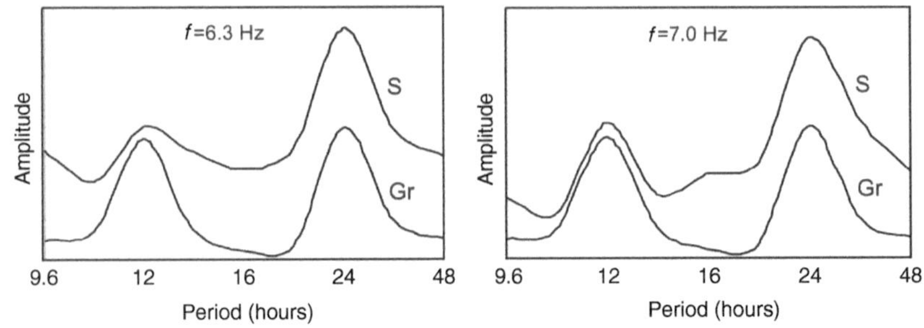

Figure 8 Average spectra of variations in the parameters of vibrosignals (S) and tidal variations of gravity (Gr). The signal frequency is 6.3 Hz (left), 7.0 Hz (right). The distance from the CV-100 vibrator is 356 km. Spectra have correlated maxima with periods of 12 and 24 h for seismic and gravity data.

not equal. Such accuracy is related to the signal/noise ratio in each session, and usually decreases in the afternoon, owing to the increase in human-caused microseismic noise. Therefore, in searching for latent periodicity, we used Fourier transforms for uniform time series with unequal accuracy points, also based on approximation by periodic functions with the least squares method. The amplitude of a spectrum component with period T was determined from the following minimization functional:

$$W_i = \sum_{n=1}^{N} \left(\frac{A_{in}}{\sigma_{in}} \right)^2 \left(\varphi_i(n\tau) - S_{\varphi i}(T) \sin \left(\frac{2\pi n\tau}{T} + \alpha_i \right) \right)^2, \quad (2)$$

where W_i is the functional for the time series i of variations for the phase ϕ_i, N is the number of sessions, τ is the time step between consecutive sessions, T is the period, $S_{\phi i}(T)$ is the amplitude over a spectrum of time variations in the phase for the period T, α_i is the phase of the spectral component, A_{in} is the amplitude of the monofrequency signal in this channel in the n-th session, and σ_{in} is the dispersion of microseismic noise in this channel in the n-th session.

The spectral components for all time series of amplitudes and phases were determined over an interval of the periods from 9.6 h to 48 h. Furthermore, as with the processing in the previous experiment, the spectrum of each time sequence was normalized to the maximum. The normalized spectra were averaged, and the spectrum for the time-parameter average for all time series of Eq. (1) was obtained.

Figure 8 shows the averaged spectra of time variations for the amplitudes and phases of monofrequency vibrosignals with the frequencies of 6.3 Hz and 7.0 Hz, and their comparison with the gravity spectrum. Both spectra contain maxima with periods of 12 h and 24 h, repeating the results of the previous experiment. Given the uniformity of the time series, the maxima of tidal variations for gravity were obtained. The spectral line width was determined by the overall duration of the experiment (4 days), allowing us to clearly detect daily and semidiurnal periodicity. The maxima of the variations in vibrosignal amplitudes and phases are above the general noise component of the spectrum. Estimates for 24-hour and 12-hour periodicity of variations in signal amplitudes give a value of 3%, and the corresponding value of the phase variations is approximately 1.5 degrees, which is close to the results of recording at a distance of 430 km.

That the same periodicities occur in the spectra of signal variations and in the Earth's tides is the major reason to suspect a relationship between them. This finding allows us to assume that the revealed variations in the vibroseismic field parameters are the result of deformation processes caused by tides, which change the velocities of seismic waves.

The obtained estimates for the amplitudes of 12-hour and 24-hour periodicity for the variations in seismic-signal parameters allow us to estimate the effect of Earth's tides on the arrival times of the seismic waves and their velocities at distances of 356-430 km. At frequencies of 6-7 Hz, a change in the signal phase by 1-2 degrees is associated with a time delay of 0.5-1 ms. Characteristic propagation times of P and S waves for a distance of 356 km are about 53 s and 94 s, respectively. Therefore, the influence of the Earth's tides on the velocities of seismic waves for a distance of 356-430 km may be characterized in the results of the experiments by the relative changes of velocities $\delta V/V \sim 10^{-5}$-10^{-6}. This estimate is considerably lower than was obtained in previous experiments and is an upper estimate for the velocity variations, since in the experiments involving vibroseismic monitoring, it was impossible to exclude the influence of other varying natural factors (temperature, pressure, etc.) with daily and half-daily periodicities.

Thus, the experiments allowed us to re-evaluate, by an order of magnitude accuracy, the expected crustal effects of tidal deformation on the velocity of seismic waves.

4. Vibromonitoring Experimental Systems

In 2000-2004 a full-scale experimental active vibromonitoring system was used in the regions of Novosibirsk and Lake Baikal, using stationary 100-ton seismic vibrators and mobile recording systems. In 2002, an experiment called "CIRCLE" was carried out, in which investigators recorded the wave fields of these powerful vibrators in various azimuthal directions at the same distance from the source. Nine recording points were located approximately at regular intervals along a 300 km diameter circle around the vibrator, in an attempt to determine the local azimuthal heterogeneity of Earth's crust.

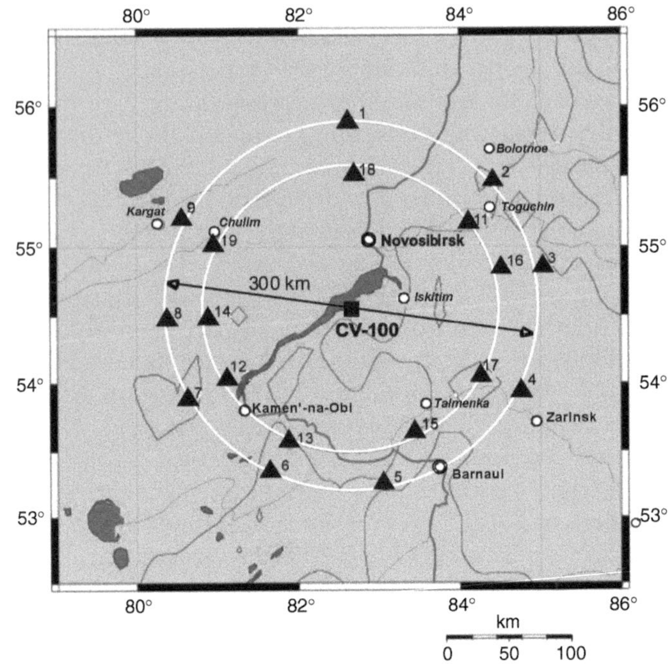

Figure 9 The scheme of the CIRCLE experiment. The CV-100 vibrator is represented by the black square at the center, and the recording points are indicated by black triangles. The goal of the experiment was full-scale modeling of a vibromonitoring system covering 100,000 km^2, using one stationary installed vibrator (CV-100) and two mobile recording complexes (VIRS, ROSA), moving from one recording point to another.

The region—a radius of 150 km around the vibroseismic test site at Bystrovka—is one with which various geological structures are in contact: the Biysk-Barnaul Depression in the South-Southwest, the Salair Range in the East-Southeast, the Tom-Kolivan folded zone in the Northeast, and the West-Siberian plate in the Northwest. The scheme of the CIRCLE experiment is given in Figure 9.

The experiment had several purposes: (1) to investigate the wave field of the vibrator in regions with various geological structures; (2) to study the structure and azimuthal heterogeneity of the Earth's crust in this region; (3) to generate a full-scale model of the vibromonitoring system over a region of 100,000 km^2.

Vibrational seismograms for every azimuthal direction, with well-defined arrivals for the main wave types, were obtained in the experiment. In the first arrivals at various azimuths from the source, Pg and Pn waves, arrived with a delay of 1.3-1.5 s from each other at a distance of 150 km and carried information about the location of the Moho boundary in this region.

In the second stage of the experiment, CIRCLE-2 conducted in 2003, recording was done at nine points of a circle with radius of 120 km at the

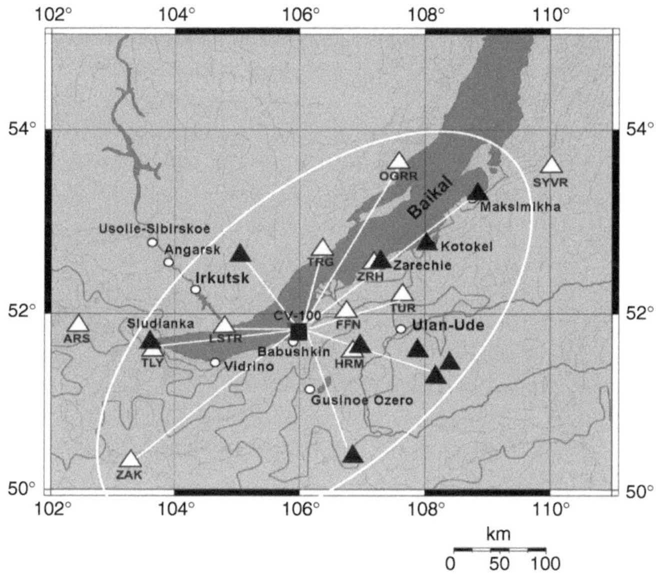

Figure 10 The scheme of the vibroseismic monitoring system in the Lake Baikal region: CV-100 vibrator (black square), mobile recording complexes (black triangles). and the local seismic stations (white triangles). The monitoring area is 550 × 250 km. The 4-6 sweep- and monofrequency sessions have been carried out by the CV-100 vibrator over night, every month since 2003.

same azimuths as in the previous experiment. The choice of a 30 km distance between recording points resulted from the need to analyze the changes in the arrival times of the Pg and Pn waves in the first arrivals. For the 120 km distance, the delay of arrival times was about 2.3-2.4 s. The variations in these delays at the 120 km and 150 km distances were used to determine the Earth's crust thickness and locate the Moho boundary at various azimuths from the source.

Active vibromonitoring, using the vibrator's monofrequency monochromatic wave fields in a seismically active zone of Lake Baikal, began for the first time in 2003. The approach was based on the use of vibrator monofrequency signals and the principles of vibroseismic interferometry. When the vibrator radiated monofrequency signals, a stationary wave field, with constant amplitudes and phases, was formed at each point of the geologic medium. Changes within the medium in a stress-concentration zone (i.e., a developing earthquake zone) led to changes in the recorded wave fields. The scheme of vibroseismic monitoring system in the Lake Baikal region is given in Figure 10.

In the Baikal Experiment, the southern region of Lake Baikal was investigated. The CV-100 vibrator used as a source was installed at the test site near the village of Babushkin. In 2003, signals were recorded at two points—at the Talaya seismic station, Sludyanka (163 km from the source), and at the Hurumsha seismic station (68 km from the source). The VIRS-M recording complex, with five three-component seismometers, was used at the Talaya seismic station; the ROSA recording complex, with four three-component seismometers, was used at the Hurumsha seismic station.

Recording of the vibrator monofrequency signals was done within a 6-10 Hz frequency range (with steps of 0.5 Hz). Frequencies with the maximal signal/noise ratio were chosen, and the spectra of monofrequency signals were obtained at each point. The repeatability of the recorded fields was checked in a series of tests. Vibrational seismograms at both recording points were also obtained, and the main types of waves were detected.

In 2004 at Lake Baikal, monitoring using monofrequency signals in seismically active zones continued. Recording points were chosen near the Zarech'e, Kotokel, and Maksimikha seismic stations, at distances from the source of 115 km, 178 km, and 248 km, respectively. VIRS-K and ROSA complexes were used. Monitoring sessions were carried out over three days, with the vibrator radiating monofrequency and sweep-signals. Four vibrator sessions were conducted in the monofrequency mode, at frequencies of 6.5, 7.0, 7.5, 8.0, 8.5, 9.0 Hz, as well as four sessions in a sweep-signal mode of 6.25-10.059 Hz, over 3272 s. Spectra for monofrequency signals and vibrograms were obtained for comparison with seismograms from the regional-network seismic stations. A database that included 432 files of wave forms was created from these seismograms. The goal of these experiments was to show the possibility of stable recording of monofrequency signals from the vibrator at the Babushkin test site, using both of the mobile recording complexes (VIRS-M and ROSA) with the permanently operating seismic stations of the regional network.

Since 2003, continuous monitoring of the Baikal seismically active zone has been done using periodic monofrequency signals recorded by seismic stations of the regional network (Figure 10). The 4-6-sweep, monofrequency sessions were generated by the CV-100 vibrator in one night, out of every month, over a monitoring area of 550 × 250 km. A database of Baikal region vibroseismic monitoring has been created at the Geological Institute of SB RAS, Ulan-Ude, Buryatia.

5. Conclusions

We have determined the sensitivity of an active seismic monitoring system using powerful (100-ton) vibrators. Our experiments have shown that such a system enables us to detect variations in seismic wave velocities on the order of 10^{-5}-10^{-6}, in a 300-400 km area around a source. We investigated the influence of the Earth's-crust tidal deformations on the order of 10^{-7} on the amplitude-phase characteristics of seismic waves radiated by the vibrators. We thereby determined that the influence of the Earth's tides on the velocities of seismic waves for a distance of 356-430 km may be characterized by the relative changes of velocities $\delta V/V \sim 10^{-5}$-10^{-6}.

The CIRCLE experiment has demonstrated the full-scale prototype of an active monitoring system for a region ~100,000 km^2, using a 100-ton vibrator, several mobile recording systems, and seismic stations within a regional network. At this time, vibroseismic monitoring of an area 550 km × 250 km (about 100 thousand sq. km) is being carried out in the Baikal seismic-prone zone using the same set of stations as in CIRCLE. This monitoring consists of periodic monofrequency sessions by the vibrator, with the aim of contributing to mid-term earthquake forecasting.

Note that beginning in 2003, the Buryat Branch of the Geophysical Survey and the Institute of Geology at SB RAS in Ulan-Ude has performed continuous seismic monitoring at Lake Baikal, using the same techniques as described above.

ACKNOWLEDGMENTS

This work was supported by Grant RFBR 07-05-00858, 07-07-00214, Programs of Basic Research, RAS 16, 16.13, 16.15.

REFERENCES

Aki, K., De Fazio, T., Reasenberg, P., Nur, A., 1970. An active experiment with earthquake fault for an estimation of the in situ stress. Bulletin of the Seismological Society of America 60, 1315–1335.

Alekseev, A.S., Chichinin, I.S., Korneev, V.A., 2005. Powerful low-frequency vibrators for active seismology. Bulletin of the Seismological Society of America 95, 1–17.

Alekseev, A.S., Geza, N.I., Glinsky, B.M., Emanov, A.F., Kashun, V.N., Kovalevsky, V.V., Manshtein, A.K., Mikhailenko, B.G., Seleznev, V.S., Serdukov, S.V., Sobisevitch, A.L., Sobisevitch, L.E., Soloviev, V.M., Khairetdinov, M.S., Chichinin, I.S., Yushin, V.I., 2004. Active Seismology with Powerful Vibrational Sources. ICM&MG SB RAS, Filial Geo publishing house of the SB RAS, Novosibirsk (in Russian).

Alekseev, A.S., Glinsky, B.M., Kovalevsky, V.V., Mikhailenko, B.G., 1997. Problems of active seismology. In: Fuch, K. (Ed.), Upper Mantle Heterogeneities from Active and Passive Seismology. In: Nato ASI Series, vol. 17. pp. 123–130.

Alekseev, A.S., Seleznev, V.S., Soloviev, V.M., Emanov, A.F., Yushin, V.I., Glinsky, B.M., Kovalevsky, V.V., Khairetdinov, M.S., Kashun, V.N., Vaptsov, K.O., Geza, N.I., 1998. Vibroseismic research of Altai-Sayan region. Bulletin of the Novosibirsk Computing Center, Series: Mathematical Modeling in Geophysics 3, 21–30.

Andreev, P.R., Grigoryuk, A.P., Shorokhov, M.N., 2002. System for reception and recording of vibroseismic signals. Bulletin of the Novosibirsk Computing Center, Series: Mathematical Modeling in Geophysics 7, 1–11.

Bungum, H., Risbo, T., Hjortenberg, E., 1977. Precise continuous monitoring of seismic velocity variations and their possible connection to solid earth tides. Journal of Geophysical Research 82, 5365.

Eisler, J.D., 1967. Investigation of a method for determining stress accumulation at depth. Bulletin of the Seismological Society of America 57, 891–911.
Fazio, T.De., Aki, K., Alba, I., 1973. Solid Earth tide and observed change in the in situ seismic velocity. Journal of Geophysical Research 78, 1319–1322.
Gamburtsev, A.G., 1992. Seismic Monitoring of the Lithosphere. Nauka Publishing House, Moscow (in Russian).
Glinsky, B.M., Kovalevsky, V.V., Khairetdinov, M.S., 2000. Vibroseismic monitoring of earthquake-prone areas. Volcanology and Seismology 21, 723–730.
Glinsky, B.M., Kovalevsky, V.V., Khairetdinov, M.S., 1999. Relationship of wave fields from powerful vibrators with atmospheric and geodynamic processes. Russian Geology and Geophysics 40 (3), 422–431.
Kovalevsky, V.V., 2006. Estimation of sensitivity of the method of active monitoring by monofrequency signals. Bulletin of the Novosibirsk Computing Center, Series: Mathematical Modeling in Geophysics 11, 43–52.
Kovalevsky, V.V., 2004. Modeling of the monitoring of anomalous zones under sounding by monofrequency vibrational signals. In: The Proceedings of 1-st International Workshop on Active Monitoring in the Solid Earth Geophysics (IWAM04), Mizunami, Japan, pp. 204–207.
Kovalevsky, V.V., 2000. Modeling of the hydroacoustic source for Earth's global tomography. Bulletin of the Novosibirsk Computing Center, Series: Mathematical Modeling in Geophysics 6, 41–48.
Kovalevsky, V.V., 1981. Vibrational source: USSR Patent 915598 (in Russian).
Popandopulo, G.A., 1982. Influence of the Earth's tides on the seismic waves velocities. Dokladi Akademii Nauk 262 (3), 580–583.
Seleznev, V.S., Alekseev, A.S., Gol'din, S.V., Emanov, A.F., Glinsky, B.M., Soloviev, V.M., Kovalevsky, V.V., Khairetdinov, M.S., Chichinin, I.S., Kashun, V.N., Danilov, I.A., 2004. Vibration geotechnologies in III millennium. In: The Proceedings of the 1st International Workshop on Active Monitoring in Solid Earth Geophysics (IWAM04), Mizunami, Japan, pp. 39–42.
Yushin, V.I., Geza, N.I., Velinsky, V.V., Mishurov, V.V., Speransky, N.F., Savvinikh, V.S., Astafiev, V.N., 1994. Vibroseismic monitoring in the Baikal Region. Journal of Earthquake Prediction Research 3, 119–134.

CHAPTER 7

Reliable Measurement Method to Reveal a Peculiar Dielectric Dispersion in Wet Rocks by Means of ACROSS

Hiroshi Matsumoto[a,b], Naotaka Shigeta[a,c],
Mineo Kumazawa[a,d], *and* Takahiro Nakajima[a]

Contents

1. Introduction

Our relevant research target is the physical states and tectonic roles of water within the solid earth's interior. The greatest potential appears in the electromagnetic sounding of rocks affected by deep-seated water, since the presence of H_2O influences the electrical properties of rocks.

[a] Tono Geoscience Center, Japan Atomic Energy Agency, 1-63, Yamanouchi, Akeyocho, Mizunami, Gifu, 509-6132, Japan
[b] Present address: EB Mask Equipment Engineering Department, Nuflare Technology Inc., 2068-3 Ooka, Numazu, Shizuoka, 410-8510, Japan
[c] Present address: Horonobe Underground Research Center, Japan Atomic Energy Agency, 432-2, Hokushin, Horonobe-cho, Teshio-gun, Hokkaido 098-3224, Japan
[d] Present address: Faculty of Science, Nagoya University, Furo-cho, Chikusa-ku, Nagoya, 464-8601, Japan

E-mail address: matsumoto.hiroshi@nuflare.co.jp (H. Matsumoto).

Handbook of Geophysical Exploration: Seismic Exploration, Volume 40

ISSN 0950-1401, DOI: 10.1016/S0950-1401(10)04012-7

The complex dielectric permittivity, defined by $\varepsilon^* = \varepsilon + \sigma/j\omega$, where ω is angular frequency, is representative of the amount and state of water contained in resistive rock; connected water serves as a path for ion conduction to raise the electrical conductivity, σ, while the accumulation of ions at the interface between water and rock-forming minerals causes the dielectric polarization to raise ε at a sufficiently low frequency.

Previous studies have shown that Debye type dielectric dispersion occurs in a frequency range around MHz, due to dielectric polarization occurring at the grain-water interface, and also in a frequency range from Hz to mHz due to the occurrence of electrochemical reactions for metallic or clayish minerals. Such a peculiar frequency dependence of ε has been studied intensively, since it is supposed to provide useful information on the polarization mechanism related to textures and constituent materials. It may depend on the shape of the grain-boundary water (De Lima and Sharma, 1992; Sen, 1981) and the presence of graphite, metallic or clay minerals causing strong dielectric dispersion (Pelton et al., 1978; Wong, 1979).

The problem arises from the phenomenon that the relative dielectric permittivity $\varepsilon/\varepsilon_0$ ($\varepsilon_0 = 8.8542 \times 10^{-12}$ F/m = the dielectric permittivity of a vacuum) in moist rocks increases in reciprocal proportion to frequency to become significantly large, 10^3-10^4 or more at 1 mHz, in comparison with 5-8 in ordinary dry rocks at high frequencies above MHz (Lockner and Byerlee, 1985; Jonscher, 1978; Olhoeft, 1985). The loss angle δ, given by $\delta = \cot^{-1}(\varepsilon\omega/\sigma)$, becomes independent of frequency, and of the order of 10^{-2} radians even at 1 mHz. Recent development of an active EM prospecting method (Nakajima et al., 2004) may attain observation of the dielectric permittivity of the earth's crust. The complex dielectric permittivity of moist rock can be an important observation in the active monitoring of underground water.

However, a critical review (Matsumoto et al., 2003) of the data reported so far leads to skepticism about the reliability and precision of the reported data; sufficient care has not been made to eliminate the polarization effect at the electrodes attached to the sample or the stray capacitance in most of the previous measurements. We suspect that observed dielectric polarization may arise from the electrode polarization. Further, the data on reported frequency dependence did not show any good systematic features.

We have developed an accurate method of measuring the complex dielectric permittivity with the least bias to provide reliable data on the peculiar dielectric dispersion by applying ACROSS (accurately controlled routinely operated signal system) (Nakajima et al., 2004), a measurement method of any linear dynamic system. On the basis of the present measurements, we shall demonstrate the reality of the '$1/f$ type' dielectric dispersion: the dielectric constant increases almost in proportion to the reciprocal frequency towards lower frequencies. Further, the importance of the geophysical significance of this phenomenon is discussed in this paper.

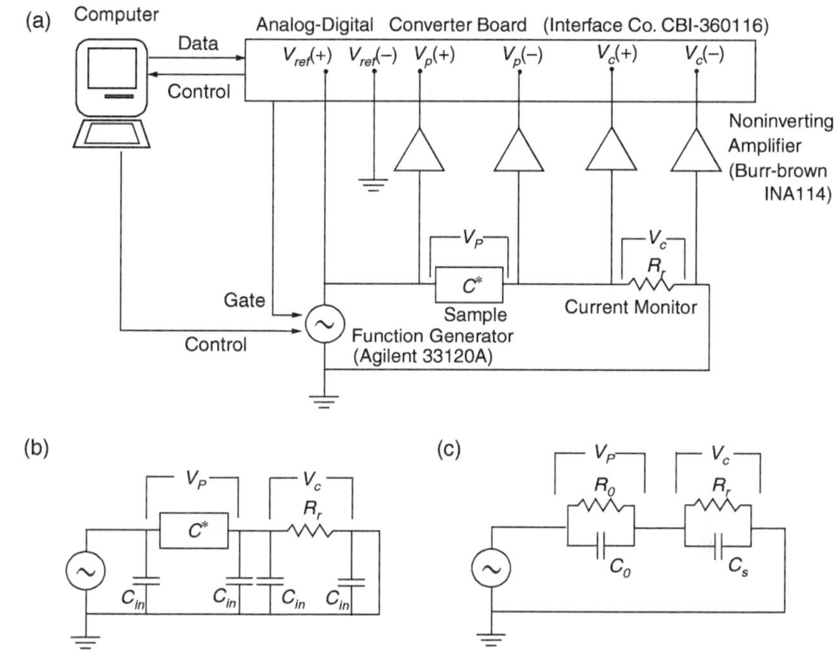

Figure 1 (a) Overview of the system of complex capacitance measurement. The sinusoidal signal is supplied from the function generator. Voltages V_p and V_c induced on the sample of complex capacitance C^* and the reference resistor of resistance R_r are measured using an amplifier and are digitized by an A/D converter and sent to the computer. (b) The equivalent circuit of the measurement circuit considering the input capacitance C_{in} of the amplifiers. (c) The model circuit used to estimate bias due to stray capacitance C_s. The sample is assumed to be a parallel circuit of resistance R_0 and capacitance C_0. The value of C_s is assumed equal to $2C_{in}$.

2. Measurement System

The measurement system is essentially the same as that of the frequency response analyzer for determining the impedance of a sample by the four-electrode method. The overview of the system is shown in Figure 1. A sinusoidal voltage signal, $V_o\,(t)$, with accurately set frequency, is fed to the sample and reference resistor R_r in series connection. We designed the simple electric circuit so that the effect of stray capacitance can be handled by a simple equivalent circuit. We used only noninverting amplifiers so that the frequency response of the amplifiers is not affected by the sample impedance or stray capacitance.

The potential difference between the two ends of the sample, $V_p\,(t)$, and those of the reference resistor, $V_c\,(t)$, are amplified by operational amplifiers having a high input impedance ($\sim$1 TΩ), converted to discrete time sequences by an A/D converter board (Interface Co. CBI-360116; 16 bits, LSD $=$ 305 μV) and then stored. The sampling condition is adjusted

Table 1 Parameters used for the measurement

Frequency (Hz)	Number of data points	Signal cycle	Data point per signal cycle
$f > 10^3$	8192	2048	4
$10 < f < 10^3$	8192	256	32
$f < 10$	8192	16	512

to the signal frequency (Table 1) so as to reduce quantization noise in the signal spectrum by increasing the number of sampling points per signal cycle and to optimize the efficiency of noise reduction by stacking the data. The signal frequency is step-scanned within a demanded frequency range (1 mHz to 100 kHz). The complex capacitance $C(\omega)$ of the sample is calculated as a function of the signal frequency as follows:

$$C(\omega) = [c(\omega)/p(\omega)]/R_r/j\omega, \tag{1}$$

where $p(\omega)$ and $c(\omega)$ are the amplitudes of the spectral lines of $V_p(t)$ and $V_c(t)$, respectively. The complex dielectric permittivity is represented by $\varepsilon^*(\omega) = C^*(\omega)/k$, where k is the cell constant given by the shape and size of the sample. The dimensionless relative permittivity is given by $\mathrm{Re}(\varepsilon^*)/\varepsilon_0$.

2.1. Bias caused by stray capacitance

Stray capacitance in the electric circuit is a major obstacle for complex dielectric permittivity measurement since the sample capacitance is quite small; a relative permittivity $\varepsilon/\varepsilon_0 = 10$ corresponds to around 1 pF with an ordinary cell constant of 10^{-2} m. We realized, after extensive analysis, that a major cause of stray capacitance is the input capacitance of the operational amplifier, which is generally 5-10 pF. When the amplifier is used as a noninverting circuit, the input capacitance C_{in} works as the capacitance connecting the input terminal and the ground (Jung, 2005), as modeled in Figure 1(b). Thus we consider that our measurement circuit is equivalent to the one shown in Figure 1(c), where C_s is the effective capacitance of the overall factors including a couple of C_{in} values in Figure 1(b) and other unspecified sources.

The equivalent circuit model indicates that the calculation of sample capacitance by Eq. (1) leads to a bias due to C_s. Instead of Eq. (1) we should use the following:

$$C(\omega) = [c(\omega)/p(\omega)]/Z_r/j\omega, \tag{1'}$$

where Z_r is the impedance of the parallel circuit of R_r and C_s. Eq. (1′) is, however, unfounded since we do not know the exact value of C_s or the validity of the model. According to the model in Figure 1(b), bias due to C_s is expected to be reduced when R_r is smaller and the measurement frequency is lower. We aim to use Eq. (1) to calculate $C(\omega)$ under the condition where the bias is insignificant. We therefore analyze the bias effect using the model to clarify both the bias effect and the conditions of bias-free measurement.

The model shown in Figure 1(c) leads to the conclusion that the sample capacitance C^* calculated by Eq. (1) is

$$C^* = (V_c/R_r)/V_p/j\omega = (Z_r/Z_0)/R_r/j\omega, \tag{2}$$

where Z_0 and Z_r are impedances of the parallel circuit of R_0 and C_0 and of the parallel circuit of R_r and C_s. At a sufficiently low frequency where $1 > \omega R_0 C_0$ and $1 > \omega R_r C_s$, we obtain $Z_0 \approx R_0(1 - j\omega R_0 C_0)$ and $Z_r \approx R_r(1 - j\omega R_r C_s)$, leading to the following:

$$C^* = C_0' + 1/R_0 j\omega \tag{3}$$

$$C_0' = C_0 - R_r C_s/R_0. \tag{4}$$

This indicates that the sample capacitance calculated by Eq. (1) has a bias of $-R_r C_s/R_0$. We can see that the bias is insignificant if $C_0 \gg R_r C_s/R_0$ with $1 > \omega R_0 C_0$, $1 > \omega R_r C_s$.

In order to verify the discussion above and to estimate C_s, we carried out a series of measurements of the parallel circuit of a 1 MΩ resistor and a 1 pF capacitor by using different values of R_r between 10^2 and 10^6 Ω. The obtained data was fitted by the model of Eq. (2) to estimate C_0 and C_s.

When an R_r smaller than 0.1-1 kΩ is used, the measured data was concordant to the model Eq. (3) with $C_s = 0$, and no bias effect was observed, as shown in Figure 2(a). $\mathrm{Re}(C^*)$ was 1.3 pF, which is considered the real capacitance of a 1 pF capacitor. At larger values of R_r, e.g., 10^6 Ω in Figure 2(b), both $\mathrm{Re}(C^*)$ and $\mathrm{Im}(C^*)$ are distorted by the bias effect. The height of the plateau of $\mathrm{Re}(C^*)$ at a lower frequency was concordant to Eq. (3) when $C_0 = 1.3$ pF and a C_s of around 20 pF are assigned, as plotted in Figure 2(c). The results summarized in Figure 2(c) show that the real value of sample capacitance C_0 is 1.3 pF and the observed capacitance $\mathrm{Re}(C^*)$ calculated by Eq. (1) is concordant to Eq. (4) for all R_r. Thus the equivalent circuit and bias estimation are valid.

Now we have clarified the bias caused by stray capacitance in our measurement circuit and the condition required for unbiased measurement; it is necessary to reduce the bias to 0.1 pF $R_r/R_0 < 0.05$ if the stray capacitance C_s equals 20 pF. In the measurements described in the rest of this paper, we followed the condition $R_r/R_0 < 0.01$ to avoid bias.

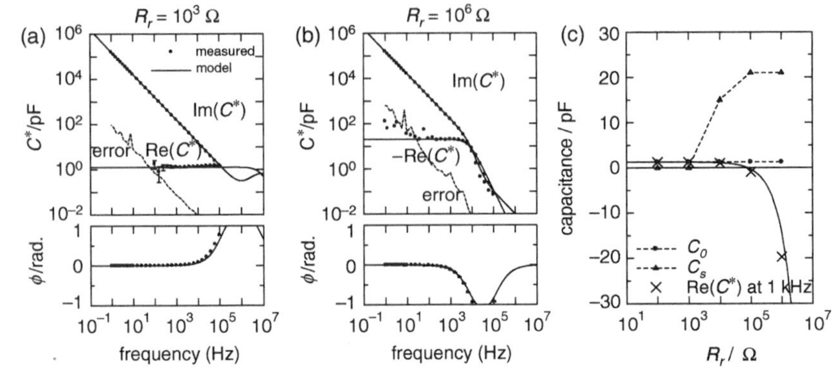

Figure 2 Analysis of the bias effect due to stray capacitance C_S (see the model shown in Figure 1 (c)) by measuring an analog sample (parallel connection of a 1 MΩ pure resistor and a 1 pF pure capacitor) with different reference resistors R_r. The complex capacitance is calculated by $C^* = (V_c/R_r)\,/\,V_p/j\omega$. Data obtained from the experiment and calculated from the model are compared. The model parameter C_S is adjusted to reproduce experimental data. (a) The bias effect is negligible at $R_r = 10^3\ \Omega$, where $\mathrm{Re}\left(C^*\right)$ was obtained as 1.3 pF by the experiment. (b) The bias at $R_r = 10^6\ \Omega$ suggested that C_S equals 20 pF. Note that $\mathrm{Re}\left(C^*\right)$ is negative. (c) The dependence on R_r of $\mathrm{Re}\left(C^*\right)$ obtained by the measurement and C_S and C_0 (sample capacitance) are assumed in the model calculation. The solid line shows $\mathrm{Re}\left(C^*\right)$ calculated by the model (see Eq. (4)).

2.2. Bias due to electrochemical reaction at the electrodes

To carry out precise and reliable measurement, we have developed the sample cell shown in Figure 3. The cell is designed for measurement of a moist sample by the four-electrode method. Elaborate care is taken in electrode installation to reduce any bias caused by electrode polarization and stray capacitance.

A metal electrode is electrically conductive, and water is ionically conductive. Thus the flow of an electric current though an interface of metal electrode and moist sample causes the polarization of charge carriers (electrons and ions) and electrochemical reactions at the interface. This causes a potential gap between the electrode and the sample, which leads to such a bias that the sample capacitance increases at low frequency. This is why the four-electrode method is necessary for measurement at frequencies lower than MHz. However, we realized, bias can be caused in the four-electrode measurement by the polarization of a potential electrode; an electric current can flow though the interface of the sample and the potential electrode attached directly to the sample causing electrode polarization.

To eliminate this bias, we installed an arc-shaped potential electrode outside the sample cell and connected to the sample by a thin buffer set in the narrow slit of the sample cell. The potential of the electrode is kept equal to the sample by the buffer, while it separates the potential electrode from

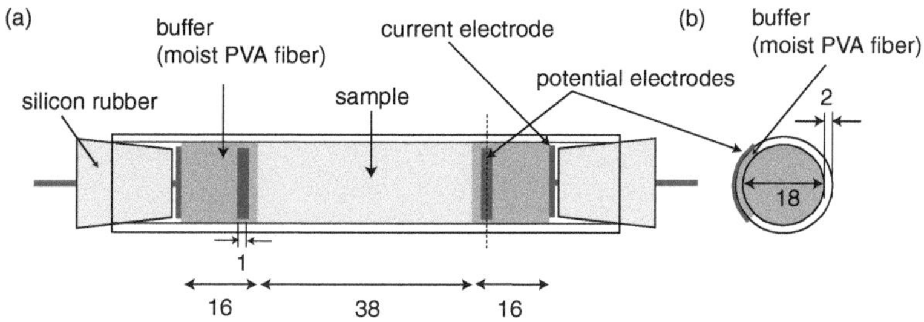

Figure 3 (a) The sample cell developed for the measurement of a moist sample by the four-electrode method. The cross section at the broken line is shown in (b). The PVA fiber is kept much more conductive than the sample by soaking with brine water so that a potential drop in the buffer is insignificant. A potential electrode attached outside of the cell is in contact with the buffer. This is for precise potential measurement by separation of the electrode from the electric current induced in the sample. The cell constant (sectional area over thickness of the sample) is 6.7×10^{-3} m.

the electric current induced in the sample. This enables us to measure the sample potential while avoiding polarization of the potential electrode.

A defect of the four-electrode method compared to the two-electrode method is the complexity in the measurement circuit, bringing about a bias due to stray capacitance. We took care of this problem by using a conductive buffer to separate the potential electrode from the current electrode; the impedance of the buffer is smaller than 10^{-2} of the sample buffer so that the measurement system is concordant with the model circuit shown in Figure 2(c). Then we can reduce the bias by using a sufficiently small R_r as discussed in the previous section.

2.3. Noise and uncertainty in measurement

The signal-to-noise ratio (SNR) is an important factor for measurement precision. However, the amplitude of the applied voltage is limited to prevent bias due to electrochemical dissociation at the current electrode; the electrochemical potential for dissociation of water to hydrogen is about 1.2 V. Whereas the larger resistance for R_r necessary to raise $V_c\,(t)$ is useful under the limited voltage value, it raises the bias due to stray capacitance, as noted above.

Substantial improvement of the SNR is made by the stacking of data acquired over a longer period of time. If the signal is accurately controlled and ideally detected, S/N is expected to increase in proportion to the square root of the time period of the data acquisition. The sampling condition of digital data is changed depending on the frequency range (Table 1) to reduce quantization noise effectively. Further, the uncertainty of the signal in terms of standard deviation is evaluated by using a noise floor (see Figure 4(a) and (b)) of $V_c\,(t)$ and $V_p\,(t)$ to calculate the error bar of the complex capacitance

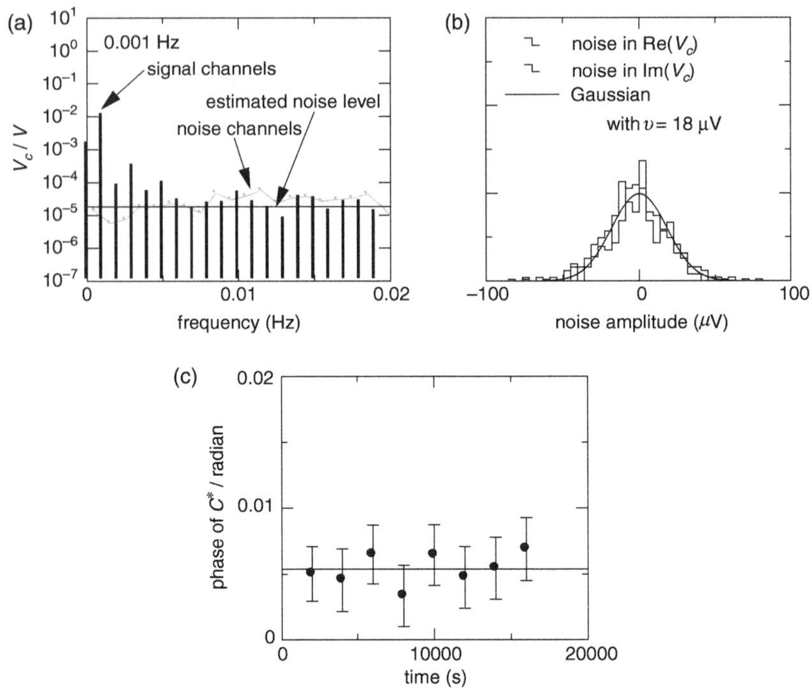

Figure 4 (a) Spectrum of V_C observed in the measurement of moist glass beads at 1 mHz. (b) Histogram of the real and imaginary parts of the noise amplitude. A Gaussian distribution with standard distribution $\upsilon = 18$ μV is plotted together to show that observed noise is Gaussian noise. (c) The phase angle calculated for each of two signal cycles showing the long-term stability during measurement. The error is estimated from the ratio of the amplitudes of signal and noise in (a).

(Figure 4(c)). Additionally the stability of the measuring system can be checked by monitoring the level and fluctuation of the phase sequence of the complex quantity (see Figure 4(c)). As a result, we can obtain reliable data with maximum SNR ($\sim 10^3$ in Figures 5 and 6) for a given measurement time (Nagao et al., 2004).

3. Measurement of Moist Grains

3.1. Sample and measurement condition

We show examples of the measurement of moist samples by our method. We measured granular samples consisting of quartz sand or glass beads of about 1 mm in grain size as moist rock analogues. The quartz sand is moistened by 10^{-3} mol/l NaCl water with 1.8% by volume and glass beads by the same

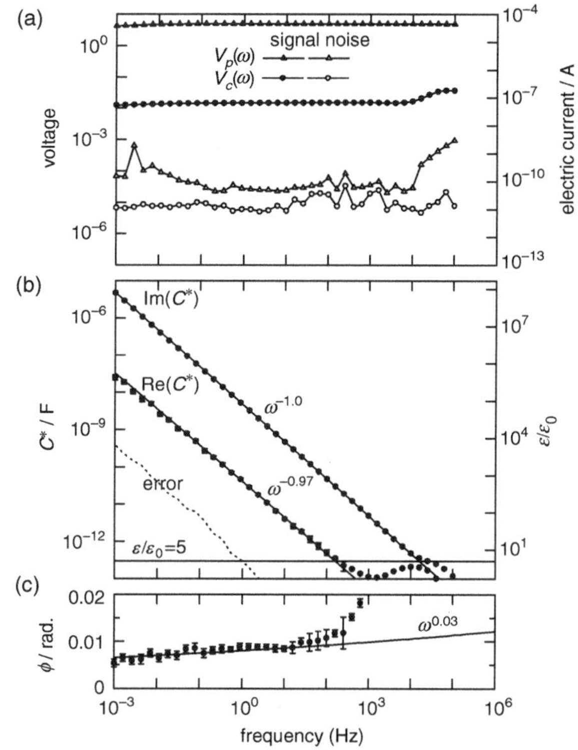

Figure 5 Data obtained for the measurement of a quartz grain of grain size about 1 mm, moistened by 10^{-3} mol/l NaCl water by 1.8% in volumetric ratio. The frequency dependency of the signal and noise in V_p and V_c (a), the complex capacitance and its estimated error (b), and the phase angle ϕ (c) are shown.

brine with 3% by volume. The output voltage of the function generator was set as 5 V. The measurement parameter was set as shown in Table 1. We checked that the voltage drop at the current electrode was smaller than 0.5 V during the entire measurement frequency (see the amplitude of V_p in Figures 5 and 6).

3.2. Result of measurement

The measured results are shown in Figure 5 for quartz sand and in Figure 6 for glass beads. The observed frequency dependence of ε^* is well represented by the following empirical formula:

$$\varepsilon^*(f) = \varepsilon_r + \alpha\,(f_0/f)^{1-p} + \sigma/j2\pi f, \tag{5}$$

where $f_0 = 1$ Hz is chosen for normalization of the frequency dependence, and $\varepsilon/\varepsilon_0 = 5$-8 is the typically expected value for the dielectric permittivity

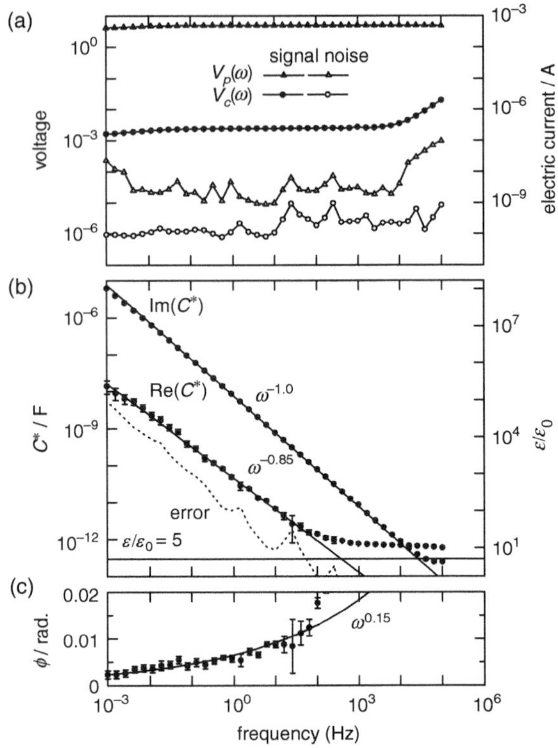

Figure 6 Data obtained for the measurement of glass beads 1 mm in diameter, moistened by 10^{-3} mol/l NaCl water by 3% in volumetric ratio. See the caption of Figure 5 for the explanation of each panel.

of dry silicate materials. While not determined well numerically by the present measurements, the data are not inconsistent with the expectation, as seen in Figures 5 and 6. The imaginary part $\mathrm{Im}\,(C^*)$ is proportional to the reciprocal frequency, and the conductivity σ is determined. It is essentially constant throughout the whole frequency range; $\sigma = 4.4 \times 10^{-6}$ Sm^{-1} for quartz sand and $\sigma = 7.2 \times 10^{-6}$ Sm^{-1} for glass beads.

The dielectric permittivity $\mathrm{Re}\,(\varepsilon^*)$ indicates that it is almost in reciprocal proportion to the frequency at levels lower than 100 kHz, whereas a small deviation represented by p in Eq. (5) is indicated to exist. The result of data fitting shows $\alpha = 6.0 \times 10^{-9}$ F m^{-1} for quartz sand and $\alpha = 6.7 \times 10^{-9}$ F m^{-1} for glass beads (α is the dielectric permittivity at 1 Hz). The relative dielectric permittivity $\varepsilon/\varepsilon_0$ reaches a level above 10^5 at 1 mHz, and it appears rise further at lower frequencies. Such a large value of dielectric constant has been found only among the synthetic ferro-electric materials for industrial use (e.g., $\sim 10^4$ for $BaTiO_3$). Therefore, one may consider it unreasonable. However, the presence of a supposedly anomalously large dielectric constant is real, as demonstrated by a very careful measurement and its estimated error. We note the small but physically important deviation from the $1/f$ law: $p = 0.03$ for quartz sand and 0.15 for glass beads.

4. Discussion

4.1. Short summary on measurement technology

Interested in the dielectric constant, or the real part of the complex dielectric permittivity in moist rocks, we have devised a reliable measurement method by analyzing the cause of unreliability recognized in the data reported so far. Uncertainty in the capacitance measurement is reduced to the order of 10^{-3} of k (σ/ω) (or $\sim 10^{-3}$ of admittance) in this work, where σ is the specific conductivity and k is the cell constant.

4.2. The $1/f$ type dielectric dispersion and its physical mechanism

Using the new measurement method to measure a moist granular material consisting of glass or quartz beads with a uniform grain size, we have demonstrated the reality of the presence of a 'f^{-1} type' dielectric dispersion, in which the relative dielectric permittivity increases nearly in proportion to the reciprocal frequency to 10^5 or more towards a lower frequency. The mechanism of dielectric dispersion is, however, not clear at the moment.

A basic idea for the cause of low-frequency dielectric dispersion is polarization of the electric double layer (EDL), which leads to a Debye type dielectric dispersion with a relaxation frequency of $f_c = 2D/a^2$, (D: diffusion constant of ions in EDL, a: grain diameter) (Schwarz, 1958). The superposition of a Debye type dispersion by means of a relaxation frequency spectrum is mathematically possible (Lesmes and Morgan, 2001). However, a specific distribution of the relaxation frequency spectrum within the five orders of frequency range below 100 kHz is demanded by the distributions of grain size and the diffusion constant in the present $1/f$ type dispersion. This appears neither feasible for the present granular samples nor adequate physically. The mechanism of a so-called Warburg impedance involving an electrochemical reaction shows that the dielectric constant is proportional to $f^{-1/2}$, instead of f^{-1} in the present case. The factor 1/2 originates from the diffusion process involving a certain finite distance with a time dependence of $t^{1/2}$. In the case of a $1/f$ type, the physical process is expected to have a linear time scale, and a diffusion-controlled mechanism like Warburg's will not provide us with a good explanation. At this moment, we postulate the presence of a certain electrochemical reaction proceeding with a constant speed under a constant electric field to result in energy storage linearly with time. However, we have not reached a detailed model of the physical mechanism yet. More data on a variety of materials to clarify any deviation (e.g., parameter p) in Eq. (5) from the '$1/f$ law', in particular, would provide us with clues to resolve the problem, since the small but finite value of p constrains the extent of energy storage reflected by the physical mechanism.

5. Conclusion

We have developed a method for complex dielectric permittivity measurement of moist samples to claim the existence of a $1/f$ dielectric dispersion. The key features are reduction of bias due to stray capacitance by careful measurement of the electric circuit, and of bias due to electrode polarization by using a newly developed method of electrode installation. Reliable data acquisition is attained by the ACROSS method which contributes features of error estimation and reliable stacking.

Our measurement method has provided experimental evidence of a $1/f$ type dielectric dispersion for samples of moist grains of quartz and glass beads having relatively uniform grain size. This indicates that the model of superposition of a Debye type dielectric dispersion caused by ion polarization at a mineral and water interface is inappropriate to account for the $1/f$ dependence. The idea of a $1/f$ dependence has been presented as an electrochemical reaction where energy is stored linearly with time.

REFERENCES

Jonscher, A.K., 1978. Low-frequency dispersion in carrier-dominated dielectrics. Philosophical Magazine B38, 587–601.

Jung, W.G., 2005. Op AMP Applications Handbook. Elsevier Inc, Newnes.

Lesmes, D.P., Morgan, F.D., 2001. Dielectric spectroscopy of sedimentary rocks. Journal of Geophysical Research 106, 13329–13346.

De Lima, O.A.I., Sharma, M.M., 1992. A generalized Maxwell–Wagner theory for membrane polarization in shaly sands. Geophysics 57, 431–440.

Lockner, D., Byerlee, J., 1985. Complex resistivity measurements of confined rocks. Journal of Geophysical Research 90, 7837–7847.

Matsumoto, H., Shigeta, N., Kumazawa, M., Nakajima, T., 2003. Study of dielectric dispersion of ionic conductors as possible observables of ACROSS. American Geophysical Union 2003 Fall Meeting, Eos Trans. AGU, 84 (46), GP11A-0253.

Nagao, H., Nakajima, T., Kumazawa, M., Kunitomo, T., 2004. Optimum weighted stacking method for acquisition of the ACROSS transfer functions having the maximum signal-to-noise ratio. In: Fujii, N., Kasahara, J., Higashihara, H., Ogawa, K. (Eds.), Proceedings of the first International Workshop on Active Monitoring in the Solid Earth Geophysics (IWAM04), Task Group for Active Monitoring, pp. 148–151.

Nakajima, T., Kunitomo, T., Nagao, H., Kumazawa, M., Shigeta, N., 2004. A proposal of imaging of the plate boundary using the active monitoring method. In: Fujii, N., Kasahara, J., Higashihara, H., Ogawa, K., (Eds.), The Proceedings of the First International Workshop on Active Monitoring in the Solid Earth Geophysics (IWAM04), Task Group for Active Monitoring, Japan, pp. 148–151.

Olhoeft, G.R., 1985. Low frequency electrical properties. Geophysics 50, 2492–2503.

Pelton, W.H., Ward, S.H., Hallof, P.G., Sill, W.R., Nelson, P.H., 1978. Mineral discrimination and removal of inductive coupling with multifrequency IP. Geophysics 43, 588–609.

Schwarz, G., 1958. A theory of the low-frequency dielectric dispersion of colloidal particles in electrolyte solution. Journal of Physics and Chemistry 66, 2636–2642.

Sen, P.N., 1981. Relation of certain geometrical features to the dielectric anomaly of rocks. Geophysics 46, 1714–1720.

Wong, J., 1979. An electrochemical model of the induced-polarization phenomenon in disseminated surface ores. Geophysics 44, 1245–1265.

CHAPTER 8

Active Geophysical Monitoring of Hydrocarbon Reservoirs Using EM Methods

Noel Black[a] *and* Michael S. Zhdanov[a]

Contents

1. Introduction

During recent years marine controlled source electromagnetic (MCSEM) surveys have become intensively used for off-shore petroleum exploration (Eidesmo et al., 2002; Ellingsrud et al., 2002; Carazzone et al., 2005). The importance of MCSEM as an exploration tool is now widely accepted. As the technology advances and the method becomes more affordable, applications other than exploration may find MCSEM data useful. In this paper we study the use of MCSEM methods for reservoir production monitoring.

Effective reservoir management requires time-lapse reservoir information throughout the interwell volume. The ability to understand and control

[a] Department of Geology and Geophysics, University of Utah

Handbook of Geophysical Exploration: Seismic Exploration, Volume 40
ISSN 0950-1401, DOI: 10.1016/S0950-1401(10)04013-9

reservoir behavior over the course of production allows for optimization of reservoir performance and production strategies. Good monitoring information makes it possible to improve the timing and location of new drilling (for both production and injection wells), to recognize flow paths, and to map oil that has been bypassed. The use of seismic data for monitoring is very challenging because of the small variation of seismic velocities over time and because of the difficulty of survey repeatability.

Early investigations of the use of cross-borehole EM data for reservoir monitoring include Malinverno and Torres-Verdín (2000), Hoversten et al. (2001), and Wilt and Morea (2004). Recent sensitivity studies (Lien and Mannseth, 2008; Wang et al., 2008) have assessed the potential of the CSEM technique for reservoir monitoring. Hu et al. (2008) have considered monitoring with array transient electromagnetic (TEM), or time-domain, sounding.

Here, we perform a numerical experiment of reservoir monitoring using MCSEM. We compute EM fields for a model of a partially depleted sea-bottom reservoir with different positions of oil-water contact over time. For a more realistic representation of the sea-bottom environment, we include a salt dome and compare models with and without a rough sea-bottom bathymetry. As an EM modeling tool we use the integral equation (IE)-based code developed at the Consortium for EM Modeling and Inversion (CEMI) at the University of Utah (Hursán and Zhdanov, 2002). We use the multiple-domain (MD) IE method (Endo et al., 2009) to efficiently model extensive anomalous regions of bathymetry, salt structures, and the reservoir itself. Maps of the EM field components clearly reflect the position of the oil-water contact. The principal advantage of the multiple-domain technique is that we are able to isolate the effect of the reservoir. This is especially important in a case where bathymetry must be considered. The oil-water interface is clearly resolved in several field components by extracting only the induction from the reservoir. Our study demonstrates that MCSEM data can accurately provide the position of the oil-water contact inside the reservoir over time.

2. Principles of Reservoir Production Monitoring using Marine EM Methods

EM methods are potentially useful as a means to monitor reservoir production because they can distinguish between hydrocarbons and saline water based on their differing resistivities. The range of resistivity variation is much greater than the range of variation for seismic velocities. Seismic data are useful primarily for estimating lithography and porosity, both of which remain essentially constant over the lifetime of a producing reservoir. Seismic velocities are different enough to discriminate fluid type (oil vs. water) only for high porosities, greater than about 30%. Since the velocity differences are small, seismic methods rely on fixing survey parameters so that differences between surveys at different points in time can be seen. The potential of seismic methods for monitoring is thus limited by the repeatability of the seismic surveys.

Enhanced oil recovery (EOR) is a prime scenario for investigating the potential use of EM for monitoring. For water injection recovery, there will be a flooding front that advances over time as the reservoir is produced. The ability to locate the position of this oil-water contact is an important test of the method. The main question with EM is resolution. It remains to be seen whether the low frequencies of these methods, which are needed to avoid attenuation to the depths of interest, can resolve the smaller-length scales associated with monitoring compared with those needed for exploration. Since the process is diffusive, there is no applicable Rayleigh criterion limiting expected resolution, and hence a numerical experiment needs to be performed.

3. Overview of the Numerical Modeling Technique

In this section we will present a short overview of the IE method with multiple domains, following Endo et al. (2009). Over the last several years CEMI has developed several computer codes for EM forward modeling based on the integral equation (IE) formulation of Maxwell's equations. In the framework of the IE method, the conductivity distribution is divided into two parts: (1) the background conductivity, σ_b, which is used for the Green's functions calculation, and (2) the anomalous conductivity, $\Delta\sigma_a$, within the domain of integration, D. One principal advantage of the IE method over the other numerical techniques is that the IE method requires discretization of the anomalous domain D only.

The inhomogeneous background conductivity (IBC) IE method (Zhdanov et al., 2006) overcomes the limitation of the conventional IE method of restriction to layered backgrounds only and allows inclusion of important geoelectrical structures like bathymetry and salt domes in the background model. Endo et al. (2009) developed an iterative inhomogeneous IE method for modeling with multiple inhomogeneous domains. By using this method we can evaluate the individual response from every domain, which includes the possible EM coupling effects between the different domains. A rigorous separate calculation of the EM fields produced by different anomalous domains representing different geological structures (e.g., bathymetry, salt structures, reservoirs) is an important practical problem of EM exploration.

The following summarizes the principles of the IE method of EM modeling with multiple inhomogeneous domains. We assume that N inhomogeneous domains ($D_i, i = 1, \ldots, N$) are located within a horizontally layered

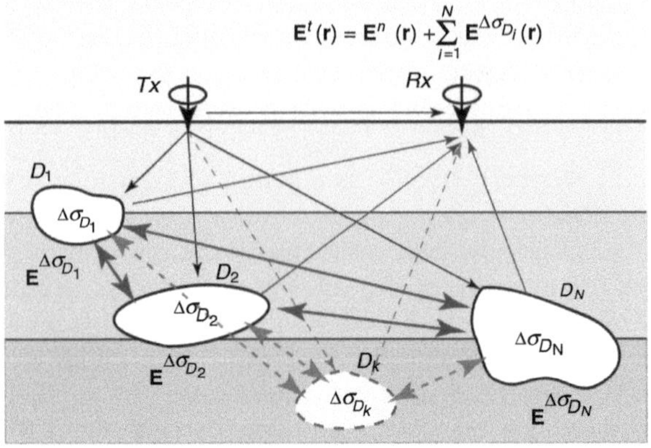

Figure 1 A sketch of a 3D geoelectrical model with horizontally layered (normal) conductivity and N inhomogeneous conductivity domains.

earth (Figure 1). The conductivity of the horizontally layered earth (normal conductivity) is σ_n, while the inhomogeneous (anomalous) conductivity within each inhomogeneous domain is denoted as $\Delta\sigma_{D_i}$, $i = 1, \ldots, N$. The total EM fields at any point $\mathbf{r}$, $\mathbf{E}^t(\mathbf{r})$ and $\mathbf{H}^t(\mathbf{r})$, can be expressed as a sum of the normal fields $\mathbf{E}^n(\mathbf{r})$, $\mathbf{H}^n(\mathbf{r})$, and the EM fields induced by every inhomogeneous domain $\mathbf{E}^{\Delta\sigma_{D_i}}(\mathbf{r})$, $\mathbf{H}^{\Delta\sigma_{D_i}}(\mathbf{r})$ $(i = 1, \ldots, N)$:

$$\mathbf{E}^t(\mathbf{r}) = \mathbf{E}^n(\mathbf{r}) + \sum_{i=1}^{N} \mathbf{E}^{\Delta\sigma_{D_i}}(\mathbf{r}) = \mathbf{E}^n(\mathbf{r}) + \sum_{i=1}^{N} G_E^{D_i}[\Delta\sigma_{D_i}\mathbf{E}^t], \tag{1}$$

$$\mathbf{H}^t(\mathbf{r}) = \mathbf{H}^n(\mathbf{r}) + \sum_{i=1}^{N} \mathbf{H}^{\Delta\sigma_{D_i}}(\mathbf{r}) = \mathbf{H}^n(\mathbf{r}) + \sum_{i=1}^{N} G_H^{D_i}[\Delta\sigma_{D_i}\mathbf{E}^t], \tag{2}$$

where $G_E^{D_i}$ and $G_H^{D_i}$ are electric and magnetic Green's operators acting within domain D_i, respectively. Then the EM modeling problem is reduced to the calculation of the total electric fields inside each inhomogeneous domain.

Rearranging Eq. (1) for the electric field induced in the inhomogeneous domain D_N, we have the following:

$$\mathbf{E}^{\Delta\sigma_{D_N}}(\mathbf{r}) = \mathbf{E}^t(\mathbf{r}) - \mathbf{E}^n(\mathbf{r}) - \sum_{i=1}^{N-1} \mathbf{E}^{\Delta\sigma_{D_i}}(\mathbf{r}). \tag{3}$$

In practice, at the first step of the field calculation, we do not know the values of any electric fields in Eq. (3). We thus first calculate the electric field in domain D_1 without taking into account the induction effect from any other domains:

$$\mathbf{E}^{\Delta\sigma_{D1}}(\mathbf{r}) = G_E^{D_1}\left[\Delta\sigma_{D_1}\mathbf{E}^t\right] = \mathbf{E}^t(\mathbf{r}) - \mathbf{E}^n(\mathbf{r}). \tag{4}$$

Equation (4) can be written as an integral equation with respect to the field $E^{\Delta\sigma_{D1}}$:

$$\mathbf{E}^{\Delta\sigma_{D1}}(\mathbf{r}) = G_E^{D_1}\left[\Delta\sigma_{D_1}(\mathbf{E}^n + \mathbf{E}^{\Delta\sigma_{D1}})\right]. \tag{5}$$

This integral equation is solved using the contraction form of the integral equations (Hursán and Zhdanov, 2002) and the complex generalized minimal residual (CGMRES) method (Zhdanov, 2002).

In the calculation of the field due to the currents induced in the next domain (2), we take into account the electric field induced from the inhomogeneous domain D_1, $\mathbf{E}^{\Delta\sigma_{D1}}(\mathbf{r})$:

$$\mathbf{E}^{\Delta\sigma_{D2}}(\mathbf{r}) = G_E^{D_2}\left[\Delta\sigma_{D_2}\mathbf{E}^t\right] = \mathbf{E}^t(\mathbf{r}) - \mathbf{E}^n(\mathbf{r}) - \mathbf{E}^{\Delta\sigma_{D1}}(\mathbf{r}). \tag{6}$$

The last equation is equivalent to the following integral equation:

$$\mathbf{E}^{\Delta\sigma_{D2}}(\mathbf{r}) = G_E^{D_2}\left[\Delta\sigma_{D_2}(\mathbf{E}^n + \mathbf{E}^{\Delta\sigma_{D2}} + \mathbf{E}^{\Delta\sigma_{D1}})\right], \tag{7}$$

which is solved again by the CGMRES method.

Finally, for the last inhomogeneous domain D_N, we already know the electric fields in all the other inhomogeneous domains and thus we can calculate the electric field $\mathbf{E}^{\Delta\sigma_{DN}}(\mathbf{r})$ as described by Eq. (3).

To improve the accuracy, we can use this scheme iteratively. In the subsequent iterations, we use the fields obtained in the previous iteration to calculate the induced fields in the given domain. For example, in the second iteration, the calculation of the electric fields from the inhomogeneous domain D_N will use the electric fields from other domains obtained in the first iteration as follows:

$$\mathbf{E}_{(2)}^{\Delta\sigma_{D1}}(\mathbf{r}) = \mathbf{E}_{(1)}^t(\mathbf{r}) - \mathbf{E}^n(\mathbf{r}) - \sum_{i=2}^{N}\mathbf{E}_{(1)}^{\Delta\sigma_{Di}}(\mathbf{r}), \tag{8}$$

where the numerical field subscripts denote the iteration number. The electric fields from the other inhomogeneous domains are calculated similarly, always using the latest obtained electric fields for the given domain. For example, for the electric fields due to domain D_2 at the second iteration, we calculate the following:

$$\mathbf{E}_{(2)}^{\Delta\sigma_{D_2}}(\mathbf{r}) = \mathbf{E}_{(1)}^{t}(\mathbf{r}) - \mathbf{E}^{n}(\mathbf{r}) - \sum_{i=3}^{N} \mathbf{E}_{(1)}^{\Delta\sigma_{D_i}}(\mathbf{r}) - \mathbf{E}_{(2)}^{\Delta\sigma_{D_1}}(\mathbf{r}). \tag{9}$$

This process is repeated until the electric fields within all the inhomogeneous domains reach self-consistency, i.e., the norm of difference between the electric fields in any domain at iterations i and $(i-1)$ is less than a certain threshold ε. In the kth inhomogeneous domain, for example, the electric fields satisfy the following inequality:

$$\frac{\left\| \mathbf{E}_{(i)}^{\Delta\sigma_{D_k}}(\mathbf{r}_j) - \mathbf{E}_{(i-1)}^{\Delta\sigma_{D_k}}(\mathbf{r}_j) \right\|_2}{\left\| \mathbf{E}_{(i)}^{\Delta\sigma_{D_k}}(\mathbf{r}_j) \right\|_2} < \varepsilon, \quad r_j \in D_k. \tag{10}$$

4. Computer Simulation of HC Reservoir Monitoring using EM Methods

We will perform a numerical experiment of forward modeling for several stages in the production of a reservoir by water injection. We consider a realistic geoelectric model of an HC reservoir and accompanying salt dome, first without and then with a sea-bottom bathymetry. The reservoir is filled from one end with injected saline water. For simplicity we model the oil-water contact as a sharp vertical interface. The EM field is generated by horizontal electric bipoles near the seafloor and detected by a line of receivers at the seafloor. The models are described in more detail in the following sections.

4.1. Model 1: HC reservoir and a salt dome structure

A vertical section of the geoelectric structure of Model 1 is shown in Figure 2. This figure shows a resistive HC reservoir with a resistivity of 100 Ohm m and a salt dome with a resistivity of 30 Ohm m located within conductive sea-bottom sediments whose resistivity is 1 Ohm m. The resistivity of the seawater layer is 0.33 Ohm m, and the depth of the seafloor is 1350 m below sea level. The reservoir is flooded from the right (positive x direction) by water having a resistivity of 0.5 Ohm m. We model four positions (x_0 = 14.0, 12.8, 11.6, and 10.4 km) of the (vertical) oil-water interface to simulate production.

The EM field in this model is excited by an x-directed horizontal electric bipole of length 1 m and current 1000 A, which is located at the points with horizontal coordinates from 0 to 20 km (every 200 m) in the x direction and

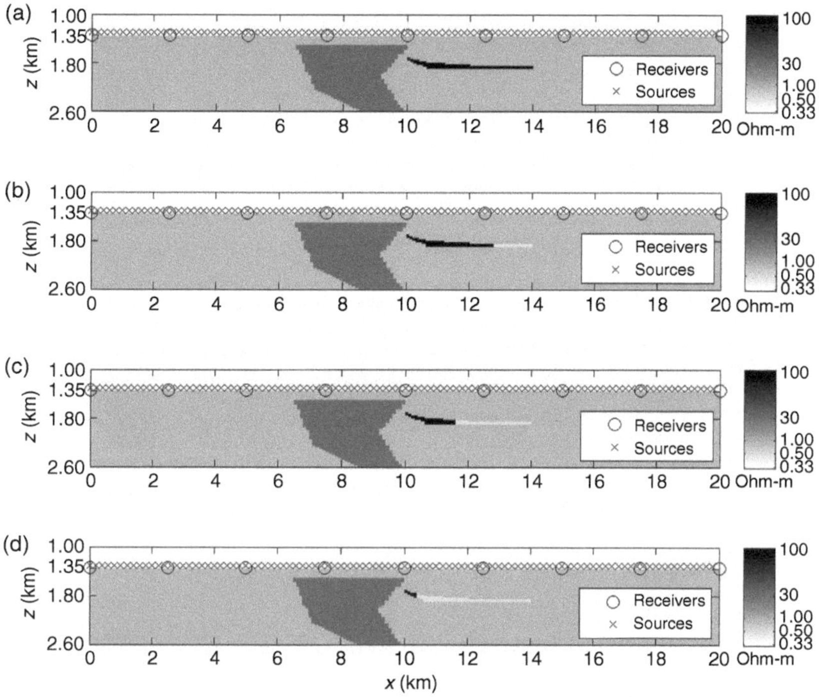

Figure 2 A vertical geoelectrical section of Model 1, containing a salt dome and a reservoir.

from −3 km to 3 km (every 200 m) in the y direction. The elevation of the transmitter bipole is 50 m above the sea bottom. The transmitter generates the frequency-domain EM field at frequencies of 0.01 Hz and 0.3162 Hz (we studied a range of frequencies between 0.01 and 10.0 Hz, and present results for these two). We modeled nine electric field receivers, equally spaced (2.5 km spacing) along the y axis between 0 and 20 km. To narrow the scope we confine our analysis in this paper to the first source at the far left of Figure 2 ($x = 0$).

Following the main principles of the MD IE method for multiple inhomogeneous domains, the modeling area is divided into two modeling domains, D_1 and D_2, corresponding to the locations of the salt dome and HC reservoir, respectively. Domain D_1 for the salt dome area is discretized into $35 \times 60 \times 10 = 36{,}960$ cells with a cell size of $100 \times 100 \times 25$ m^3, and domain D_2 for the HC reservoir area is discretized into $40 \times 40 \times 8 = 12{,}9800$ cells with the same cell size, $100 \times 100 \times 25$ m^3.

4.2. Model 1: Forward modeling results

Figures 3 and 4 show the x (in-line) component of the total electric field normalized by the fields induced in the layered background and the salt

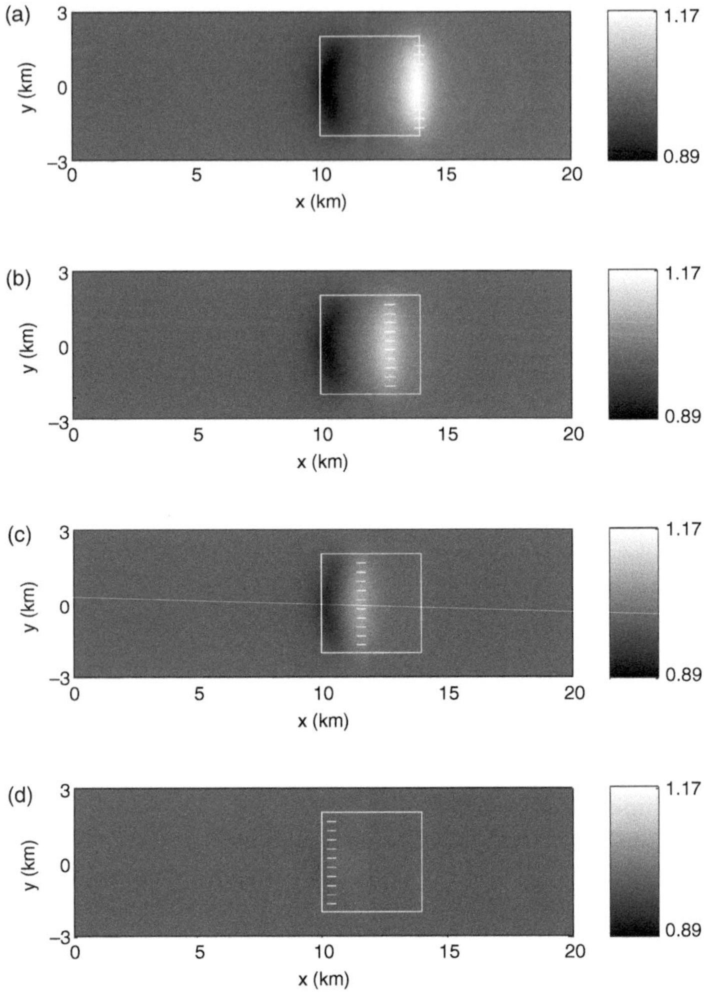

Figure 3 $|E_x^{\text{tot}}|$ normalized by background and salt dome fields for frequency = 0.01 Hz.

dome, $|E_x^{\text{tot}}| / (|E_x^b| + |E_x^{sd}|)$, for the sea-bottom receiver #1 (located at $x = 0$ km) and for the transmitter frequencies 0.01 Hz (Figure 3) and 0.3162 Hz (Figure 4). The higher frequency produces greater resolution and contrast but a smaller total field. The solid white box indicates the horizontal position of the reservoir, and the dashed white line indicates the horizontal position x_0 of the oil-water interface for four different positions (i.e., stages of production). Water fills from the right (positive x direction), forcing the oil to the left (negative x direction). The interface is clearly seen for all four positions as an enhancement of the normalized field component.

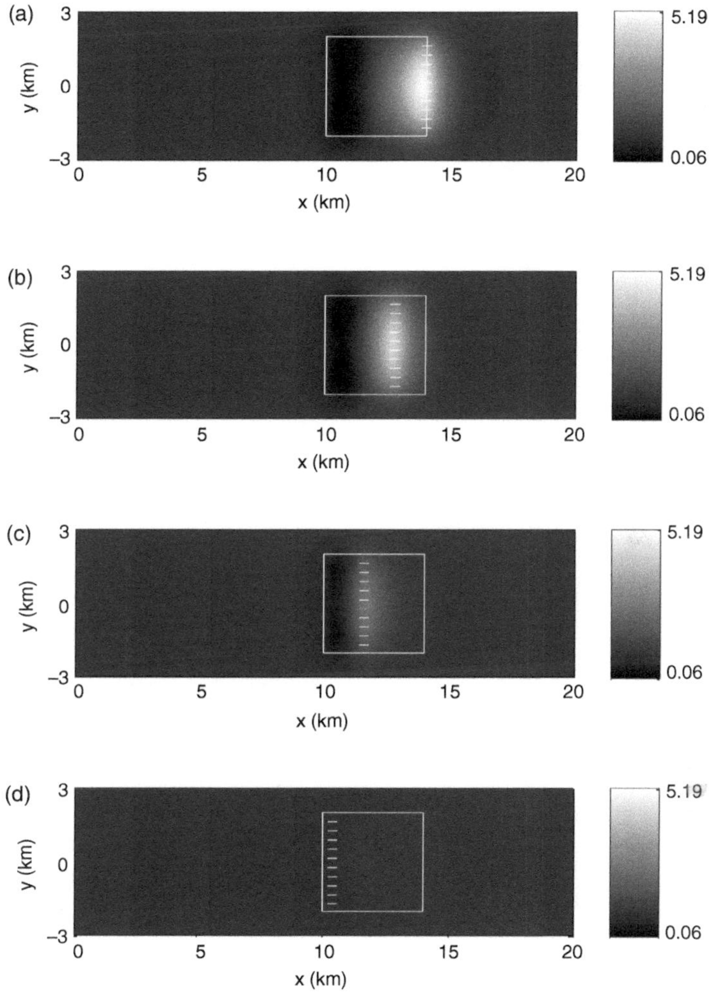

Figure 4 $|E_x^{\text{tot}}|$ normalized by background and salt dome fields for frequency $= 0.3162$ Hz.

Figures 5 and 6 show the same but normalized by the layered background field only $\left(|E_x^{\text{tot}}|\,/\,|E_x^b|\right)$. The salt dome location is now seen, as is the left (negative x) boundary of the HC reservoir.

Figures 7 and 8 show the z (vertical) component of the total electric field normalized by the fields induced in the layered background and the salt dome, $|E_z^{\text{tot}}|\,/\,\left(|E_z^b| + |E_z^{sd}|\right)$, for the sea-bottom receiver #1 (compare with Figures 3 and 4 for the x component). The location of the HC reservoir and the oil-water interface is clearly seen for the lower frequency as a negative anomaly. The unproduced portion of the reservoir is nicely

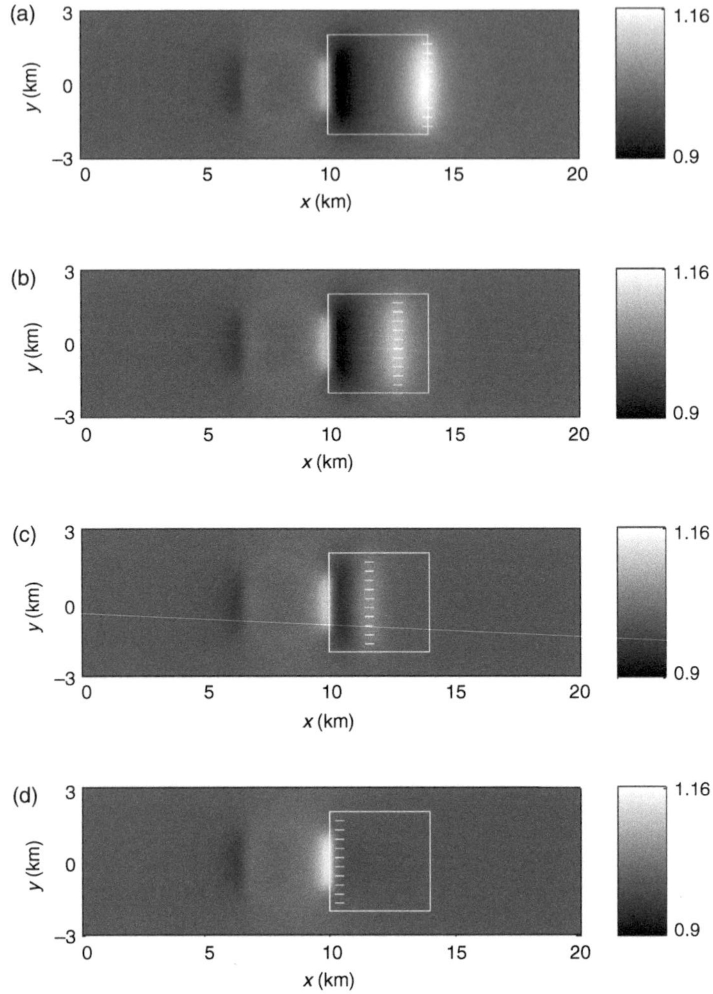

Figure 5 $|E_x^{tot}|$ normalized by background field for frequency = 0.01 Hz.

outlined in the higher frequency plot by a positive anomalous field along its boundary. For both frequencies, we can clearly see not only the oil-water contact but also the location of the remaining hydrocarbon.

The magnitude of the x and z electric field components of the source-moment-normalized anomalous reservoir electric field is in the range of 10^{-14}-10^{-15} V/m. This is slightly above, or approximately at, the noise level of 10^{-15} V/m set by Um and Alumbaugh (2007). This noise estimate is slightly higher than that given by Constable and Weiss (2006). The anomalous reservoir response is therefore expected to be detectable.

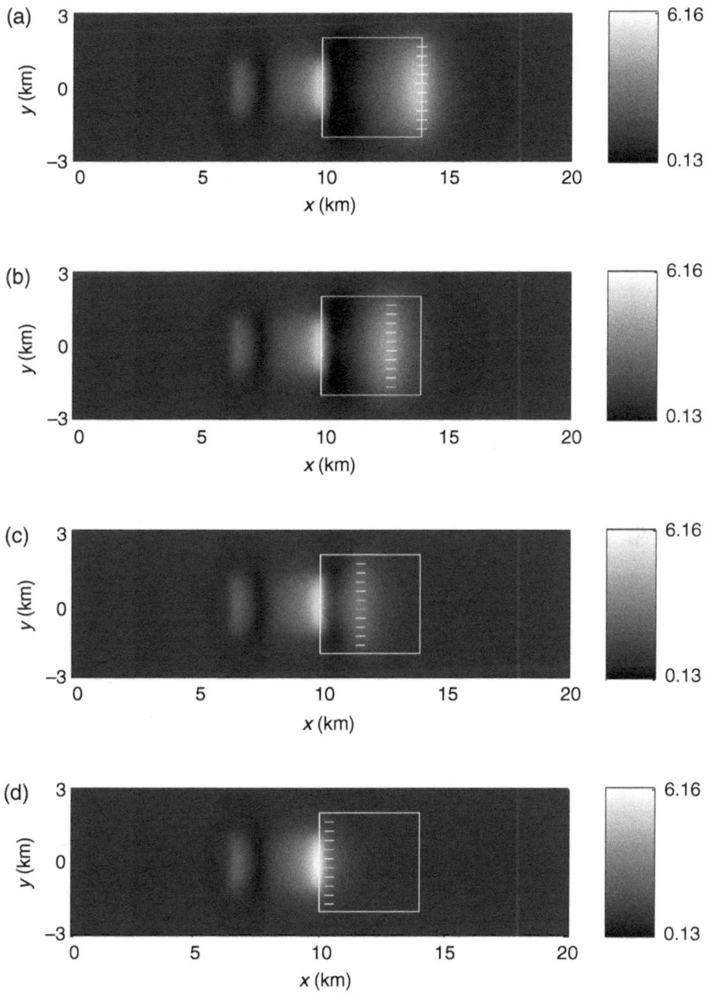

Figure 6 $|E_x^{tot}|$ normalized by background field for frequency = 0.3162 Hz.

4.3. Model 2: HC reservoir and a salt dome in an area with a rough sea-bottom bathymetry

In this section we study the feasibility of identifying the location of the oil-water interface from MCSEM data in a case where there is a rough sea-bottom bathymetry. This is very important from a practical standpoint because the effect of the bathymetry can significantly distort the EM response from an HC reservoir, and therefore the oil-water interface position in production monitoring can also be affected by the bathymetry. As a prototype of the bathymetry structure we use a simplified model of the known

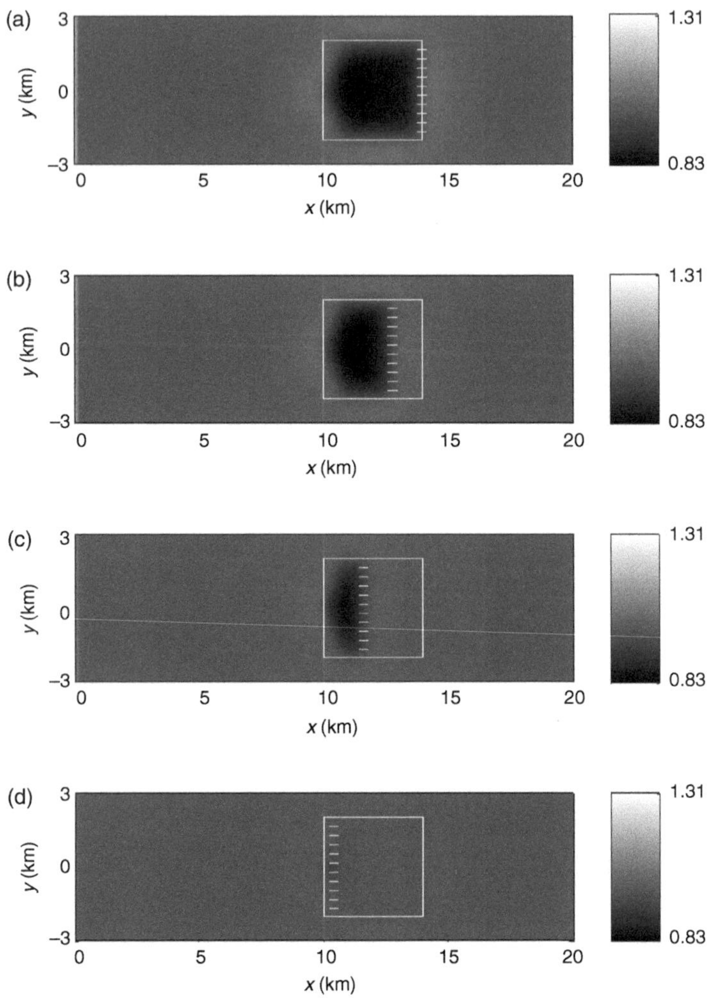

Figure 7 $|E_z^{tot}|$ normalized by background and salt dome fields for frequency = 0.01 Hz.

bathymetry of the Sabah area, Malaysia (provided by Shell International Exploration and Production, acquired in 2004). A vertical section of the model is shown in Figure 9. It includes three domains: D_1 for the bathymetry discretized as $200 \times 60 \times 10 = 120{,}000$ cells, and domains D_2 for the salt dome and D_3 for the HC reservoir, discretized as in Model 1.

4.4. Model 2: Forward modeling results

Figures 10 and 11 show the x (in-line) component of the total electric field normalized by the fields induced in the layered background, bathymetry, and

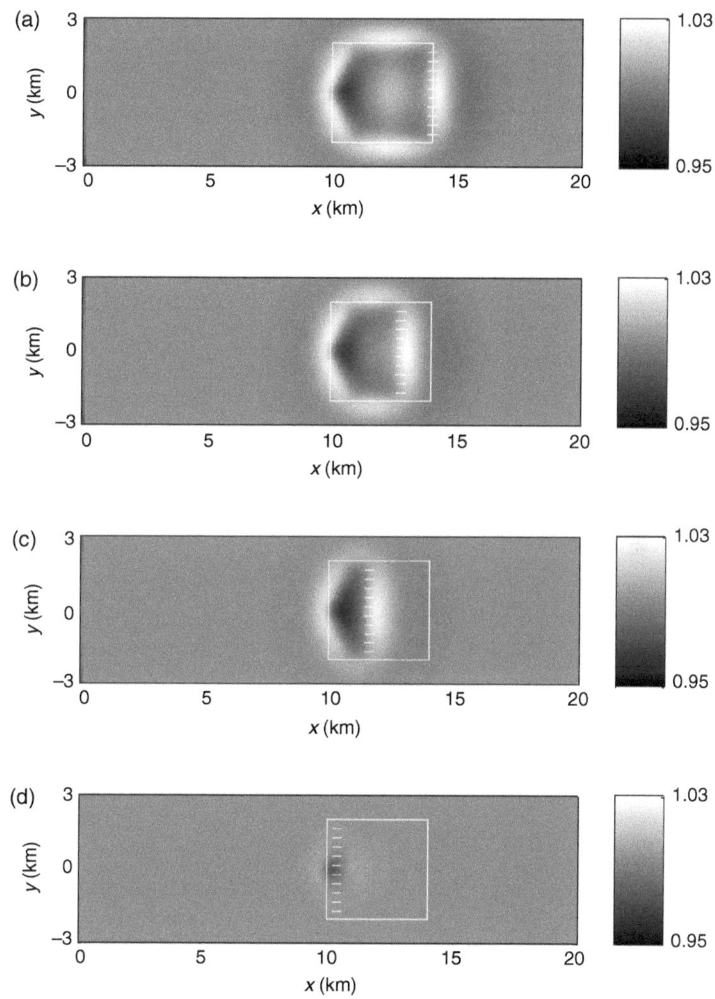

Figure 8 $|E_z^{\text{tot}}|$ normalized by background and salt dome fields for frequency = 0.3162 Hz.

the salt dome $|E_x^{\text{tot}}| / (|E_x^b| + |E_x^{\text{bath}}| + |E_x^{sd}|)$ for the sea-bottom receiver #1 and for both frequencies. The position of the oil-water contact is again seen as an enhancement in the field. The anomalous field from the reservoir is slightly less pronounced when a bathymetry is present, but this is not a large effect.

Figures 12 and 13 show the same electric field component normalized by the background and bathymetry fields. Again, the location of the salt dome is now revealed; the oil-water contact is still clearly seen although its anomalous response is somewhat suppressed compared to the first normalization, which includes the salt dome.

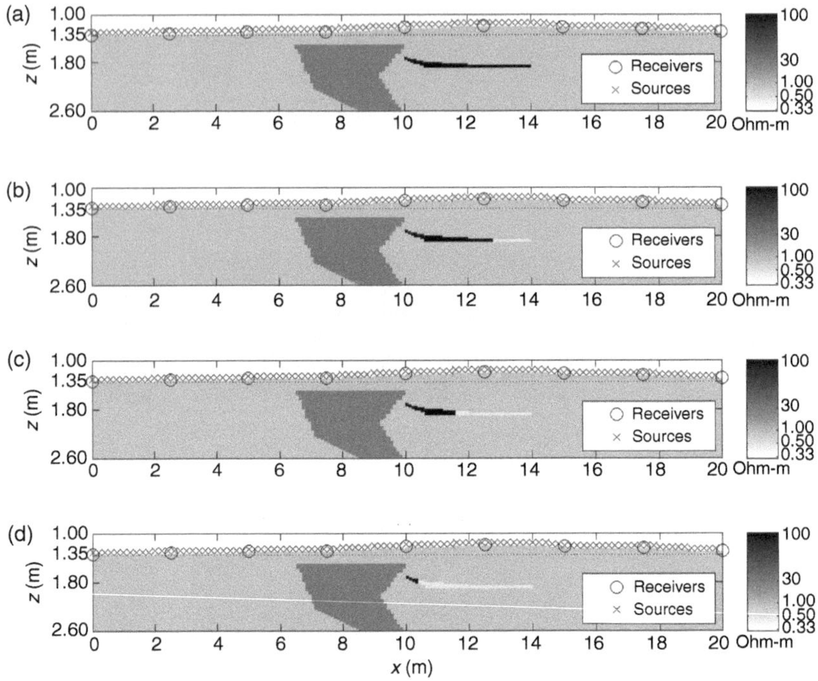

Figure 9 A vertical geoelectrical section of Model 2, containing a salt dome and a reservoir with a sea-bottom bathymetry.

Similar results were obtained for the *z* component of the electric field (Figures 14 and 15) to those obtained without a sea-bottom bathymetry. The unproduced portion of the reservoir and in particular the oil-water contact are clearly seen with only a slightly less pronounced anomalous field compared to the case without bathymetry.

Figures 16 and 17 show the *y* component of the magnetic field for the two frequencies under consideration. The results are similar to those for the *x* component of the electric field, showing the oil-water contact position very nicely. Figures 18 and 19 show the *x* component of the magnetic field, with results similar to the *z* component of the electric field. The remaining hydrocarbon can be seen, and in particular the oil-water contact location is clearly demarked.

The magnitude of the *x* and *y* magnetic field components of the source-moment-normalized anomalous reservoir magnetic field is in the range 10^{-8}-10^{-11} nT/m. The noise level quoted by Um and Alumbaugh (2007) is 10^{-10} nT/m. We therefore expect that at least some of these components will be detectable for the survey parameters modeled here.

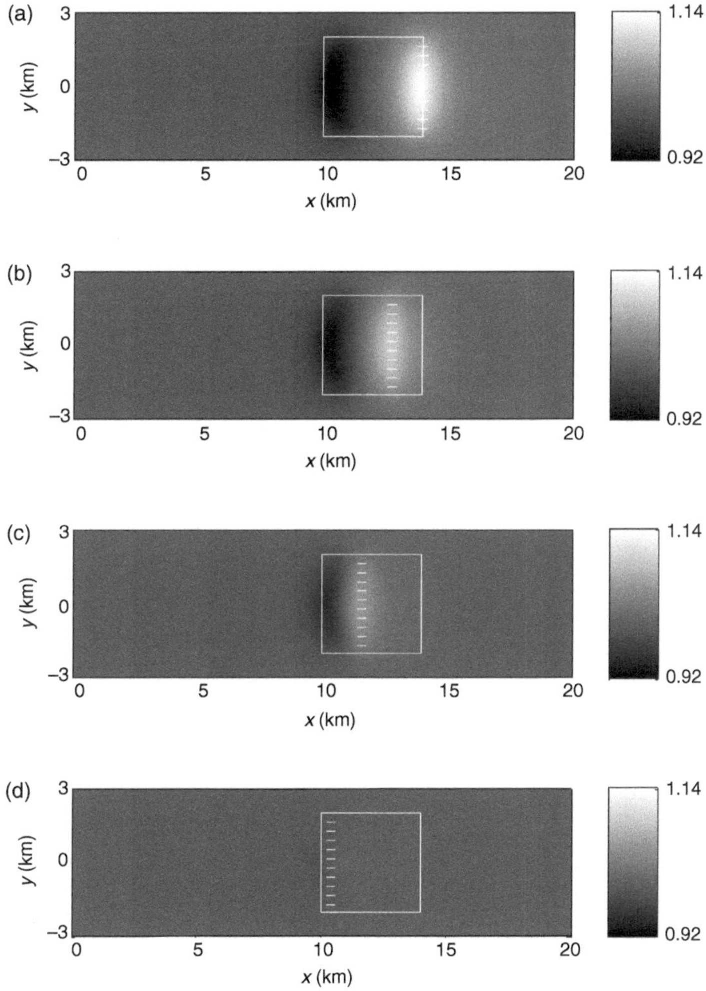

Figure 10 $|E_x^{tot}|$ normalized by background, bathymetry, and salt dome fields for frequency $= 0.01$ Hz.

5. Conclusions

The difference in the resistivities of the oil and the water filling the reservoir during production allows the possibility of monitoring the flooding front by EM methods. The oil-water interface can be clearly seen in several field components, even with a sea-bottom bathymetry. Induction due to the different anomalous domains can be extracted separately with the IBC method, allowing for different normalizations of the total field. The reservoir, and therefore the oil-water contact, can be seen most clearly when normalized by the background, bathymetry, and salt dome fields.

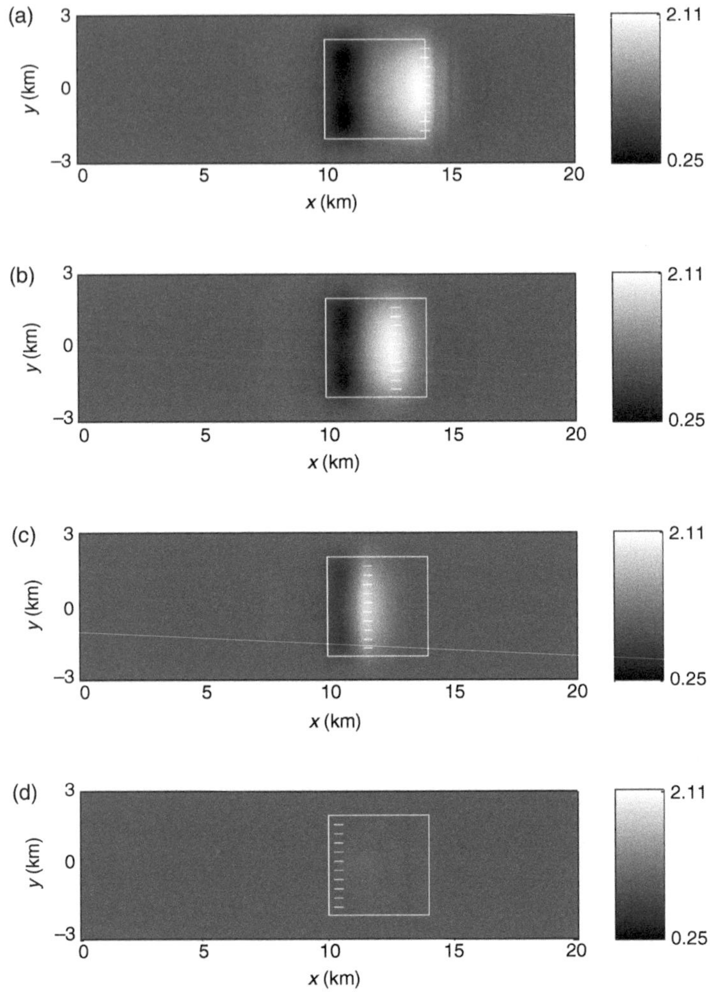

Figure 11 $|E_x^{\text{tot}}|$ normalized by background, bathymetry, and salt dome fields for frequency = 0.3162 Hz.

Our study demonstrates that measurable MCSEM data can provide an accurate position of the oil-water contact inside the reservoir over time. Future work should study more realistic models of the flooding front in the water injection recovery method. The oil-water interface can be nonvertical (angled), it can be nonsharp (occurring as a transition zone of up to tens of meters thick), and we should consider flooding from the bottom. Injection water of a range of resistivities should also be studied. The positive results of our experiment here, with a simplified flooding scheme (but a realistic geological model), warrant these further investigations into the use of EM methods for reservoir monitoring during production.

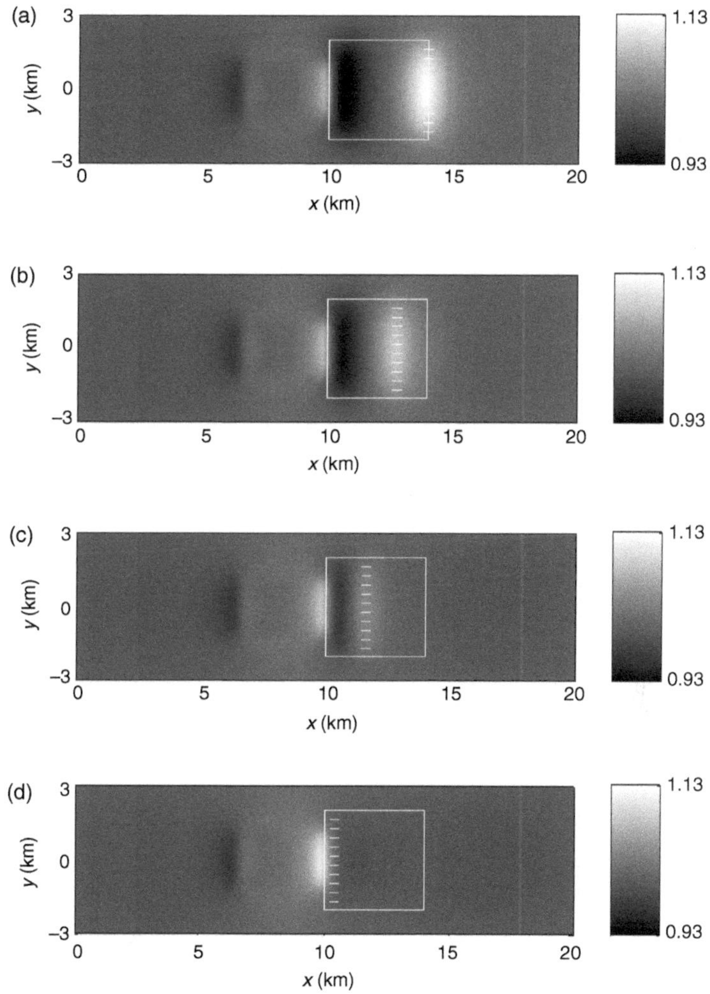

Figure 12 $\left|E_x^{tot}\right|$ normalized by background and bathymetry fields for frequency = 0.01 Hz.

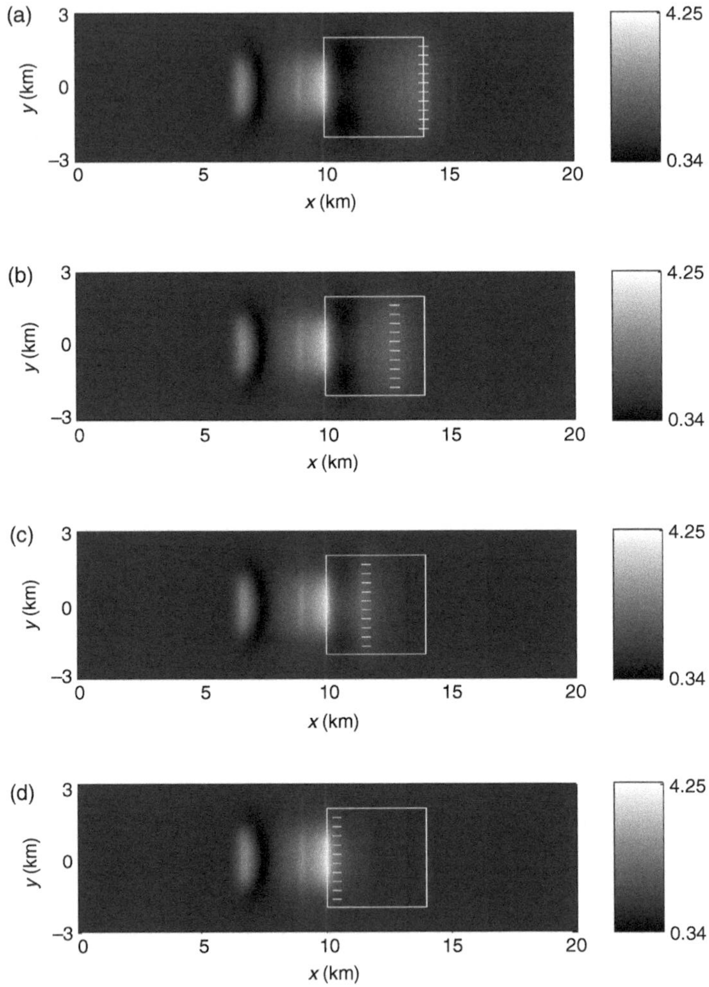

Figure 13 $|E_x^{tot}|$ normalized by background and bathymetry fields for frequency = 0.3162 Hz.

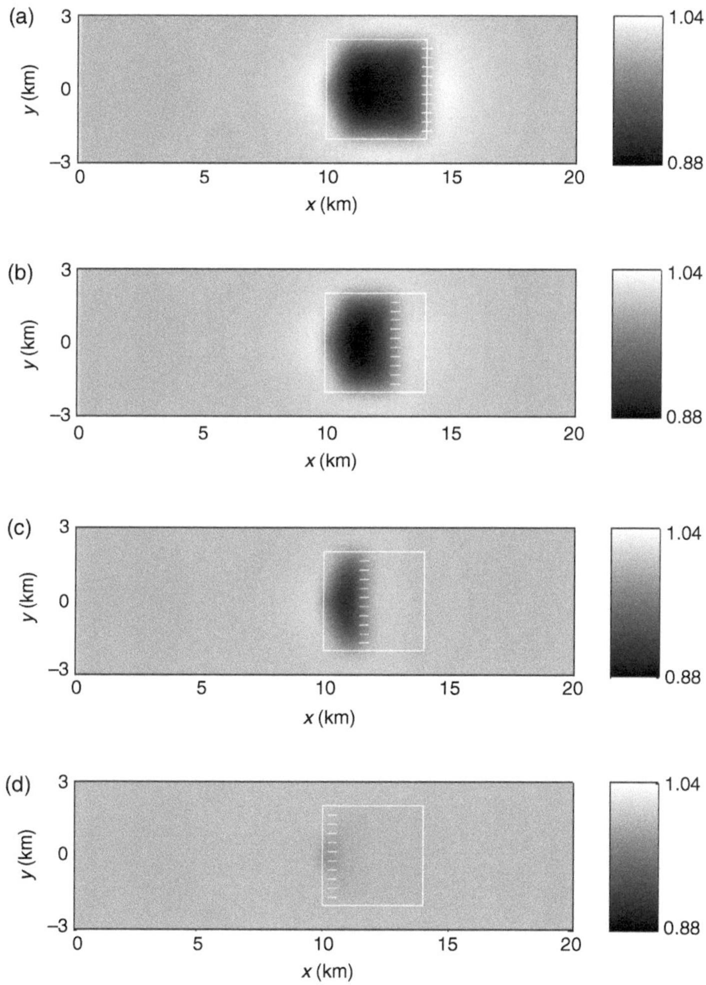

Figure 14 $|E_z^{\mathrm{tot}}|$ normalized by background, bathymetry, and salt dome fields for frequency $= 0.01$ Hz.

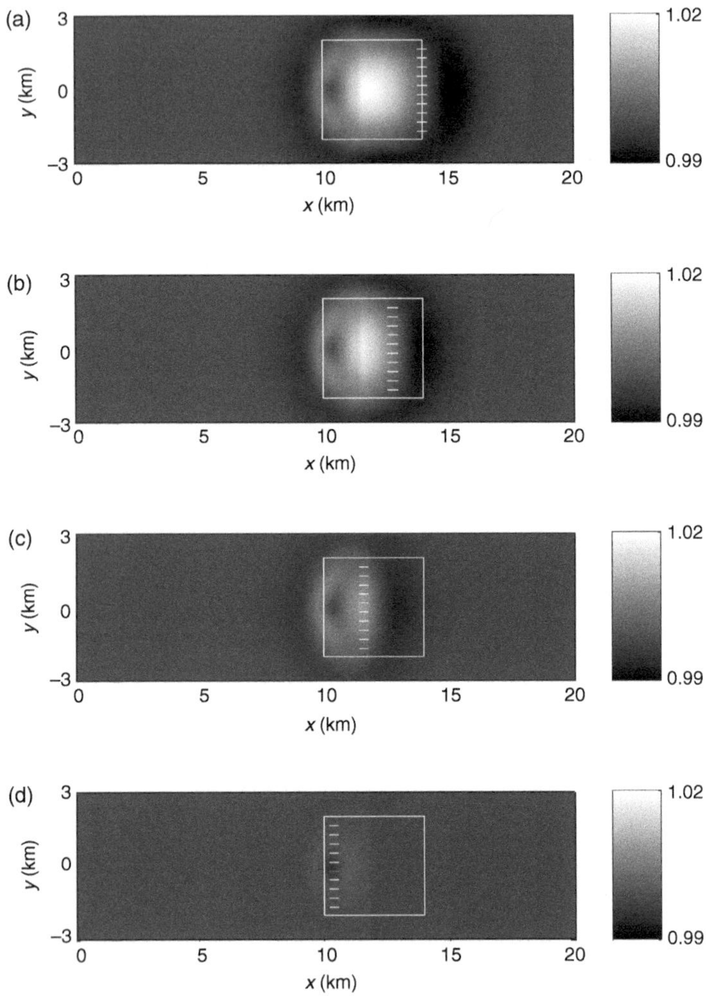

Figure 15 $\left|E_z^{\text{tot}}\right|$ normalized by background, bathymetry, and salt dome fields for frequency $= 0.3162$ Hz.

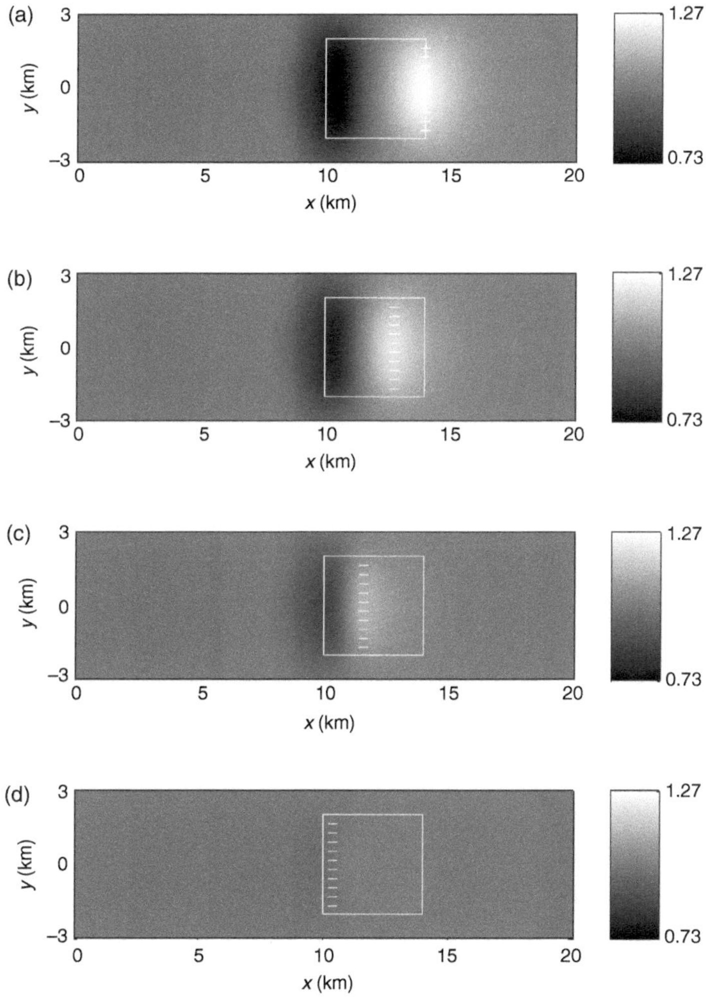

Figure 16 $\left|H_y^{\mathrm{tot}}\right|$ normalized by background, bathymetry, and salt dome fields for frequency = 0.01 Hz.

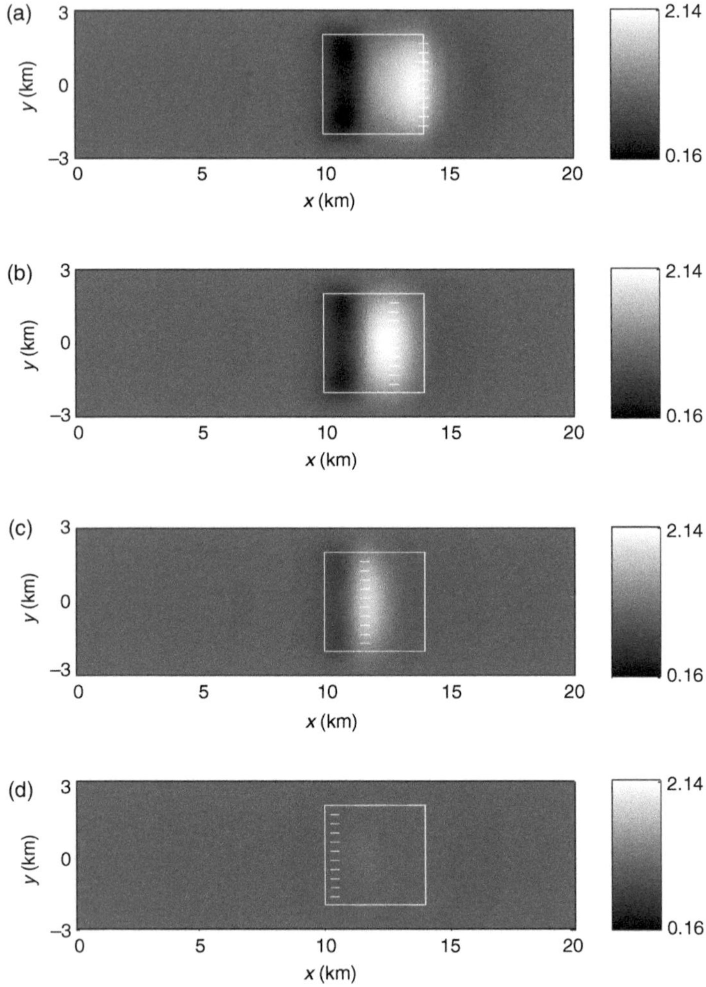

Figure 17 $\left|H_y^{\text{tot}}\right|$ normalized by background, bathymetry, and salt dome fields for frequency $= 0.3162$ Hz.

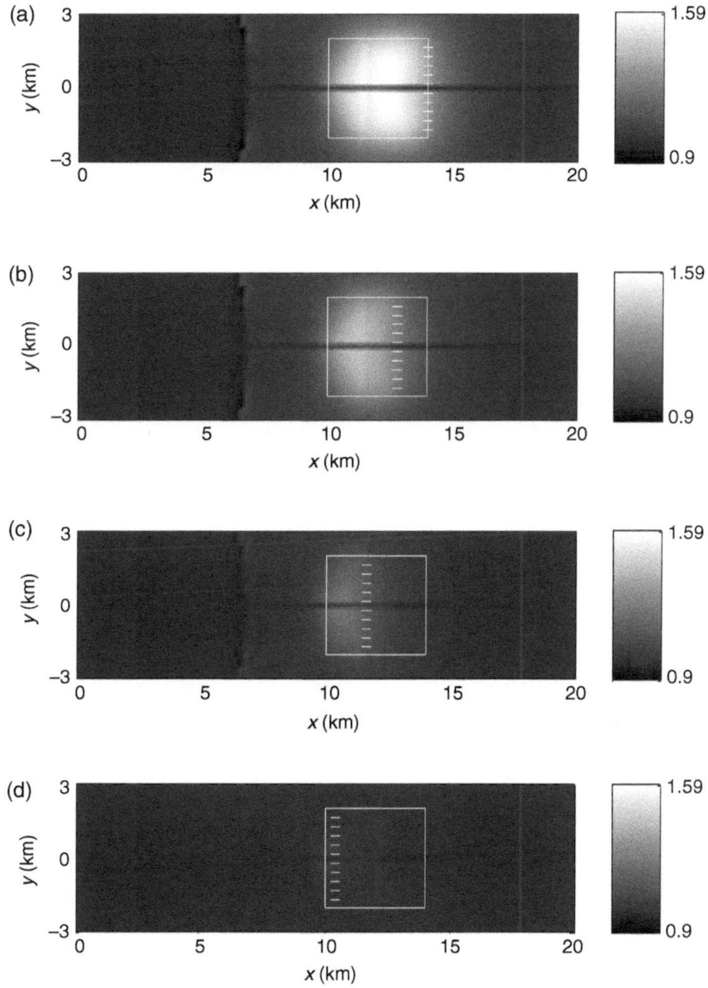

Figure 18 $|H_x^{\text{tot}}|$ normalized by background, bathymetry, and salt dome fields for frequency = 0.01 Hz.

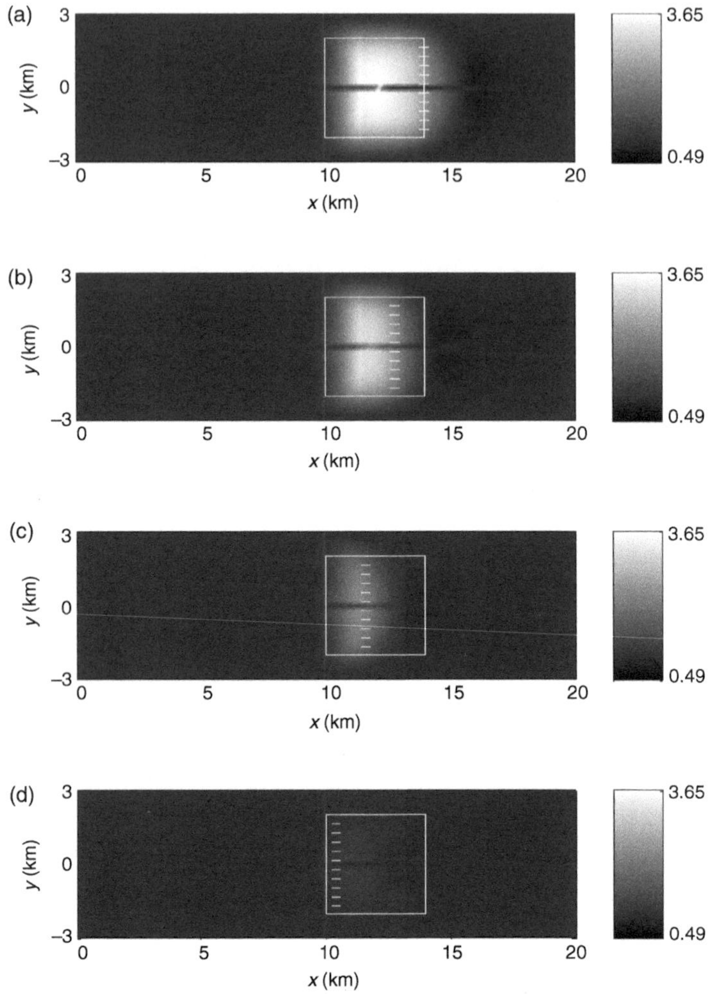

Figure 19 $|H_x^{\mathrm{tot}}|$ normalized by background, bathymetry, and salt dome fields for frequency = 0.3162 Hz.

ACKNOWLEDGMENTS

The authors acknowledge the support of the University of Utah Consortium for Electromagnetic Modeling and Inversion (CEMI), which includes BAE Systems, Baker Atlas Logging Services, BGP China National Petroleum Corporation, BHP Billiton World Exploration Inc., BP, EMGS, ENI S.p.A., ExxonMobil Upstream Research Company, Fugro, Halliburton Energy Services, Information Systems Laboratories, Newmont Mining Co., OHM, Petrobras, PGS, Rio Tinto - Kennecott, Rocksource, Russian

Research Center Kurchatov Institute, Schlumberger, Science Applications International Co., Shell International Exploration and Production Inc., StatoilHydro, Sumitomo Metal Mining Co., Woodside Energy, and Zonge Engineering and Research Organization.

REFERENCES

Carazzone, J.J., Burtz, O.M., Green, K.E., Pavlov, D.A., 2005. Three dimensional imaging of marine CSEM data. In: 75th Annual International Meeting, SEG, Expanded Abstracts, pp. 575–578.

Constable, S., Weiss, C.J., 2006. Mapping thin resistors and hydrocarbons with marine EM methods: Insight from 1D modeling. Geophysics 71 (2), G43–G51.

Eidesmo, T., Ellingsrud, S., MacGregor, L.M., Constable, S., Sinha, M.C., Johansen, S., Kong, F.N., Westerdahl, H., 2002. Sea bed logging (SBL), a new method for remote and direct identification of hydrocarbon filled layers in deepwater areas. First Break 20, 144–152.

Ellingsrud, S., Eidesmo, T., Johansen, S., Sinha, M.C., MacGregor, L.M., Constable, S., 2002. Remote sensing of hydrocarbon layers by seabed logging (SBL): Results from a cruise offshore Angola. The Leading Edge 21, 972–982.

Endo, M., Cuma, M., Zhdanov, M.S., 2009. Large-scale electromagnetic modeling for multiple inhomogeneous domains. Communications in Computational Physics 6, 269–289.

Hoversten, G.M., Newman, G.A., Morrison, H.F., Gasperikova, E., 2001. Reservoir characterization using crosswell electromagnetic inversion: A feasibility study for the Snorre field, North Sea. Geophysics 66, 1177–1189.

Hu, W., Yan, L., Su, Z., Zheng, R., Strack, K., 2008. Array TEM sounding and application for reservoir monitoring. In: 78th Annual International Meeting, SEG, Expanded Abstracts, pp. 634–638.

Hursán, G., Zhdanov, M.S., 2002. Contraction integral equation method in three-dimensional electromagnetic modeling. Radio Science 37, 1089–2002.

Lien, M., Mannseth, T., 2008. Sensitivity study of marine CSEM data for reservoir production monitoring. Geophysics 73, F151–F163.

Malinverno, A., Torres-Verdín, C., 2000. Bayesian inversion of DC electrical measurements with uncertainties for reservoir monitoring. Inverse Problems 16, 1343–1356.

Um, E.S., Alumbaugh, D.L., 2007. On the physics of the marine controlled-source electromagnetic method. Geophysics 72, WA13–WA26.

Wang, Z., Gelius, L., Kong, F., 2008. A sensitivity analysis of the sea bed logging technique with respect to reservoir heterogeneities. In: 78th Annual International Meeting, SEG, Expanded Abstracts, pp. 711–715.

Wilt, M., Morea, M., 2004. 3D waterflood monitoring at Lost Hills with crosshole EM. The Leading Edge 23, 489–493.

Zhdanov, M.S., 2002. Geophysical inverse theory and regularization problems. Elsevier.

Zhdanov, M.S., Lee, S.K., Yoshioka, K., 2006. Integral equation method for 3D modeling of electromagnetic fields in complex structures with inhomogeneous background conductivity. Geophysics 71, G333–G345.

PART II

THEORY AND TECHNOLOGY OF ACTIVE MONITORING

TECHNOLOGY OF ACTIVE MONITORING

9. Acquisition of Crosswell Seismic Monitoring Data
10. EM-ACROSS System and Corresponding Tensor Transfer Functions in Diffusion Field Region
11. Design and Workout of ACROSS Transmitters Using Air Bearings

SIGNAL PROCESSING AND ACCURACY CONTROL IN ACTIVE MONITORING

12. Stacking Strategy for Acquisition of an ACROSS Transfer Function
13. Secondary Calibration Method of Seismometers by Utilizing ACROSS Signal
14. Signal Discrimination of External Geomagnetic Effects Using the Transfer Function Approach with Continuous Wavelet Transform
15. Development of Simulation Algorithm for Analyzing Temporal Stability of ACROSS Seismic Source
16. Nonlinear Processes in Seismic Active Monitoring

THEORY OF DATA ANALYSIS AND INTERPRETATION

17. Maxwell's Equations and Numerical Electromagnetic Modeling in the Context of the Theory of Differential Forms
18. 3D Electromagnetic Holographic Imaging in Active Monitoring of Sea-Bottom Geoelectrical Structures
19. Foundations of the Method of EM Field Separation into Upgoing and Downgoing Parts and its Application to MCSEM Data
20. Automatic Travel Time Determination from a Frequency-domain Transfer Function: The Sompi Event Analysis
21. Effect of Spatial Sampling on Time-lapse Seismic Monitoring in Random Heterogeneous Media
22. Singular Value Decomposition in Active Monitoring Data Analysis

Technology of Active Monitoring

CHAPTER 9

Acquisition of Crosswell Seismic Monitoring Data

Thomas M. Daley[a], Fenglin Niu[b], Paul G. Silver[c], *and* Ernest L. Majer[a]

Contents

1. Introduction

Seismic velocity (both compressional and shear) is a fundamental property of materials, a property that varies with changes in conditions both external (stress, temperature) and internal (fluid saturation, crack density). Monitoring of changes in these external or internal conditions is a goal of geophysical investigations such as earthquake prediction (via stress change monitoring) and reservoir exploitation (via fluid saturation monitoring). Using an active source and fixed receivers, the simplest means of observing a time-dependent stress change is to measure the delay time between subsequent source pulses for the same path. The precision at which this measurement can be made determines the precision to which stress changes can be detected, and the interval between measurements determines the temporal resolution. Active source, continuous monitoring of seismic velocity can provide high-precision *in situ* measurement of temporal changes. Using a crosswell geometry for acquisition (i.e., having source and sensors, at depth, in separate boreholes) provides a methodology

[a] Lawrence Berkeley National Laboratory
[b] Rice University
[c] Carnegie Institute

Handbook of Geophysical Exploration: Seismic Exploration, Volume 40 — 2010 Published by Elsevier Ltd.
ISSN 0950-1401, DOI: 10.1016/S0950-1401(10)04014-0

for spatially localizing the monitored region. Ideally, the source and sensors would both be in a formation of interest, and only changes in that formation would affect the data. This minimizes the effects of near-surface changes, which are often unwanted "noise", and simplifies data analysis.

An important component of *in situ* monitoring with active-source seismic measurement is calibrating the change in the seismic P- or S-wave velocity with a known stress variation, such as earth tides or atmospheric pressure change. This would allow the identification of that part of seismic velocity changes that is due to stress change in a given rock volume, and allow an *in situ* determination of the stress-velocity relationship. Changes in seismic velocity induced by solid earth tidal loading and variations in barometric pressure have been estimated in some studies. The fractional change in seismic velocity with respect to stress change (i.e., the pressure sensitivity) is reported to be in the range of 10^{-9}/Pa to 10^{-6}/Pa (De Fazio et al., 1973; Yukutake et al., 1988; Sano et al., 1999)

Previous active source monitoring efforts had variable resolution of *in situ* velocity changes; however, recent work has shown dramatic improvement. The major advances, compared to the earlier studies, are twofold. First, the characteristics of available seismic sources have significantly improved, primarily in repeatability. Detection of travel-time changes is maximized for highly repeatable sources that can operate at high frequency and for extended periods of time. Previous work used sources such as surface vibrators or air guns, which were difficult to keep running for extended periods and had limited repeatability. Modern piezoelectric borehole sources are highly repeatable and dependable for millions of source excitations. Second, the precision in measuring differential travel time has increased, made possible in large part by the greatly increased sample rates available with modern high-speed data acquisition systems using high-speed internal clocks, and by the computational capability of performing massive waveform stacking. Additionally, the development of seismic sensors, both coil/magnet type geophones and piezoelectric-type hydrophones, has improved the sensitivity and dependability over recent decades. Studies with permanent installations have demonstrated the improvement in monitoring resolution.

A logical next step is to take continuous monitoring measurements to field sites and use equipment designed for temporary deployment in wells of opportunity, allowing *in situ* monitoring to assess the sensitivity of a given site and then redeployment when and if desired. In this paper, we present techniques used for *in situ* travel-time monitoring using shallow (10–100 m) field site boreholes, with equipment designed for crosswell seismic acquisition, recorded for time periods of 6 and 35 days at two separate sites. The analysis is focused on correlation of barometric pressure changes to velocity changes, allowing calibration of the stress sensitivity.

2. Optimal Acquisition Parameters

For a given acquisition geometry, we would like to optimize acquisition parameters such as frequency content and source-receiver distance, which tend to scale (i.e., lower frequencies at larger distances due to attenuation). In travel-time monitoring, we measure τ, the lag time between an arriving wavelet (usually the first) on two seismograms, with a total travel-time T, and a center or characteristic frequency f_0. If we express both τ and T in terms of the characteristic frequency, f_0, $\tau = \varepsilon/f_0$, $T = N/f_0$, where N is the number of wavelengths between the source and receiver and ε is a normalized time delay. Then the velocity perturbation, $\delta\hat{v} = \delta V/V$ can be simply rewritten as the ratio of two dimensionless parameters:

$$\delta\hat{v} = \varepsilon/N. \tag{1}$$

This particular parameterization provides a simple way of scaling experiments both in terms of frequency and spatial extent. From Eq. (1), it is clear that achieving the lowest possible detection threshold for $\delta\hat{v}$ can be broken into two separate tasks: (1) maximizing the number N of wavelengths between the source and receiver, and (2) increasing the precision in measuring the normalized delay time, ε (effectively the phase). The first task, maximizing N, is a property of the medium, the geometry, and the source characteristics, since N can be written as $N = Lf_0/V$. Thus, the acquisition geometry (path length $= L$) and maximum frequency (a function of source output and the medium's attenuation) will factor into determining N. The second task is a function of frequency content and signal-to-noise. The attenuation, Q (Toksoz and Johnston, 1981) will affect the signal-to-noise and therefore the precision. It can be shown that the optimal $N \sim Q/\pi$ (Silver et al., 2007).

For $Q = 60$, a typical upper crust value, we have $N \sim 20$, and assuming the P velocity is 1.5 km/s, the optimal frequencies for source-receiver distances of 3 m, 30 m, 100 m and 1,000 m (corresponding to the distance range that we expect to encounter) are 10 kHz, 1.0 kHz, 330 Hz, and 33 Hz, respectively. To the extent possible, we use the optimal frequency in the data acquisition design. For our two test sites, we were able to use 10 kHz and 1 kHz at 3 and 30 m, respectively.

3. Acquisition Hardware

Data acquisition for continuous crosswell monitoring builds on development of crosswell seismic work of the previous 10-20 years, which was usually designed for tomographic imaging (e.g., Daley et al. (2004),

Majer et al. (1997), Rector III (1995)), but more recently has been used for time-lapse monitoring (e.g., Vasco (2004), Hoversten et al. (2003)). Three major data acquisition components are involved: seismic source, seismic sensor, and the recording system. Our acquisition was conducted with a combination of commercial and special-built equipment. The special-built components are the piezoelectric source and the high-voltage amplifier used to power it. The source is made up of piezoelectric-ceramic (lead zirconate titanate) cylindrical rings epoxied together and wired for positive and negative voltage on the inner and outer surfaces. This type of source has been used in many seismic crosswell surveys (e.g., Majer et al. (1997), Daley et al. (2004)) and is known to be repeatable and dependable. The tests reported here used a single-cycle square wave generated by a source waveform generator (in this case, a programmable analog signal generator), which also sends a trigger signal to the recording system. Other source waveforms are known to provide greater signal-to-noise ratio. The sensors were commercial hydrophones designed for large bandwidth and high sensitivity, using piezoelectric film and built-in amplifiers with analog data transmission to the surface recording system.

The commercial recording system, a "Geode" manufactured by Geometrics, has a 24-bit analog-to-digital converter using a sigma-delta conversion algorithm. Our data was collected at a sampling rate of 48,000 samples per second. The triggering of the recording system for each shot is an important consideration. The digitizer is continually sampling the data, and receives a trigger that will generally be between two digitized samples. Including a section of pre-trigger data, the time series is interpolated and resampled, so that the time series begins at the time of the trigger. This start time is not exact, and, at a sampling rate of 48,000/s, this time is computed to within a 20th of a sample (Geometrics engineering, personal communication). Thus, there is a delay-time measurement error that will be a 40th of a sample (halfway between samples), and the average error will be an 80th of a sample, assuming that the errors are uniformly distributed. This corresponds to an average error of 260 ns per trigger. The error in the stacked data is smaller by a factor of $N^{1/2}$, assuming they are uncorrelated. For $N = 36{,}000$ traces, as in the case of our experiments, the error is thus 1.4 ns for one-hour sampling, which is below other sources of error.

4. Field Tests

Our first test was conducted at a distance of about 3 m between two 15-m deep holes at the Lawrence Berkeley National Laboratory (LBNL). The piezoelectric source was placed at the bottom of one hole. The source power supply for this test was a commercial high-voltage pulser, made by Cober Electronics. The sensor cable had 24 sensors at 0.5 m spacing, spanning most

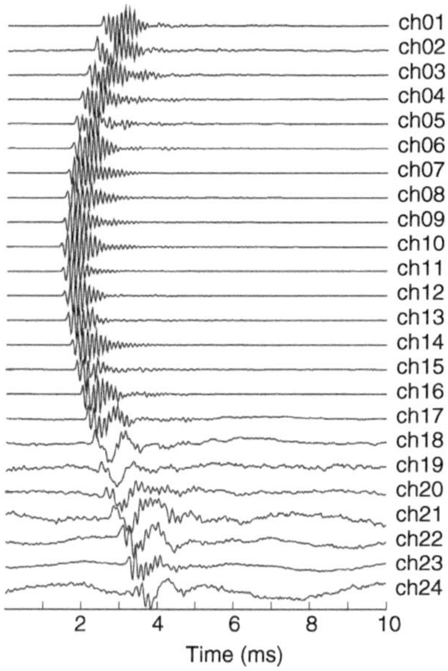

Figure 1 A single recording (no stacking) for the 24 sensors at 0.25 m spacing from the LBNL data set. The well spacing is 3 m. The dominant frequency is about 10,000 Hz.

of the second hole. All 24 sensors were recorded. The source was pulsed every 100 ms with a record length of 10 ms, which yields 10 traces per sensor per second. These individual traces were stacked in the acquisition system, with a stacked set of 24 traces output every minute. A one-minute record is thus a stacked recording consisting of a real-time stack of 600 single traces for each sensor. An example of a single shot record (no stacking) for all 24 sensors is shown in Figure 1. The signal-to-noise ratio is very high even without any stacking, especially for those sensors in the middle of the receiver string. The sensors near the surface (channels 18-24) had a much lower signal level, because they were above the water table in a much more attenuative medium. One advantage of using multiple receivers is that we can select the channels to be analyzed to obtain ε as low as possible.

In this 3 m experiment, we continuously recorded a total of 160 h starting from November 12, 2003. The SNRs average was ~4,000 for the stacked one-minute records. Figure 2(a and b) shows that SNR for this experiment continues to increase for up to 10^4 stacks, implying that a nonrandom noise "floor" has not been reached.

Figure 2(c) shows the histogram of the measured delay time between two adjacent one-minute records, which follows a normal distribution with a standard deviation of ~50 ns. Since the P-wave velocity at the test site is ~1.5 km/s, the travel time, T, is ~2 ms. For one-hour stacked records, we can achieve a standard deviation, or precision, of up to 6 ns, which

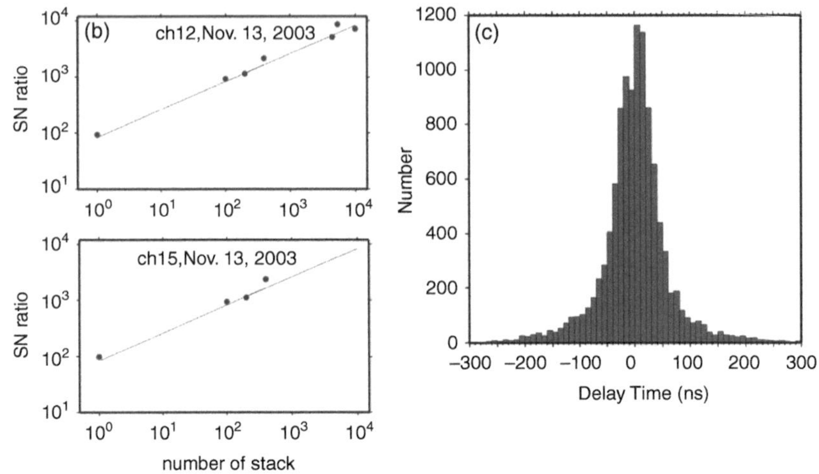

Figure 2 (a and b) (left) shows the increase in signal-to-noise ratio versus the number of stacked recordings for two data channels (ch12 and ch15 from Figure 1). (c) (right) A histogram of the delay time measurements between adjacent one minute stacks (600 records) with a standard deviation of about 50 ns (10^{-9} s).

corresponds to a resolution of 3×10^{-6} on the velocity perturbation, $\delta\hat{v}$. Our actual precision is possibly better than this, since there is a contribution from actual stress-induced velocity perturbations in the histogram. This nanosecond precision is the key to successful stress monitoring.

In this experiment, we measured the delay time using two different time windows: a short time window, which includes the first arrival only (1.4 cycles), and a long time window, which contains both the first arrival and coda (10 cycles). The measured delay times with respect to the trace using the two time windows are plotted in Figure 3(a) and b. Both show a time variation of ~3 μs (for a travel time ~2 ms), corresponding to $\delta\hat{v}$ of 10^{-3}. We also show the variations in barometric pressure (Figure 3(c)) and dilatation (Figure 3(d)) recorded by a nearby strain meter. During the period of observation, we see a large pressure excursion (~10^3 Pa) that continues for 3 days (Figure 3(c)). The delay times closely track this barometric pressure change. Using the variations in delay time and the barometric pressure change, we obtain a value of stress sensitivity of ~10^{-6}/Pa. Cross correlations between delay time with barometric pressure look very similar to the autocorrelation of the delay time with itself, demonstrating that at least over a long period, there is an excellent correlation (Figure 3(e-g)).

The sign of the travel-time fluctuation relative to the barometric pressure (increasing travel time for increasing pressure) is the opposite of the standard expectation and requires some explanation. An increase in barometric pressure acting directly on the borehole will raise pore pressure and decrease

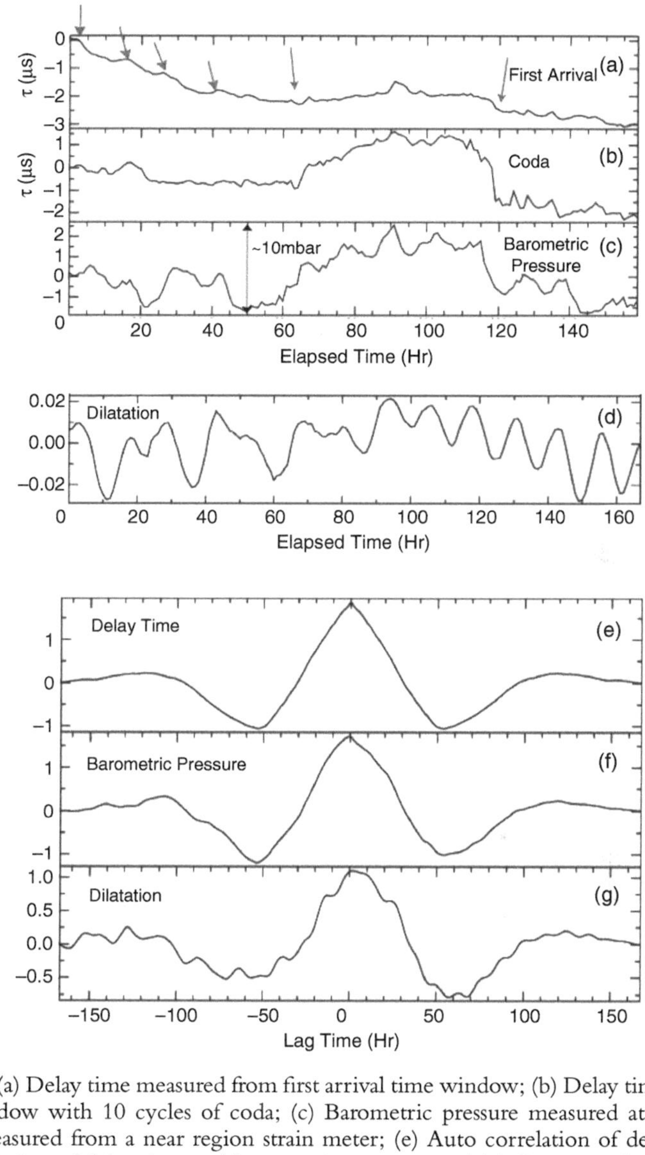

Figure 3 (a) Delay time measured from first arrival time window; (b) Delay time measured from a window with 10 cycles of coda; (c) Barometric pressure measured at surface; (d) Dilation measured from a near region strain meter; (e) Auto correlation of delay time; (f) Cross correlation of delay time and barometric pressure; and (g) Cross correlation of delay time with dilation.

the effective stress in the near-field region next to the borehole. In the far field, the same increase in barometric pressure will act on unchanged pore pressure and increase the effective stress. The change in seismic velocity along a particular ray path thus depends on the net change in effective stress. When two boreholes are relatively close, near-field effective stress can decrease and cause a decrease of seismic velocity.

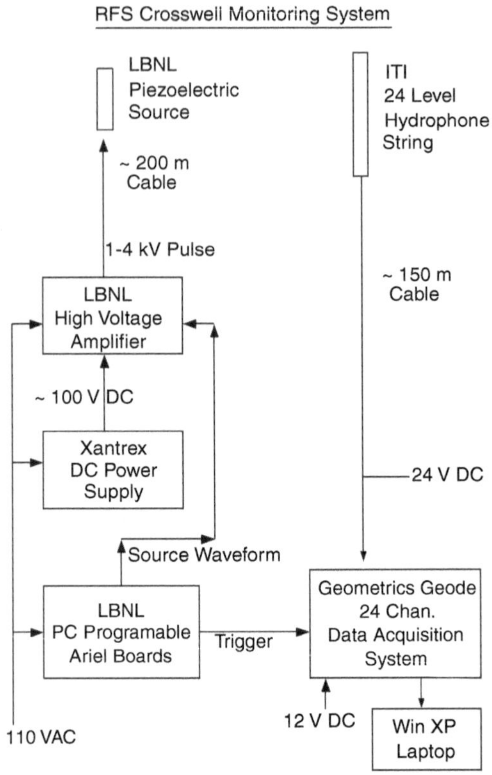

Figure 4 Schematic of data acquisition equipment used in the 36 day monitoring experiment at RFS.

After the LBNL site work, we wanted to deploy in deeper wells and at a longer interwell distance, hopefully moving out of the near-well conditions at the LBNL site. We chose to deploy in wells at the Richmond (California) Field Station (RFS), a test site on San Francisco Bay near the University of California, Berkeley. The RFS test site has several boreholes approximately 70 m deep and 30 m apart from each other (Daley and Gritto, 2001).

At RFS, using equipment shown in Figure 4, the source waveform was a 1 ms wide pulse (half cycle of a square wave). The output waveform had 1.2 kV peak-to-peak amplitude. The source was repeated 10 times per second, with a 50 ms record length, and these pulses were stacked by the recording system to give recordings every 1 minute (600 pulses). For this study, we attempted to reduce any movement of the sensors with "centralizers" (flexible bands of metal acting like springs), which kept three of the sensors fixed, and with hard foam attached to the source. Figure 5 shows the sensors with centralizers and the source with foam. Unfortunately, at some point during deployment, the source foam was compressed, apparently by the water pressure.

Figure 5 (left) Seismic sensor (hydrophone) with centralizer as used at the RFS experiment. (Right) Piezoelectric seismic source with foam centralizer, as used at the RFS experiment.

The dominant frequency observed in the RFS data was about 1 kHz; the travel time was about 20 ms (a Vp of 1.5 km/s). Compared to the LBNL test, the data frequency is about 1/10, but the travel time is 10 times as large; therefore, we achieved a resolution for $\delta\hat{v}$, essentially the same as the value we obtained at LBNL, $\sim 3 \times 10^{-6}$. The signal-to-noise ratio of 1-minute stacks is 604, with a standard deviation of 39.2 ns, while 1-hour stacks have a SNR of 3185 with 5.2 ns standard deviation. We recorded 35 days of continuous data in the RFS test.

The RFS data, shown in Figure 6(a), required some initial preprocessing. Bad shots were deleted and the long-term trend was removed. The cause of the long-term trend is not known, but it has a period larger than our signals of interest. During the 30 m experiment, we found that ambient surface temperature (Figure 6(b)) had a measurable effect on the data. We observed that delay time correlates very strongly with temperature. Recording of temperature inside the recording container allowed correction for temperature variation. The scaling is about 0.1 s/°C (as shown in Figure 6(d)) where swings of 10 °C correspond to oscillations in delay time of about a microsecond.

This is a large error, which we treated in two ways. First, we can solve for a temperature correction coefficient and subtract out this effect, assuming it is linear. A linear scaling constant between the measured temperature and the delay time variation was estimated, and the scaled temperature data was then subtracted from the delay time data. This was effective in minimizing the 24-hour variation in delay time caused by temperature. Second, we can minimize the effect by maintaining the electronics within

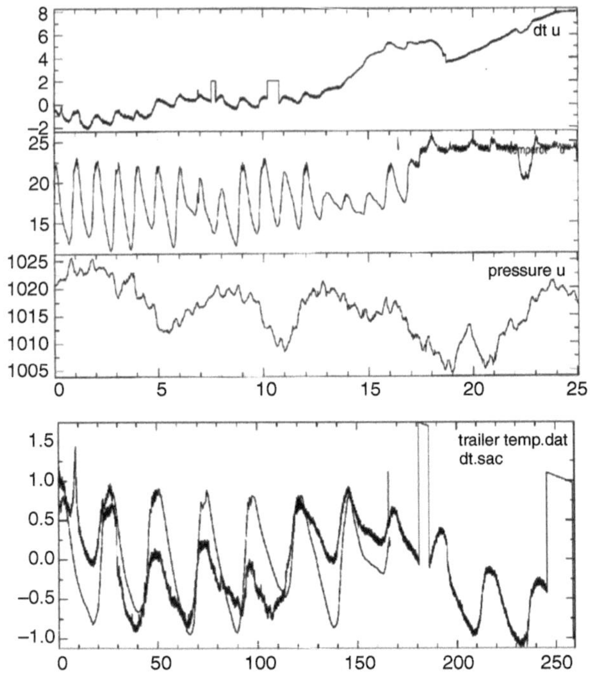

Figure 6 (a) (top) Measured raw delay time in microseconds for the RFS experiment for first 25 days. Excursion from 14 to 20 days is during heavy rainfall. (b) Measured temperature, degrees centigrade, inside the recording container. (c) Measured barometric pressure, millibars. (d) Comparison of raw delay time (microseconds) and temperature for initial 250 h at RFS.

a temperature range, which was done midway through the experiment by installing thermostat-controlled heating and air conditioning in the recording container, which stabilized the temperature. The temperature control began on Day 17 and the minimization of temperature variation can be seen in Figure 6(b). We were successful in maintaining temperature variation within ~2 °C using standard commercial thermostats to control heating and cooling.

The data collection coincided with a period of heavy rain, and we found that the correlation to barometric pressure had changed following this rainfall event. The data from the rainfall period is difficult to interpret. At least two possible effects are involved, an increase in surface load (increasing effective stress) resulting from the mass of the rainwater, and an increase in pore pressure (decreasing effective stress) resulting from the infiltration of rain water. The effect on travel time due to rain is therefore difficult to interpret and we do not attempt to do so here.

Figure 7 shows the delay time and barometric pressure for two time periods, before and after the rain, corresponding to days 1-14 and 24-36. In the first time period, a delay time change of 1 microsecond approximately

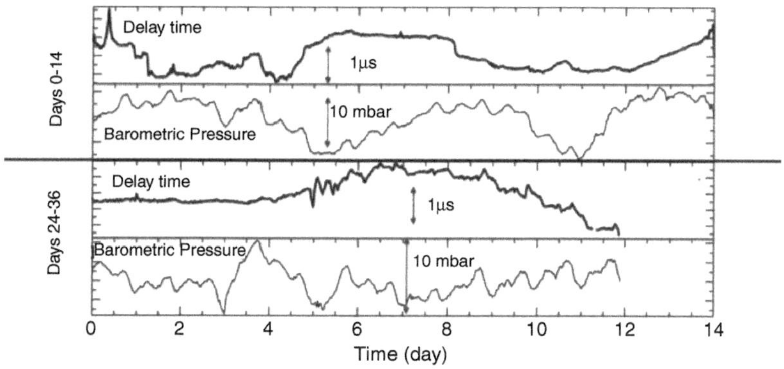

Figure 7 Delay times after temperature correction and removal of the linear trend from the data in Figure 6(a) are shown with barometric data for days 1-14 (top) and 24-36 (bottom).

corresponds to a barometric pressure change of 15 millibar, giving a stress sensitivity of about 5×10^{-8}/Pa. The second time period had a stress sensitivity of about 12.5×10^{-8}/Pa. Importantly, the sign of travel-time fluctuation relative to the barometric pressure is now the expected decrease in travel time for increasing pressure, indicating that we are measuring the "far field" effect away from the boreholes.

5. Conclusions

Modern crosswell data acquisition systems are capable of high-precision subsample delay time measurement. We have acquired two crosswell monitoring data sets over time periods of 7 days and 36 days. A large sensitivity to surface temperature in the recording instruments was observed and corrections were applied. In the second experiment, centralizers were used to stabilize the position of the source and sensors. In these two separate experiments, we demonstrated travel-time measurement precision on the order of 10^{-7} to 10^{-8} s. This precision allows us to monitor barometric pressure as a calibration signal to measure the stress sensitivity of velocity at each field site. Near-field effects dominated our first test at 3 m spacing between wells. These effects caused the velocity to decrease with increasing pressure, with a sensitivity of 10^{-6}/Pa. In a larger-scale 30-m experiment, the velocity was observed to increase with increasing barometric pressure, with a sensitivity of 5-12.5 $\times 10^{-8}$/Pa.

Addendum

Since the completion of this manuscript, we have successfully deployed a crosswell stress monitoring experiment at 1 km depth, for multiple weeks in two deployments, using the San Andreas Fault Observatory at Depth (SAFOD) site boreholes. As well as measuring the stress sensitivity, pre- and coseismic stress changes were observed. This work is described in Niu et al. (2008), and the reader is referred to this recent experiment. In 2009, our coauthor Paul Silver died unexpectedly. We would like to acknowledge his friendship and leadership in this work.

Acknowledgments

The authors would like to thank Rob Trautz of LBNL for supplying the barometric pressure logger and helping with field work. This work was supported by NSF Grants EAR-0352134 (FN), EAR-0453471(FN), EAR-0352119 (PGS) and EAR-0453638 (PGS), and the Carnegie Institution of Washington, DTM, and by the US Department of Energy, under Contract No. DE-AC02-05CH11231.

REFERENCES

Daley, T.M., Gritto, R., 2001. Field test of INEEL tube-wave suppressor and LBNL borehole seismic system at Richmond Field Station, Lawrence Berkeley National Laboratory Report, LBNL-49015.

Daley, T.M., Majer, E.L., Peterson, J.E., 2004. Crosswell seismic imaging in a contaminated basalt aquifer. Geophysics 69, 16–24.

De Fazio, T.L., Aki, K., Alba, J., 1973. Solid earth tide and observed change in the in situ seismic velocity. J. Geophys. Res. 78, 1319–1322.

Hoversten, G.M., Gritto, R., Washbourne, J., Daley, T., 2003. Pressure and fluid saturation prediction in a multicomponent reservoir using combined seismic and electromagnetic imaging. Geophysics 68, 1580–1591.

Majer, E.L., Peterson, J.E., Daley, T.M., Kaelin, B., Myer, L., Queen, J., D'Onfro, Peter, Rizer, W., 1997. Fracture detection using crosswell and single well surveys. Geophysics 62, 495–504.

Niu, F., Silver, P.G., Daley, T.M., Cheng, X., Majer, E.L., 2008. Preseismic velocity changes observed from active source monitoring at the Parkfield SAFOD drill site. Nature 454, 204–208. doi:10.1038/nature07111.

Rector III, J.W., 1995. Crosswell methods: Where are we, where are we going? Geophysics 60, 629–630.

Sano, O., Hieda, K., Hirano, K., Hirano, T., Ishii, H., Hirata, Y., Matsumoto, S., Yamauchi, T., 1999. Stress-sensitivity of the sound velocity at Kamaishi mine. Paper presented at the Seismological Society of Japan 1999 Fall Meeting, Sendai, Japan.

Silver, P.G., Daley, T.M., Niu, F., Majer, E.L., 2007. Active source monitoring of crosswell seismic travel time for stress induced changes. Bulletin of Seismological Society of America 97 (1B), 281–293.

Toksoz, M.N., Johnston, D.H. (Eds.), 1981. Seismic Wave Attenuation. In: Geophysics Reprint Series, vol. 2. Society of Exploration Geophysicists.

Vasco, D.W., 2004. Seismic imaging of reservoir flow properties: Time-lapse pressure changes. Geophysics 69, 511–521.

Yukutake, H., Nakajima, T., Doi, K., 1988. In situ measurements of elastic wave velocity in a mine, and the effects of water and stress on their variation. Tectonophysics 149, 165–175.

CHAPTER 10

EM-ACROSS System and Corresponding Tensor Transfer Functions in Diffusion Field Region

Takahiro Nakajima[a], Takahiro Kunitomo[a], Hiromichi Nagao[a,b], Mineo Kumazawa[a,c], *and* Naotaka Shigeta[a,d]

Contents

1. Introduction

Following the suggestion by Ogawa and Kumazawa (1996), Kumazawa et al. (2000), an electromagnetic method was developed for ACROSS (EM-ACROSS). The propagation of an electromagnetic disturbance into the earth's crust is mostly dependent on the electrical conductivity of

[a] Tono Geoscience Center, Japan Atomic Energy Agency, 1-63, Yamanouchi, Akeyocho, Mizunami, Gifu, 509-6132, Japan

[b] Present address: Institute for Research on Earth Evolution, Japan Agency for Marine-Earth Science and Technology, 3173-25, Showa-machi, Kanazawa-ku, Yokohama, 236-0001, Japan

[c] Present address: Faculty of Science, Nagoya University, Furo-cho, Chikusa-ku, Nagoya, 464-8601, Japan

[d] Present address: Horonobe Underground Research Center, Japan Atomic Energy Agency, 432-2, Hokushin, Horonobe-cho, Teshio-gun, Hokkaido 098-3224, Japan

E-mail address: nakajima.takahiro@jaea.go.jp (T. Nakajima).

Handbook of Geophysical Exploration: Seismic Exploration, Volume 40

ISSN 0950-1401, DOI: 10.1016/S0950-1401(10)04015-2

the ground. The electrical conductivity is expected to depend on the H_2O content and the physicochemical state of the crust. Kasahara et al. (2001) discussed the relationship of free water, hydrous minerals, and the seismic characteristics in three typical subduction zones, and suggested the possibility that earthquake occurrence is controlled by differences in the free-water content of the rocks and hydrous minerals along subduction zone boundaries. If H_2O is sufficiently abundant in the seismogenic zone to be a factor in earthquake generation, it could be detected as an electrical conductivity anomaly. In order to investigate the presence of water and any change in the physicochemical state due to H_2O flow, continuous monitoring using electromagnetic fields is important. Most electromagnetic methods utilize natural disturbances such as magnetic storms, because natural signals are much larger than the signals that can be generated from artificial electromagnetic sources. However, natural disturbances are irregular and their incident direction and signal form are usually unknown. They are not suitable for continuous monitoring.

Therefore, for accurate monitoring, it is essential to use an artificial signal. EM-ACROSS uses a time base synchronized to a GPS clock similar to the seismic ACROSS (Kunitomo and Kumazawa, 2004), to obtain signal enhancement by stacking. The transfer function obtained for the electromagnetic field propagated from the source dipole to the receiver is essentially a Green's function.

In this paper, we present the theoretical background of EM-ACROSS in the diffusion field region by showing the basic characteristics of the electromagnetic waves. Next, we briefly describe the observation system implemented at the test area, and show the data obtained from the trial observations. The results are decomposed into the tensor transfer function to investigate the heterogeneity around the test site. Tentative interpretations are made of the transfer function obtained. We also investigate temporal variations of the transfer function.

2. Basic Equations for the Electromagnetic Field and the Transfer Function in EM-ACROSS

The basic equation for electromagnetic phenomena with a time dependency of $\exp(i\omega t)$ is a Maxwell's equation in the frequency domain:

$$\nabla \times \mathbf{H} = \mathbf{J}_0 + \mathbf{J} + i\omega\mathbf{D}, \qquad \nabla \times \mathbf{E} = -i\omega\mathbf{B},$$

where the physical parameters are related by the state equations,

$$\mathbf{J} = \sigma\mathbf{E}, \qquad \mathbf{D} = \varepsilon\mathbf{E}, \qquad \mathbf{B} = \mu\mathbf{H}.$$

In the above equations, $\mathbf{J}$, $\mathbf{J}_0$, $\mathbf{D}$, $\mathbf{E}$, $\mathbf{B}$, and $\mathbf{H}$ are electric current density, source current density, electric displacement, electrical field intensity, magnetic flux density, and magnetic field intensity, respectively. The material parameters are electrical conductivity σ (inverse of resistivity), dielectric permittivity ε, and magnetic permeability μ. Generally these parameters have heterogeneity, anisotropy, dispersion, nonlinearity and time dependency. In most electromagnetic explorations, all medium properties are assumed to be linear, and the magnetic permeability is the same as that of free space in order to simplify the data analysis (e.g., Ward and Hohmann (1987)).

Now we investigate the properties of an electromagnetic field in a uniform medium. The propagation of an electromagnetic disturbance in a dielectric and semiconductive medium is obtained by solving

$$\nabla^2\mathbf{A} + k^2\mathbf{A} = -\mathbf{J}_0,$$
$$k^2 = \omega^2\mu\varepsilon^*,$$
$$\varepsilon^* = \varepsilon + \sigma/\mathrm{i}\omega,$$

and the electromagnetic field is represented as

$$\mathbf{H} = \nabla \times \mathbf{A}, \qquad \mathbf{E} = -(\mathrm{i}\omega\mu)\mathbf{A} + \frac{1}{\mathrm{i}\omega\varepsilon^*}\nabla(\nabla \cdot \mathbf{A}),$$

where $\mathbf{A}$ is the Schelkunoff potential and k is the wave number as a function of complex dielectric permittivity ε^*. The real part of ε^* denotes an effective dielectric permittivity, and the imaginary part is an effective electrical conductivity contributing to energy dissipation. The square of the wave number shows that the propagation of the disturbance has two different features depending on the frequency range. When the real part is dominant, at an angular frequency higher than the critical value $\omega_c = \sigma/\varepsilon$, the electromagnetic field appears to behave as a "traveling wave field". When the angular frequency is lower than ω_c, the electromagnetic field is described as propagation by "diffusion". Whereas the diffusion phenomenon is not usually regarded as a "wave", it is well registered as a group of traveling waves along their ray paths with a group velocity given by $1/(\partial k/\partial\omega)$. We designate such a field an "EM diffusion wave". This is one of the most important points of our approach, as noted earlier by Yokoyama et al. (2002). The group velocity in the "diffusion" region is given by $2(2\omega/\mu\sigma)^{1/2}$, which is proportional to the square root of the frequency (See Nakajima et al. (2004a) for more details).

We use the source current (essentially a current in a grounded wire with a dipole moment $\mathbf{m}_0 = I_0\ell \cdot \hat{\mathbf{e}}$ in the direction $\hat{\mathbf{e}}$) located at a particular site and observe the induced electromagnetic fields, $\mathbf{E}$ and $\mathbf{H}$ (or $\mathbf{B}$). We can define two types of transfer functions:

$$\mathbf{E} = Z\mathbf{H} \quad \text{or} \quad \mathbf{H} = A\mathbf{E},$$

and

$$\mathbf{E} = T^E\mathbf{m}_0 \quad \text{or} \quad \mathbf{H} = T^H\mathbf{m}_0.$$

The impedance Z or admittance $A(=1/Z)$ defined above is the transfer function between the electric and magnetic fields. They are dependent on the electromagnetic properties of the medium nearby as indicated by the Maxwell's equations. Impedance is the familiar quantity utilized in the magnetotelluric method (e.g., Vozoff (1972)). In our approach, we take $T^E(=ZT^H)[(\text{V/m})/\text{Am}]$, and $T^H(=AT^E)[(\text{A/m})/\text{Am}]$ or $T^B(=\mu AT^E)[T/\text{Am}]$ as sets of two observed Green's functions in tensor form in the frequency domain. The transfer functions T^E and T^H carry information on the electromagnetic properties of the medium accommodating the EM fields induced by the sources. The transfer functions T^E and T^H can be used to calculate the impedance $Z = T^E/T^H$ or the admittance $A = T^H/T^E$.

3. The EM-ACROSS Observation System

In this section we explain the measurement system used to achieve electrical conductivity monitoring of the earth by EM-ACROSS. We developed a transmitter that operates in a frequency range below 10 kHz and used it at the test sites around the Tono Mine campus of JAEA (Nakajima et al., 2000, 2004b). We used electrical current dipoles for transmitting the electromagnetic field. We positioned three grounded electrodes separated by 150 m (Figure 1). The three electrodes can be used to generate three horizontal current dipole sources with different azimuths in order to obtain essential information on anisotropy and lateral heterogeneity.

The transmitting signal can be designed appropriately for the ACROSS measurements. The signal is fed through a power amplifier to a pair of electrodes. Using the transmitter with a function generator and power amplifier, we can suppress the effect of higher harmonics of the transmitting signal in comparison with the rectangular wave produced by switching to DC current. It is also preferable to concentrate the energy into a limited number of spectral lines because the total energy of the transmitting signal is limited in practice. We continuously recorded the potential difference between the electrodes and the transmitting current, and the time signal of the GPS clock, to monitor the operational stability of the system and environmental changes near the transmitter.

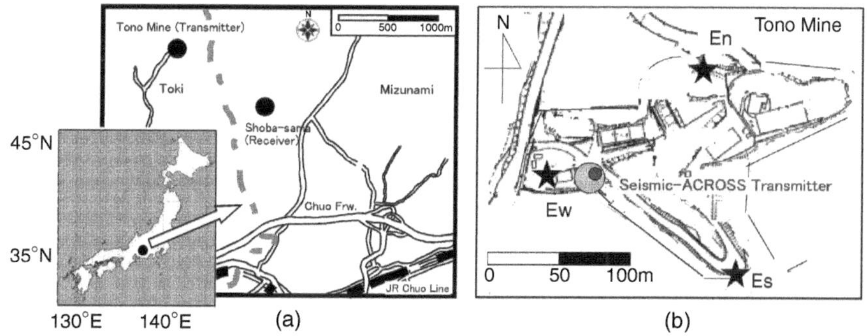

Figure 1 Tono test site. (a) Arrangement of transmitter and receiver, (b) Layout of transmitting electrodes.

The separation between the source and the receiver sites was approximately 950 m. The three-component magnetic field was measured by a fluxgate magnetometer (Bartington MAG-03) and the two horizontal components of the electric field by a couple of dipoles 33 m long with grounded Pb-$PbCl_2$ electrodes. The recording system was the same as that for the transmitter monitor. We recorded the time sequence data first, and then we calculated the amplitude and phase of the signal in its frequency domain. All the data were corrected by the frequency response characteristics of the measuring devices after verification and calibration. The spectrum data with reliability (noise level) were stacked for a longer period of time by means of the "optimum weight stacking method" (Nagao et al., 2004) to obtain better signal-to-noise ratio (SNR) data.

4. RESULTS OF THE TRIAL OBSERVATION

In this section we show the results for the static properties obtained in the trial observations.

First we explain the transmitting signal used for the observations. The transmitting signal was designed by selecting an appropriate set of continuous sinusoids with different frequencies and adjusting each phase so as to minimize the maximum instantaneous amplitude. To determine the characteristics of the transfer function in a wide frequency band, and the detailed characteristics in a narrow frequency band with a limited signal power, we adopted a signal, for which 13 sinusoids were distributed in three frequency bands as follows:

Band L : 3 sinusoids = 35.0, 37.5, 40.0 Hz
Band M : 5 sinusoids = 125.0, 127.5, 130.0, 132.5, 135.0 Hz
Band H : 5 sinusoids = 362.5, 365.0, 367.5, 370.0, 372.5 Hz.

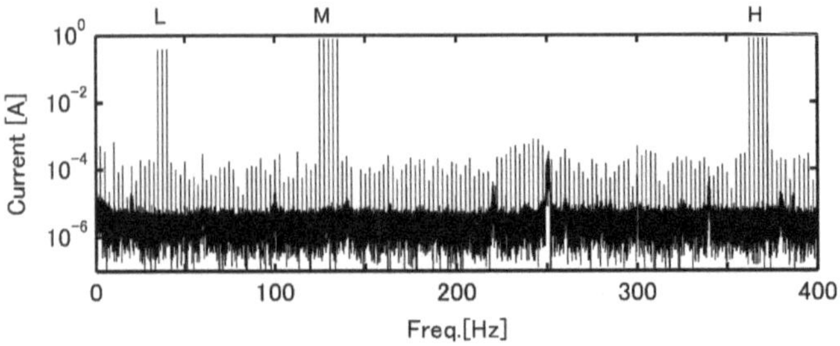

Figure 2 Amplitude spectrum of the transmitting signal showing the 13 sinusoids distributed in three frequency bands: L, M, and H.

From the consideration of the dispersion of the diffusion field, band M was set around the geometric mean of bands L and H. With the expectation that the higher-frequency signal may be more attenuated, we took the amplitude of the sinusoids in bands M and H to be double (Figure 2). The largest common divider of the 13 signal frequencies was 2.5 Hz, so the synthesized signal had a repetition period of 0.4 s.

The stacked data for 120 hours (5 days) in the entire frequency range are shown in Figure 3. The 13 signal spectral lines are identified well with a very large SNR. The peaks near 60 Hz and higher modes came from power line noise, which was larger than the transmitted signal. The results of the electric field have better SNR than those of the magnetic field. One of the reasons is that the noise level from the instruments used for the electrical measurements was lower.

The transfer function was calculated after the observed electromagnetic fields were divided by the transmitting dipole moment. Because both the observed values and the current dipole source are vectors, the transfer function obtained can be represented as second rank tensors. The tensor relationship from the current dipole to three components of the electric and magnetic fields can be written as

$$\begin{pmatrix} E_r \\ E_t \\ E_z \end{pmatrix} = \begin{pmatrix} T^E_{rr} & T^E_{rt} & T^E_{rz} \\ T^E_{tr} & T^E_{tt} & T^E_{tz} \\ T^E_{zr} & T^E_{zt} & T^E_{zz} \end{pmatrix} \begin{pmatrix} m_r \\ m_t \\ m_z \end{pmatrix}, \qquad \begin{pmatrix} B_r \\ B_t \\ B_z \end{pmatrix} = \begin{pmatrix} T^B_{rr} & T^B_{rt} & T^B_{rz} \\ T^B_{tr} & T^B_{tt} & T^B_{tz} \\ T^B_{zr} & T^B_{zt} & T^B_{zz} \end{pmatrix} \begin{pmatrix} m_r \\ m_t \\ m_z \end{pmatrix},$$

where the subscripts r, t, z represent the radial, transverse and downward directions respectively, and m_r shows the radial component of the dipole moment. There are 18 independent components, but we could obtain only 10 elements of the transfer function in our observations, because of the lack of a vertical current dipole and vertical components of the electric field measurements. The observed components of the tensor transfer functions are shown in Figure 4, in which the average and its slope (first derivatives in

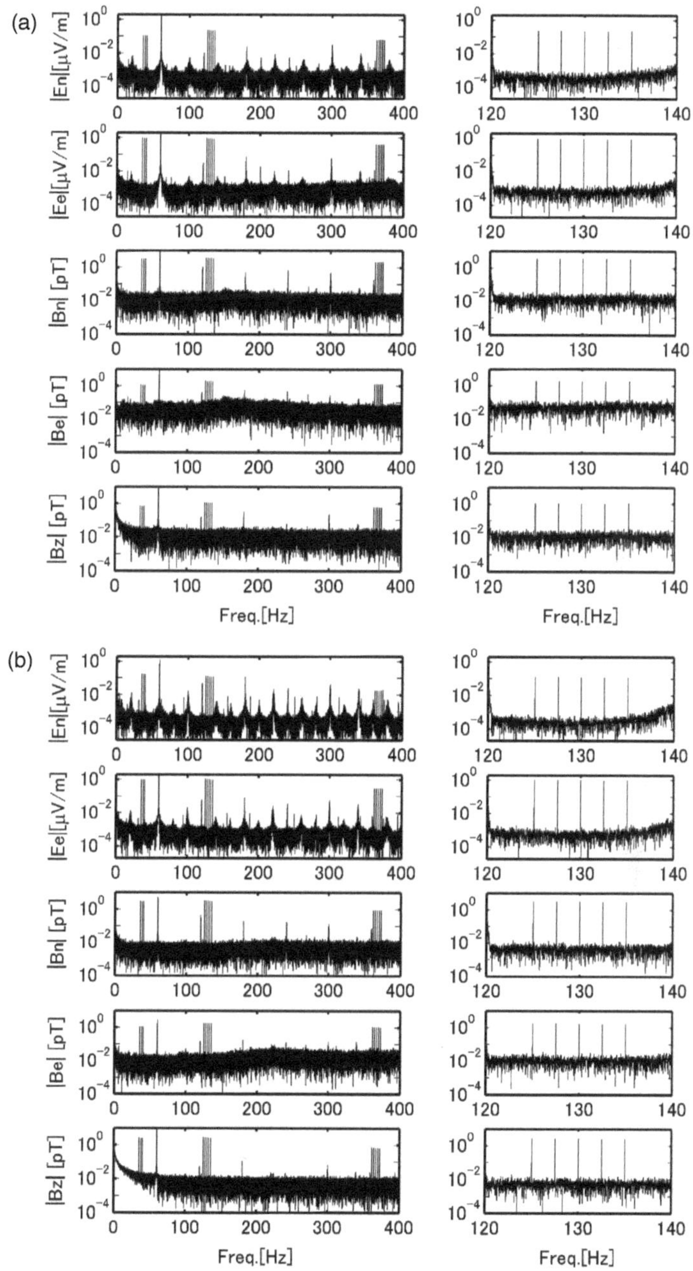

Figure 3 The amplitude spectrum of observed signals (two electric fields and three magnetic fields with a stacking period of 120 hours) transmitted by a dipole source at Tono Mine (see Figure 1). The source dipole electrodes are "En" and "Ew" for (a), and "Es" and "Ew" for (b). The right-hand column show the data in band M. Peaks at 60 Hz and the higher modes originated from a commercial power line.

(a)

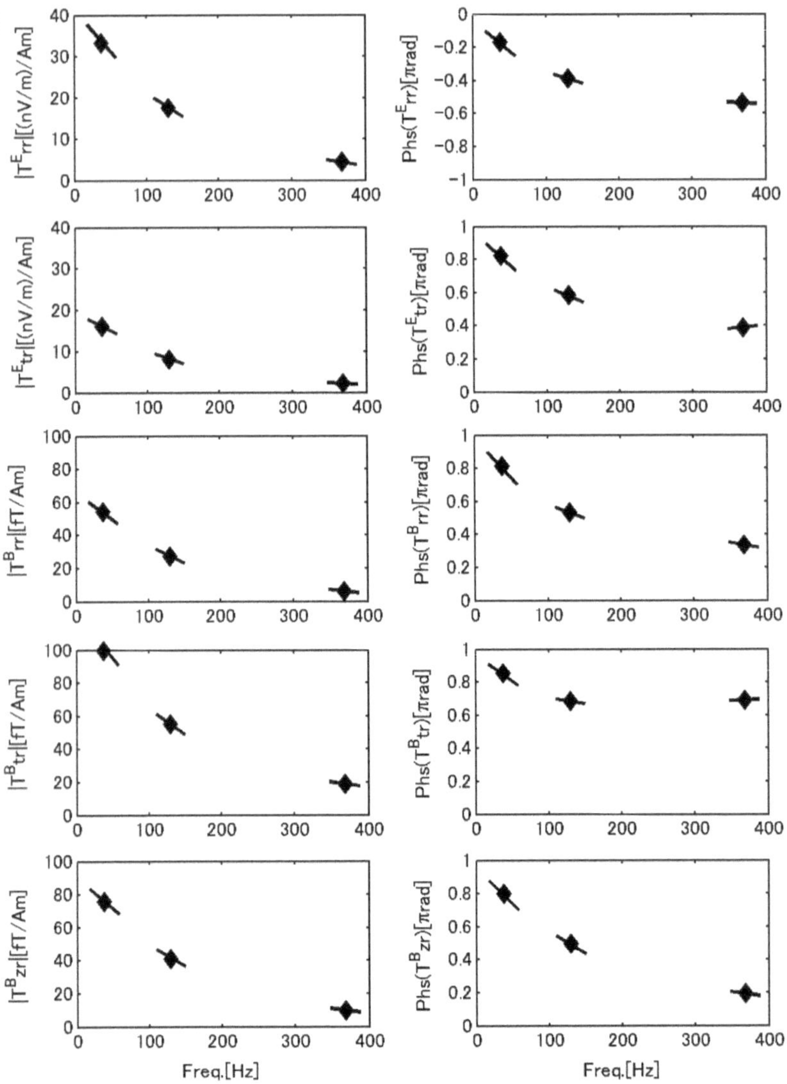

Figure 4 The ten components of the observed transfer function. The diamonds indicate the average and short segments are the slope of the transfer function in each frequency band. (a) shows the results of radial excitation, and (b) shows the case of transverse excitation. The right- and left-hand sides of each panel show the amplitude and phase of the transfer function.

the frequency domain) in each of the three frequency bands are expressed. The errors of the transfer functions were small because their noise levels were small after stacking. We could image the tendency around each frequency band from the slope of the transfer function.

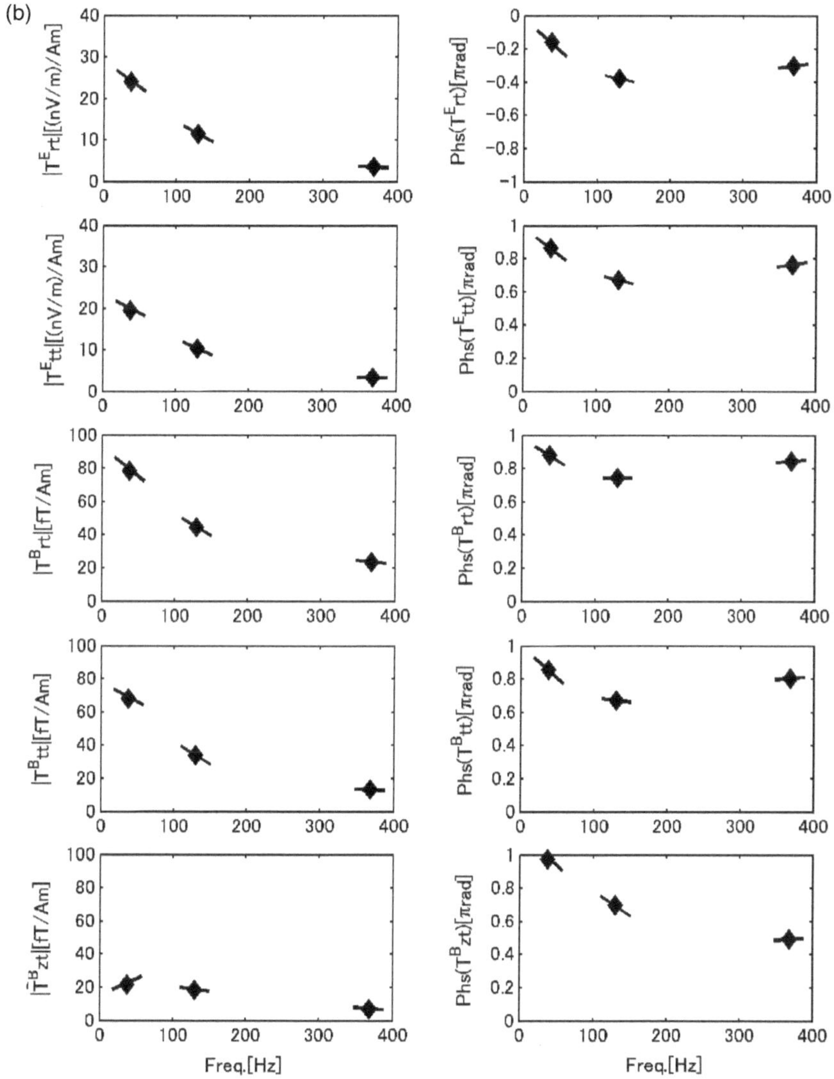

Figure 4 *(continued)*

5. Properties of the Transfer Function

5.1. Static tensor properties of the transfer function and the information on the underground conditions

In Figure 4 we show the observed transfer function with their averages and slopes in each of the three frequency bands. The terms T_{rr}^{B}, T_{zr}^{B}, T_{tr}^{E}, T_{tt}^{B},

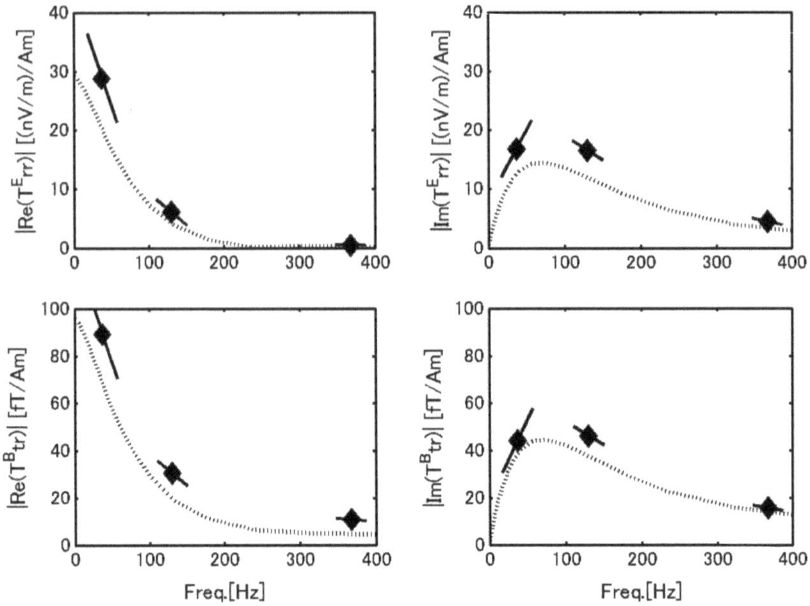

Figure 5 Comparison between the observed transfer functions (diamonds and short solid line segments) and the calculated frequency dependency (dashed line) from radial excitation in the horizontal layered earth model (Figure 6). The upper panels show the real and imaginary parts of the electric field, and the lower panels show those of the magnetic field.

and T_{rt}^{E} are expected to be zero for any layered earth models. However, the observed values are not zeros, rather their amplitudes are of the same order of magnitude, indicating that there must be significant lateral heterogeneity and/or anisotropy around the test site.

Although we presume the presence of lateral heterogeneity or anisotropy, we do not yet have enough observation points to describe them at this time. Therefore, we computed the electromagnetic fields using a previously reported ground model (Ogata et al., 1998) and compared them with the observed values. As an approximation, the conductivity of the top layer was set at about 0.1 S/m, and it became smaller than 0.001 S/m in deeper parts where the granite is present at about 100 m depth. We used their model as a horizontal layer model and as a first approximation for the computation of the electromagnetic field excited by the horizontal current dipoles. We changed the depth and the conductivity of the approximated model to fit the observed data. Figure 5 shows a comparison between the observed transfer function and the computed value derived from the model in Figure 6. We cannot say that the model is good, since the magnetic field cannot be explained well. In order to investigate the conductivity structure in this region, we need more observation points and also a tool to calculate the propagation of the electromagnetic field in 3D.

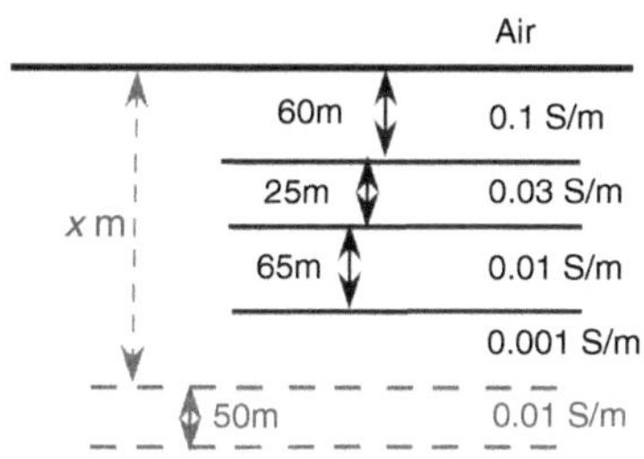

Figure 6 Horizontal layered earth model to compare the observed transfer functions shown in Figure 5. The layer indicated by the broken lines is added for the discussion on the high-conductivity layer (Figure 7).

Next, we examine the potential of our system for exploration depth. The test site is on granite, so there is less information on any deeper layers. Because the amount of H_2O in the seismogenic zones is thought to be a key factor with respect to earthquake activity (e.g., Kasahara et al. (2001)), the electrical conductivity in such zones should also be higher. The occurrence and the magnitude of a conductivity change in the layers is most important in electromagnetic monitoring. As a simple estimation, we used the model in Figure 6, but we added a high conductivity (0.01 S/m) layer with a thickness of 50 m. We changed the depth of the layer from 500 m to 2000 m and checked the difference in the observed electromagnetic field. The relationship between the source and receiver was set the same as for our trial observation of radial excitation. Figure 7 shows the result. The lines in the figure indicate the difference in the electromagnetic field between the base model and the high-conductivity layer at some depth. It is clear that the signal weakens when the high-conductivity layer is deeper. The frequency where the peak occurs decreases when a high-conductivity layer is deeper. This corresponds to a longer travel time of the reflecting electromagnetic field between the surface and the high-conductivity layer.

From Figure 7, the differences at a depth of 2000 m are 0.01 [(nV/m)/Am] in the electric field and 0.5 [fT/Am] in the magnetic field, which is in a frequency range of 10-30 Hz. These signals can be amplified by using the source dipole with a moment 150 m times 1 A; then the signals become 1.5 [nV/m] and 75 [fT]. Assuming that the electromagnetic noise is at the same level as the test site, the electric and magnetic noise are 0.6 [nV/m] and 30 [fT] respectively after 120 hours of observation (Figure 3). The noise is less than the signal from the high-conductivity layer, so the signal could be detectable. When we observe with better SNR or detect the signal from the deeper part, we will need more stacking time.

5.2. Temporal variations

In order to investigate temporal variations of the transfer functions, we examined the data obtained every 2 hours for the electric field and every

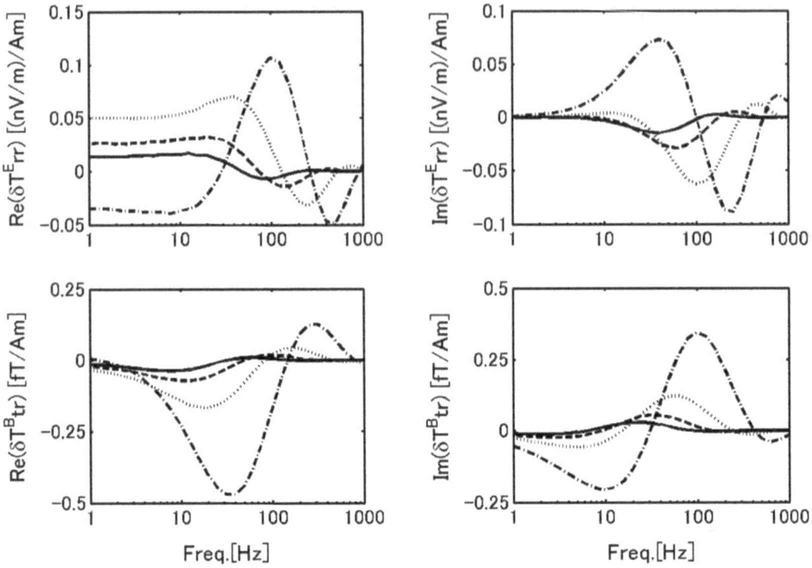

Figure 7 The frequency dependency of the difference in the signal between the model in Figure 6 and that with a high-conductivity layer of a thickness of 50 m. The upper panels show the real and imaginary parts of the electric field, and the lower panels show the magnetic field. The differences in the lines represent the depth of the high-conductivity layer: the line-dot is 500 m, the dotted line is 1000 m, the broken line is 1500 m, and the solid line represents 2000 m.

6 hours for the magnetic field. Figure 8 shows one of the examples in the "frequency band M" between September to November of 2003 acquired by transmitted signals from the "En" and "Ew" electrodes. The data plotted are the amplitudes and the phases with confidence limits given as standard deviation. The important point to note is that the magnitude of the temporal variation is greater than the confidence limit, especially in the electric field, indicating the presence of meaningful variation.

To investigate the correlations among the temporal variations and other natural phenomena, we compared them with the pressure and rainfall at the test site. As shown in Figure 8, the amplitude and average phase of the electric fields decrease just before rainfall events and start to recover during rainfall. There are also slow recoveries after rainfall events. The decreases might be related to the onset of low atmospheric pressures, but the overall changes in the electric field do not look like pressure-related responses. In the observations of the magnetic field, such changes are uncertain. The reasons why the electric field measurements were sensitive are thought to be related to (1) electric field measurements having higher SNR, (2) magnetic fields not being sensitive to local changes of conductivity, and (3) fast changes being possibly related to the electrokinetic phenomenon around the observation

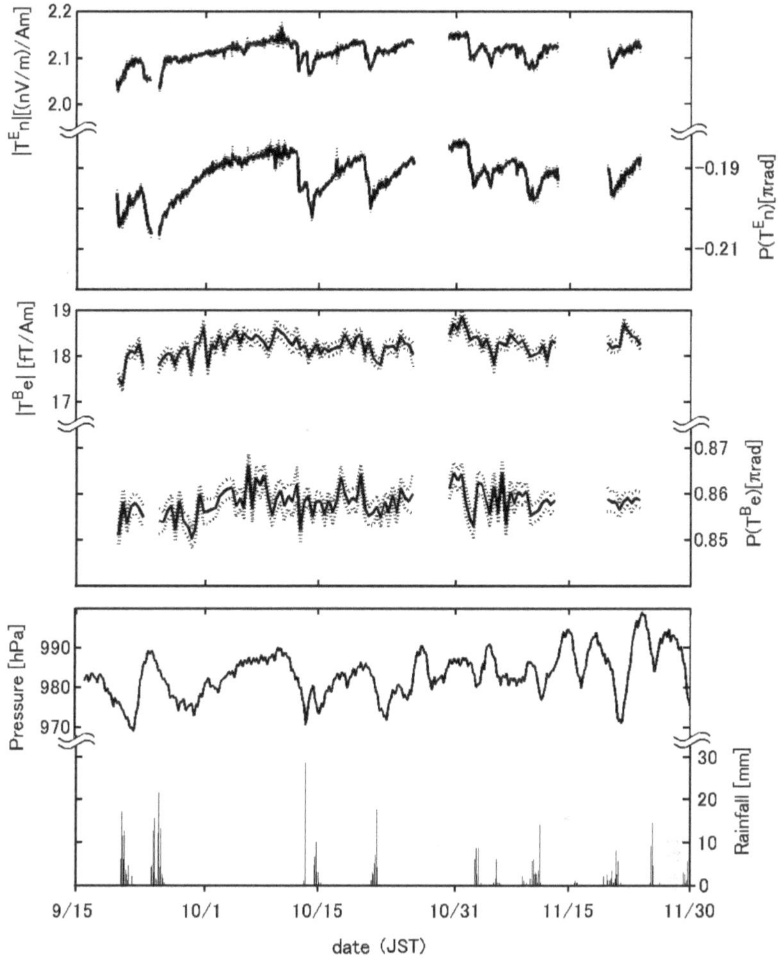

Figure 8 Examples of temporal variations of transfer functions in band M. The top panel shows the amplitude (upper part) and phase (lower part) of the east component of the electric field, the middle panel shows the same information for the north component of the magnetic field, and the bottom panel show the atmospheric pressure and the rainfall at the site. The solid lines are obtained from a 2-hour stacking of the electric field and from a 6-hour stacking of the magnetic field. The dotted lines indicate the reliabilities of the stacking results. There is no data from the period 9/24, from the period 10/27 to 10/30, or from the period 11/14 to 11/18, 2003, due to power failures.

electrodes. This environmental influence will be noise in the detection of small changes in the deeper underground. Therefore the investigation of corrections needed for the removal of environmental effects is an important subject for the detection of the subtle signal changes in the deep crust.

6. Conclusions

We carried out operational tests of the underground monitoring system using an electromagnetic field. We applied EM-ACROSS for this purpose. We acquired the tensor transfer function with good SNR and with confidence limits. The tensor transfer function has information on the lateral heterogeneity and the anisotropy of the ground. After three months of operation, we saw temporal variations in the transfer function phase. We found these variations correlated with rainfall. Although the trial observations were limited and on a small scale, the results indicate that EM-ACROSS could be a tool to provide information on the conductivity of the earth's crust.

REFERENCES

Kasahara, J., Kamimura, A., Fujie, G., Hino, R., 2001. Influence of water on earthquake generation along subduction zones. Bulletin of Earthquake Research Institute University of Tokyo 76, 289–301.

Kumazawa, M., Kunitomo, T., Yokoyama, Y., Nakajima, T., Tsuruga, K., 2000. ACROSS: Theoretical and technical developments and prospect to future applications. JNC Technical Review 9, 115–129 (in Japanese).

Kunitomo, T., Kumazawa, M., 2004. Active monitoring of the Earth's structure by the seismic ACROSS –Development of the seismic ACROSS and continuous transmission at Tono Mine ACROSS station. In: Proceedings of the "1st International Workshop on Active Monitoring in the Solid Earth Geophysics (IWAM04)".

Nagao, H., Nakajima, T., Kumazawa, M., Kunitomo, T., 2004. Optimum weighted stacking method for acquisition of the ACROSS transfer functions having the maximum signal-to-noise ratio. In: Proceedings of the "1st International Workshop on Active Monitoring in the Solid Earth Geophysics (IWAM04)".

Nakajima, T., Kumazawa, M., Shigeta, N., Kunitomo, T., Nagao, H., Matsumoto, M., 2004a. Overview of EM-ACROSS and its development. In: Proceedings of the "1st International Workshop on Active Monitoring in the Solid Earth Geophysics (IWAM04)".

Nakajima, T., Kunitomo, T., Nagao, H., Kumazawa, M., Shigeta, N., 2004b. Long-term operation of the EM-ACROSS and derived transfer function in the diffusion field region. In: Proceedings of the "1st International Workshop on Active Monitoring in the Solid Earth Geophysics (IWAM04)".

Nakajima, T., Kunitomo, T., Kumazawa, M., Yokoyama, Y., 2000. Development of EM-ACROSS and its field test. Bulletin of Earthquake Research Institute University of Tokyo 75, 413–428.

Ogata, N., Ota, K., Hama, K., Kubota, K., 1998. Regional hydrogeological study in the Tono area. JNC Technical Review 1, 41–50 (in Japanese with English abstract).

Ogawa, K., Kumazawa, M., 1996. Towards the continuous remote sensing of H_2O, tectonic stress and physical states in the Earth's crust by means of acoustic and electromagnetic ACROSS. Abstracts of the Seismological Society of Japan, 45 (in Japanese).

Vozoff, K., 1972. The magnetotelluric method in the exploration of sedimentary basins. Geophysics 37, 98–141.

Ward, S.H., Hohmann, G.W., 1987. Electromagnetic theory for geophysical applications. In: Nabighian, M.N. (Ed.), Applied Geophysics, Vol.1. Society of Exploration Geophysicists, Oklahoma, pp. 131–312.

Yokoyama, Y., Kumazawa, M., Nakajima, T., 2002. Transfer function measured by electromagnetic sounding with an accurately controlled signal. Earth Planets Space 54, 459–472.

CHAPTER 11

Design and Workout of ACROSS Transmitters Using Air Bearings

Tomohiko Ise[a], Takao Torii[b], Nobuyoshi Morita[b], Takahiro Kunitomo[c], Shinichi Togo[d], *and* Makoto Okano[e]

Contents

1. Introduction

The seismic ACROSS transmitters of the rotary type currently operating generate their periodic single force by means of rotation of an eccentric mass (Kumazawa et al., 2000; Kunitomo and Kumazawa, 2004a,b). In these transmitters, rolling contact type bearings are used to support an eccentric rotor. In practice, the frictional coefficient of the bearing is 0.003 at best and significant frictional heating takes place in the current transmitters.

[a] Department of Mechanical and System Engineering, University of Hyogo, 2167 Shosha, Himeji, Hyogo 671-2280, Japan

[b] Department of Mechanical Engineering, Shizuoka University, 3-5-1 Jouhoku, Naka-ku, Hamamatsu-shi, Shizuoka, 432-8561, Japan

[c] Japan Atomic Energy Agency, 1-63, Yamanouchi, Akeyocho, Mizunami, Gifu, 509-6132, Japan

[d] Department of Mechanical Engineering, Tohoku-Gakuin University, 1-13-1 Chuo, Tagajo-shi, Miyagi, 985-8537, Japan

[e] National Institute of Advanced Industrial Science and Technology, Tsukuba Central 2, 1-1-1 Umezono, Tsukuba-shi, Ibaraki, 305-8568, Japan

E-mail address: ise@eng.u-hyogo.ac.jp (T. Ise).

Handbook of Geophysical Exploration: Seismic Exploration, Volume 40

ISSN 0950-1401, DOI: 10.1016/S0950-1401(10)04016-4

In the case of a force of the order of 10^5 N supported by a bearing of diameter 0.20 m at a frequency of 25 Hz, for example, the frictional energy loss amounts to 3.8 kW whereas the elastic wave energy transmitted toward far fields is in the order of 10 W. In order to remove the generated heat in the transmitter, further energy is necessary to circulate the cooling fluid and to run the other peripheral devices. The operation of the ACROSS transmitter is continuous over years, and we have problems not only with significant energy consumption and heat generation in the transmitter but also with the durability and stable maintenance of the transmitter as a whole.

In contrast to ordinary bearings utilizing balls or rollers to support the force, hydrostatic gas bearings support the force by hydrostatic pressure of a gas without any solid mechanical contact, so that there is no friction at all and the pressurized gas itself works as a coolant for the viscous heating and other sources of heat within the transmitter. The durability is expected to be very high because there is no possibility of either flaking or any other wear from solid contact. In addition it is economical to run when using air as the working gas, whereas additional equipment such as gas compressors and related peripherals are needed. Air bearings have been utilized only to support a well-balanced precise rotor and no experience has been reported so far on the support of an eccentric mass rotor, where they are thought theoretically to have a large allowance of radial force above 10^5 N. These features raise the necessity of a technical evaluation of the advantages and disadvantages of using hydrostatic air bearings for supporting rotary type ACROSS transmitters in comparison with the current rolling contact bearing. The purpose of this study is to find out whether or not the use of an eccentric rotor with air bearings is promising for the ACROSS transmitter as a maintenance-free routine device of the next generation.

In this paper we carried out operating experiments on air bearings to support the centrifugal force generated by an eccentric mass. In the first preliminary experiment, pressurized air was supplied through restrictors from the stator to the ordinary symmetric air bearings which support the balanced rotor. The eccentric moment was devised by adding an external mass to this rotor and measurements of the stiffness and other characteristics were successfully made. On the basis of this preliminary test, we designed and made a bearing test rig for the application of high stiffness bearings expected to have a force allowance of 2000 N at 50 Hz if it was rigidly installed to a sufficiently heavy table. This rig was tested up to 30 Hz without any problem, various mechanical characteristics were measured and an analysis was made to extrapolate various factors up to the larger force and higher frequency. We have confirmed that the air bearing appears ideal for rotary type ACROSS transmitters except for one problem. This is the consumption of too much pressurized air, resulting in an energy loss larger than roller bearings, if the traditional symmetric air bearing for balanced rotor is used as it is. Since we have an idea of a counter plan to suppress the air supply rate, we have come

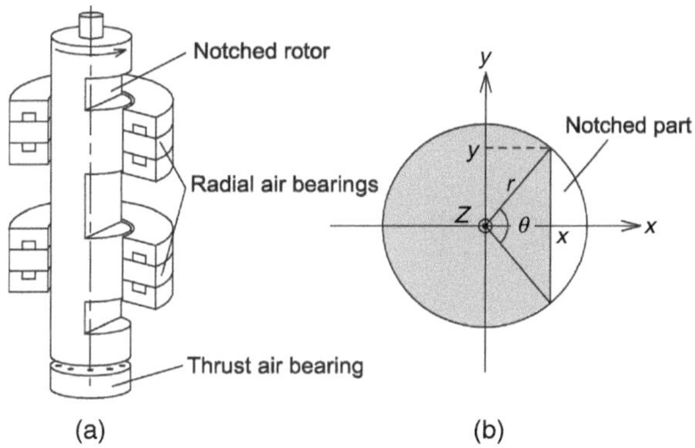

Figure 1 Schematic construction of an eccentric rotor and air bearings.

to believe that the air bearing is an important technique to realize a stable ACROSS transmitter useful for the active monitoring work over years.

2. Construction of an ACROSS Transmitter Using Air Bearings

Figure 1(a) shows one of the constructions of an eccentric rotor supported by air bearings to generate a sinusoidal centrifugal force. Two radial bearings are separately positioned and evenly support the centrifugal force. The rotor mass is supported vertically by a one thrust air bearing positioned at the bottom. This design is a typical design for an air bearing for a balanced rotor which may sustain a radial force. When a mass is added to the balanced rotor and the radial force is generated by itself, the force F is given as a function of rotational frequency f in Hz by the following equation:

$$F = M_1(2\pi f)^2 \tag{1}$$

where M_1 is the first rotational moment, or eccentric moment of the mass. When a mass deficit, by notching the cylindrical rotor, is used to generate eccentricity, as in Figure 1(b), M_1 is given by $(2L\rho r^3/3)\sin^3(\theta/2)$, where ρ and r are the density and the radius of the rotor, and L is the sum of the notch length.

3. Working Mechanism of a Hydrostatic Air Bearing

Figure 2 shows the schematic illustration of a hydrostatic air bearing and its working principle by an equivalent model of an electrical circuit.

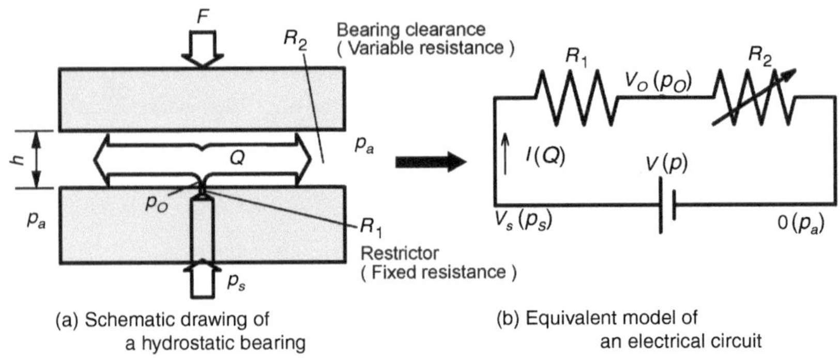

Figure 2 Working mechanism of a hydrostatic air bearing.

Pressurized air (pressure p_s) is supplied to the bearing through a restrictor from the outer source. At the outlet of the restrictor, the pressure of the air will drop from p_s to p_o caused by the flow resistance (R_1 in Figure 2(a)) of the restrictor. Then the supplied air flows between the narrow clearance of the bearing and goes to the outside of the bearing where the pressure is the ambient pressure p_a. In this process, the bearing clearance acts as a variable flow resistance (R_2 in Figure 2(a)). Since the flow resistance R_2 depends on the bearing clearance, it works as a variable resistance, whereas the flow resistance R_1 of the restrictor works as a fixed resistance. As a result, the outlet pressure of the restrictor p_o changes with a change in the bearing clearance. This change in the outlet pressure p_o of the restrictor can be easily recognized by the analogy of an electric circuit, as shown in Figure 2(b). The outlet pressure p_o of the restrictor gives the load capacity of the bearing, and the change in p_o with a change in the bearing clearance gives the bearing stiffness.

4. Preliminary Test

Up to now, hydrostatic air bearings have been designed to support a well-balanced precise rotor only and have no experience of supporting unbalanced rotors such as an eccentric mass rotor. Thus, the first trial was to verify the possibility of supporting an eccentric mass rotor with air bearings. Preliminary tests were carried out using trial air bearings.

Figure 3 shows the preliminary test rig for the air bearing to support an unbalance rotor. The radial air bearings are inherent orifice type and their diameter and length are 50 mm respectively. The rotor has maximum diameter of 74 mm and its total mass is 3.39 kg. The spin axis of the rotor is positioned horizontally and an eccentric mass of 0.29 kg is clamped by a screw on the outer surface at the maximum diameter of the rotor. The unbalanced rotor is driven by an impulse turbine placed at the end of the

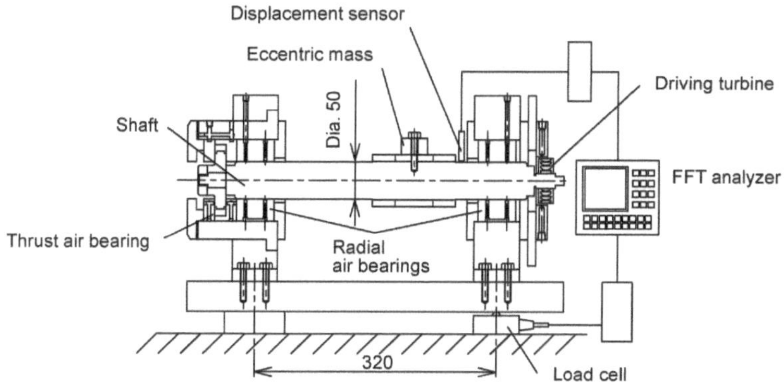

Figure 3 Preliminary test system and the hydrostatic air bearing test rig with a notched rotor of $M_1 = 1.07 \times 10^{-2}$ kg m.

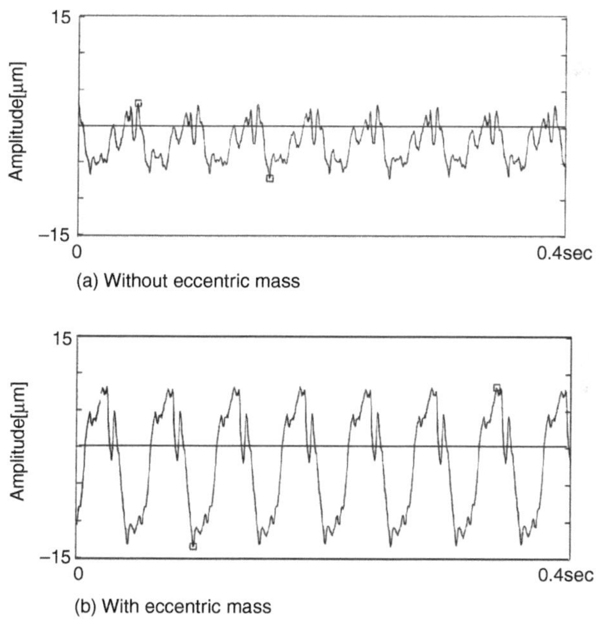

Figure 4 Amplitude of the eccentric mass on the rotor vibrating at a rotational frequency of 20 Hz.

rotor. The amplitude of the rotor vibration is measured by an eddy current type non contact displacement sensor. The centrifugal force generated by the unbalanced mass is measured by a strain gauge type load cell placed under the test rig.

Figure 4 shows the measured vibrations of the rotor at a rotational frequency of 20 Hz. The amplitude includes the effect of the roundness and surface roughness of the rotor. Figure 4(a) is the vibration without the

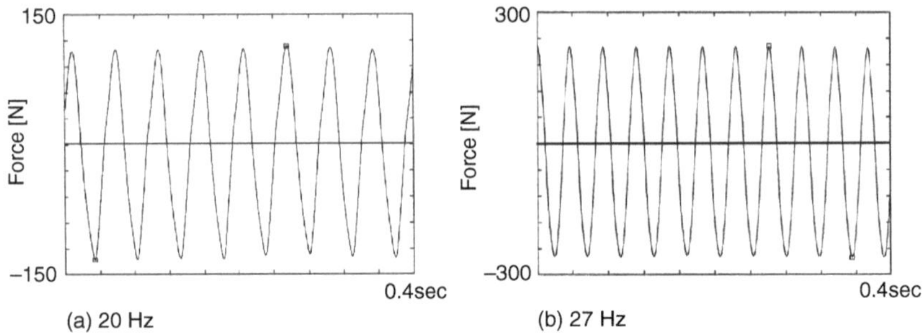

Figure 5 Measured centrifugal force generated by the eccentric mass.

eccentric mass and (b) is the vibration with the eccentric mass. The increase in the amplitude of vibration with the unbalanced mass is about 11.3 μm. This increase in amplitude corresponds to the dynamic displacement e of the rotor by the centrifugal force.

The radial clearance C_r of the bearings is 20 μm. One of the important parameters to express the operational condition of bearings is an index called the eccentricity ratio ε, defined by the ratio of rotor displacement e from the bearing center to the radial clearance C_r, i.e., $\varepsilon = e/C_r$. The eccentricity ratio of the radial air bearings at a rotational frequency of 20 Hz with the eccentric mass corresponds $\varepsilon = 11.3/20 = 0.57$. As $\varepsilon = 1$ means a touch down of the bearings, there is enough margin to avoid this. The possibility of supporting an eccentric mass rotor with an air bearing is verified.

Figure 5 shows the time history of the measured centrifugal force generated by the unbalanced rotor. Figure 5(a) represents a rotational frequency of 20 Hz and (b) represents a rotational frequency of 27 Hz. The calculated centrifugal force generated by the unbalanced rotor is 187 N at a rotational frequency of 20 Hz and 342 N at a rotational frequency of 27 Hz. The measured centrifugal forces are 118 N at a rotational frequency of 20 Hz and 232 N at 27 Hz. The reason for these differences can be explained by the setting method of the load cell. It is set under the test rig at the bottom of the driving turbine side radial bearing. So that the sharing rate of the load by the load cell is estimated as 66 percent of the total force. The centrifugal forces measured by the load cell should be corrected by the load sharing rate. The corrected values are 123 N (=187 N × 0.66) at a rotational frequency of 20 Hz and 226 N (=342 N × 0.66) at 27 Hz. These values agree well with the measured values.

5. Test Trial of Air Bearings for Generation of 2000 N

On the basis of the preliminary test, a 2000 N class trial ACROSS transmitter using air bearings was made for test operation. As shown in

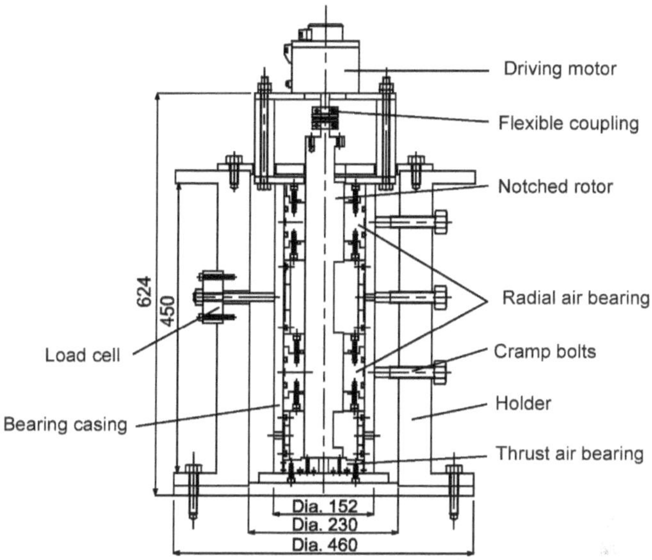

Figure 6 Cross section of the 2000 N class test trial rig using air bearings to support a notched rotor with $M_1 = 2.12 \times 10^{-2}$ kg m.

Figure 6, the transmitter is composed of two main parts, i.e., the part for the generation of the centrifugal force and the casing part. The force generation part is composed of a notched rotor, a driving motor, two radial air bearings and a thrust air bearing. The casing part is composed of a holder and clamp bolts.

The spin axis of the notched rotor is positioned vertically and driven by a variable frequency motor connected through a flexible coupler. The notched rotor is made of stainless steel (in JIS SUS420) and its diameter, length and mass are 60 mm, 520 mm and 12 kg respectively. The radial bearings and spacers are fitted in the bearing casing with a clearance of 0.1 mm. Gum O-rings are used to prevent air leaks. To align the centers of the two radial bearings and to fix them in the bearing casing, they are pressed against the bearing casing by bolts from the outside. The bearing casing is set at the center of a holder and clamped from the outside of the holder in three right angle directions with clamp bolts and in one direction with a strain gauge type load cell.

The main specifications of the air bearings are listed in Table 1. The radial clearance C_r is set to be 30 μm and an allowable eccentricity ratio of 0.4 at the rotational frequency of 50 Hz was chosen for safety operation. The centrifugal force generated by the notched rotor is supported by two radial bearings. Slot restricted double admission type radial bearings, as shown in Figure 7(a), (Togo, 2002) are adopted to obtain a high load capacity and stable operation. The inherent orifice type is adopted for the thrust

Table 1 Specifications of the air bearings

	Radial	Thrust	Note
Dimensions (mm)	$D = 60$ $L = 120$ $C_r = 0.03$ $h_{sl} = 0.015[0.022]$	$r_o = 29.5$ $r_g = 23$ $r_i = 7.5$	
Load capacity (N)	1000 [890] × 2 at $\varepsilon = 0.4$	120 [120] at $h = 30$ μm	$p_s = 0.69$ MPa (gauge)
Friction loss (W)	1.3 × 2	0.06	at $f = 50$ Hz
	Total 2.72		
Air flow rate (liter/min)	76 [173] × 2	28 [28]	
	Total 180 [374]		

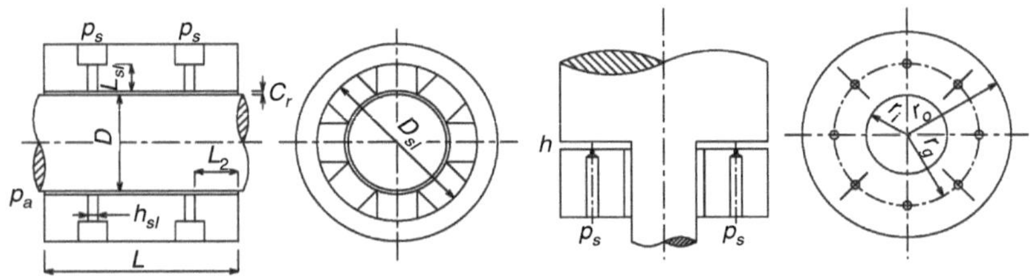

(a) Slot restriction type double admission radial bearing

(b) Inherent orifice type circular thrust bearing

Figure 7 Construction of air bearings for the test rig.

bearing because of its simple configuration, as shown in Figure 7(b). In Table 1, values without brackets are designed values and the values in brackets are actual values. Unfortunately, as the actual slot clearance h_{sl} differed from the design value, the air flow rate of the radial bearings increased and the load capacity decreased from the target values. In hydrostatic air bearings, frictional loss depends mainly on the viscosity of air and the radial clearance C_r. Though the loss changes proportionally to the square of the rotational frequency f, it does not change the centrifugal force F. Figure 8 shows a photograph of the outside view of the 2000 N class trial ACROSS transmitter using air bearings and Figure 9 show its main parts.

6. Workout of the Trial Air Bearings

A testing system for the 2000 N class trial ACROSS transmitter using air bearings is shown in Figure 10. The notched rotor is driven by a variable

Figure 8 Photograph of the 2000 N class air bearing test rig.

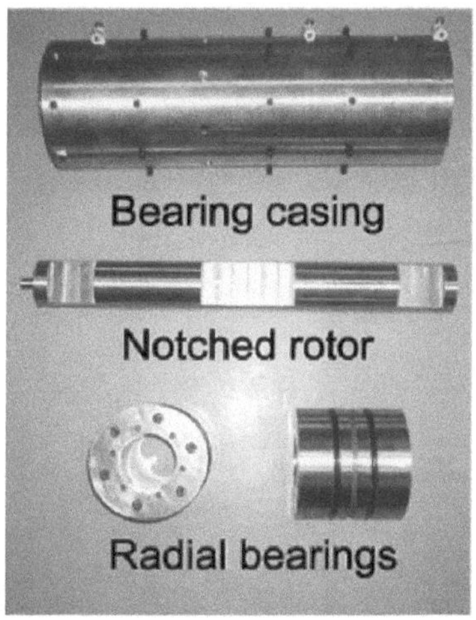

Figure 9 Main parts of the 2000 N class air bearing test rig.

frequency motor connected with a flexible coupling. A digital power meter is connected to the motor cable to measure the driving power.

The amplitude of the rotor vibration is measured at the upper part of the rotor by an eddy current type displacement sensor. The centrifugal force generated by the notched rotor is measured by a strain gauge type load cell clamped to the holder. As mentioned above, the bearing casing is clamped to the holder with bolts, the force is transmitted from the bearing casing to the load cell through the bolts. Then, the outputs of the load cell are

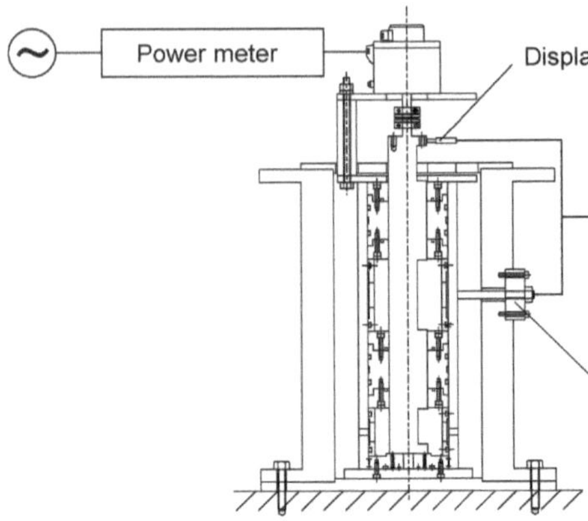

Figure 10 Cross section of the 2000 N class trial test rig using air bearings to support a notched rotor with $M_1 = 2.12 \times 10^{-2}$kg m.

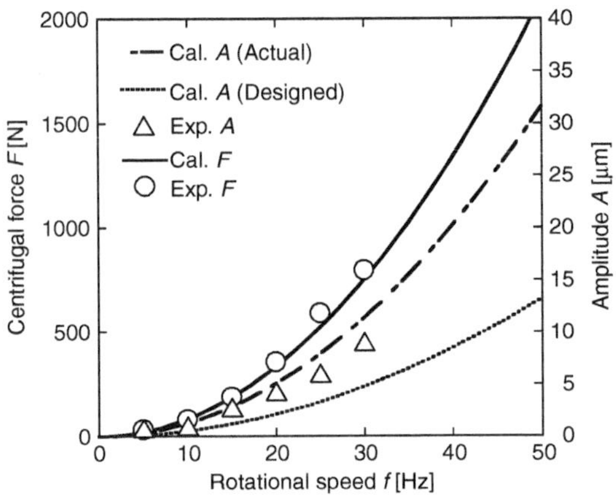

Figure 11 Rotational frequency versus the generated force and the amplitude of the shaft vibration.

calibrated to the load by every assembly of the transmitter. Outputs of the displacement sensor and the load cell are analyzed by FFT. In the test, the maximum rotational frequency is limited to 30 Hz because of the vibrations of the test stand and the floor. The air supply pressure to the bearings is limited to 0.59 MPa (gauge) by the capacity of the air compressor.

Figure 11 shows the relationship of the generated force and the vibration amplitude of the notched rotor to the rotational frequency. In Figure 11, the solid line represents the calculated force and points plotted by open circles are the measured values. The measured force agrees well with the calculated one. In the same figure, points plotted by triangles represent the measured amplitude of the vibration of the notched rotor. The dash-dotted line represents the theoretical one calculated from the load capacity of the

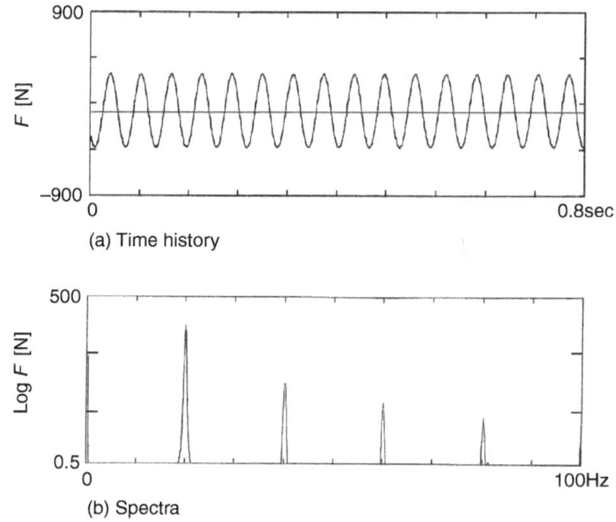

Figure 12 Time history and spectrum of the centrifugal force generated by the notched rotor (20 Hz).

radial bearings with actual dimensions and the dotted line is with design dimensions. The measured values of the vibration amplitude overrun the design values.

The measured amplitude of the notched rotor at a rotational frequency of 30 Hz reaches 9 μm. At this frequency, the generated force is about 750 N. The force is supported with two radial bearings, one radial bearing shares the load of 375 N. When the stiffness of the radial bearing is assumed to be constant and the load capacity is linear to the rotor displacement, the amplitude of the rotor vibration will reach 24 μm with a centrifugal force of 2000 N at a rotational frequency of 50 Hz. As the radial clearance is 30 μm, the maximum eccentricity ratio will reach about 0.8. This value is much higher than the allowable target eccentricity ratio of $\varepsilon = 0.4$. This discrepancy in the eccentricity ratio of the bearings is caused by manufacturing errors in the dimension of the slot clearance.

The measured time history of the centrifugal force and its spectrum at a rotational frequency of 20 Hz are shown in Figure 12. Whereas the generated force appears as a pure sinusoidal form, as in Figure 12(a), it contains a higher order vibration, as revealed by spectrum in Figure 12(b). This may be caused by a non-linear effect on the rotor motion due to a rocking vibration of the transmitter body and also to a nonlinear response of the load cell. To avoid the higher order vibration, the transmitter body should be firmly fixed to a more rigid body with a larger mass, the Earth itself.

Table 2 shows the various factors measured by the present experiments and those estimated on the basis of various backgrounds. The values in brackets of rotational frequency represent the theoretical centrifugal force.

Table 2 Driving power and losses of the 2000 N class trial air bearings

Rotational frequency (Force in N)	Motor input (Measured)	Motor loss (Measured)	Loss of gas bearings ($2L_r + L_t$) (Calculated)	Windage loss L_A (Calculated)	Loss of other elements (Calculated)	Loss of rolling contact bearings L_B (Estimated)
[Hz]				[W]		
5 (21)	5.5	5.1	0.030	0.004	0.366	0.523
10 (83)	7.3	7.0	0.120	0.037	0.143	2.092
15 (188)	10.4	9.4	0.271	0.126	0.603	4.708
20 (334)	13.0	11.5	0.481	0.299	0.720	8.369
25 (522)	16.3	14.0	0.752	0.584	0.964	13.076
30 (753)	21.4	16.9	1.083	1.009	2.408	18.830
35 (1025)	26.2	19.8	1.470	1.602	3.328	25.630
40 (1338)	31.8	22.9	1.920	2.392	4.588	33.476
45 (1694)	38.0	26.3	2.430	3.406	5.864	42.368
50 (2091)	44.8	29.8	3.000	4.672	7.328	52.306

The driving power (energy loss rate in Watt) is measured electrically from the energy consumption of the motor with and without the rotor at several rotational frequencies. The difference of the two is attributed to the sum of bearing loss, windage loss, other mechanical elements in the transmitter, and also the radiated kinetic energy of vibration towards the outside. As shown in Table 2, the largest fraction of energy consumption is in the motor itself, and bearing loss + windage loss is really a small fraction as estimated by the following formulas for the radial bearing L_r and the thrust bearing L_t:

$$L_r = \frac{1}{\sqrt{1-\varepsilon^2}} \frac{\eta\pi^3 D^3 L f^2}{C_r} \tag{2}$$

$$L_t = \frac{\eta\pi^3 f^2 (r_o^4 - r_i^4)}{h} \tag{3}$$

where η is the viscosity coefficient. The windage loss of the notched rotor L_A is calculated by following equation (Yoshizawa, 1974):

$$L_A = C_M \rho r^4 L (2\pi f)^3 \tag{4}$$

where ρ is the gas density of the atmosphere and C_M (=$4.07/N$) is a drag coefficient with $N = 3$ given for the notched shape, as determined experimentally by (Yoshizawa, 1974). If we are to support the force by a rolling contact bearing with the same diameter of 60 mm and a frictional coefficient $\mu = 0.0025$, then the loss of the rolling contact bearings L_B can be calculated from the following equation:

$$L_B = M_1 \mu r (2\pi f)^3 \tag{5}$$

since M_1 of the rotor is 2.12×10^{-2} kg m and $r = 30$ mm, L_B is estimated as given in Table 2, in which the estimated values are given from 30 to 50 Hz. The energy loss from the roller bearing is extremely large at higher frequencies, in particular, in comparison with that of gas bearings. The use of gas bearings is certainly effective to support an eccentric rotor with the least energy loss. The reason for this is twofold: not only is the friction associated with mechanical rollers larger than that of a gas, but the loss is proportional to the frequency cubed for a roller bearing, in contrast to frequency squared in a gas bearing.

7. Conceptual Design of a 2×10^5 N Class ACROSS Transmitter Using Air Bearings

The transmitter (FIT-1) now in operation since 1996 at the Tono transmitting station has a rotor of eccentric moment 8.0 kg m and the

Table 3 Specifications of air bearings for the 2×10^5 N class ACROSS transmitter

	Radial/1 bearing	Thrust	Note
Dimensions (mm)	$D = 400$ $L = 1000$ $C_r = 0.07$	$r_o = 200$ $h = 0.07$	
Load capacity (N)	0.6×10^5 ($\varepsilon = 0.4$ at 19.2537 Hz)	1.0×10^4	$p_s = 0.69$ MPa (gauge)
Friction loss (kW)	0.206 Total 0.428	0.016	
Flow rate (liter/min)	911.4 Total 1990.2	167.6	
Air compressor (Air flow rate-power)	1990.2 liter/min—13.59 kW		Screw type Including air dryer power
Total power (kW)	14.018		

generated force is 2×10^5 N at 25 Hz. The energy balance has been measured carefully under the operational conditions to transmit elastic waves of frequency modulation within a range between 10.2537 Hz and 19.2537 Hz at a precise carrier frequency of 13.0050 Hz. The average force generated is 0.54×10^5 N, nearly one quarter of the force sustainable by the bearings for the expected life time of 1 year. The total power consumption of the transmitter is 5.59 kW in this situation. The energy loss directly related to friction at the bearings amounts to 3.28 kW. Additional energy, not evaluated well, is expended in circulating the coolant and other peripherals to extract the heat generated by friction.

On the basis of the technical experience and considerations presented above, a conceptual design was made for the equivalent transmitter with gas bearings. The eccentric rotor was designed to be slim so as to suppress the rotational moment of inertia. The diameter of the notched rotor with an eccentric moment of 8.0 kg m was set to be 400 mm, and the length was then determined as 1000 mm to generate a force of 2×10^5 N at 50 Hz, as shown in Figure 13. The estimated energy balance at the same operational condition as for FIT-1 is shown in Table 3. The total frictional loss 0.428 kW in Table 3 is substantially smaller than the case with roller bearings. However, it requires a tremendous flow rate of compressed air to consume the energy of 13.59 kW. This energy consumption is significantly larger, even in comparison with the use of roller bearings, as theoretically expected in advance. This is the result of the symmetric gas bearing having been designed

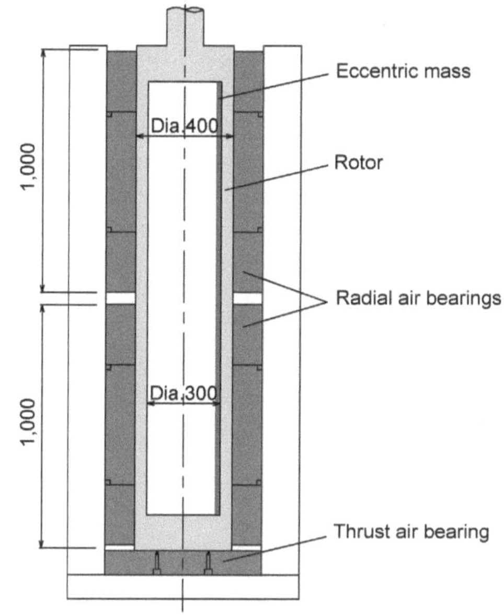

Figure 13 Schematic cross section of the 2×10^5 N class transmitter using air bearings.

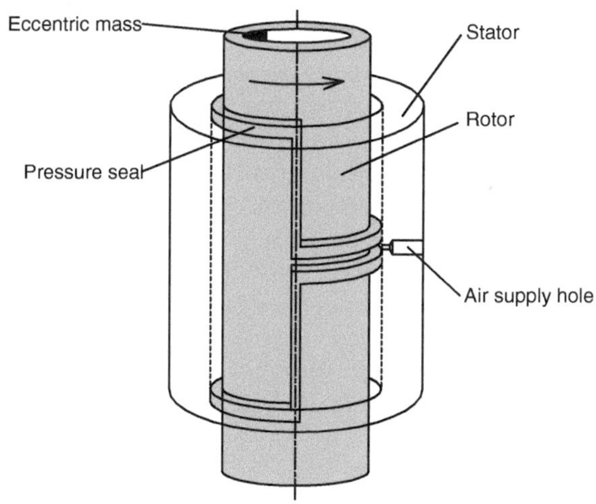

Figure 14 Conceptual design of the load supporting mechanism by a pressure seal.

for a balanced rotor sustaining only a limited level of radial force. However, heat generation within the transmitter is drastically suppressed and the technical problems arising from friction, involving possible flakes and wear, which could threaten the continuous and stable operation are all resolved. The remaining problem is the reduction of the supply rate of compressed air.

Next we propose a new design policy in which the gas consumption is minimized by an approach shown in Figure 14. Pressurized air is always supplied in rotation to a chamber surrounded by a pressure seal. The seal is constructed asymmetrically on the rotor surface, i.e., the eccentric mass side (loading side) is high in pressure and that of counter-loading part is subjected almost to open. Any possible variation of radial clearance is sealed by some flexible and lubricating element. A subordinate device to realize this design as a practical one is the feedback control of pressure to keep the force balanced so as to minimize the force and consequent abrasion of the seal elements. Further we propose the parallel use of roller bearings with substantially large rigidity, which works (as an insurance) to prevent the machine destruction only at a time when the gas control system fails or a significant ground acceleration (e.g., an earthquake) takes place. This design policy will be a practical one for a rotary type ACROSS transmitter supported by air bearings.

8. Conclusion

The possibility of substituting a hydrostatic air bearing for ball bearings was explored to support an eccentric mass rotor in order to solve the tribological problems of the current ACROSS transmitter. An eccentric rotor to sustain a 2000 N force at 50 Hz was designed using symmetric air bearings, and tested up to 30 Hz successfully. It was demonstrated that a large force was supported with very small frictional loss. The use of air bearings for the ACROSS transmitter was shown to be useful and advantageous in the suppression of heat generation inside the transmitter itself and the elimination of peripheral cooling devices. However, the traditional design of symmetric air bearings for a well balanced rotor is associated with a large air consumption, i.e., a large capacity air compressor is required. Next, we proposed a new approach to reduce the air consumption by the introduction of (1) a pressure seal with an air supply only in the direction of the generated force rotating with the rotor and also (2) feedback control of the gas pressure to keep a balance with the generated force. Our preliminary experiments on pressure control show that the recent digital controller is accurate enough to warrant the demanded control of gas pressure with a sufficiently short time constant. We claim, on the basis of this study, that an air bearing, appropriately designed, is the basic technology for rotary type ACROSS transmitters with stable and routine operation useful for active monitoring over years.

Acknowledgments

A number of colleagues have generously cooperated in this study. In particular, T. Higashidate, Y. Yamada, Y. Endo and K. Hikichi. We are grateful to Dr. M. Kumazawa for his discussions on the present work.

REFERENCES

Kumazawa, M., Kunitomo, T., Yokoyama, Y., Nakajima, T., Tsuruga, K., 2000. ACROSS: Theoretical and technical developments and prospect to future applications. JNC Technical Review 9, 115–129 (in Japanese with English abstract).

Kunitomo, T., Kumazawa, M., 2004a. Active Monitoring of the Earth's Structure by the Sesmic ACROSS -Transmitting and Receiving Technologies of the Seismic ACROSS. In: The Proceedings of 1st International Workshop on Active Monitoring in the Solid Earth Geophysics (IWAM04), Task Group for Active Monitoring, Mizunami, Japan, pp. 181–184.

Kunitomo, T., Kumazawa, M., 2004b. Active monitoring of the earth's structure by the sesmic ACROSS -development of the seismic ACROSS and continuous transmission at tono mine ACROSS station. In: The Proceedings of 1st International Workshop on Active Monitoring in the Solid Earth Geophysics (IWAM04), Task Group for Active Monitoring, Mizunami, Japan, pp. 218–222.

Togo, S., 2002. Gas Bearing Design Guide Book. Kyoritsu Shuppan Co., Ltd (in Japanese).

Yoshizawa, T., 1974. Reserch on drag of multi-faced mirrors at high speed rotation. Journal of the Japan Society of Precision Engineering 40 (2), 162–166 (in Japanese with English abstract).

Signal Processing and Accuracy Control in Active Monitoring

CHAPTER 12

Stacking Strategy for Acquisition of an ACROSS Transfer Function

Hiromichi Nagao[a,b], Takahiro Nakajima[a],
Mineo Kumazawa[a], *and* Takahiro Kunitomo[a]

Contents

1. Introduction

A stacking technique is important for active monitoring in order to enhance the signal-to-noise ratio (SNR) of observed data. However, the expected SNR cannot be obtained when a simple stacking method, which averages whole data, is applied to a noisy data set. Typical sources of such contamination are local earthquakes in the case of seismic data, passing electric trains near the observation sites in the case of electromagnetic data, and daily and weekly variations due to human activity in highly populated regions in both cases. The truncation of data in the noisy time period is often used to avoid such reduction of the SNR. For example, only data obtained at nighttime are used in the magnetotelluric (MT) or controlled signal magnetotelluric (CSMT) method. However, the needed information is included even in the noisy data, and we are trying to utilize them all. Some

[a] Tono Geoscience Center, Japan Atomic Energy Agency, 1-63, Akeyo-cho, Yamanouchi, Mizunami, Gifu 509-6132, Japan
[b] The Institute of Statistical Mathematics, 10-3, Midori-cho, Tachikawa, Tokyo 190-8562, Japan

E-mail address: hnagao@ism.ac.jp (H. Nagao).
Handbook of Geophysical Exploration: Seismic Exploration, Volume 40
ISSN 0950-1401, DOI: 10.1016/S0950-1401(10)04017-6

weighted stacking methods have been proposed so far to make good use of the noisy data (e.g., Robinson (1970), Anderson and McMechan (1990)), in which the optimum weight is set to be in reciprocal proportion to the square of the noise level. We remind the reader that there is a significant problem in the weighted stacking method; the difficulty is in the reliable evaluation of noise level in the data, which is necessary to determine the optimum weight. Here we propose the optimum weighted stacking method (OWSM) for data obtained by the Accurately Controlled Routinely Operated Signal System (ACROSS) with a reasonable method for noise level evaluation (Kumazawa, 1998). An essential point of the OWSM is that the SNR would be enhanced by data stacking in reciprocal proportion to the square root of the time period of observation. The OWSM enables us to obtain a better SNR from observed ACROSS data even if the noise level changes with time and frequency, assuming that the temporal variation of the signal is small. We show an application of the OWSM to real seismic and electromagnetic data obtained at our observation test site of ACROSS.

2. Methodology

2.1. Properties of signals and noise in ACROSS data

We give a model of an observed ACROSS data sequence assuming that properties of signal and noise are included in the data and that the optimum method of data stacking is derived to attain a maximum SNR. A discrete time series of the ACROSS data in a segment $x(j)$ ($j = 1, 2, \ldots, N_s$; N_s is the number of data in the segment) acquired continuously with a sampling time interval Δt is given by a sum of periodic signals $s(j)$ and observation noise $n(j)$ at the discrete time $j\Delta t$, i.e.,

$$x(j) = s(j) + n(j). \tag{1}$$

The time length T_s of this segment is an integer multiple of this period. The noise may be an ideally incoherent and stationary random sequence with mean zero and mean power $\sigma^2(j)$ at one data point, i.e.,

$$E(n(j)) = 0, \tag{2}$$

$$E(n(j - j')n(j - j'')) = \sigma^2(j)\delta_{j'j''}, \tag{3}$$

where $\delta_{j'j''}$ is the Kronecker delta, i.e., $\delta_{j'j''} = 0$ if $j' \neq j''$ and $\delta_{j'j''} = 1$ if $j' = j''$. The noise level $\sigma(j)$ is usually larger than the signal amplitude $|s(j)|$, and an appropriate method to eliminate the noise is essential for effective extraction of the signal components. Whereas the noise model described as Eqs. (2) and (3) is simple to deal with, the actual noise usually contains some coherent but non-stationary components. Such quasi-coherent components can be removed by some means, and we, therefore, consider that the noise is practically regarded as incoherent, as specified by Eq. (3), after the removal of quasi-coherent noise components. The remaining fraction of noise is significantly non-stationary, in particular, in highly populated regions and according to our experience the daytime noise level is much higher than at night because of human activities. Furthermore, there is sporadic noise such as local earthquakes and passing electric trains. The appropriate evaluation of noise level and its time variation is necessary in data processing to extract the signal with higher reliability. The data acquisition system for ACROSS has been devised accordingly to make such evaluation possible.

The Fourier transform of $x(j)$ shows a set of spectral lines at the frequencies

$$f_k = k\Delta f_s \quad (k = 0, 1, 2, \ldots, N_s/2) \tag{4}$$

from 0 up to the Nyquist frequency, with a frequency spacing $\Delta f_s = 1/N_s\Delta t$. Since the ACROSS signal has an accurate fundamental period and the time length of a segment is an integer multiple of the fundamental period, the discrete frequency series of spectra (complex numbers) at the frequencies in Eq. (4) have the form

$$X(k_0) = S(k_0) + N(k_0), \tag{5}$$

$$X(k_1) = 0 + N(k_1), \tag{6}$$

where we call k_0 the "signal channel", which accommodates the ACROSS signal frequency, and k_1 the "noise channel", which is a frequency other than the signal channels. We define the discrete frequency sequences of noise power

$$P(k_0) = N(k_0)^* N(k_0) = (X(k_0) - S(k_0))^*(X(k_0) - S(k_0)) \tag{7}$$

$$P(k_1) = N(k_1)^* N(k_1) = X(k_1)^* X(k_1) \tag{8}$$

where $*$ denotes a conjugate complex number. An estimation of $P(k_0)$ from k_0 only is impossible because the signal channel contains not just the noise, but also the signal, while $P(k_1)$ can be obtained directly from data. One of the features in the design of the ACROSS signal is that the noise power in a signal channel $P(k_0)$ can be estimated quantitatively from the noise power in noise channels $P(k_1)$. An essential point is that the respective values of $P(k_1)$ involve a statistical fluctuation. What we would like to have is not the actual value but the expectation of $P(k_0)$, the square root of which is identified with the estimated error of $S(k_0)$ due to noise at the signal channel.

The problem arising here is the method of estimating the expectation of $P(k_0)$ from $P(k_1)$ on the basis of an appropriate model of noise. The random noise (or slightly colored noise) shows a smooth continuous function of frequency, so the appropriate model for $P(k_1)$ appears to be low order polynomials as a function of frequency. In actual cases, however, the noise level depends on frequency in more complicated ways (probably due to the presence of weak unknown resonators amplifying or absorbing the random noise). In order to treat these cases, local smoothing and interpolation appear to be better. There are several methods for smoothing and interpolating such discrete sequence with missing elements such as $P(k_0)$. The simplest filter for this purpose is a moving average of the noise power in discrete noise channels from $k - k'$ to $k + k'$ defined by

$$P'(k_0) = \frac{1}{K'} \sum_{j=-k'}^{k'}{}' P(k_0 + j), \tag{9}$$

where $\sum'$ denotes the summation of the noise channels only, and K' is the number of noise channels used for the estimation. Note that the estimated value $P'(k_0)$ for $P(k_0)$ can be obtained by this filtering. Then the uncertainty σ_s^2 of $S(k_0)$ can be expressed with $P'(k_0)$, i.e.,

$$\sigma_s^2 = \sqrt{P'(k_0)}. \tag{10}$$

Further the uncertainty $\Delta\sigma_s^2$ of σ_s^2 originating from the statistical fluctuation is theoretically evaluated by

$$\Delta\sigma_s^2 = \frac{\sigma_s^2}{2\sqrt{K'}}. \tag{11}$$

2.2. Derivation of optimum weight

Following the considerations in the previous subsection, the probability density function (PDF) of spectrum $X(k_0)$ (complex number) at a signal channel k_0 in one time segment is represented by a normal distribution with variance σ_s^2, estimated by Eq. (10),

$$p(X(k_0)) = \left(\frac{1}{\sqrt{2\pi\sigma_s^2}}\right)^2 \exp\left[-\frac{(X(k_0) - S(k_0))^*(X(k_0) - S(k_0))}{2\sigma_s^2}\right]. \tag{12}$$

Note that this PDF is a two-dimensional normal distribution function with zero covariance between the real and imaginary parts. The time segment T_s in the current relevant ACROSS work is usually of the order of 10 to 100 seconds. If we consider this situation, the effect of noise level on the uncertainty is given at each of T_s, which is short enough to take a good account of the temporal variation of noise.

Suppose that we have a sequence of time segments and obtain a set of data, $X(s; k_0)$ and $\sigma^2(s)$ $(s = 1, 2, \ldots, M)$, where s denotes the segment number. The likelihood function for a set of observed data for M segments is given by

$$L = \prod_{s=1}^{M} p_s(X(s; k_0)). \tag{13}$$

Assuming that the data set ($X(s; k_0)$ and $\sigma^2(s)$) is the sampled data from a uniform group, we maximize Eq. (13) to obtain the most likely value of $S(k_0)$ as given by

$$S_{uM}(k_0) = \sum_{s=1}^{M} w(s) X(s; k_0), \tag{14}$$

$$\sigma_{uM}^2(k_0) = 1 \Bigg/ \sum_{s=1}^{M} \frac{1}{\sigma^2(s)}, \tag{15}$$

$$w(s) = \frac{1}{\sigma^2(s)} \Bigg/ \sum_{s'=1}^{M} \frac{1}{\sigma^2(s')} \tag{16}$$

where $S_{uM}(k_0)$ is an averaged data and $\sigma_{uM}^2(k_0)$ is the error level of $S_{uM}(k_0)$ obtained by an optimum weighted stacking of M segments. Finally, the SNR of the stacked data is evaluated by the formula

$$\mathrm{SNR}_{uM}(k_0) = \sqrt{S_{uM}(k_0)^* S_{uM}(k_0)} / \sigma_{uM}(k_0). \tag{17}$$

Eqs. (14)–(17) mean that the optimum estimation of the signal level is given by an optimum weighted stacking method (OWSM), and the optimum weight for each signal channel in each segment is in reciprocal proportion to the noise power. This weight is designed to be $0 < w(s) < 1$, providing that the SNR of stacked data always increases, even if any segment having a large noise level is added to the data set. Suppose there is a time segment, in which, due to some reasons, no data was acquired or an extreme outlier took place to give a significantly large value for $\sigma^2(s)$. Such bad data is automatically set to be least-weighted to provide no influence within the whole data according to Eq. (16).

2.3. Successive stacking for larger SNR

Continuous observation leads to the successive accumulation of data, and this situation provides us with an uncertainty of data describing the stationary target of observation which increases normally in reciprocal proportion to the square root of time. We can detect a temporal variation of the target when the temporal variation exceeds the uncertainty of the data. There is a trade-off relation between the detectable minimum temporal variation of a target underground and the time resolution of the monitoring. In this sense, the quantitative evaluation of the uncertainty (Eq. (10)) and the temporal variation of these parameters play an important role in reliable monitoring. Consequently, a small temporal variation would be detectable after a certain period of continuous observation using ACROSS.

3. Application of Optimum Weighted Stacking Method

In this section, several numerical experiments of data stacking are made using ACROSS data. Usually, noise level shows a significant daily variation both in seismic and electromagnetic data; whereas the nighttime is quiet, the daytime from early morning to midnight is noisy due to human activities. The daytime is longer while noise level is larger, so an appropriate method of noise suppression is important. The results of the numerical experiments are compared for the different stacking methods; (1) the simple stacking method (SSM), which is an ordinary averaging of whole data, (2) the nighttime stacking method (NSM), in which SSM is applied only to the data at nighttime by discarding the noisy daytime data, and (3) the OWSM derived in the previous section.

A long-term ACROSS observation test started in October 2002 at Tono, Central Japan. The seismic ACROSS transmitter is located in Toki City, and the seismic ACROSS signal is received at Hi-net observatories operated by the National Research Institute for Earth Science and Disaster Prevention, Japan. The ACROSS signal is designed as a frequency-modulated (FM) signal changing temporally with a period of 20 seconds, and in the 18.52 ± 0.50 Hz frequency range. Signal channels are produced in every 0.05 Hz, i.e., 18.52 Hz, 18.52 ± 0.05 Hz, 18.52 ± 0.10 Hz, ... (Kunitomo and Kumazawa, 2003). We select the vertical component obtained by the velocity seismometers at Yaotsu, Kasugai, Hiraya, Takagi, and Fujimi observatories. The location of the transmitter and receivers is shown in Figure 1. Figure 2(a)-(c) shows the spectral amplitudes of stacked data

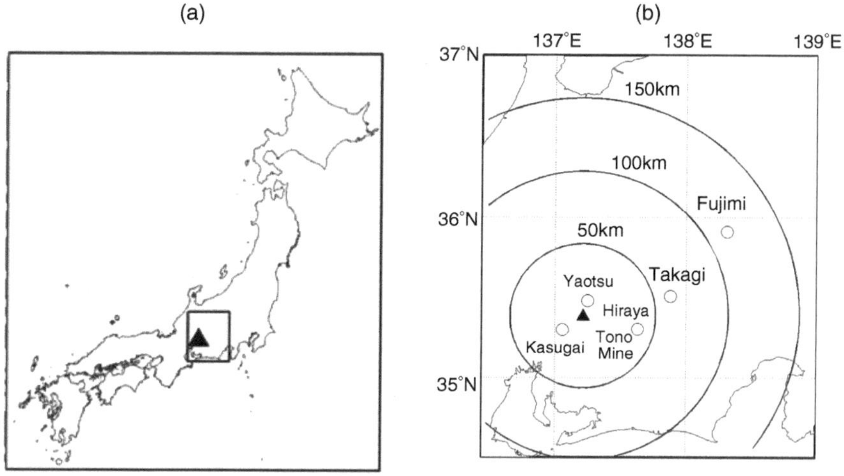

Figure 1 (a) Location of the Tono mine (solid triangle) and Tono area (square) in Japan. Both seismic and electromagnetic ACROSS transmitters are at the Tono mine. (b) Enlargement of the square box in the left panel, and the location of the Tono mine and five Hi-net observatories (circles) are shown. The distance to each observatory from the transmitter is 11.1, 17.0, 38.6, 61.7, and 115 km, respectively.

obtained by the three stacking methods raised above. The observation period of this seismic ACROSS test is 11 days. We adopt here $T_S = 100$ seconds as the time length of a segment and the sampling time is $\Delta t = 0.01$ second. Figure 2(a)-(c) confirms that the noise level obtained by the OWSM is clearly smaller than those obtained by other stacking methods for each observatory. The signal amplitudes are much larger than the noise levels obtained by the OWSM at the nearest four observatories.

Figure 2(d) shows the signal amplitude for each signal channel and noise level obtained by the three stacking methods overlapping the noise level obtained by the OWSM using frequency-dependent weights determined by the noise power estimated for each signal channel. According to this, the resultant noise level is not so improved in spite of the usage of weights that depend on the signal channel; it is found that the handling of temporally changing noise is more significant than that of frequency-dependent noise in this case. The signal amplitudes at some signal channels are larger than the noise level in Fujimi, whereas the signal amplitudes at most of signal channels are below the noise levels obtained by the SSM and NSM at Hiraya, Takagi, and Fujimi. That is, the seismic ACROSS signal is detected by the OWSM at Fujimi, >100 km from the transmitters, while it is not detected by other stacking methods. Figure 3 shows the time series of noise levels at one of the signal channels 18.37 Hz estimated by Eq. (10). A clear daily variation and sudden large noise related to local earthquakes are found. Figure 4 shows

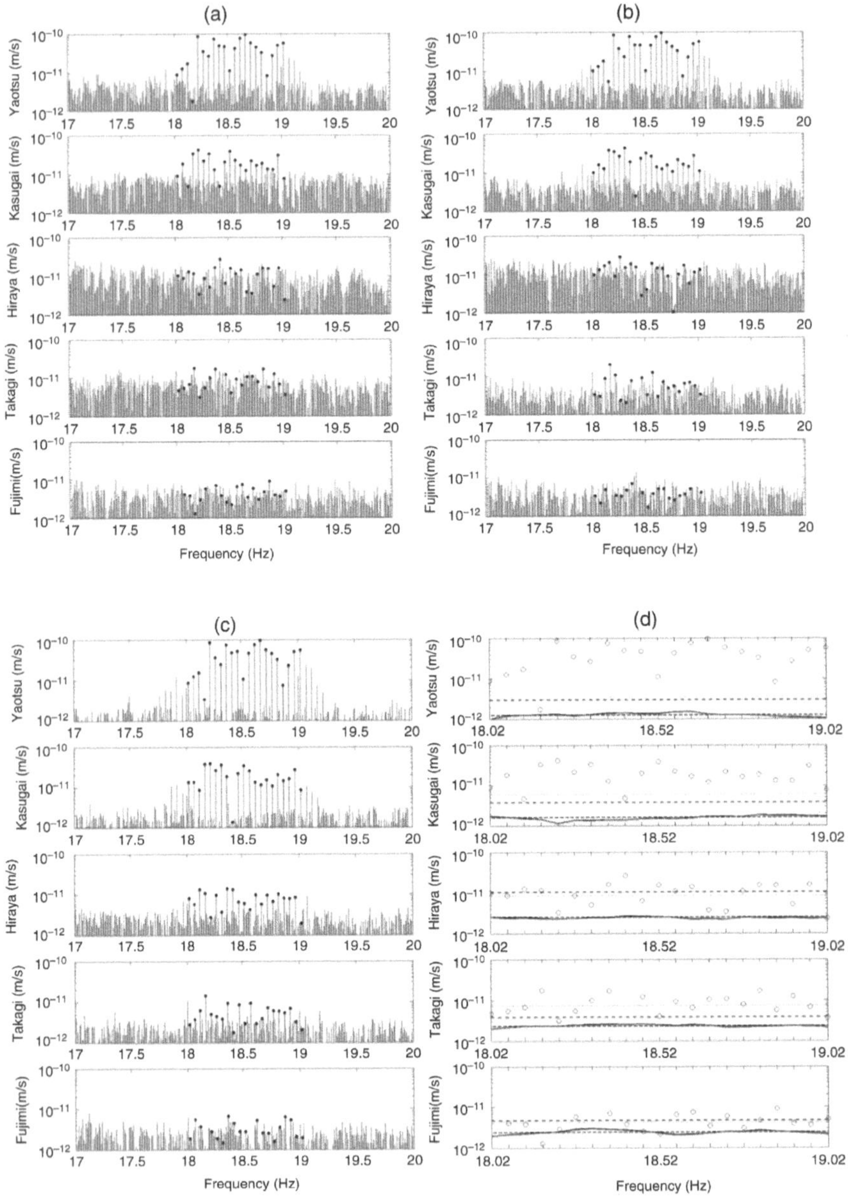

Figure 2 The stacked amplitude spectrum obtained by applying (a) SSM, (b) NSM, and (c) OWSM to the seismic data (the vertical component of the velocity seismometer) at each Hi-net observatory. The solid circles indicate the signal channels. (d) Signal and noise levels of the stacked seismic data. Each open circle indicates the signal level at each signal channel, and the dotted, dot-dashed, and dashed lines indicate the noise levels obtained by the SSM, NSM, and OWSM, respectively. The solid line indicates the noise level obtained by the OWSM using the frequency-dependent weights.

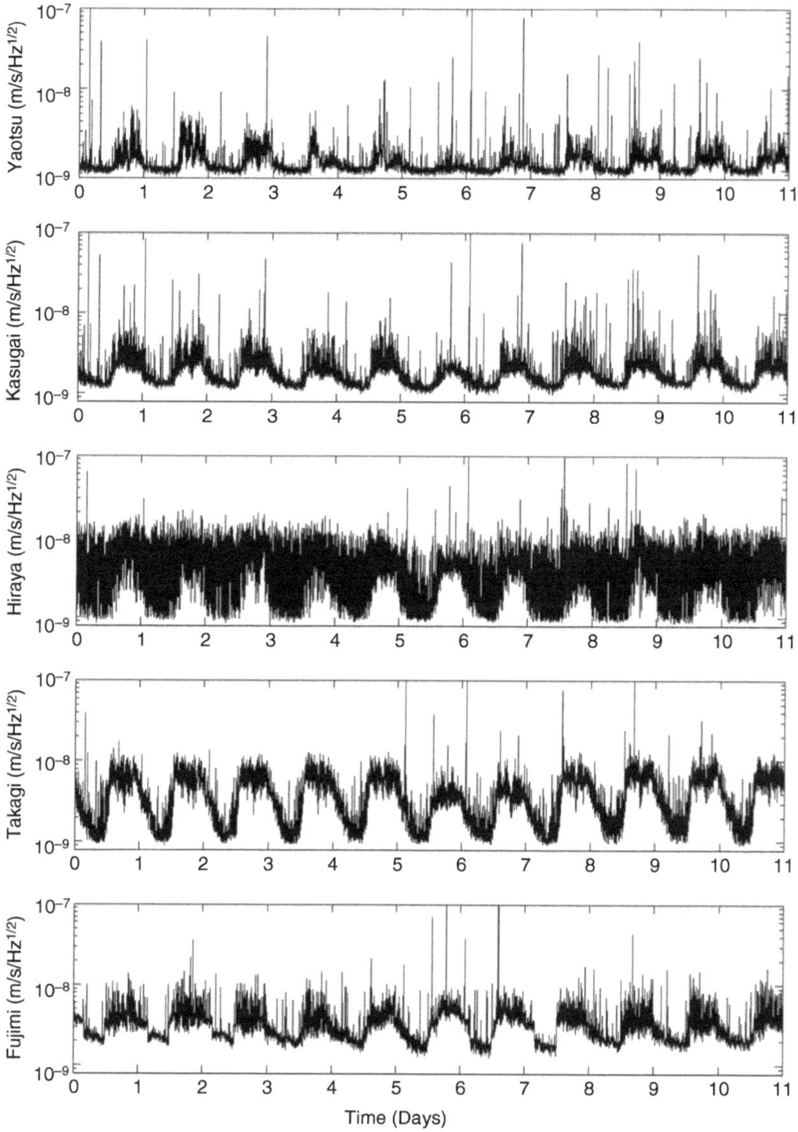

Figure 3 Time-dependent noise level at a signal channel of 18.37 Hz at each Hi-net observatory.

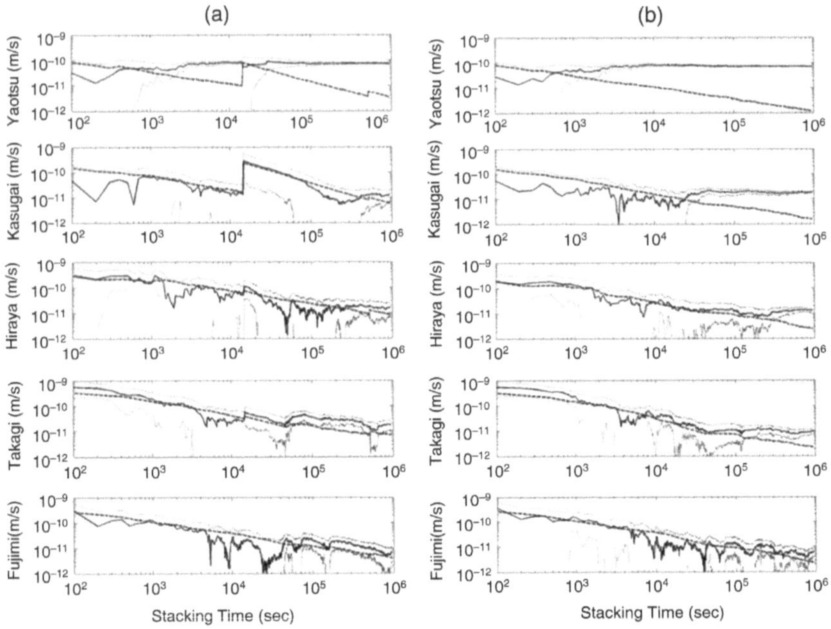

Figure 4 Variation of the signal level (solid line) and noise level (dashed line) with the stacking time at a signal channel of 18.37 Hz at each Hi-net observatory. (a) and (b) are obtained by the SSM and OWSM, respectively. The estimated error levels (dotted lines) are also indicated for both signal and noise levels.

the variation of the signal and noise levels at 18.37 Hz obtained by the SSM and OWSM, and Figure 5 shows the growth curve of the SNR at this signal channel. The SNR obtained by the SSM decreases suddenly when stacking segments of daytime and those including the earthquakes. It is worth noting that the SNR obtained by OWSM is not affected by these events, and is certainly larger than those obtained by the other stacking methods.

Figures 6–8 are the results obtained from electromagnetic ACROSS data. The source signal consists of three sinusoidal components with frequencies of 0.9375, 1.2500, and 1.5625 Hz, and the source dipole moment is 3.8 Am for each component. It is necessary to use very low frequencies ($\ll$1 Hz) when we explore the electrical conductivity distribution in the Earth's crust and mantle in the future with an electromagnetic controlled source. See Nakajima et al. (2003) for a detailed introduction of the electromagnetic ACROSS. A receiver is located at a distance of 7.7 km from the transmitter, and three components of the magnetic field and two components of the electric field are observed. The observation period is 11 days, the time length of a segment is set to $T_S = 400$ seconds, and the sampling time is $= 0.001$ second. Figure 6 shows the time series of noise level estimated at

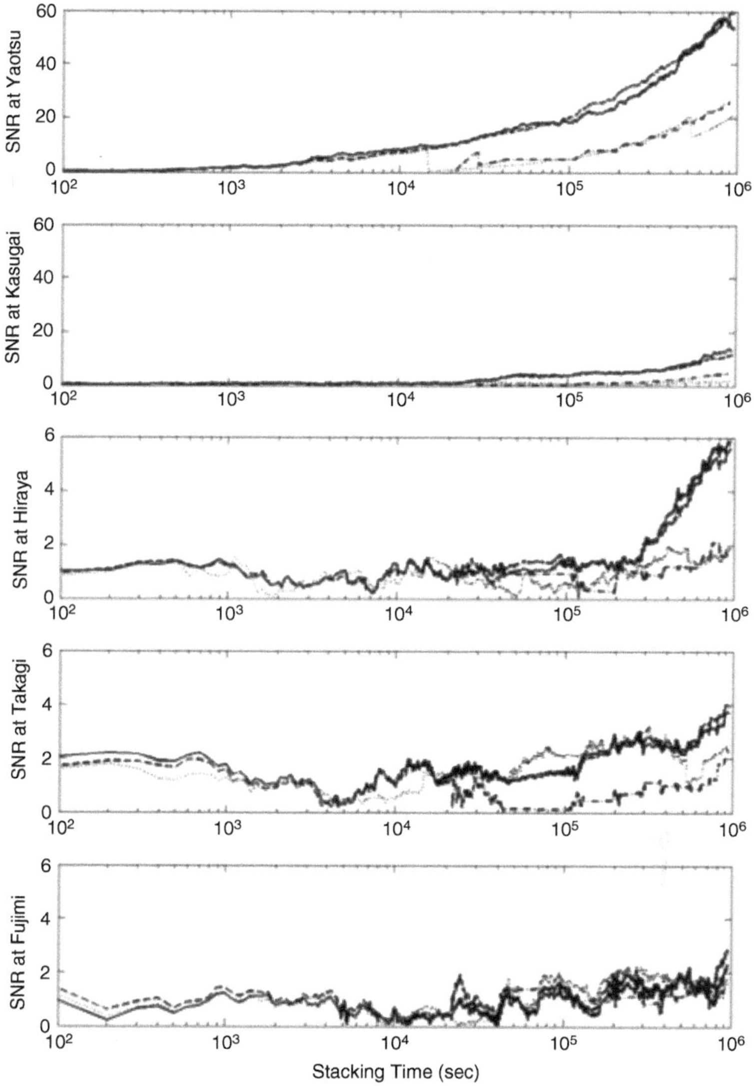

Figure 5 The growth curves of the SNR at the signal channel of 18.37 Hz obtained by the SSM (dotted line), NSM (dot-dashed line), OWSM using non-frequency-dependent weights (dashed line), and OWSM using frequency-dependent weights (solid line).

a signal channel of 1.2500 Hz using Eq. (10). The noise level at daytime is several times larger than that at nighttime, which is the same as the case of the seismic data. Figures 7 and 8, respectively, show the temporal variation of the signal and noise levels of stacked data, and the growth curve of the SNR at one of the signal channels, 1.2500 Hz. It is confirmed that the SNR

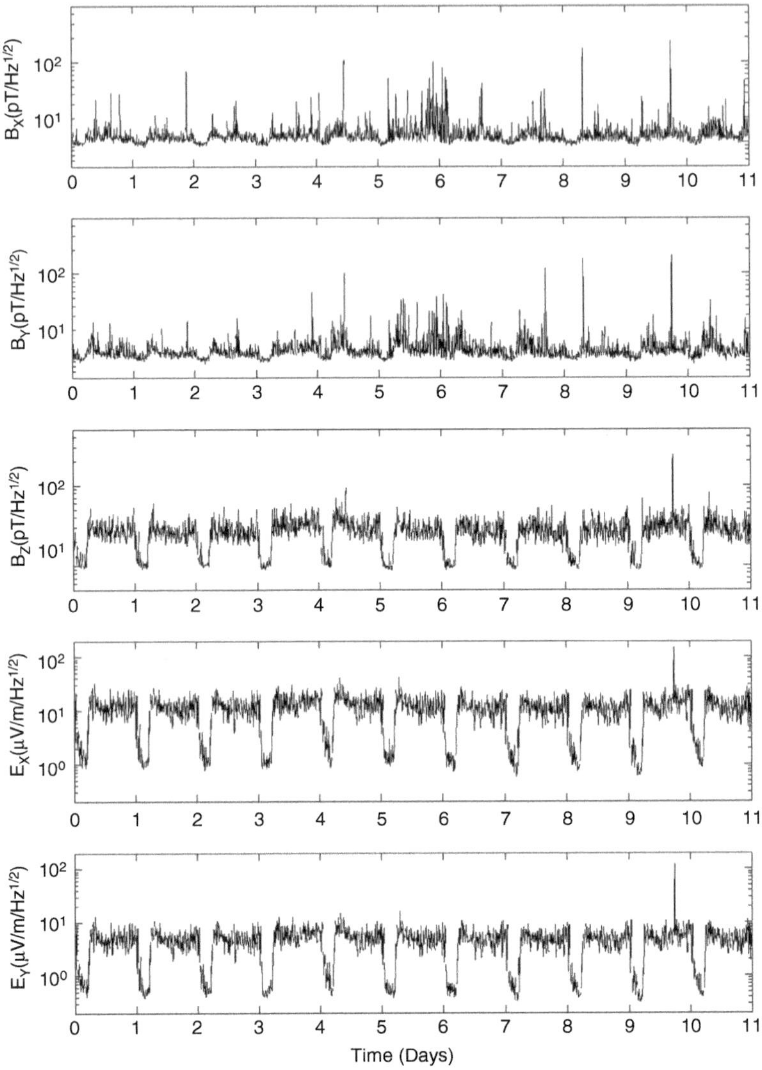

Figure 6 Time-varying noise level at a signal channel of 1.25 Hz in the electromagnetic ACROSS data observed at a receiver 7.7 km from the transmitter. The data are three components of the magnetic field (geomagnetic northward B_X, eastward B_Y, and downward B_Z) and two components of the electric field (geomagnetic northward E_X and eastward E_Y.)

obtained by the SSM or NSM does not exceed that obtained by the OWSM. However, there is a difference from the case of the seismic data; the SNR obtained by the OWSM is not so different from those obtained by the SSM for B_X and B_Y components. The reason for this is the noise level at daytime is

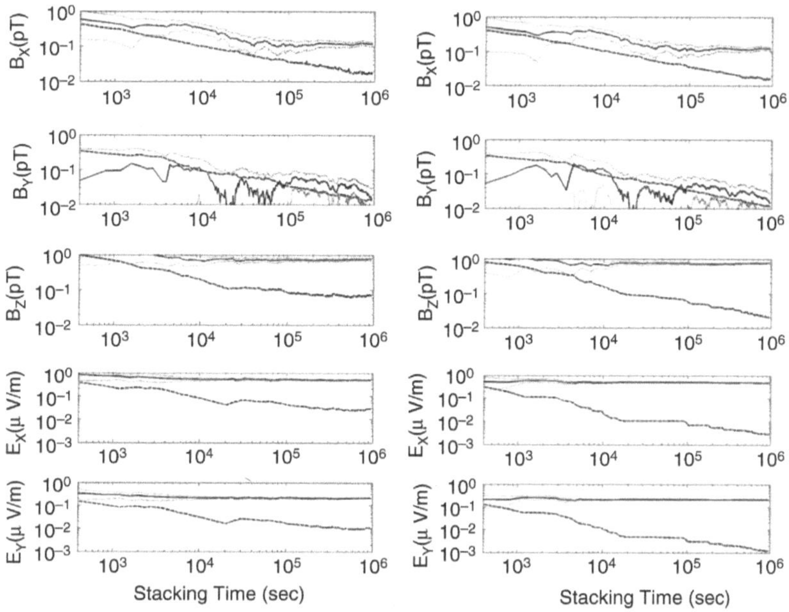

Figure 7 Variation of the signal level (solid line) and noise level (dashed line) with stacking time in the case of the signal channel of 1.25 Hz in the electromagnetic ACROSS data. (a) and (b) correspond to SSM and OWSM, respectively. The error level is indicated by a dotted line for each signal and noise level.

not so large in these components compared with that at nighttime, although the difference is remarkable in other components.

4. Conclusions

We showed that the best estimation method is the OWSM for the signal in continuously acquired ACROSS data, and the optimum weight is in reciprocal proportion to the noise power. It is necessary to estimate the noise level at each signal channel in order to determine the optimum weight, and we suggested a new method for the noise level estimation in a signal channel. The noise level is estimated from noise channels produced around the signal channel, and the optimum estimation is shown to be the mean of the noise power at the noise channels, as described in Eq. (9). The OWSM was applied to real ACROSS data, and the obtained SNR is confirmed to be larger than the cases of the other stacking methods used so far. The observation of ACROSS in daytime is as invaluable as in nighttime because the daytime data contribute to increase the SNR of a transfer function when using the OWSM.

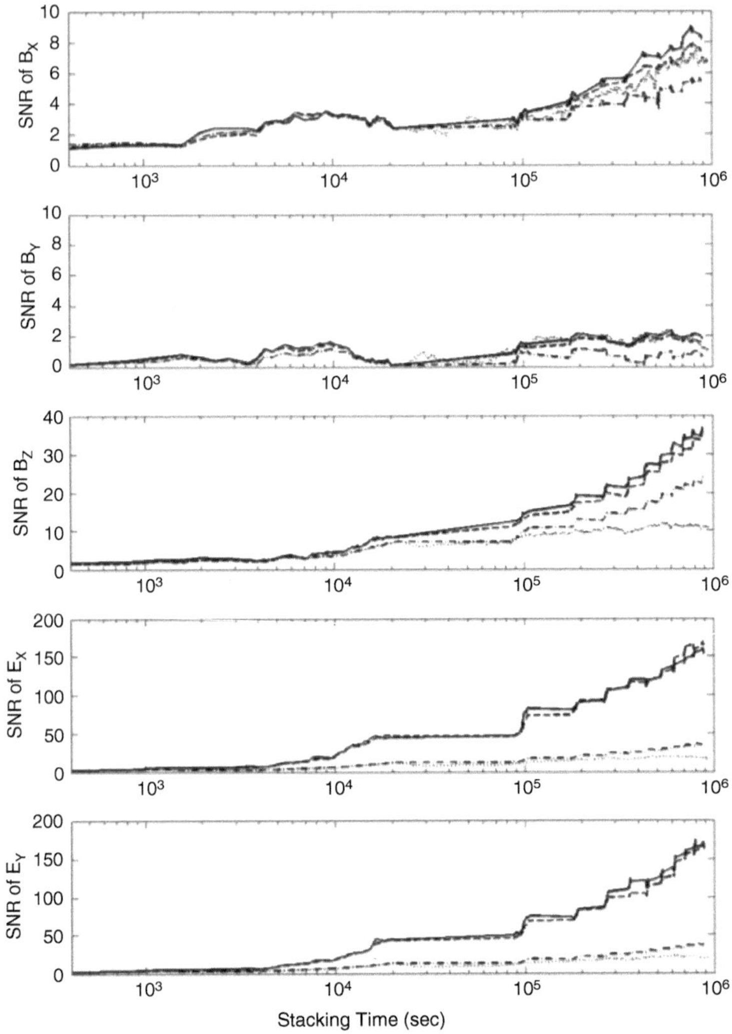

Figure 8 The growth curves of the SNR at a signal channel of 1.25 Hz of the stacked electromagnetic ACROSS data. The dotted, dot-dashed, dashed, and solid lines indicate SSM, NSM, OWSM using non-frequency-dependent weights, and OWSM using frequency-dependent weights, respectively.

ACKNOWLEDGMENTS

We used in this paper the seismic data obtained from Hi-net observatories operated by the National Research Institute for Earth Science and Disaster Prevention, Japan.

REFERENCES

Anderson, R.G., McMechan, G.A., 1990. Weighted stacking of seismic data using amplitude-decay rates and noise amplitudes. Geophysical Prospecting 38, 365–380.

Kumazawa, M., 1998. A new light and new eye to look into the solid Earth and a potential monitoring methodology of geodynamic states; Introduction to ACROSS, Abstracts of International Workshop on Frontiers in monitoring science and technology for earthquake environments, A1.

Kunitomo, T., Kumazawa, M., 2003. Optimum FM signal for monitoring the underground states by means of the seismic ACROSS, Abstracts of IUGG Meeting 2003, Sapporo.

Nakajima, T., Kunitomo, T., Kumazawa, M., Shigeta, N., Nagao, H., 2003, Observations of the transfer function using an electromagnetic sounding system, EM-ACROSS, Abstracts of IUGG Meeting 2003, Sapporo.

Robinson, J.C., 1970. Statistically optimal stacking of seismic data. Geophysics 35, 436–446.

CHAPTER 13

Secondary Calibration Method of Seismometers by Utilizing ACROSS Signal

Kayoko Tsuruga[a,d], Mineo Kumazawa[a,b], Takahiro Kunitomo[a,c], Naotaka Shigeta[a], *and* Junzo Kasahara[a,c,d]

Contents

1. Introduction

One of major tasks for active seismic monitoring is to detect temporal changes in the Earth's interior because the precursory phenomena in the seismogenic zones may be caused by any physical property change. In order to detect subtle changes, a precise description of the measurement system is mandatory. A seismometer is one of the most important parts of a measurement system, and calibration of seismometers is extremely important

[a] Tono Geoscience Center, Japan Atomic Energy Agency, Japan
[b] Graduate School of Environmental Study, Nagoya University, Japan
[c] Faculty of Science, Shizuoka University, Japan
[d] JGI, Inc., Japan

E-mail address: tsuruga.k@jgi.co.jp (K. Tsuruga).

Handbook of Geophysical Exploration: Seismic Exploration, Volume 40
ISSN 0950-1401, DOI: 10.1016/S0950-1401(10)04018-8

for the above objectives because a large number of reliable seismometers are necessary for the array observations.

"Calibration of a seismometer" is a determination of the frequency response for velocity seismometers or accelerometers to the input signal. The "primary calibration" is to determine the response characteristics of sensors referring to "primary standard units" such as length, time and weight specified by an international organization or the standard of units. The "secondary calibration" is to obtain the response characteristics of sensors by comparison with more reliable easy-to-use measurement tools and/or reference sensors which are already calibrated through the primary calibration.

The major conventional calibration methods of seismometers are to vibrate testing sensors on a vibration table directly by sinusoidal seismic waves or impulses, or to move a coil inside the sensor by an electric signal; this is the so called "indirect calibration method". The accuracy of the input signals made by conventional equipment, however, is not high enough and the uncertainty of a measurement is almost ambiguous because the noise factors accompanying the measurement are difficult to avoid, and also because the characteristics depend upon several factors such as temperature, minimum resolution, and self-mechanical and/or electrical noise levels etc. The cross-coupling sensitivity and nonlinearity of a sensor are not fully evaluated in this paper.

It is possible to estimate a highly reliable frequency response for several seismometers relatively simple and simultaneously, by utilizing the Accurately-Controlled Routinely-Operated Signal System (ACROSS) (Kumazawa et al., 2000) and a vibration table using reference accelerometers. One of the most important advantages of using ACROSS signals is that both the transmitted signal and receiving system are accurately controlled in phase (/frequency) and time by reference to a GPS clock (Kunitomo and Kumazawa, 2004a).

In the following, we describe the basic concept and analysis method for the calibration of seismometers using the ACROSS methodology.

2. Basic Concept and Method of Analysis

2.1. Basic concept of the current method

It is appropriate for the primary calibration of physical quantities (such as acceleration) to use a laser wavelength provided by a cesium clock and the velocity of light in an inertia system as the international primary standard of time and length. However such a primary calibration is not yet established or familiar to general users of seismometers because of the high costs and the shortage of primary standard facilities (Japan Machinery Federation and Japan Industrial Technology Association, 2003).

We, here, describe an indirect calibration method as a secondary calibration, noting that reference accelerometers, which are calibrated accurately by primary units, already exist. The method is based on the simultaneous determination of the frequency response characteristics for a number of seismometers placed on the same vibration table, which is a little different from the common usage. In our method, the "vibration table" is a plate exciting the same vibration by the external vibrator (here, the seismic ACROSS). The motion of the table should be determined by a set of feedback-type reference accelerometers.

2.2. Major features

The following are the major features of the current calibration method of seismometers using the ACROSS methodology (e.g., Kunitomo and Kumazawa (2004a,b)).

(1) An analysis method which can be properly used according to the required calibration accuracy is given.
(2) The vibration table has sufficiently high rigidity such that the effect of its elastic strain may become smaller than 10^{-4}-10^{-5}.
(3) Reference seismometers are installed surrounding the test seismometers on the vibration table with an azimuth error of 10^{-4} radian or less. They are also required having low-self noise
(4) The number of reference seismometers is required to be larger than the number of degrees for freedom of the kinematics model for the vibration table with high redundancy in order to estimate several unknown parameters. The required degrees for freedom is six for describing the model as a rigid body motion.
(5) A sufficient signal-to-noise ratio (i.e., S/N) can be obtained by stacking data according to environmental noise. In the experiment, S/N will be attained to be 10^3 to 10^4.
(6) An optimized algorithm which can check the internal consistency of the kinematics model of a vibration table is adopted.
(7) The vibration table is oscillated by a ground motion excited by ACROSS seismic waves.
(8) ACROSS signals enable us to precisely determine the transfer function between the calculated input vibration and a target seismometer with the reliability of estimation by the maximum likelihood method.

2.3. Model for the motion of the vibration table and the frequency response model of the seismometer

On the basis of data processing for ACROSS signals, we here introduce a method to characterize the frequency response of a target sensor which

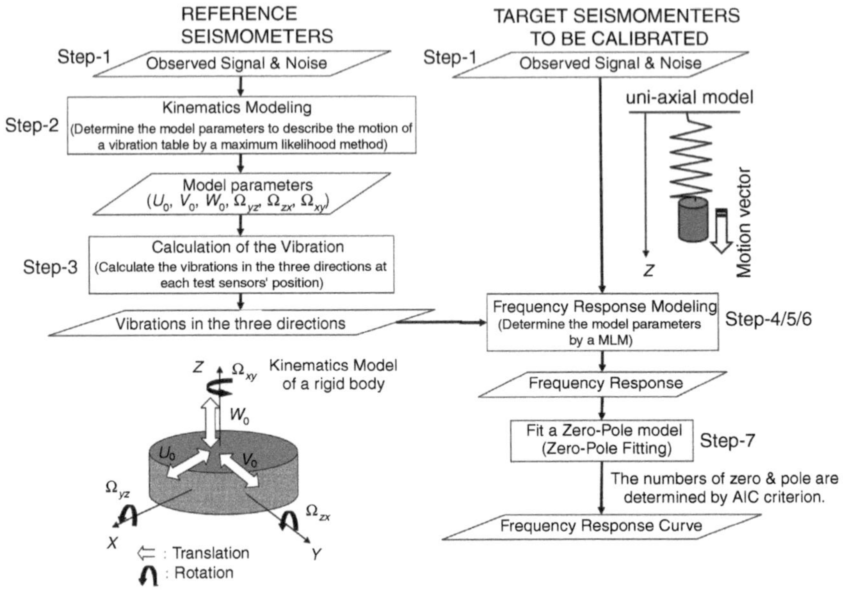

Figure 1 Flowchart of the analyses for the secondary calibration of seismometers by using the ACROSS methodology.

has an unknown characteristic of the frequency response function (FRF). We measure the spectral amplitude and phase responses to several ACROSS signals by target test seismometers on a vibration table, and determine the frequency response in velocity through kinematic modeling for a rigid body.

Figure 1 shows the flowchart of the current secondary calibration method by using ACROSS signals. Here we described six parameters of the rigid-body motion model, that is, three translations, U_0, V_0 and W_0, and three rotations, Ω_{yz}, Ω_{zx} and Ω_{xy}, as the kinematics model. There are six major steps in this method as follows:

1. Observe the seismic ACROSS signal, which is transmitted at a frequency f_k (where k is a discrete frequency number, $k = 1, 2, \ldots, K$) of the m-th different vibration modes (mode index, $m <= M$), which have different directions of vibration to each other, using both the target sensors and the reference accelerometers placed on the vibration table. According to the ACROSS methodology (Kumazawa et al., 2000), an observation error, σ_k, for each signal frequency is estimated by a root-mean-square of noise levels nearby each signal. The reliability of the data can be represented by the Signal-to-Noise ratio (i.e., S/N) while its uncertainty is a reciprocal of S/N (i.e., N/S).
2. Determine the parameters of the kinematics model for the vibration table at each frequency, f_k, by referring the data to the reference accelerometers. The unknown parameters can be estimated by a weighted

maximum likelihood method using the observed noise level, σ_k, as a weight, $1/\sigma_k^2$. The validity of the kinematics model of the vibration table is evaluated by comparison with the measurement error (i.e., noise level), and the residue between the restoration values calculated from the model and the measurement data of the reference accelerometers.

3. Calculate three components of the inputs (e.g., displacement, velocity or acceleration) through the vibration table to each target sensor, assuming the rigid-body motion model determined at Step 2. The inputs, E_{km}, at the frequency, f_k, of the m-th vibration mode are represented by

$$\mathbf{E}_{km} = [E_{km1}, E_{km2}, E_{km3}], \tag{1}$$

where the subscripts, 1, 2 and 3 denote the x, y and z directions.

4. Describe the output voltage, $\mathbf{V}_{km}$, of the target sensor by both the sensitivity of the target sensors and $\mathbf{E}_{km}$. We suggest the "uni-axial response model" and the "tri-axial response model" for estimating the frequency response characteristics of a seismometer. A tri-axial response model has sensitivity in three directions including the directions perpendicular to the principal and the number of degrees for freedom is three. The sensitivity of two directions perpendicular to a principal direction is called the "cross-coupling sensitivity". On the other hand, a uni-axial model has sensitivity only in one direction assuming the sensitivity on the crossing directions to be zero and the number of degrees for freedom is one. The frequency response characteristic is defined as a transfer function from an input vibration to a target sensor to its output in voltage. The output voltage, $\mathbf{V}_{km}$, can be expressed with a sensitivity in three directions represented by $\mathbf{R}_k = [R_{k1}, R_{k2}, R_{k3}]^{\mathrm{T}}$ as follows:

$$\mathbf{V}_{km} = \mathbf{E}_{km}\mathbf{R}_k = E_{km1}R_{k1} + E_{km2}R_{k2} + E_{km3}R_{k3}. \tag{2}$$

5. Perform the procedures of 1-4 for all the vibration modes. In order to have a redundancy in the estimation, the number of vibration modes, M, should be larger than 3 for a tri-axial response model and M should be larger than 2 for a uni-axial response model.
6. Estimate the sensitivity, $\mathbf{R}_k$ $(k = 1, 2, \ldots, K)$ of the target sensor by solving M simultaneous equations. The observed noise level of the target sensor can be used as a weight in the estimation by a weighted maximum likelihood method. The reliability of this estimation for a target sensor can be also represented in the same way as in estimating the model at Step 2.
7. Describe the frequency response function of each target sensor by expressing it as a Zero-Pole (Z-P) curve for easy use. The kinematics model of the vibration table is validated by comparison with the measurement error (i.e., noise level) of eight reference accelerometers, and the residue between the restoration values calculated from the model

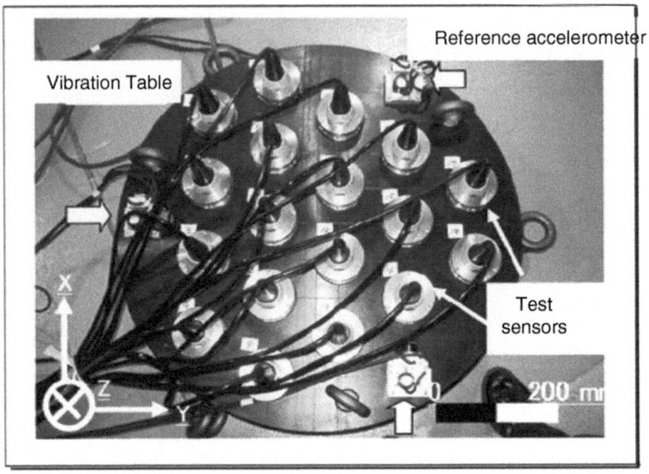

Figure 2 Reference accelerometers and velocity seismometers installed on a cylindrical vibration table made from about 450 kg iron with a diameter of 660 mm. Eight directional components of reference accelerometers (Japan Aviation Electronics Industry, Ltd., JA-5) are located at three surface areas around its perimeter. Eighteen 3-component seismometers (e.g., Seismic Nail sensors by OYO GeoSpace) were set into the holes inside the vibration table.

and the measurement data. Here, the "variation" of frequency response among the target sensors is expressed by the ratio of the standard deviation to the amplitude average estimated at each frequency.

3. Experimental Measurements

Here we demonstrate an example of an experiment, using some feedback-type accelerometers as the reference seismometers, in order to find the theoretical and technical problems in the evaluation and treatment of the statistical errors existing in a measurement, and to investigate the appropriate logical descriptions of a frequency response characteristic in this calibration method.

Figure 2 shows our rigid vibration table made from cylindrical steel, having a diameter of 660 mm, thickness of 190 mm and weight of about 500 kg, with 18 holes for mounting the sensors inside. Eight feedback-type accelerometers, JA-5 (made by Japan Aviation Electronics Industry, Ltd.) were used as the reference seismometers at three locations on the surface circumference of the vibration table. The frequency response characteristics of these accelerometers were given, based on absolute sensitivity measurements, with temperature coefficients of about 0.004–0.007%/°C according to manufacturer.

Moving-coil type velocity seismometers, Seismic Nail (X-phone, made by OYO GeoSpace Co. Ltd.) were investigated as the target sensor to be calibrated. The X-phone used has three uni-axial sensors with a natural frequency of 10 Hz, is a relatively inexpensive and small enough to be installed into the ground. The outputs of 54 uni-axial sensors were measured simultaneously by a time-segment-stacking recorder, MD-8464 (made by PAVEC), which has $\Delta\Sigma$ type 24-bit A/D converters (Kunitomo and Kumazawa, 2004a).

To avoid the temperature influence during calibration measurement, the vibration table was installed in a heat insulation box made of polystyrene foam and the rest of the recording systems were in an air-conditioned room. The following results were obtained at a temperature of 19-20 °C.

The motions of the vibration table were excited by ground motions at the known frequencies by seismic ACROSS transmitters located approximately 30 m apart. Eight vibration modes were excited by two types of transmitters: one with a rotation axis in the horizontal plane and another with an axis in the vertical direction. Each transmitter generated two rotational vibrations in the clockwise and anticlockwise directions. The vibration table was placed at two different x and y directions. Thus we could excite the eight different vibration modes.

In this experiment, we obtained vibration data at 21 discrete frequencies from 9 to 29 Hz with 1 Hz interval. Because of the comparable Signal-to-Noise ratio (S/N) of the measurement data in the target frequency range, we investigated the characteristics of background noise before the experiment and then optimized both the transmitting duration and the stacking time period for each signal frequency. For example, we stacked the data for 4800 and 400 s periods while the signals were transmitted at frequencies of 9 and 22-29 Hz, respectively.

4. Results and Interpretation

Here we show the results of the experiment using 8 reference seismometers and 52 target velocity seismometers, and interpret their frequency response characteristics.

Figure 3 shows the amplitude and phase spectra observed by one of the reference accelerometers for one of eight vibration modes. The 21 discrete frequencies were measured in the range of 9 to 29 Hz (i.e., $f_k = 9, 10, \ldots, 29$). As a result of optimizing the transmission duration as described above, the S/N of the observed spectral amplitude was almost constant with a very high value of 10^3-10^5 at its maximum. The reliability of the phase in dimensionless units is also extremely high, with a level of 10^4 or more.

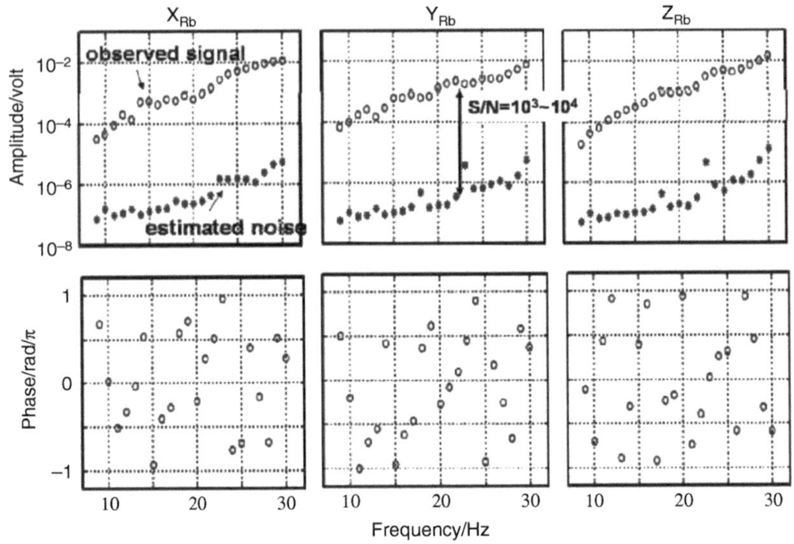

Figure 3 Spectral data in the three components observed by the reference accelerometers. ACROSS signals, which were transmitted by FIT2, exist in the frequency range from 9 to 29 Hz with an interval of 1 Hz. The noise is defined as a root-mean-square amplitude calculated from the observed noise amplitudes around each signal. The signal-to-noise ratio (S/N) attained a nearly uniform reliability up to 104 since the stacking time length changed with respect to the selected frequencies (e.g., 2880 s for 9 Hz, and 240 s for 29 Hz, respectively).

At Step 2, the rigid-body motion model of a vibration table was determined by the maximum likelihood method using the data recorded by eight reference accelerometers. Figure 4 shows the amplitude spectra of the six components (i.e., 3 translations and 3 rotations) together with their estimation errors. The uncertainty of the estimated parameters at each signal frequency was smaller than 10^{-2}–10^{-3}. It is found that three translation components have the dominant contribution to the motions of the vibration table. The displacement excited by the rotation components became approximately several tens % of the translation components in the higher frequency range while it reached several % in the lower frequency range. It was found that the residues between the restoration values calculated from the model and the measurement data of the reference seismometers were almost below or equivalent to the noise level. Since the reliability of estimations (i.e., signal-to-noise ratio in amplitude) was to be 10^2–10^4 (with a mean of 10^3) in this experiment, although the values depended on the vibration modes and the frequencies, these results indicated that the assumption of a rigid-body motion model for the vibration table was mostly appropriate in this experiment.

According to step 3, Figure 5 shows the calculated amplitude spectra of the input vibration through the vibration table to the target sensor and the

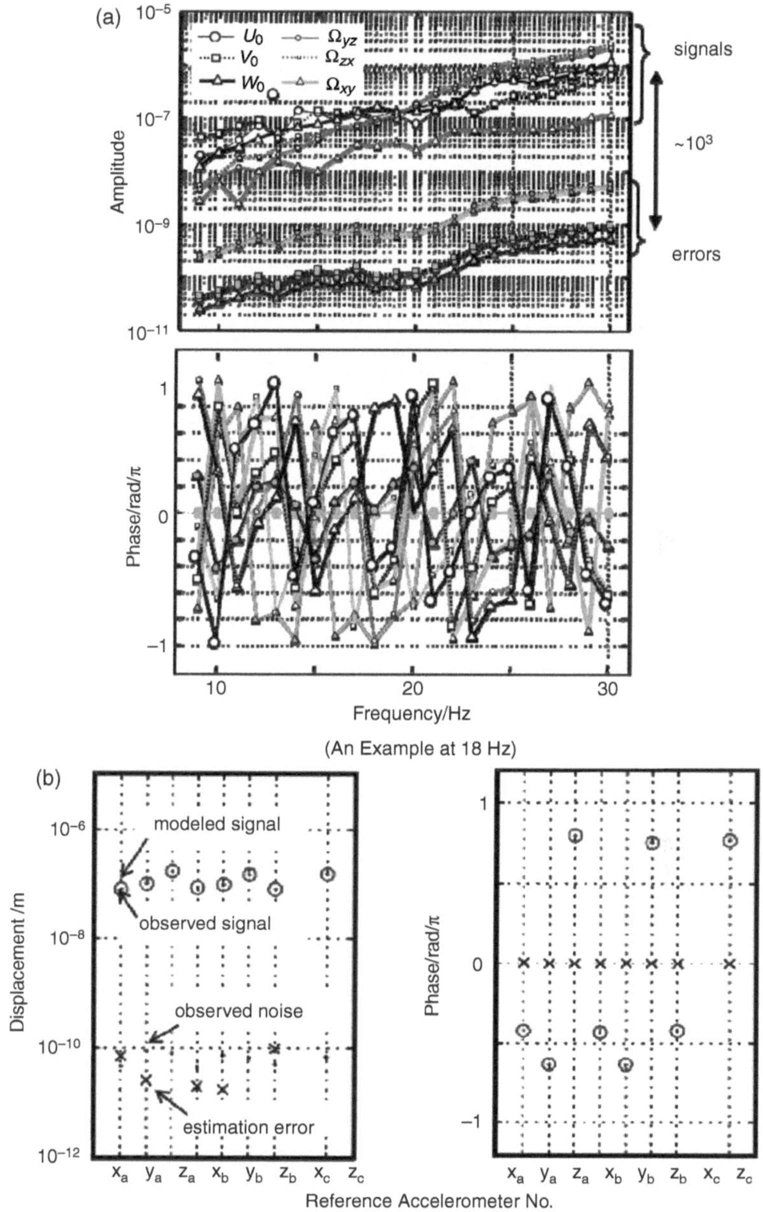

Figure 4 The vibrations of the table were determined by the eight reference accelerometers at each frequency under the assumption of a rigid vibration table. (a) Black solid, dot, and thick lines with open symbols represent the following kinematics model parameters: translations (i.e., U_0, V_0 and W_0), and gray solid, dot, and thick lines with open symbols represent the rotations (i.e., Ω_{yz}, Ω_{zy}, and Ω_{xy}), respectively. The gray solid symbols represent their estimation errors, respectively. (b) An example of the observed signals and the estimated signals at the frequency of 18 Hz.

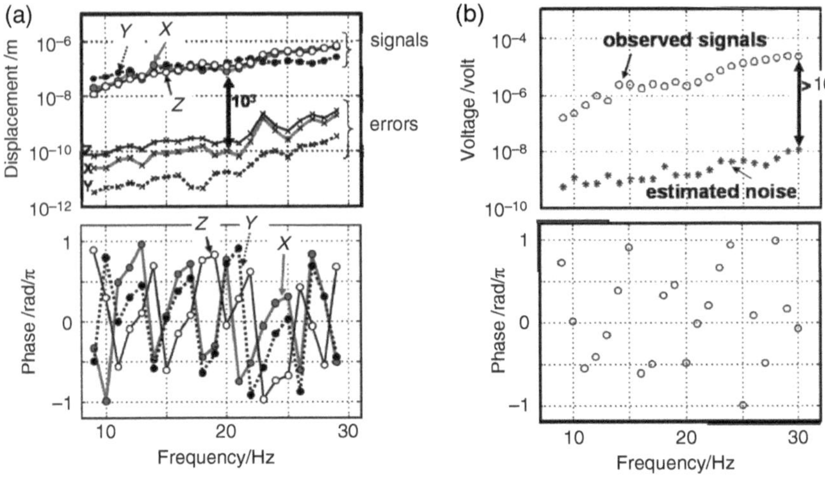

Figure 5 (a) The input vibrations in the three components (i.e., x, y and z) were calculated from the kinematics model of the vibration table. (b) ACROSS signals and estimated noise observed in the x component of a test sensor.

output voltage recorded by the target sensor. The estimation errors of input vibrations depend on the directions and frequencies. For example, in the lower frequency range of 22 Hz, the ratio of the estimation error to the calculated input vibration is 10^{-3}-10^{-4} for the horizontal components and 10^{-2}-10^{-3} for the vertical component.

Through steps 4 to 7, we obtained the frequency response characteristics of a target sensor assuming a uni-axial response model as shown in Figure 6. We also show the Zero-Pole (Z-P) curves estimated from both response models are consistent with each other within their estimation errors. The ratio of the estimation error to response amplitude is smaller than 10^{-2} and the reliability of estimation is to be of the order of 10^2-10^3 in the frequency range from 9 to 29 Hz, except for 13, 18 and 29 Hz. Such a reliability was expected from the qualities of the calculated input vibration and observed data with a S/N ratio of 10^2-10^3. On the other hand, we could obtain a similar reliability for the frequency response of the characteristic in the case of a tri-axial model. The response values of the perpendicular directions may be less than the measurement error of about 10^{-2} so that it may be discarded in this calibration experiment. It follows that the uni-axial model is enough for the calibration with a reliability up to 10^2-10^3. However, it is necessary to consider the cross-coupling effects if the required accuracy is higher than 10^{-4} (i.e., reliability of 10^4) by measuring precise data and by eliminating the bias which is inherent in the process of measurement and/or analyses. It is also needed to examine another frequency response model.

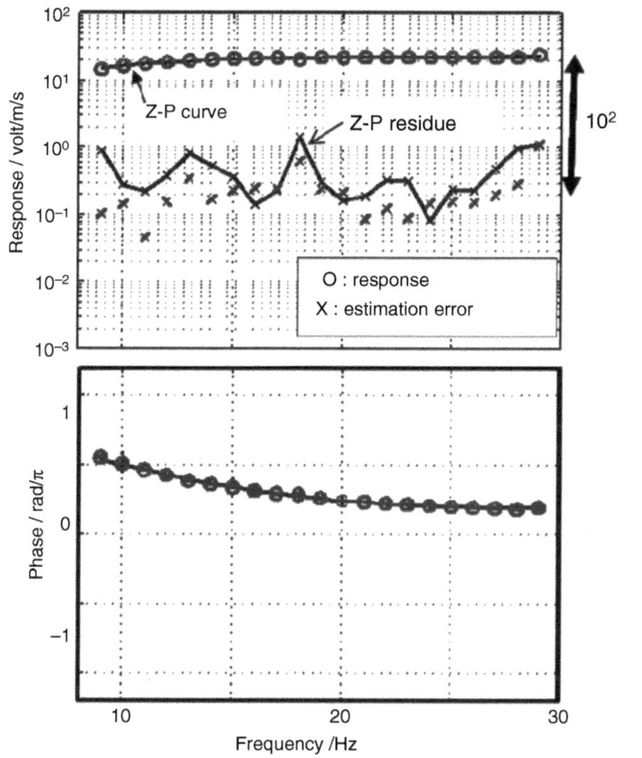

Figure 6 Frequency response estimated from the 8 vibration-mode data at each frequency from 9 to 29 Hz for one of the test sensor in the x direction on the basis of the uni-axial response model in the x direction. Solid lines represent the curve determined by a Zero-Pole model with an optimum order by the AIC criterion.

Figure 7 summarizes the frequency response characteristics for 52 target sensors of three directional components on the assumption of a uni-axial response model. The values of frequency response and their estimation errors depend on the sensor components and frequency. Variations of the response amplitude among the target sensors are less than 2% and are almost within the estimation error range of 1-2% in the frequency range from 9 to 29 Hz in the horizontal components, while some singularities (where the variation exceeded their estimation errors) existed around particular frequencies such as the natural frequency of the seismometer (i.e., 10 Hz) and 17 Hz, and higher frequency range up to 29 Hz. In the vertical component, the variations of the response among seismometers were to be 5 to 10% and their estimation errors to be about 2-3%. Comparing the Z-P curve with the theoretical response curve, it is found that it is mostly in agreement within the estimation error band, but the approximate value was also found to deviate by +2% in amplitude and by +0.02 radian in phase

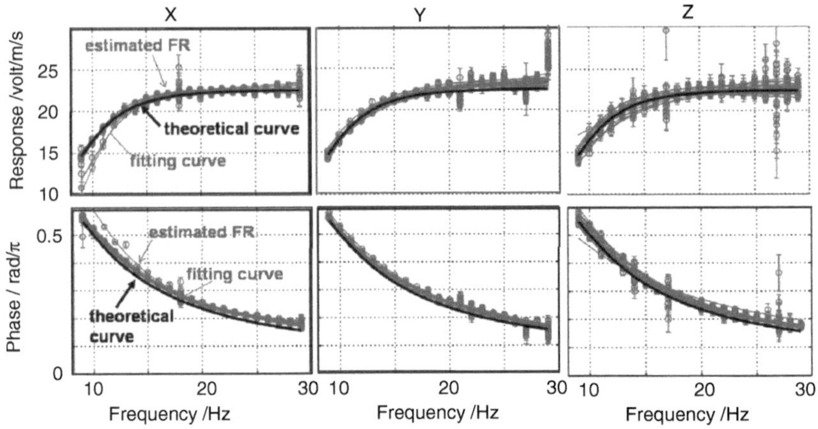

Figure 7 Frequency response characteristics of 52 test sensors. Gray circles with an error-bar represent the frequency response values estimated from observed data. Gray lines show the curve determined by a Zero-Pole model. The solid line shows the theoretical frequency response of a velocity-type seismometer with a natural undamped frequency of 10 Hz, an intrinsic sensitivity of 22.5 volt/m/s, and an open circuit damping coefficient of 0.68 (c.f., http://www.geospacelp.com).

from the theoretical values. Some other sources of deviation might be in the measurement system and/or the analyzing procedures.

5. Discussion

The demonstration of the secondary calibration method for seismometers by ACROSS methodology showed each of 52 target seismometers could have a reliable frequency response and its estimated response curve in the frequency range from 9 to 29 Hz with an estimation error. Here, we discuss how the obtained results relate to the requirements for the calibration of seismometers to be applied for active monitoring of the Earth. In any observations to detect subtle changes of underground media, a few errors in a measurement system may significantly affect the signal detection and evaluation of the changes in the targeted region. It is, therefore, necessary to quantitatively evaluate not only the characteristics of the seismometer sensitivity, the amplitude and phase responses at different frequencies, seismometer stability and low self-noise, but also the combined error of all these factors.

The major requirements for seismometers used for active monitoring of underground dynamic structures are as follows:

(1) Record the time series of background noise in an acceleration or velocity unit without a bias at the receiver site.

(2) Give values and/or equations in order to correct the amplitude and phase response characteristics in the frequency domain and their temperature dependency, the cross-coupling characteristics of seismometers, the temperature dependency and drift and/or annual change of the sensor's sensitivity, the accuracy and minimum resolution, etc. We recommend an accuracy up to 10^{-4} for active monitoring.
(3) Attain an accuracy of 10^{-4} for the amplitude and 10^{-4} radian for the phase after various kinds of noise levels have been decreased by stacking and by correcting the systematic bias and/or characteristics.
(4) Have small enough sizes to be installed at the subsurface, and also be inexpensive to use a large number of seismometers for array observation.

The change of the acoustic impedance ratio at the interface between the different geological layers, for example, a basement rock and sedimentary layer, a fault fracture zone and a hard rock, a crack within a granite rock, seems to be from several tens of % to several tenths of % approximately. If the P-wave velocity of the medium is assumed to be about 5 km/s and its change along the propagation path is to be about 0.1%, the travel time is approximately of the order of 10 s and its variation is 10 ms along a 50 km offset distance. A much smaller change of travel time should be expected for active monitoring the underground state. Considering the accuracy (< 1 μs) of a GPS clock and the synchronous performance of the present ACROSS transmitting system recently, a detection level of travel time change of 10 μs can be detected if the detection limit is caused by velocity seismometers.

6. Conclusions

In this paper, we introduced an indirect calibration method by using ACROSS signal. The method is based on the simultaneous determination of the frequency response characteristics of a set of seismometers placed on the same vibration table which are vibrated by the seismic ACROSS signal. One of the essential points of the current calibration method is the data acquisition strategy. The sinusoidal sweep signals from the seismic ACROSS transmitter are used to oscillate the table, and the outputs of calibrated accelerometers and target seismometers are recorded according to the ACROSS methodology. This allows an accurate stacking of data with a sufficient redundancy. Another important point is the data analysis method. The approach used, based on a maximum likelihood optimization, enables us to obtain not only accurate parameters of the models, such as a kinematics model for the vibration table and a physical model describing the seismometer's sensitivity, but also their estimates of reliability.

ACKNOWLEDGMENTS

This study was supported by the Japan Nuclear Cycle Development Institute (now Japan Atomic Energy Agency).

REFERENCES

Japan Machinery Federation and Japan Industrial Technology Association, 2003. The research report of the present situation and the trend of 2003 of accelerometer standardization and the needs of the industrial world in Japan.

Kumazawa, M., Kunitomo, T., Yokoyama, Y., Nakajima, T., Tsuruga, K., 2000. ACROSS: Theoretical and technical developments and prospect to future applications. Technical report, Japan Nuclear Cycle Development Institute, 9, 115–129.

Kunitomo, T., Kumazawa, M., 2004a. Active monitoring of the Earth's structure by the seismic ACROSS –Transmitting and receiving technologies of the seismic ACROSS. In: Proc. 1st International Workshop on 'Active Monitoring in the Solid Earth Geophysics' in Mizunami, Japan, 30 June–2 July, pp. 181–184.

Kunitomo, T., Kumazawa, M., 2004b. Active monitoring of the Earth's structure by the seismic ACROSS? Development of the seismic ACROSS and continuous transmission at Tono mine ACROSS station. In: Proc. 1st International Workshop on 'Active Monitoring in the Solid Earth Geophysics' in Mizunami, 30 June–2 July, pp. 218–222.

CHAPTER 14

Signal Discrimination of External Geomagnetic Effects Using the Transfer Function Approach with Continuous Wavelet Transform

Makoto Harada[a], Katsumi Hattori[b], *and* Nobuhiro Isezaki[b,a]

Contents

1. Introduction

In order to study changes in the deep electrical structure by using controlled electromagnetic waves, it is desirable to apply periodic signals in the ultra-low-frequency (hereafter ULF) band ($f < 10\,\text{Hz}$) from the viewpoint of the skin effect. However, it is well known that observed ULF electric and magnetic fields on the ground are susceptible to contamination by various artificial and natural noises. Such noises have a bad influence

[a] Tokai University, Institute of Oceanic Research and Development
[b] Chiba University, Graduate School of Science

E-mail address: mharada@scc.u-tokai.ac.jp (M. Harada).

Handbook of Geophysical Exploration: Seismic Exploration, Volume 40
ISSN 0950-1401, DOI: 10.1016/S0950-1401(10)04019-X

upon accuracy in transfer-function estimation, leading to a decline in time resolution.

Several methods have been suggested for the reduction of artificial noise (e.g., Nagao et al. (this monograph)). Since their method is based on the arithmetic mean, it should be effective for random noise, which can be considered as following the Gaussian distribution.

Concerning natural sources, the most intense signals are from geomagnetic storms and pulsations, which are associated with solar-terrestrial activities. They appear simultaneously on a global scale. Such geomagnetic field fluctuations have $1/f$ properties; therefore, simple stacking or averaging would not be appropriate for their reduction. To estimate the desired transfer functions these periodic signals should be eliminated beforehand.

We describe a method based on the interstation transfer-function (hereafter ISTF) approach for the elimination of external-source geomagnetic fields and related induced electric fields (Harada et al., 2004a,b).

2. Formulation of Signal Discrimination

2.1. Interstation transfer function (ISTF)

ULF magnetic field fluctuations have wavelengths ranging from several hundred to several thousand kilometers. In the geomagnetic mid-latitude region, one may assume approximately vertically incident plane waves propagating through the atmosphere (Schmucker, 1970). Under this assumption, the three observed components of the magnetic field on the ground, $\mathbf{H}_s(\omega) = (\Delta X_s(\omega), \Delta Y_s(\omega), \Delta Z_s(\omega))$, are considered as the superposition of the following inputs: [1] a "normal field" $\mathbf{H}_n(\omega)$, which is the sum of the contributions from the external field and the internal field induced by a regional horizontally layered electrical conductivity structure; [2] an "anomalous field" $\mathbf{H}_a(\omega)$, which is due to a locally varying conductivity structure; and [3] another source field, $\delta\mathbf{H}(\omega)$, including artificial noise (Gough and Ingham, 1983; Beamish and Banks, 1983). Thus, the observed magnetic field can be written as follows:

$$\mathbf{H}_s(\omega) = \mathbf{H}_n(\omega) + \mathbf{H}_a(\omega) + \delta\mathbf{H}(\omega), \tag{1}$$

where $\mathbf{H}_n(\omega)$ consists of the two horizontal components $(\Delta X_n(\omega), \Delta Y_n(\omega), 0)$ because of the source-field assumption that plane waves penetrate to the earth vertically. Since $\mathbf{H}_a(\omega)$ is considered as the secondary field of the input of $\mathbf{H}_n(\omega)$,

$$\mathbf{H}_a(\omega) = \mathbf{T}(\omega) \cdot \mathbf{H}_n(\omega), \tag{2}$$

where $\mathbf{T}(\omega)$ is the transfer-function, 3×2 complex matrix.

Substituting Eq. (2) into Eq. (1),

$$\mathbf{H}_s(\omega) = \mathbf{T}'(\omega) \cdot \mathbf{H}_n(\omega) + \delta\mathbf{H}(\omega), \tag{3}$$

where $\mathbf{T}'(\omega) = \mathbf{I} + \mathbf{T}(\omega)$ ($\mathbf{I}$ is a unit matrix). To get around some difficulties in solving Eq. (3), the remote reference data with sufficiently small artificial noise, $\mathbf{H}_r(\omega)$, are applied instead of $\mathbf{H}_n(\omega)$ (Egbert, 1997). Thus,

$$\mathbf{H}_s(\omega) = \mathbf{T}_{is}(\omega) \cdot \mathbf{H}_r(\omega) + \delta\mathbf{H}(\omega). \tag{4}$$

When $\mathbf{H}_r(\omega)$ contains no artificial noise and the vertical component at the reference site is negligible, it consists only of the horizontal component $(\Delta X_r(\omega), \Delta Y_r(\omega), 0)$. $\mathbf{T}_{is}(\omega)$ is the ISTF, 3×2 complex matrix. The ISTF should be estimated by using the data when the external source field is much greater than the uncorrelated part.

2.2. ISTF approach to the magnetotelluric (MT) response function

The principal component of electric field variation on the ground surface is induced by geomagnetic fluctuations from an external source. The relationship between the electric and magnetic field is written as follows (e.g., Vozoff (1972)):

$$\mathbf{E}_s(\omega) = \mathbf{Z}_s(\omega) \cdot \mathbf{H}_s(\omega), \tag{5}$$

where $\mathbf{E}_s(\omega)$ and $\mathbf{H}_s(\omega)$ indicate the two horizontal components of the electric and magnetic fields, respectively. $\mathbf{Z}_s(\omega)$ is a complex response function, which is defined as a 2×2 matrix. Since $\mathbf{Z}_s(\omega)$ contains information on the underground electrical structure, a number of geophysical applications have been carried out so far.

Several articles have been devoted to eliminating the electric field variations related to the external-source geomagnetic fluctuations (Chouliaras and Rasmussen, 1988; Hadjiannou et al., 1993). They were based on the magnetotelluric expression (5), and used simultaneous magnetic and electric data at the same site. Their results are effective for the reduction of an induced field under favorable conditions. However, problems arise when undesired correlated magnetic and electric components are not negligible. Therefore, in this case, adopting the ISTF approach is essential for the calculation of the MT response function.

The observed electric field is represented as follows:

$$\mathbf{E}_s(\omega) = \mathbf{Z}_s(\omega) \cdot \mathbf{H}_s(\omega) + \delta\mathbf{E}(\omega), \tag{6}$$

where $\delta\mathbf{E}(\omega)$ is the uncorrelated part, which is expected to include the transmitted signal. If a sufficiently large external-source field is loaded,

$\delta\mathbf{E}(\omega)$ could be negligible. In practice, $\mathbf{T}_{is}(\omega)$ in Eq. (4) is estimated with a high multiple correlation value ($\gamma^2 \geq 0.9$) between $\mathbf{H}_s(\omega)$ and $\mathbf{H}_r(\omega)$. Taking into account this condition and assuming $\mathbf{T}_{is}(\omega)$ to be time-invariant, Eq. (6) could be given as follows:

$$\mathbf{E}_s(\omega) = \mathbf{Z}_s(\omega) \cdot \mathbf{T}_{is}(\omega) \cdot \mathbf{H}_r(\omega) + \delta\mathbf{E}(\omega). \tag{7}$$

Then, the relationship between the electric field at a certain station, $\mathbf{E}_s(\omega)$, and the magnetic field at the remote reference station, $\mathbf{H}_r(\omega)$, is described by using the interstation MT transfer function $\mathbf{Z}_{is}(\omega)$:

$$\mathbf{E}_s(\omega) = \mathbf{Z}_{is}(\omega) \cdot \mathbf{H}_r(\omega) + \delta\mathbf{E}(\omega), \tag{8}$$

where

$$\mathbf{Z}_{is}(\omega) = \mathbf{Z}_s(\omega) \cdot \mathbf{T}_{is}(\omega). \tag{9}$$

2.3. Elimination of the external geomagnetic effects

The estimated interstation transfer-functions are assumed to be time invariant, because the electric property under the ground is supposedly stable when there is no large crustal change (Honkura, 1979). Taking this assumption into account, we can estimate the possible magnetic ($\mathbf{H}_{sg}(\omega)$) and electric ($\mathbf{E}_{sg}(\omega)$) fields at a site by using the remote reference data $\mathbf{H}_r(\omega)$:

$$\mathbf{H}_{sg}(\omega) = \mathbf{T}_{is}(\omega) \cdot \mathbf{H}_r(\omega), \tag{10}$$

$$\mathbf{E}_{sg}(\omega) = \mathbf{Z}_{is}(\omega) \cdot \mathbf{H}_r(\omega). \tag{11}$$

Therefore, if one subtracts the estimated electric/magnetic field from the observed data, one can be expected to remove the effects of the global electromagnetic field changes and their secondary fields:

$$\Delta\mathbf{H}(\omega) = \mathbf{H}_s(\omega) - \mathbf{H}_{sg}(\omega), \tag{12}$$

$$\Delta\mathbf{E}(\omega) = \mathbf{E}_s(\omega) - \mathbf{E}_{sg}(\omega), \tag{13}$$

where $\Delta\mathbf{H}(\omega)$ and $\Delta\mathbf{E}(\omega)$ represent the residual components.

3. Application of the Wavelet Transform

3.1. Continuous wavelet transform

Geomagnetic fluctuation is generally nonstationary, and the observed data contain many transient signals. The wavelet transform is useful in analyzing

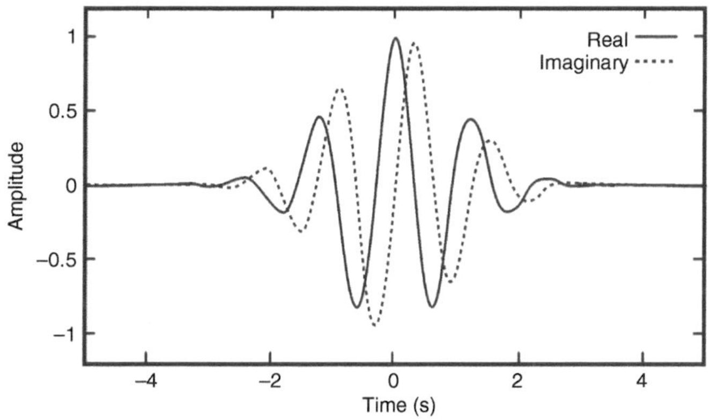

Figure 1 The waveforms of the Morlet wavelet.

time series data which contain nonstationary power at many different frequencies (Farge, 1992).

The mother wavelet is usually chosen depending on the signal properties and the purpose of analysis. In estimating the transfer functions, the mother wavelet must be complex-valued, and the relationship between the scale and frequency should be simple. The wavelet transform should characterize the amplitude and phase information of the electromagnetic field in both time and frequency domains. These requirements are satisfied by the complex-valued Morlet wavelet (Grossman and Morlet, 1984), which consists of a plane wave modulated by a Gaussian window (see Figure 1):

$$\psi_0(t) = \pi^{-1/4} e^{i\omega_0 t} e^{-t^2/2}, \tag{14}$$

where t is a nondimensional time parameter, and ω_0 is a nondimensional frequency, here taken to be 6 to satisfy the admissibility condition.

The relationship between the wavelet scale a and Fourier frequency ω is given as follows (Meyers et al., 1993):

$$a = \frac{\omega_0 + \sqrt{2 + \omega_0^2}}{2\omega}. \tag{15}$$

The continuous wavelet transform of the real square integrable function $f(t) \in L^2(\mathfrak{R})$ at the location b, relative to a real integrable mother wavelet $\psi(t)$ at scale a, can be defined as:

$$Wf(b, a) = \langle f, \psi_{b,a} \rangle = \frac{1}{\sqrt{a}} \int_{-\infty}^{+\infty} f(t) \overline{\psi\left(\frac{t-b}{a}\right)} dt, \tag{16}$$

where $Wf(b, a)$ is called the wavelet coefficient, and – stands for the complex conjugate. In the Fourier space, the wavelet transform can be expressed as:

$$Wf(b,a) = \frac{1}{2\pi}\int_{-\infty}^{+\infty} \hat{f}(\omega)\hat{\psi}_{b,a}(\omega)\mathrm{e}^{\mathrm{i}b\omega}\mathrm{d}\omega, \tag{17}$$

where $\hat{f}$ and $\hat{\psi}$ are the Fourier transforms of f and ψ, respectively. The inverse wavelet transform is represented as follows:

$$f(t) = \frac{1}{C_\psi}\cdot\frac{1}{a^2}\int_{0+}^{+\infty}\int_{-\infty}^{+\infty} Wf(b,a)\psi_{b,a}(t)\mathrm{d}b\mathrm{d}a, \tag{18}$$

where C_ψ is the integral of $\|\hat{\psi}(\omega)\|^2\omega^{-1}$ overall frequencies.

Because of the redundancy in time and scale in the case of a continuous wavelet transform, the reconstruction of the time series using a completely different function from the mother wavelet is possible (Farge, 1992; Torrence and Compo, 1998). Here we choose the delta function as the kernel for the reconstruction. In this case, the reconstructed time series is just the sum of the real part of the wavelet coefficients over all scales (Torrence and Compo, 1998):

$$f(t) = \frac{\mathrm{d}j\mathrm{d}t^{1/2}}{C_\delta\psi_0(0)}\sum_{j=0}^{J}\frac{\Re\{Wf(b,a_j)\}}{a_j^{1/2}}, \tag{19}$$

where j indicates the scale index and J yields the maximal number of resolvable scales. The factor $\psi_0(0)$ removes the energy scaling, while the $a_j^{1/2}$ converts the wavelet coefficients to an energy density. $\psi_0(0)$ and C_δ are constants of $\pi^{-1/4}$ and 0.776 for the Morlet wavelet in this study, respectively (Torrence and Compo, 1998).

3.2. Auto- and cross-correlation functions based on the wavelet transform

The wavelet-based auto-correlation function, $WC(a,\tau)$, of wavelet coefficients yielded by Eq. (16), is defined by the following formula (Li, 1998).

$$WC_f(a,\tau) = \lim_{T\to\infty}\frac{1}{T}\int_{-2/T}^{2/T}\overline{Wf(b,a)}Wf(b+\tau,a)\mathrm{d}b, \tag{20}$$

where τ is the time delay of wavelet coefficients in the wavelet space, or the time delay of the signal $f(t)$. The wavelet auto-correlation function

$WC(a,\tau)$ can also be written in Fourier space:

$$WC_f(a,\tau) = \frac{1}{2\pi}\int_{-\infty}^{\infty} P_f(\omega)\left|\hat{\psi}(a\omega)\right|^2 e^{i\tau\omega}d\omega. \tag{21}$$

In analogy with the wavelet auto-correlation, the wavelet cross-correlation function of two different signals, $f(t)$ and $g(t)$, can be defined as follows:

$$WC_{fg}(a,\tau) = \lim_{T\to\infty}\frac{1}{T}\int_{-2/T}^{2/T} \overline{Wf(b,a)}Wg(b+\tau,a)db, \tag{22}$$

where $Wf(b,a)$ and $Wg(b,a)$ are wavelet coefficients, and τ indicates the time delay between two signals. The wavelet cross-correlation function $WC_{fg}(a,\tau)$ can also be written in Fourier space:

$$WC_{fg}(a,\tau) = \frac{1}{2\pi}\int_{-\infty}^{\infty} P_{fg}(\omega)\left|\hat{\psi}(a\omega)\right|^2 e^{i\tau\omega}d\omega, \tag{23}$$

where $P_{fg}(\omega)$ is the traditional cross-spectrum. Equation (23) represents the relationship between the wavelet and Fourier transforms, and describes the characteristics of $WC_{fg}(a,\tau)$ as a local filter, in which the Fourier power spectrum $P_{fg}(\omega)$ is filtered by $|\hat{\psi}(a\omega)|^2$ in Fourier space.

3.3. Wavelet-based transfer function and correlation function

The transfer functions used in this study are defined by the following complex linear equation, which is formulated between 2-input and 1-output components:

$$y^j(\tau) = h_1^j \cdot x_1^j(\tau) + h_2^j \cdot x_2^j(\tau), \tag{24}$$

where x_1^j, x_2^j, and y^j represent the wavelet coefficients of the 2-input and 1-output components. The letters j and τ indicate the scale index and the translation of wavelet coefficients, respectively. h_1^j and h_2^j are the transfer functions which are calculated from the conventional power-spectrum approach (Vozoff, 1972).

The transfer functions in each time segment, k, the length of which is determined in each scale, j, are represented as follows:

$$h_1^{j,k} = \frac{\left\{\begin{array}{l}\sum_k WCx_1^j y^j\left(\tau_k\right)\cdot\sum_k WCx_2^j x_2^j\left(\tau_k\right)\\ \quad -\sum_k WCx_1^j x_2^j\left(\tau_k\right)\cdot\sum_k WCx_2^j y^j\left(\tau_k\right)\end{array}\right\}}{\sum_k WCx_1^j x_1^j\left(\tau_k\right)\cdot\sum_k WCx_2^j x_2^j\left(\tau_k\right)-\left|\sum_k WCx_1^j x_2^j\left(\tau_k\right)\right|^2}, \quad (25)$$

$$h_2^{j,k} = \frac{\left\{\begin{array}{l}\sum_k WCx_2^j y^j\left(\tau_k\right)\cdot\sum_k WCx_1^j x_1^j\left(\tau_k\right)\\ \quad -\sum_k WCx_2^j x_1^j\left(\tau_k\right)\cdot\sum_k WCx_1^j y^j\left(\tau_k\right)\end{array}\right\}}{\sum_k WCx_1^j x_1^j\left(\tau_k\right)\cdot\sum_k WCx_2^j x_2^j\left(\tau_k\right)-\left|\sum_k WCx_1^j x_2^j\left(\tau_k\right)\right|^2}, \quad (26)$$

where $WCx_1^j x_1^j\left(\tau_k\right)$ and $WCx_2^j x_2^j\left(\tau_k\right)$ are wavelet auto-correlation functions, and $WCx_1^j x_2^j\left(\tau_k\right)$, $WCx_2^j x_1^j\left(\tau_k\right)$, $WCx_1^j y^j\left(\tau_k\right)$, and $WCx_2^j y^j\left(\tau_k\right)$ are wavelet cross-correlation functions. For example, $WCx_1^j y^j\left(\tau_k\right)$ can be derived from $WCx_1^j x_2^j\left(\tau_k\right) = \overline{x_1^j\left(\tau_k\right)}\cdot x_2^j\left(\tau_k\right)$. Here, input components should be independent of each other. The multiple correlation function between the input $(x_1^j(\tau_k), x_2^j(\tau_k))$ and output $y^j(\tau_k)$ parameters of each segment k is represented as follows:

$$R^{j,k}\gamma^2 = \mathrm{Re}\left\{\frac{\begin{array}{l}h_1^j\cdot\overline{h_1^j}\cdot WCx_1^j x_1^j\left(\tau_k\right)+h_1^j\cdot\overline{h_{21}^j}\cdot W\underline{C}x_1^j x_2^j\left(\tau_k\right)\\ \quad +h_2^j\cdot\overline{h_1^j}\cdot WCx_2^j x_1^j\left(\tau_k\right)+h_2^j\cdot\overline{h_2^j}\cdot WCx_2^j x_2^j\left(\tau_k\right)\end{array}}{WCy^j y^j\left(\tau_k\right)}\right\}, \quad (27)$$

where $R^{j,k}\gamma^2$ consists of the real component and satisfies $0\le R^{j,k}\gamma^2\le 1$. Here, input components should be independent of each other.

4. Application to Observed Data

In this subsection, we show the performance and capability of the proposed ISTF approach for reducing geomagnetic fluctuation and the related secondary fields.

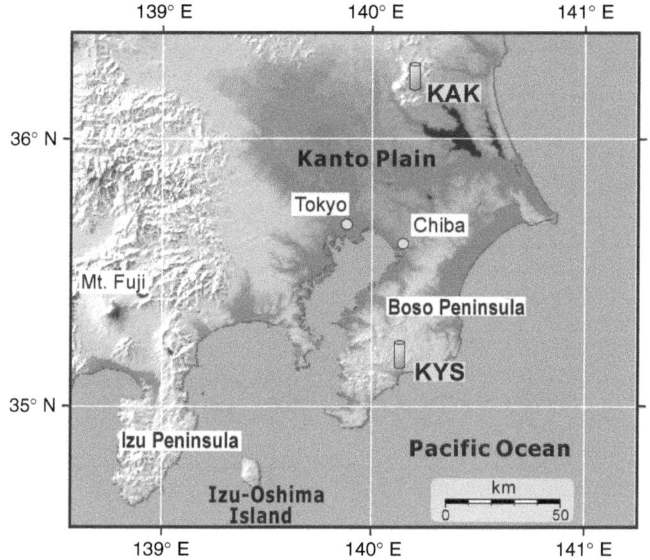

Figure 2 Configuration of the electric and magnetic stations. *KYS* is one of the electric and magnetic stations to monitor earthquake-related electromagnetic phenomena. *KAK* indicates the Kakioka Magnetic Observatory, Japan Meteorological Agency. The magnetic data at *KAK* are used as a remote reference.

Figure 2 illustrates the configuration of the ULF geomagnetic stations in this study. Kiyosumi (hereafter *KYS*) is one of our observation points (Hattori et al., 2004). We use the magnetic field data at the Kakioka Magnetic Observatory, Japan Meteorological Agency (hereafter *KAK*) as the remote reference. *KAK* is located at about 100 km almost due north of *KYS*.

Figure 3 shows the observed magnetic data at the *KAK* and *KYS* stations, and the electric data at *KYS* in the 4-hour interval of 01:00-05:00 (LT) on 16 July 2000, when the geomagnetic activity was extremely high ($Kp = 9-$). The proposed method was applied to these time series.

Figure 4 shows the estimated ISTF between the magnetic data at *KYS* and the referenced magnetic data at *KAK*, which were estimated from the midnight time data. Figure 5 shows the estimated $\mathbf{Z}_{is}$ calculated between the electric data at *KYS* and the magnetic data at *KAK*. The time interval used for the estimation is the same as for ISTF. For the accurate estimation of ISTF, data segments whose multiple correlation ($R\gamma^2$) is greater than 0.9 should be chosen and averaged in both the time and scale spaces. The transfer functions for periods longer than 8 seconds seem to be well estimated.

Figures 6 and 7 show the results of elimination of the external source effects from the magnetic and electric data at *KYS*. The reconstruction of the time series was practiced by using Eq. (19). These panels indicate that the global geomagnetic fluctuation and related induced electric/magnetic field at *KYS* were well estimated and eliminated effectively.

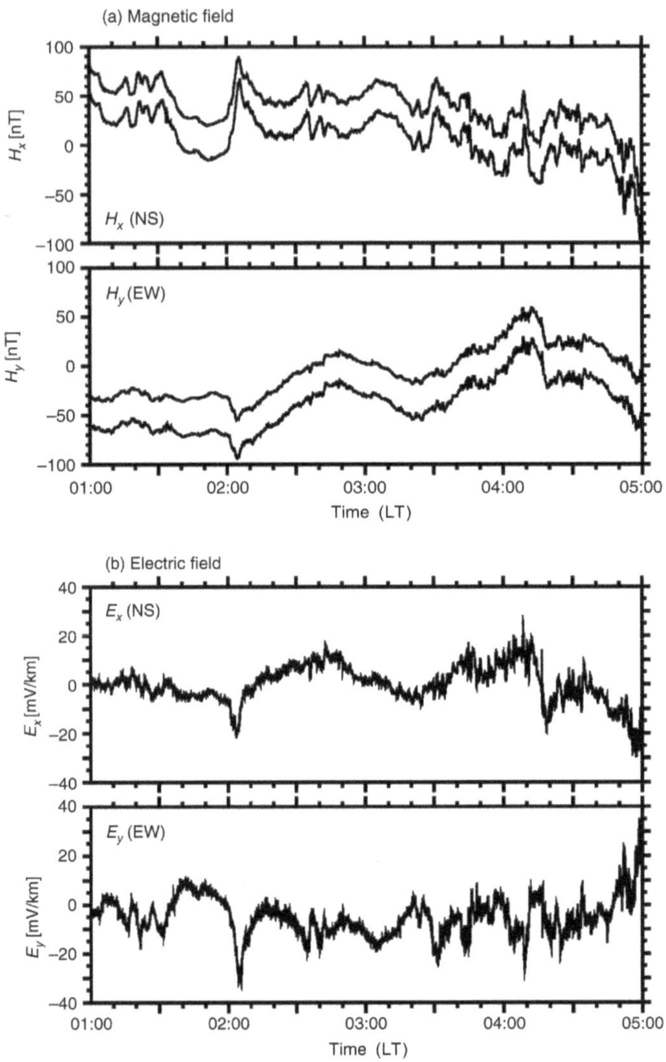

Figure 3 Magnetic and electric field variations in the period of 01:00-05:00 on 16 July 2000 (LT). (a) Horizontal components of the magnetic field at the *KAK* (upper line in each component) and *KYS* (lower line) stations. (b) Horizontal components of electric field data at *KYS*.

5. Verification of the Detectability of Artificial Signals

In the previous subsection, we performed a reduction of the external-source fields from the magnetic and electric data at *KYS*. In this subsection,

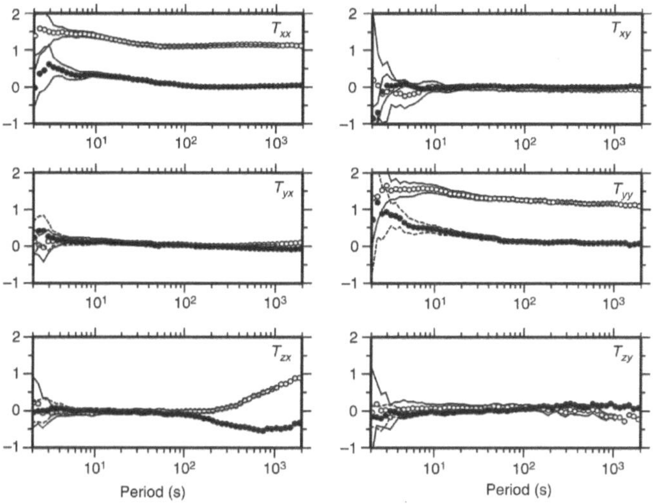

Figure 4 Estimated interstation magnetic transfer functions between the KAK and *KYS* stations. Open and solid circles indicate the real and imaginary parts of the transfer functions. Upper and lower boundaries of the 95% confidence limits are also illustrated.

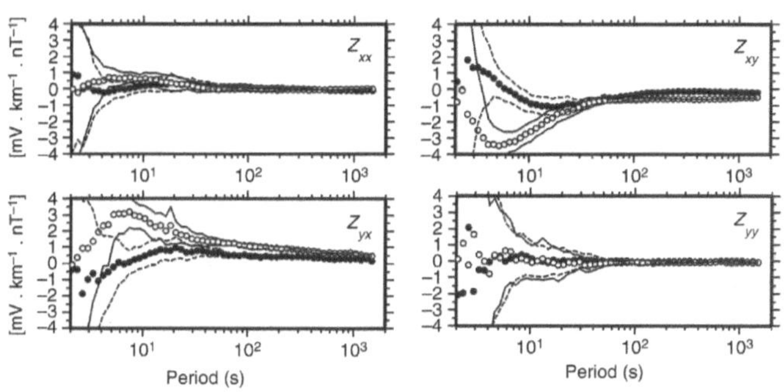

Figure 5 Estimated interstation electromagnetic transfer functions between the *KAK* and *KYS* stations. Open and solid circles indicate the real and imaginary parts of the transfer functions. Upper and lower boundaries of the 95% confidence limits are also shown.

we verify whether the waveforms of the detected signals keep their original features after this operation. Imposed artificial sets of signals for this examination were (a) cosine bell and (b) rectangular in shapes, as illustrated in Figure 8. The amplitude of these signals was set to 0.5 nT. The durations were set to 100 seconds and 10 seconds. These simulated signals were added to the actual magnetic data at *KYS* at 02:30:00 (LT).

Figure 9 represents the result of the signal discrimination test. The left and right panels indicate the artificial signal of cosine bell and the rectangle, respectively. Panel (a) illustrates the time series which contains the signal.

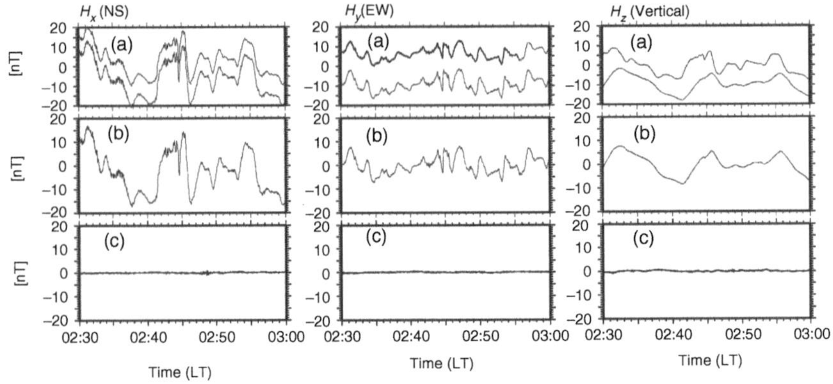

Figure 6 Results of the elimination of external source effects at the *KYS* station. (a), (b), and (c) illustrate the high-pass filtered ($T < 1024$ s) original data at KAK (upper line) and KYS (lower line), the magnetic field at KYS estimated from the reference data at *KAK* using ISTF, and the residual components of (a)-(b).

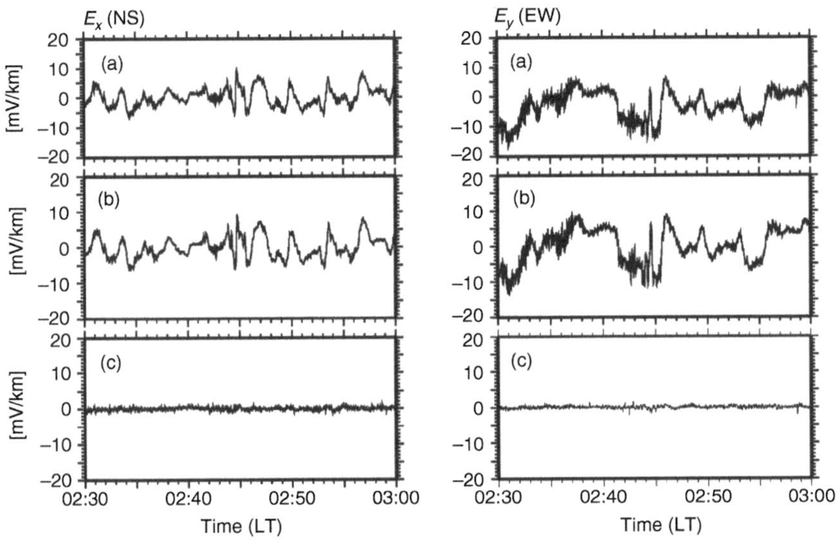

Figure 7 Results of the elimination of the induced electric field associated with external source effects at the *KYS* station. (a), (b), and (c) illustrate the high-pass filtered ($T < 1024$ s) original data at *KYS*, the electric field estimated from *KAK* using $\mathbf{Z}_{is}(\omega)$, and the residual components (a)-(b).

One cannot recognize the existence of an added signal by eye. Panel (b) shows the results of the signal discrimination when no signal was imposed. Panel (c) represents the case when the artificial signals were imposed. We can recognize the added signals clearly. Panel (d) shows the differences of (c)-(b).

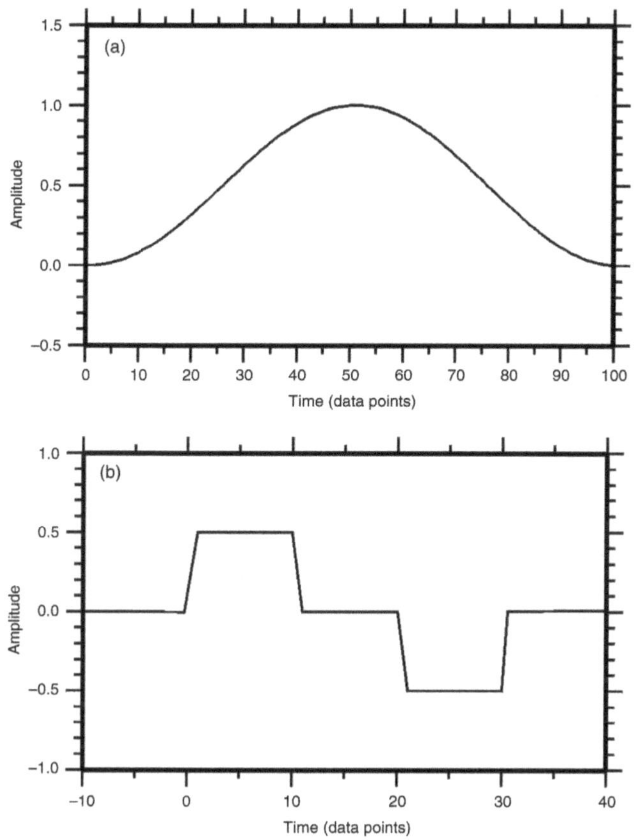

Figure 8 Simulated signals used for verification of signal detectability. (a) Cosine bell, and (b) rectangle type. The amplitude of the signals is set to 0.5 (nT).

We can see that the imposed signals have been restored in keeping with their waveform.

In order to confirm the validity of the signal reconstruction, we show the differences between the restored signals in Figure 9(d) and the original added signals in Figure 8.

Since the wavelet transform is essentially a band-pass filter of uniform shape and varying location and width, the reconstructed signal should have some differences in principle. In addition, the performance of the wavelet transform is dependent on the choice of mother wavelet. In this study, the Morlet wavelet was chosen, and the reconstruction of the time series was practiced by using scales of less than 1024 seconds (i.e., a high-pass filter).

Figure 10(a) and (b) both show that the differences in their amplitude are less than 0.5% of the input signal and are very small indeed.

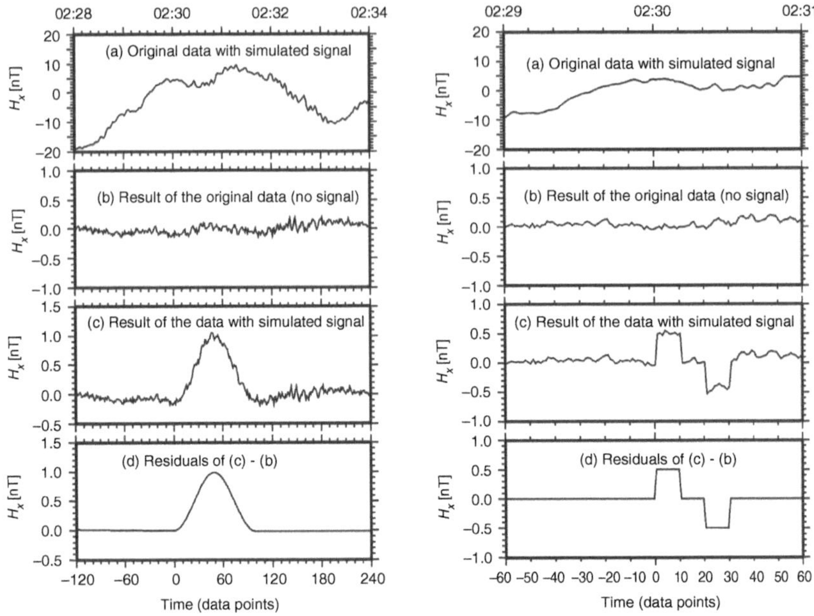

Figure 9 Results of the verification of signal discrimination. The left and right panels illustrate the results of the cosine bell and rectangle signals, respectively. (a) Time series of H_x, which contains the simulated signal shown in Figure 8. (b) Result of external source field reduction when no simulated signal was imposed. (c) The same when the simulated signal was imposed. (d) The residual (c)-(b).

From the above verification of the detectability of imposed simulated signals, we can safely say that the proposed method is effective for the detection of desired signals.

6. Conclusion

An effective method of eliminating the contaminant signals originating from external magnetic fluctuation in the frequency range of 0.0001 to 1.0 Hz has been developed. This method is based on the interstation transfer-function approach with the application of a continuous wavelet transform.

It is important to estimate the transfer functions with good accuracy for effective signal discrimination. Furthermore, signal discrimination in the time-scale (frequency) domain is also of essential importance. We formulated all of the processes of the spectral analysis in the wavelet domain.

We examined the performance of the proposed method by using the actual electric and magnetic data observed at Kiyosumi, Central Japan. We also used the magnetic data at Kakioka, JMA, as the remote reference.

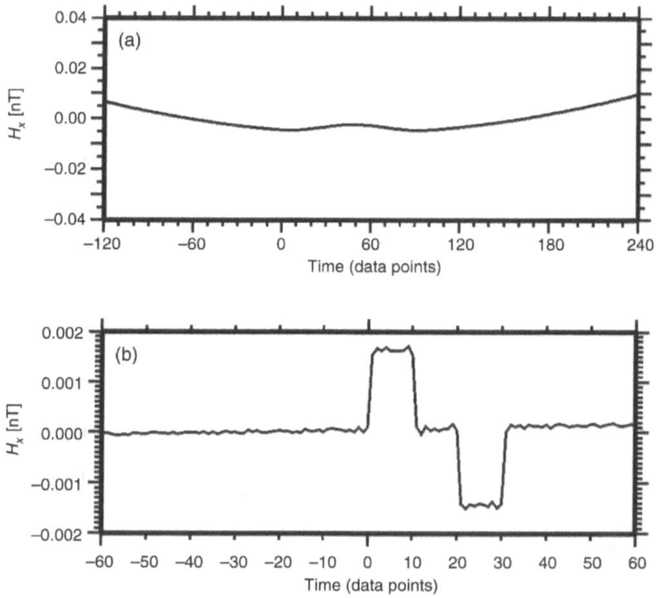

Figure 10 The differences of the original signals in Figure 8 and the residual component in Figure 9 (d). (a) and (b) illustrate the results of the cosine bell and rectangle type added signals.

The results were satisfying in both the elimination of external source effects and the detection of imposed synthetic signals. This implies that the detection of desired electromagnetic signals will be possible even during a period of severe geomagnetic storms. Therefore, this method will also be practical for electromagnetic sounding using a controlled source, including monitoring the time changes of underground electrical structures by using continuously transmitted source signals (Kumazawa et al., 2000).

REFERENCES

Beamish, D., Banks, R.J., 1983. Geomagnetic variation anomalies in northern England: Processing and presentation of data from a non-simultaneous array. Geophysical Journal of the Royal Astronomical Society 75, 513–539.

Chouliaras, G., Rasmussen, T.M., 1988. The application of the magnetotelluric impedance tensor to earthquake prediction research in Greece. Tectonophysics 152, 119–135.

Egbert, G.D., 1997. Robust multiple-station magnetotelluric data processing. Geophysical Journal International 130, 475–496.

Farge, M., 1992. Wavelet transforms and their applications to turbulence. Annual Reviews in Fluid Mechanism 24, 395–457.

Gough, D.I., Ingham, M.R., 1983. Interpretation methods for magnetometer arrays. Reviews of Geophysics and Space Physics 21, 805–827.

Grossman, A., Morlet, J., 1984. Decomposition of Hardy functions into square integrable wavelets of constant shape. SIAM Journal on Mathematical Analysis 15, 723–736.

Hadjiannou, D., Vallianatos, F., Efraxias, K., Hadjincotis, V., Nomikos, K., 1993. Subtraction of the telluric inductive component from VAN measurements. Tectonophysics 224, 113–124.

Harada, M., Hattori, K., Isezaki, N., 2004a. Transfer function approach to signal discrimination of ULF geomagnetic data. Physics and Chemistry of the Earth 29, 409–417.

Harada, M., Hattori, K., Isezaki, N., 2004b. Reduction of geomagnetic effects (periods $T < 940$ s) from geoelectric potential difference data. IEEJ Transaction of Fundamentals and Materials 124 (A12), 1245–1250.

Hattori, K., Takahashi, I., Yoshino, C., Isezaki, N., Iwasaki, H., Harada, M., Kawabata, K., Kopytenko, E., Kopytenko, Y., Maltsev, P., Korepanov, V., Molchanov, O.A., Hayakawa, M., Noda, Y., Nagao, T., Uyeda, S., 2004. ULF geomagnetic field measurements in Japan and some recent results associated with Iwateken Nairiku Hokubu earthquake in 1998. Physics of the Chemistry of the Earth 29, 481–494.

Honkura, Y., 1979. Observations of short-period geomagnetic variations at Nakaizu (2): Changes in transfer functions associated with the Izu-Oshima-Kinkai earthquake of 1978. Bulletin of the Earthquake Research Institute, University of Tokyo, 54, 477–490.

Kumazawa, M., Kunitomo, T., Yokoyama, Y., Nakajima, T., Tsuruga, K., 2000. ACROSS: Theoretical and technological developments and prospect to future applications. Journal of Nuclear Science and Technology 9, 115–129 (in Japanese).

Li, H., 1998. Identification of coherent structure in turbulent shear flow with wavelet correlation analysis. Journal of Fluid Engineering 120, 778–785.

Meyers, S.D., Kelley, B.G., O'Brien, J.J., 1993. An introduction to wavelet analysis in oceanography and meteorology: With application to the dispersion of Yanai wave. Monthly Weather Review 121, 2858–2866.

Nagao, H., Nakajima, T., Kumazawa, M., Kunitomo, T., Stacking strategy for acquisition of an ACROSS transfer function with large signal-to-noise ratio, this monograph.

Schmucker, U., 1970. Anomalies of geomagnetic variations in the southwestern United States. Bulletin of the Scripps Institution of Oceanography 13, 1–165.

Torrence, C., Compo, G.P., 1998. A practical guide to wavelet analysis. Bulletin of the American Meteorological Society 79, 61–78.

Vozoff, K., 1972. The magnetotelluric method in the exploration of sedimentary basins. Geophysics 37, 98–141.

CHAPTER 15

Development of Simulation Algorithm for Analyzing Temporal Stability of ACROSS Seismic Source

Masayuki Saeki[a], Koshun Yamaoka[b], *and* Ryoya Ikuta[c]

Contents

1. Introduction

One of the objectives of this research is to estimate the effects of temporal variations of the surface ground properties on the seismic waves excited by the ACROSS source. In this research, we develop a simulation

[a] Department of Civil Engineering, Tokyo University of Science
[b] Research Center for seismology, Volcanology and disaster Mitigation, Graduate School of Environmental Studies, Nagoya University
[c] Faculty of Science, Shizuoka University

E-mail address: saeki@rs.noda.tus.ac.jp (M. Saeki).

Handbook of Geophysical Exploration: Seismic Exploration, Volume 40

ISSN 0950-1401, DOI: 10.1016/S0950-1401(10)04020-6

program of ACROSS seismic waves and investigate the effects of temporal variations of the surface ground. At first, we explain the backgrounds of this research here.

In active geophysical monitoring, signals are continuously (or repeatedly) generated by an artificial source. The excited signals propagate through the Earth's crust and mantle and they are monitored at observation points. If some temporal variations in the observed signals are detected, those variations are analyzed to estimate the changes of crustal model parameters. The changes of signals corresponding to the changes of physical properties related to a target region (like a fault zone) are considered to be very small and they might be easily contaminated by noise. Therefore, a highly stable signal source is needed if there are some limitations on excitation size.

The ACROSS seismic source is designed to generate a very stable force. This source uses the centrifugal force that is generated by rotation of an eccentric mass. The centrifugal force depends only on the rotational frequency of the eccentric mass. Since the phase of rotation of eccentric mass is controlled using a GPS clock that is synchronized to the UTC (Coordinated Universal Time) with an accuracy to 10^{-6}, the strength and the phase of excited force are exceptionally stable.

Because of the temporal stability of the force, one may expect that the ACROSS seismic source generates very stable seismic waves if it is settled on the ground firmly. However, the measurements obtained in site experiments show that the responses observed near to the ACROSS seismic source have changed temporally over a day or seasons (Yamaoka et al., 2001). This means that the excited waves are not necessarily stable even if the force of seismic source is stable. This temporal change seems to make it difficult to detect the small changes of signals corresponding to the changes of physical properties related to a deep target region. Therefore, specifying the cause of temporal variations of excited signals and removing the variations from the observed signals are very important tasks for the active geophysical monitoring.

The cause of these temporal changes has not been made clear yet. The largest possible cause of this change is considered to be a change in the physical properties of the surrounding ground to which the ACROSS seismic source is fixed. Even if the force of the source is stable, the responses could be changed due to a change of ground properties such as a shear modulus. This results in changes of the excited signals. This problem was also pointed out by Alekseev et al. (2005). They conducted an experiment in Russia using their powerful low-frequency vibrator and observed the refracted waves at 80 km distance. The spectra of the observed waves were different in summer and winter. They pointed out that the surrounding ground was frozen in winter and that the freezing ground changed the spectral properties of the ground. These facts stimulated us to investigate the effects of changes of shallow ground properties on the excited seismic waves. If the ground properties of the near field change, the motion of source also changes, and this results in a change of the excited waves. One of the other possible causes is the change of coupling condition between the source and the ground. It is a very important problem how to fix the source on the ground firmly. If the source is not in good contact with the ground, the coupling condition might be affected by environmental conditions. This is an engineering problem and the change might be controllable. Thus, we will not discuss this coupling problem any further here.

In this paper, we focus on the effects of changes of shallow ground properties on the near field responses. First, we review the results of site experiments and discuss the required properties of the simulation algorithm. Next, we try to develop the algorithm of ACROSS seismic waves. This algorithm should be able to simulate the effects of possible mechanisms of the temporal changes. Finally, we numerically investigate the effects of changes in the physical properties of shallow ground on the excited waves using a simple velocity structure model.

2. Brief Review of Observed Temporal Variations of Near Field Responses

In this section, we briefly review two examples of the observed temporal variations of near field responses and discuss the possible causes of the temporal variation. The simulation model is determined according to this discussion.

2.1. Temporal variations observed at Awaji Island

One of the reports related to the temporal variations of near field responses due to the ACROSS seismic source was made by Yamaoka et al. (2001). Here, we briefly explain their experimental conditions and the observed temporal variations.

In their experiment, two ACROSS seismic sources were settled on the surface of granite rock near to the Nojima fault. These sources were driven with center frequencies of 18.0 Hz and 27.5 Hz, respectively. The ACROSS signals were continuously observed by means of three borehole seismometers which were installed at depths of 10 m, 800 m and 1600 m. The observed signals were divided into shorter segments with a period of 100 seconds and these segments were stacked 18 times. Then, one record was made every 30 minutes. A Fourier Transform of each stacked data was performed and the spectral peaks corresponding to ACROSS signals were picked up. The signal levels of the spectral peak corresponding to the ACROSS signals were about 100 times larger than the noise level in the frequency range.

They showed the time series of the amplitudes and the phases of the spectral peak at 18.0 Hz observed at 800 m borehole seismometer. In the time series, temporal variations appeared in the amplitudes and phases, and the daily variations were predominant. The variations of phases were almost similar for all components, even though those of amplitudes were different from each other. It was also reported that similar temporal variations appeared in the time series observed at the 10 m and 1600 m borehole seismometers and that the temporal variations seemed to correlate to the temporal variations of the atmospheric temperature. These two facts indicated that some changes should occur in the shallower part of the ground or in the coupling condition between the source and the ground. However, the mechanism of how the change of atmospheric temperature affected the response was not clarified.

They also tried to explain the temporal variations of responses using a rigid motion of the source in order to remove the temporal variations from the measurements. They observed the vibrations of the source foundation using three seismometers mounted on the source, determined the rigid motion of source, and estimated a linear relationship between the responses and the rigid motion of source. From this analysis, they showed that 90% of the temporal variations of the responses could be explained from the source's rigid motion.

In other experiments carried out in the same experimental site, seasonal variations were also found. For example, from July to September, the phase of responses decreased with time up to 1 ms and suddenly recovered to the usual value after heavy rainfall. From this fact, it was considered that the near field responses were largely affected by the condition of the surrounding rocks around the source. Ikuta et al. (2002) pointed out the possibility that rainfall softened the shallow part of the ground and delayed the phase of the response.

2.2. Relatively stable responses in a large tunnel at the Kazunogawa site

The experimental results of Awaji Island suggest that the near field responses are affected by the change of physical properties of shallow ground. Here, we show the other example in which the ACROSS seismic source is fixed in a deep tunnel.

Higashihara et al. (2002) deployed the ACROSS seismic source in a deep tunnel to avoid the problems of surface ground and Saeki et al. (2003) conducted experiments to investigate the temporal stability of the source. In their experimental site, there is a huge network of tunnels that is constructed for the Kazunogawa power station of the Tokyo Electric Power Company. A large pumped storage power station exists at the bottom of the tunnels at a depth of 500 m. One ACROSS seismic source and three seismometers are deployed in the same tunnel. Figure 1 shows the overview of the tunnels and the deployment of the ACROSS seismic source. The three seismometers are deployed about 3 m, 260 m and 1000 m away from the source, respectively.

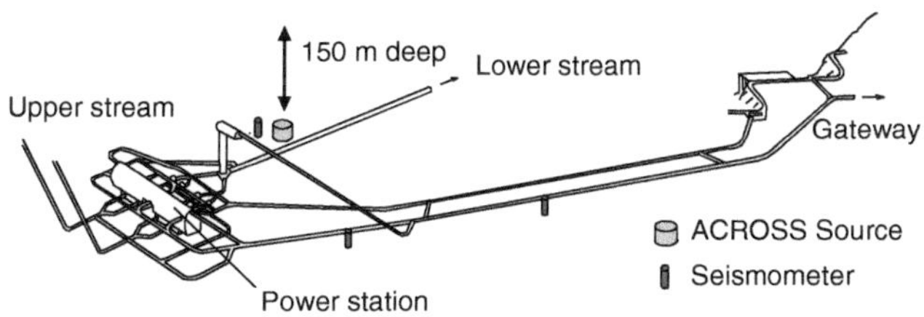

Figure 1 Overview of the tunnels and the deployment of the ACROSS seismic source at the Kazunogawa site.

The source and the seismometers are connected to each other by means of optical fiber cables for synchronizing the source rotation and the data sampling.

Temporal stability of the near field responses was investigated using the ACROSS seismic source and the deployed seismometers. In the observation, the source was continuously driven for 24 hours with a constant frequency. The responses were logged with 1 kHz sampling and saved on a hard disk. The observed acceleration seismogram was divided into shorter segments of 100 s. Each segment was Fourier Transformed and the amplitude and phase of the spectral peak corresponding to the ACROSS signal was estimated. The time series of the estimated amplitudes and phases are plotted in Figure 2. In the figure, the upper three graphs are for the amplitudes of each component, and the lower three graphs are for the phases. The horizontal axis represents the accumulation time [hours] from the start of observation.

In this experiment, the fluctuations of phases are less than 100 ms. The accuracies that are defined as the standard deviation of the amplitudes divided by the averaged amplitude are estimated at about 0.1%. These values are relatively smaller than those of the results at Awaji Island.

2.3. Discussions about the observed results

The results of the two experiments might be summarized as follows:

Temporal variations appear in the time series of amplitudes and phases of the observed ACROSS signals in the case that the ACROSS seismic source is settled on the ground surface;

The temporal variations are similar in the observations which are observed at the depth of 10 m, 800 m and 1600 m on Awaji Island;

About 90% of the temporal variations can be explained by means of rigid motions of the source;

At the Kazunogawa site, the ACROSS seismic source is settled on the ground about 150 m deep. The temporal variations observed at the Kazunogawa site are suppressed compared to the results on Awaji Island.

As mentioned before, the ACROSS seismic source uses a centrifugal force. The excited force is calculated by $mr\omega^2$ where m is the weight of the eccentric mass, r is the radius of rotation and ω is the angular frequency.

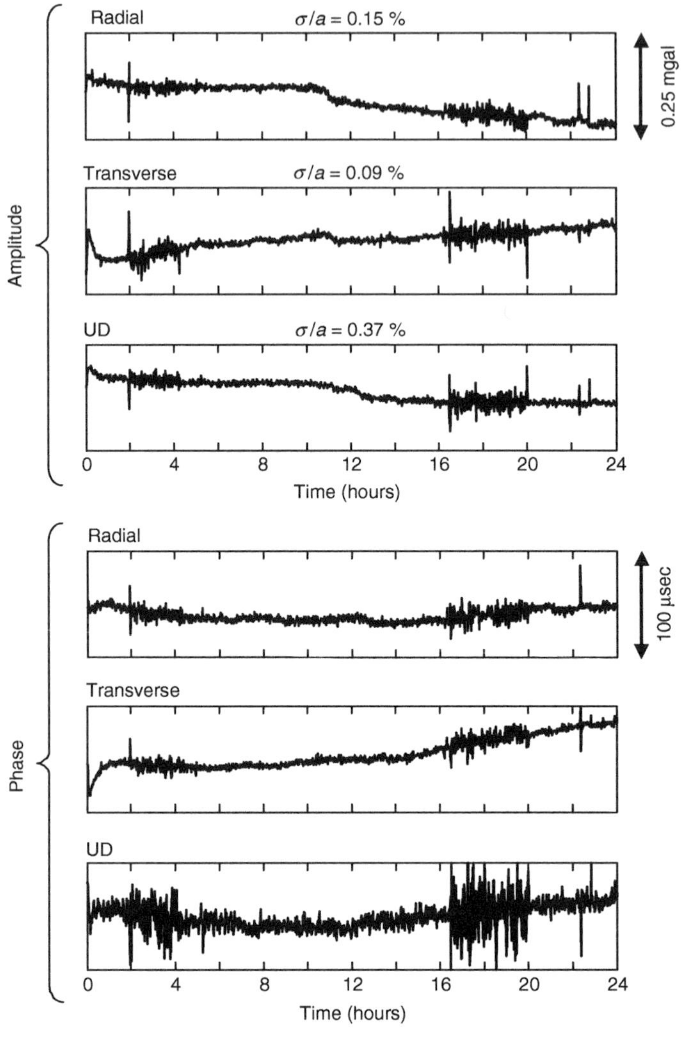

Figure 2 Temporal variations of the amplitudes and phases which are estimated from the data continuously observed at the Kazunogawa site.

Since the ACROSS seismic source is able to control and monitor the phase of rotation very accurately, ω is controllable and the force only depends on the rotational radius r if m does not change. Also the change of radius r is negligibly small because the displacement of source itself is small compared to the rotational radius. Therefore, the force of the ACROSS seismic source is very stable. However, the near field responses have a temporal change. This means that the observed temporal variations are caused due to some changes in the physical properties of the ground. In addition, the above results indicate that the changes must occur in a shallow part of the ground.

One of the significant differences between the two experiments, at Awaji Island and Kazunogawa, is the state of the surrounding rocks at the source. In general, rocks near to the surface of the ground are weathered and under less effective stress. Such rocks might be relatively easier to be affected by weather conditions such as rainfall. Therefore, in the experiment at Awaji Island, it can be considered that the physical properties of the surrounding rocks at the source have a temporal change due to the environmental conditions. This may result in a change of the rigid motion of source and then the near field responses may change temporally.

One of our objectives is to detect the temporal variations of the transfer function corresponding to the change of physical properties around a target region by active geophysical monitoring. The temporal variations of the seismic waves related to the shallow ground to which the ACROSS seismic source is fixed are noise. Therefore, we have to remove the cause of temporal changes in a hardware manner or to develop some algorithm for deleting the temporal changes. However, the mechanism of the temporal variation is not made clear yet. Therefore, we should try to find out the mechanism of the temporal change with the help of numerical simulation.

3. Development of Simulation Algorithm of ACROSS Seismic Waves

In this section, we present the algorithm for calculating the rigid motion of the ACROSS seismic source and the seismic waves generated by the source. In this algorithm, the ACROSS seismic source and the ground are modeled as a horizontally forced rigid circular disk and a multi-layered elastic half space, respectively. We derive the solution of the wave field by applying the explicit Green's function approach (Higashihara, 1984) to the solution of a horizontal point force in a multilayered media.

3.1. Simulation model

First, we consider the required properties of the simulation algorithm. As mentioned in the previous section, the temporal variation of observed signals can be estimated by using the information of the source's rigid motion. In order to study the mechanism of the temporal variation, investigating the relationship between the source's rigid motion and the excited waves seems very helpful. Therefore, the simulation algorithm should be able to calculate the source's rigid motion as well as the generated waves. Also, physical properties such as S wave velocities in the shallow ground should be changeable. The other constraint (or required property) of the simulation algorithm is computational accuracy. The magnitude of the observed temporal changes of signals is very small and it is considered that

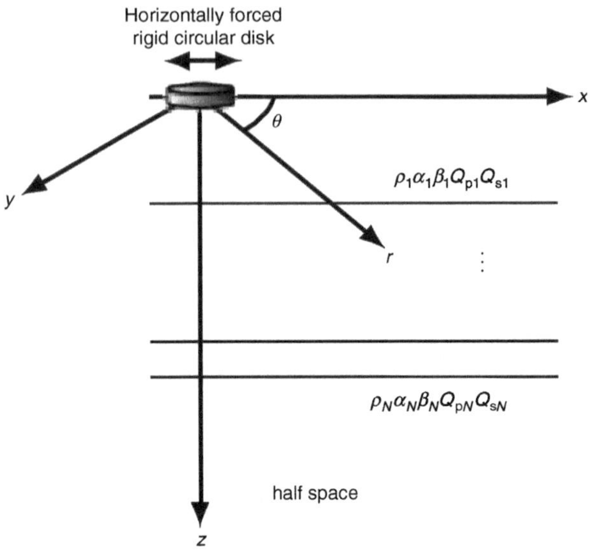

Figure 3 Schematic view of the simulation model and the coordinate system. A horizontally forced rigid circular disk is fixed on the surface of multilayered elastic media.

the corresponding change of physical properties of the model parameters is also small. Therefore, computational accuracy is very important. According to the above considerations on the constraints and required properties, in this research, we model (i) the ACROSS seismic source as a horizontally forced rigid circular disk and (ii) the ground as a multilayered elastic half space.

Figure 3 shows the schematic view of the simulation model and the coordinate system. In this model, the horizontally forced rigid circular disk is deployed on the free surface of multilayered media and the force is applied in the x-direction. Each layer is homogeneous and has its own ground properties. In the figure, ρ_n is the density, α_n is the P wave velocity, β_n is the S wave velocity, Q is the quality factor.

The displacement vector $\mathbf{u}$ and the stresses, σ_{zz}, τ_{rz} and $\tau_{r\theta}$, satisfy the following governing equation and the boundary conditions. The time dependence is assumed to be $\mathbf{e}^{i\omega t}$.

Governing equation:

$$-\omega^2\rho\mathbf{u} = (\lambda+\mu)\nabla(\nabla\cdot\mathbf{u}) + \mu\nabla^2\mathbf{u} + \mathbf{f}, \tag{1}$$

where ω is the angular frequency, λ and μ are Lame constants and $\mathbf{f}$ is the source function.

$$\rho = \rho_n, \quad \lambda = \lambda_n, \quad \mu = \mu_n, \quad \text{if } z_n < z < z_{n+1}. \tag{2}$$

The boundary conditions of the multilayered media are:

$$\sigma_{zz} = 0, \quad \tau_{zr} = 0, \quad \tau_{z\theta} = 0 \tag{3}$$

at the free surface of ground are;

$$\mathbf{u}(z^-) = \mathbf{u}(z^+), \quad \sigma_{ij}(z^-) = \sigma_{ij}(z^+) \tag{4}$$

and at the internal surface of ground are;

$$\mathbf{u}(z) \rightarrow 0, \quad \sigma_{ij}(z) \rightarrow 0 \quad \text{as } z \rightarrow \infty \tag{5}$$

in the half space.

Boundary conditions of the contact plane between the source and the multilayered media are:

$$u_x(0) = \Delta, \quad u_y(0) = 0, \quad \sigma_{zz}(0) = 0, \tag{6}$$

where Δ is the resultant displacement of a rigid circular disk due to the horizontal force.

There is much research on algorithms for calculating the wave field generated by a point source in a multilayered elastic half space. For example, Harkrider (1964) describes the details of a semi-analytical solution of the wave field using the propagator matrix. In his formulation, there is the problem that the components of propagator matrix become very large when high frequency waves propagate through a very low speed media and the large numbers are numerically canceled out. This cancelation decreases the computational accuracy and sometimes makes it difficult to calculate the solution. This problem is serious in the computation of the ACROSS wave field because the frequency band of the ACROSS seismic source is about 5-30 Hz and the source is generally settled on a soft ground surface. Kennett (1974, 1980) overcomes this cancelation problem and presents the details of his formulation. In our simulation program, the Kennett algorithm is employed.

In Kennett's (or other researchers') algorithm for calculating the wave field in multilayered media, the source is assumed to be a point source. In this case, the displacements diverge at the source location and it is impossible to calculate the source's rigid motion. Thus, we need to improve the algorithm so that the rigid motion of source can be calculated. As mentioned above, we model the ACROSS seismic source as a horizontally forced rigid circular disk instead of a point source. The modeling of a rigid circular disk makes it possible to calculate the motion of source. The dynamic response of a rigid circular disk on an elastic half space is well studied and the solution is derived from the solution of a point source in the elastic half space.

The dynamic response of a rigid circular disk due to a horizontal force has been numerically investigated by Luco and Westmann (1971). He obtains the numerical solution of the compliance function by solving Fredholm integral equations of the second kind. Higashihara (1987) presents a different approach called the explicit Green's function approach and checks the validity of the assumptions which are made in Luco's and the other methods by comparing their accuracies. In the explicit Green's function approach, the solution of the dynamic response of a rigid circular disk on the elastic half space is derived from the solution of the point source. In this paper, we apply the explicit Green's function approach (Higashihara's approach) to the problem of a horizontally forced rigid circular disk on the multilayered media.

3.2. Displacements excited by a horizontally forced rigid circular disk deployed on a multilayered elastic half space

We start from the solution of a point source in a multilayered media. The displacements and stresses of a point force are expressed as follows (Harkrider, 1964):

$$u(r,\theta,z) = \int_0^\infty \left[U(z,k)\frac{\mathrm{d}J_1(kr)}{\mathrm{d}kr} + V(z,k)\frac{J_1(kr)}{kr} \right] \mathrm{d}k \cos\theta \tag{7}$$

$$v(r,\theta,z) = -\int_0^\infty \left[U(z,k)\frac{J_1(kr)}{kr} + V(z,k)\frac{\mathrm{d}J_1(kr)}{\mathrm{d}kr} \right] \mathrm{d}k \sin\theta \tag{8}$$

$$w(r,\theta,z) = \int_0^\infty W(z,k)J_1(kr)\mathrm{d}k \cos\theta \tag{9}$$

$$\tau_{rz}(r,\theta,z) = \int_0^\infty \left[\tau_R(z,k)\frac{\mathrm{d}J_1(kr)}{\mathrm{d}kr} + \tau_L(z,k)\frac{J_1(kr)}{kr} \right] \mathrm{d}k \cos\theta \tag{10}$$

$$\tau_{\theta z}(r,\theta,z) = -\int_0^\infty \left[\tau_R(z,k)\frac{J_1(kr)}{kr} + \tau_L(z,k)\frac{\mathrm{d}J_1(kr)}{\mathrm{d}kr} \right] \mathrm{d}k \sin\theta \tag{11}$$

$$\sigma_{zz}(r,\theta,z) = \int_0^\infty \sigma_R(z,k)J_1(kr)\mathrm{d}k \cos\theta \tag{12}$$

where k is the wave number in the radial direction, $J_1(kr)$ is the Bessel function of first kind of order 1. $u(r,\theta,z)$, $v(r,\theta,z)$ and $w(r,\theta,z)$ represents the radial, transversal and vertical components of displacements, respectively. $U(z,k)$, $V(z,k)$ and $W(z,k)$ are the kernels of the wave number integral of displacements. $\tau_R(z,k)$, $\tau_L(z,k)$ and $\sigma_R(z,k)$ are the kernel of the wave number integral of stresses.

Now, we suppose that the integral kernels of stresses, $\tau_R(z,k)$ and $\tau_L(z,k)$, are expressed as the product of the integral kernels of displacements, $U(z,k)$ and $V(z,k)$, and some explicit functions $a(z,k)$ and $b(z,k)$.

$$\tau_R(z,k) = a(z,k)U(z,k) \tag{13}$$

$$\tau_L(z,k) = b(z,k)V(z,k). \tag{14}$$

In the case of the elastic half space, the kernels of wave number integrals are very simple and one can easily find out the explicit relations between the integral kernels of displacements and stresses. If we can successfully obtain the function $a(k)$ and $b(k)$, the explicit Green's function approach can be applied to Eqs (7)–(12). As the result, the integral equations of displacements and stresses are derived as follows;

$$2\Delta = \int_0^R sA(r,s)\tau_x(s)\mathrm{d}s + \int_0^R sC(s,r)\tau_y(s)\mathrm{d}s \tag{15}$$

$$0 = \int_0^R sC(r,s)\tau_x(s)\mathrm{d}s + \int_0^R sB(s,r)\tau_y(s)\mathrm{d}s \tag{16}$$

where $\tau_x(s)$ and $\tau_y(s)$ are the stress distributions on the contact plane of the source and the multilayered media, Δ is the displacement of the rigid circular disk in the x-direction, R is the radius of rigid circular disk. The functions $A(r,s)$, $B(r,s)$ and $C(r,s)$ are expressed as follows:

$$A(r,s) = \int_0^\infty \left(\frac{k}{a(k)} + \frac{k}{b(k)}\right) J_0(kr)J_0(ks)\mathrm{d}k \tag{17}$$

$$B(r,s) = \int_0^\infty \left(\frac{k}{a(k)} + \frac{k}{b(k)}\right) J_2(kr)J_2(ks)\mathrm{d}k \tag{18}$$

$$C(r,s) = \int_0^\infty \left(-\frac{k}{a(k)} + \frac{k}{b(k)}\right) J_2(kr)J_0(kr)\mathrm{d}k. \tag{19}$$

In this research, we assume the stress in the y-direction to be zero for simplicity:

$$\tau_y(s) = 0. \tag{20}$$

The validity of this assumption is studied by Higashihara (1987). The integral Eq. (15) is numerically solved with Eq. (20) and the discrete values of stresses over the rigid circular disk $\tau_x(s)$ are obtained. Using this obtained stress distribution $\tau_x(s)$ as the source function, the integral kernels $U(z,k)$, $V(z,k)$ and $W(z,k)$ in Eqs (7)–(9) are calculated. Next, the displacements excited by the horizontally forced rigid circular disk on a multilayered elastic half space are estimated. On the other hand, the force of source is calculated by integrating the obtained stress over the rigid circular disk.

$$F = \int_0^R 2\pi r \tau_x(r)\mathrm{d}r. \tag{21}$$

We show the procedure to calculate the stress distribution over the contact plane of the rigid circular disk and the displacement field excited by the stress distribution. This procedure can be applied if the functions $a(k)$ and $b(k)$ are known. Next, we try to apply this procedure to the case of a multilayered elastic half space. Now, our problem is how to find out the functions $a(k)$ and $b(k)$.

In Kennett's formulation, the kernels of wave number integrals in in-plane problem are expressed as

$$\begin{bmatrix} U(x,k) \\ W(z,k) \end{bmatrix} = (M_D + M_U R_D^{SN}) V_D(z,k) \tag{22}$$

$$\begin{bmatrix} \sigma_R(z,k) \\ \tau_R(z,k) \end{bmatrix} = (N_D + N_U R_D^{SN}) V_D(z,k) \tag{23}$$

where M_D and M_U are $2*2$ matrices of displacement transformations, N_D and N_U are $2*2$ matrices of stress transformations, R_D^{SN} is the $2*2$ matrix of reflection coefficients and $V_D(z,k)$ is the potential vector of the downward traveling plane wave in Kennett (1979). Deleting $V_D(z,k)$ from Eqs (22) and (23) gives the following equation.

$$\begin{bmatrix} \sigma_R(z,k) \\ \tau_R(z,k) \end{bmatrix} = (N_D + N_U R_D^{SN})(M_D + M_U R_D^{SN})^{-1} \begin{bmatrix} U(x,k) \\ W(z,k) \end{bmatrix}. \tag{24}$$

Solving Eq. (24) yields the functions $a(k)$ and $b(k)$. In the case of elastic half space, the functions $a(k)$ and $b(k)$ are explicitly obtained. On the other hand, in the case of a multilayered half space, the functions $a(k)$ and $b(k)$ are numerically obtained because the matrix R_D^{SN} is very complicated in this case.

3.3. Compliance function

One of the possible mechanisms that explain the temporal variations of near field responses is the temporal change of the compliance function. The compliance function $Ce^{i\phi}$ is the function that determines the relationship between the force and the displacement of the source.

$$Ce^{i\phi} = \Delta/F. \tag{25}$$

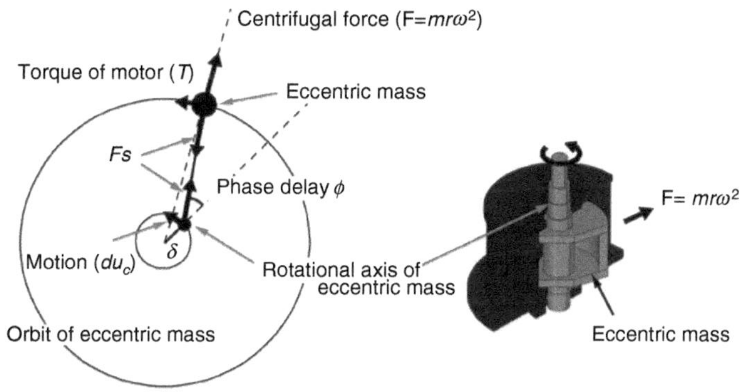

Figure 4 Schematic view of the horizontal motion of the ACROSS seismic source (left) and the source's vertical section (right).

This compliance function includes some parameters related to the ground properties. Thus, the response of the source can be changed due to a change of ground properties through the compliance function.

First, we graphically explain the relationship between the force and the response of the ACROSS seismic source. Figure 4 shows the schematic view of the horizontal motion of the ACROSS seismic source (left). The source's vertical section (right) is also shown in the same figure.

Now, we suppose that the eccentric mass rotates counter-clockwise with a constant angular frequency ω around the rotational axis of the eccentric mass. This rotation is driven by the torque of motor $\boldsymbol{T}$. The rotation generates a centrifugal force $\boldsymbol{F}$ which displaces the rotational axis of the eccentric mass by Δ. The rotational axis moves along a circle of radius Δ. $\boldsymbol{F}_s$ is the force that balances with $\boldsymbol{F}$ and $\boldsymbol{T}$. In the dynamic case, the displacement of rotational axis of eccentric mass has a phase delay ϕ to the force. If $\phi = 0$ were true, the source could not emit any energy to the ground because the orbit of the eccentric mass and the centrifugal force would always be perpendicular to each other. Therefore, ϕ should not be zero in the dynamic case. This phase delay determines the amount of energy that is emitted into the ground. The amount of this energy is equal to the amount of work done by the force $\boldsymbol{F}_s$ that is defined as

$$W = \int \boldsymbol{F}_s \cdot \mathrm{d}\boldsymbol{u}_c \tag{26}$$

where $\mathrm{d}\boldsymbol{u}_c$ is the vector representing the motion of the rotational axis of eccentric mass. As you can see from Figure 4, a larger phase delay ϕ yields a larger work for the force $\boldsymbol{F}_s$. The phase delay ϕ and the displacement Δ can be obtained by solving the dynamic response of the ACROSS seismic source.

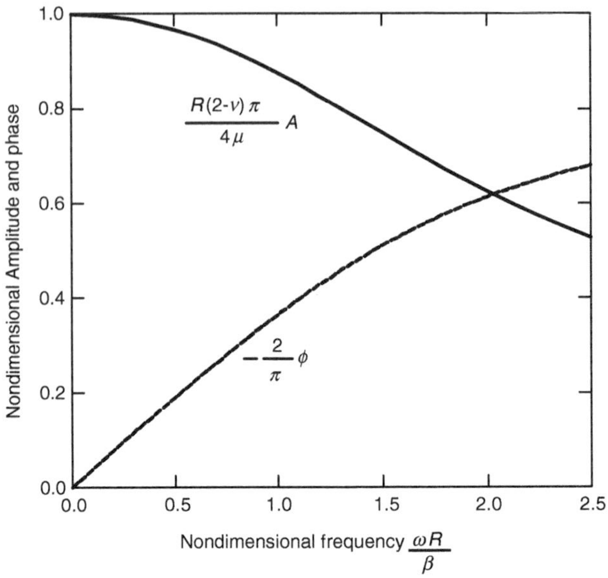

Figure 5 Compliance function $Ce^{i\phi}$ of a rigid circular disk on an elastic half space.

Figure 5 shows the compliance function $Ce^{i\phi}$ of a rigid circular disk on an elastic half space that is calculated by using the program developed in this research. The vertical axis represents the non-dimensional amplitude and phase of the compliance function. The horizontal axis represents the non-dimensional frequency, defined as $\omega R/\beta$, where β is S wave velocity. The non-dimensional frequency corresponding to the general use of the ACROSS seismic source is roughly less than 1.0. In this frequency range, the phase delay is almost linear with the non-dimensional frequency. If the ground becomes softer, the non-dimensional frequency becomes larger and this change makes the phase delay larger.

4. Numerical Simulation

To investigate the effects of shallow ground properties on the source's rigid motion and the waves excited by the ACROSS seismic source, we carry out some numerical simulations using a simple velocity structure model. Here, we show some results of this simulation.

In the numerical simulation, the two layered elastic half space model is assumed. The values of the parameters of the velocity structure model are shown in Table 1. The horizontally forced rigid circular disk is assumed to be fixed on the surface of the layered media. The radius of the rigid circular

Table 1 Numerical simulation model of multilayered elastic half-space

	ρ (kg/m^3)	α (m/s)	β (m/s)	Qp	Qs
Layer 1	2000	1732	1000	100	100
Layer 2	2000	3464	2000	100	100

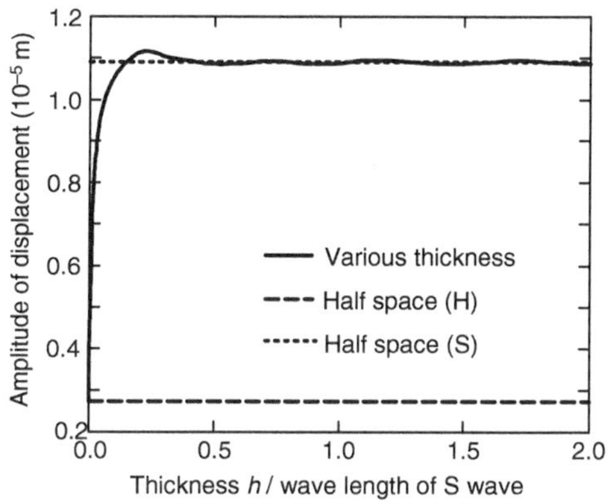

Figure 6 The amplitude of displacement of a rigid circular disk mounted on the free surface of an elastic multilayered half space.

disk is 1 m, the driving frequency of the source is 20 Hz and the force of the source is 100 kN.

4.1. Effect of the thickness of the surface layer on the source's rigid motion

First, we show the effect of the thickness of the surface layer on the source's rigid motion. The source's rigid motion is calculated for various thicknesses (h) of the surface layer (layer 1) from 0 m to 100 m. The amount of displacement of the rigid circular disk and the phase delay are plotted in Figures 6 and 7, respectively. In Figure 6, the horizontal axis represents the thickness of surface layer normalized by the wavelength of the S wave. The vertical axis represents the amplitude of displacement. In Figure 7, the horizontal axis is the same as that of Figure 6 and the vertical axis is the phase delay of the displacement. As a reference, the displacements excited by the rigid circular disk fixed on the elastic half space are also plotted in both figures. In the figures, half space (S) represents the result of the elastic half space whose parameters are the same as layer 1 and half space (H) represents the result of the elastic half space whose parameters are the same as layer 2.

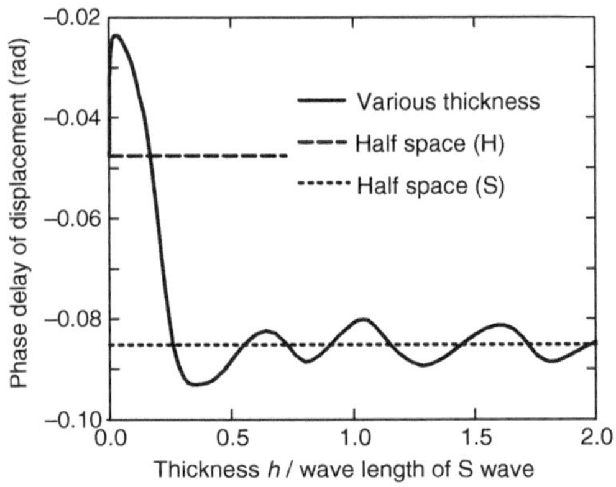

Figure 7 The phase delay of displacement of a rigid circular disk mounted on the free surface of an elastic multilayered half space.

In Figures 6 and 7, the displacement is the same as that of elastic half space (H) when $h = 0$. On the other hand, as h becomes larger, the displacement seems to converge to the same response as in the case of the elastic half space (S). These facts are consistent with our intuition and the numerical results seem reliable.

It is shown that the response of the rigid circular disk drastically changes with the thickness of the soft surface layer, especially when the thickness h is less than a quarter of the wavelength of the S wave $\lambda_\beta/4$.

4.2. Effect of the thickness of the surface layer on the waves excited by the ACROSS seismic source

Next, we investigate the response beneath the ACROSS source at a depth of 1 km. The amplitude and phase of the displacement are plotted in Figures 8 and 9, respectively. The vertical axis and the horizontal axis are the same as those of Figures 6 and 7.

In Figure 8, it is shown that the displacement is amplified at $h/\lambda_\beta = 3n/4$, $(n = 1, 2, \ldots)$. When the condition $h/\lambda_\beta = 3n/4$ is satisfied, the generated waves propagating in the surface layer and the waves reflected at the internal surface between layer 1 and 2 have the same phase. Thus the displacement is amplified. This behavior of displacement at a depth of 1 km is very different from the motion of a rigid circular disk. If there is a thin surface layer, which means the thickness of the surface layer is less than a quarter of the wavelength of the S wave, the source's rigid motion is largely affected by the properties of the surface layer but the responses at depth are not so sensitive to the change of surface ground properties. A similar behavior is also shown in the graph of phase.

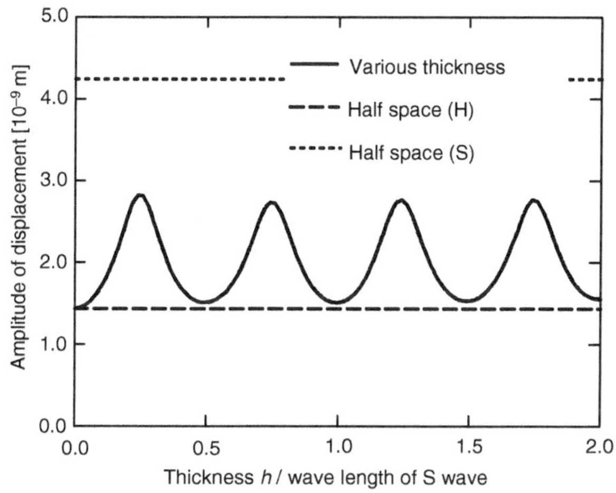

Figure 8 The amplitude of displacement at a point 1 km deep beneath the rigid circular disk.

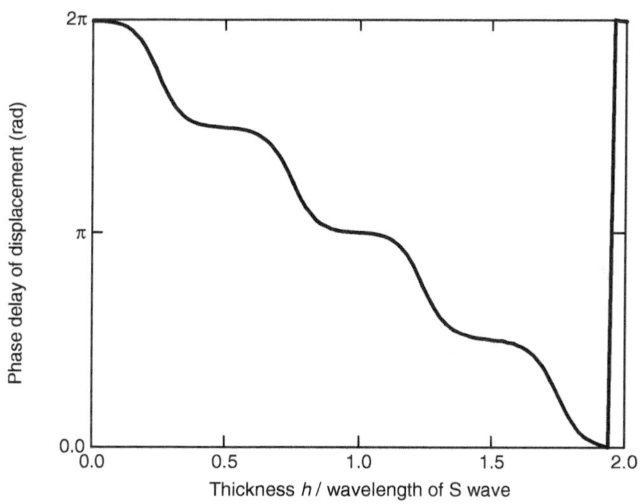

Figure 9 The phase of displacement at a point 1 km deep beneath the rigid circular disk.

5. Summary

The temporal stability of a seismic source is a very important problem for active geophysical monitoring. One of the main causes of the temporal variation could be a change of shallow ground properties. However, the mechanism of the temporal variation is not clarified yet. Therefore, we developed a simulation program of ACROSS seismic waves to estimate the effects of the shallow ground properties on the excited waves.

In this paper, we review the experimental results of temporal variations and discuss the required properties of the simulation algorithm. According to the results of the discussion, we develop the algorithm for calculating the rigid motion of the ACROSS seismic source and the wave field generated by the source. In this algorithm, the ACROSS seismic source and the ground are modeled as a horizontally forced rigid circular disk and a multilayered elastic half space, respectively. We also simulate the source's rigid motion and the generated waves by using the program developed in this research. The velocity structure model in this simulation is very simple. We consider a two layered media and investigate the effect of thickness of the surface ground on the source's rigid motion and the generated waves at a deep point beneath the source.

Numerical simulation based on a more realistic velocity structure should be carried out as future work. Also the relationship between the source's rigid motion and the responses at a deep point should be investigated in detail, since the signals observed at a far field point might be corrected using the information of the source's rigid motion or of near field responses. The simulation program developed in this research seems a useful tool for studying the temporal stability of the ACROSS seismic source.

REFERENCES

Alekseev, A.S., Chichinin, I.S., Korneev, V.A., 2005. Powerful low-frequency vibrators for active seismology. Bulletin of the Seismological Society of America 95, 1–17.

Harkrider, D.G., 1964. Surface waves in multilayered elastic media I. Rayleigh and love waves from buried sources in a multilayered elastic half space. Bulletin of the Seismological Society of America 54, 627–679.

Higashihara, H., 1984. Explicit Green's function approach to forced vertical vibrations of circular disk on semi-infinite elastic space. Journal of Engineering Mechanics Division of the American Society of Civil Engineering 110 (EM10), 1510–1523.

Higashihara, H., 1987. Coupled horizontal and rocking vibration of circular disk on elastic half space. Journal of the Japanese Society of Civil Engineering 386 (I-8) (in Japanese).

Higashihara, H., Saeki, M., Weijian, Zeng, 2002. High-resolution seismic tomography based on coherent wave technologies: Proceedings of the 15th Engineering Mechanics Conference of the American Society of Civil Engineering, CD-ROM.

Ikuta, R., Yamaoka, K., Miyakawa, K., Kunitomo, T., Kumazawa, M., 2002. Continuous monitoring of propagation velocity of seismic wave using ACROSS. Geophysical Research Letters.

Kennett, B.L.N., 1974. Reflections, rays and reverberations. Bulletin of the Seismological Society of America 64, 1685–1696.

Kennett, B.L.N., 1979. Seismic waves in a stratified half space. Geophysical Journal of the Royal Astronomical Society 57, 557–583.

Kennett, B.L.N., 1980. Seismic waves in a stratified half space—II Theoretical seismograms. Geophysical Journal of the Royal Astronomical Society 61, 1–10.

Luco, J.E., Westmann, R.A., 1971. Dynamic response of circular footings. Journal of Engineering Mechanics Division of the American Society of Civil Engineering 110 (EM5), 1381–1395.

Saeki, M., Higashihara, H., Ohtake, Y., 2003, Deep deployment of new ACROSS seismic tomography instruments and its performance assessment. In: Proceedings of the 16th Engineering Mechanics Conference of the American Society of Civil Engineering, CD-ROM.

Yamaoka, K., Kunitomo, T., Miyakawa, K., Kobayashi, K., Kumazawa, M., 2001. A trial for monitoring temporal variation of seismic velocity using an ACROSS system. The Island Arc 10, 336–347.

CHAPTER 16

Nonlinear Processes in Seismic Active Monitoring

Marat S. Khairetdinov[a] *and* Gulnara M. Voskoboynikova[a]

Contents

1. Introduction

The purpose of active vibroseismic monitoring of seismically dangerous zones is to observe the rheological characteristics of the geological medium during the geodynamic processes that precede large earthquakes. Such monitoring is based on regular sounding of the Earth using recurrent seismic oscillations from low-frequency vibrators. Each sounding is a *vibrosession*, an analysis of the medium's response parameters extracted from the recorded data (Alekseev et al., 2004, 1998). The repeatability of vibrosessions can be disrupted by various interfering factors: seasonal factors associated with the freezing and thawing of the ground under the vibrator, instrumental errors affecting the intensity characteristics of the radiation field, etc. There is thus the difficulty of providing informative field parameters independent from these unwanted factors.

In addition, vibrosessions are accompanied by nonlinear effects that manifest themselves during the radiation and propagation of vibroseismic

[a] Institute of Computational Mathematics and Mathematical Geophysics SB RAS, Blvd. Akademika Lavrenteva, 6, Novosibirsk, 630090, Russia

E-mail address: marat@opg.sscc.ru (M.S. Khairetdinov).
Handbook of Geophysical Exploration: Seismic Exploration, Volume 40
ISSN 0950-1401, DOI: 10.1016/S0950-1401(10)04021-8

oscillations. The problem with nonlinear effects occurring during acoustic wave propagation in elastic isotropic media has been studied for several decades. In particular, Polyakova (1964) investigated those nonlinear effects that appear in an elastic isotropic medium during finite deformations and the propagation of an elastic wave through calculation of cubic terms in the elastic energy. Polyakova showed that the interaction between longitudinal and transverse waves leads to the appearance of second harmonics in the transverse wave. More recently, Korneev et al. (1998) presented results based on Murnaghan's elastic theory for elastic wave propagation in isotropic solids, as well as a summary of the major results from the nonlinear interaction of plane elastic waves. Other important studies involving the nonlinear effects of wave propagation in an elastic, isotropic medium are presented in the works of Zarembo and Krasilnikov (1966), Rudenko and Soluyan (1975), and Brekhovskich and Goncharov (1982).

Nikolaev (1987) suggests that fractured media in seismically active regions cause nonlinear effects in propagating waves. The related problems have not only scientific value, but also have practical implications, related to the sensitivity and accuracy of vibroseismic monitoring. This chapter describes some experiments aimed at estimating the nonlinear effects caused by vibrators, as well as the propagation of seismic waves over a 335 km long region of the earth's crust during periods of lunar-solar tides.

We show that taking into account the amplitude ratios between the multiple and fundamental harmonics of seismic wave fields enables monitoring results to be independent of seasonal and instrumental variations. At the same time, the high sensitivity of these ratios to small stress variations in the Earth's crust is retained.

2. Nonlinear Phenomena

Nonlinearity in seismic wave propagation is caused in two ways: by the source and by the geological medium. Source nonlinearity results from source-ground coupling and from the strong nonlinear properties of the soil under the source, leading to generation of sub- and multiple harmonics. As an illustration, Figure 1 presents the spectrum versus time oscillations radiated by two types of vibrators: the eccentric vibrator CV-100 (Figure 1(a)) and the hydroresonance vibrator HRV-50 (Figure 1(b)), with initiation forces of 100 tons and 50 tons, respectively (Alekseev et al., 2005). In the first case, the generated spectrum is associated with a linear frequency modulation (the sweep signal) of 6.25-9.5 Hz and duration of 600 s. In the second case, we have 5-7 Hz and 1.400 s, respectively. In both cases, the nonlinear radiation effect is characterized by the appearance of second and third harmonics in the spectra.

To estimate the level of nonlinear effects for both mentioned sources, in Figures 2(a) and 2(b), we plotted the histograms corresponding to the

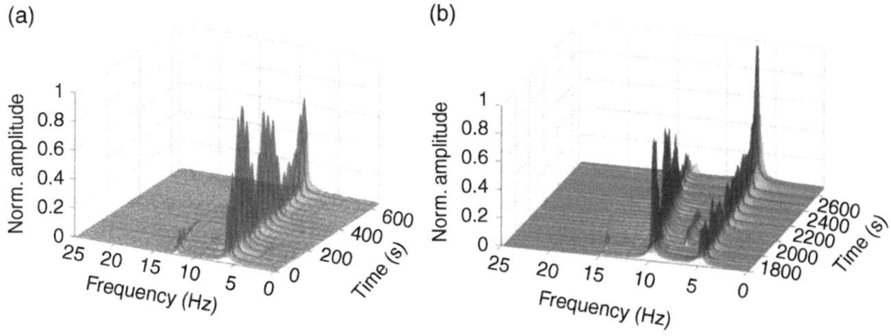

Figure 1 Amplitude spectra vs. time for linear frequency modulation radiated by two types of vibrators: (a) the centrifugal vibrator CV-100 (frequency range of mainly 5.5-8.5 Hz); (b) the hydroresonance vibrator HRV-50 (frequency range of mainly 5-7 Hz). Higher harmonics are clearly defined in both spectra.

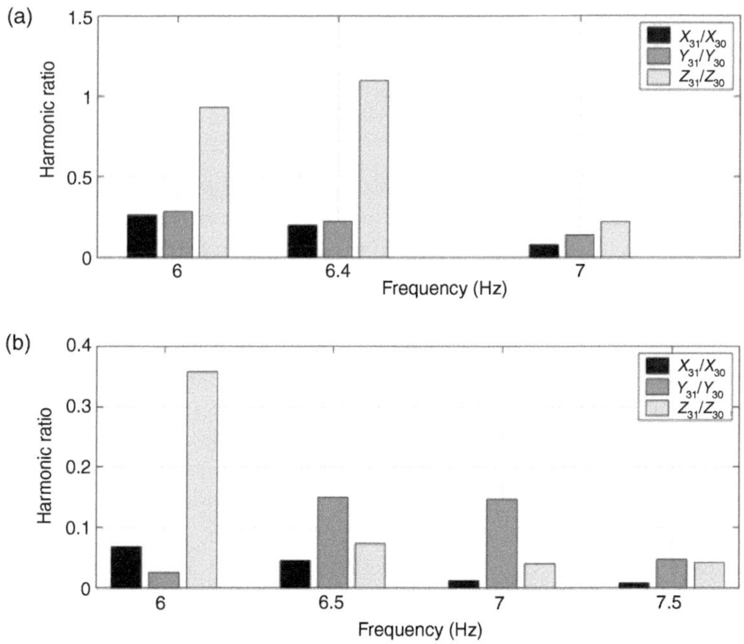

Figure 2 The histograms for the ratios between the amplitudes of the second and base harmonics of vibrators HRV-50 and CV-100 at the discrete radiation frequencies 6, 6.4, 6.5, 7.0, and 7.5 Hz and for spatial coordinates x, y, z of the wave field, where x and y are horizontal components, and z is the vertical field component. There is a considerable contribution of second harmonics to the radiation field in the harmonic regime, which reaches 100% and more for the first source type and about 35% for the second source type.

ratios (nonlinear coefficients) between the amplitudes of the second and fundamental harmonics at the discrete radiation frequencies of 6, 6.4, 6.5,

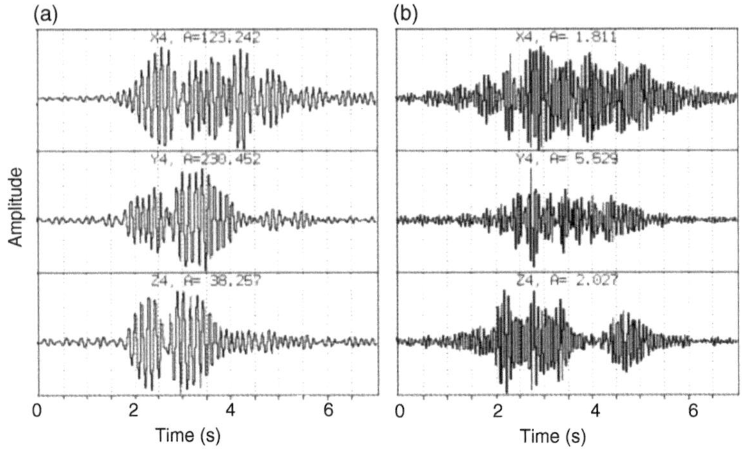

Figure 3 Vibrational seismograms for fundamental and second harmonics at a distance of 0.3 km. The ratios between the maximum values of secondary and primary waves for X, Y, and Z components are 1.5%, 2.3%, and 5.3%, respectively. A higher nonlinearity level is observed for a vertically polarized perturbing force. One can see higher wave resolution in the zone of second harmonics than in the first ones.

7.0, and 7.5 Hz. These ratios are calculated for x, y, z components of the wave field, where x and y are horizontal components, and z is the vertical component. The histograms show that nonlinearity at certain frequencies can reach 100%. This effect is associated with the vibrators' design, failure to meet the conditions of permanent contact of the source with the ground, or the ground's nonlinear properties (Khairetdinov, 2003).

Below, we present the results of data analysis, taking into account the nonlinear effects of the seismic wave field. When the source conducts sweeps in a wide frequency band, the algorithm for calculating vibrational seismograms is based on the correlation convolution between the recorded seismic signals and the reference signal, which repeats the shape of the sounding signal from the oscillation source: Here, $u(t_i)$ is the recorded seismic signal. The reference (fundamental) signal $v(t_i)$ has the form $v_1(t_i) = A\cos(\omega_0 t_i + \beta t_i^2/2)$ and the second harmonic has the form $v_2(t_i) = A\cos(2\omega_0 t_i + \beta t_i^2)$, with β the frequency sweep rate. In this case, the signal $v_1(t_i)$ is within a 5.5-8.5 Hz frequency band, and $v_2(t_i)$ has an 11-17 Hz frequency band. We recorded vibrational seismograms corresponding to different distances from the vibrator, namely 0.3 km (Figure 3), 20 km, and 50 km (Figure 4). In these figures, the amplitudes of dominant waves are additionally presented, in discrete units of the analog-to-digital converter. Figure 3(a) presents the results of convolution within the frequency bands of 5.5-8.5 Hz (base frequency); Figure 3(b) shows the results for the frequency bands of 11.0-17 Hz (second harmonics). It is clear that seismograms of the second harmonics are characterized by the sharper arrivals of first

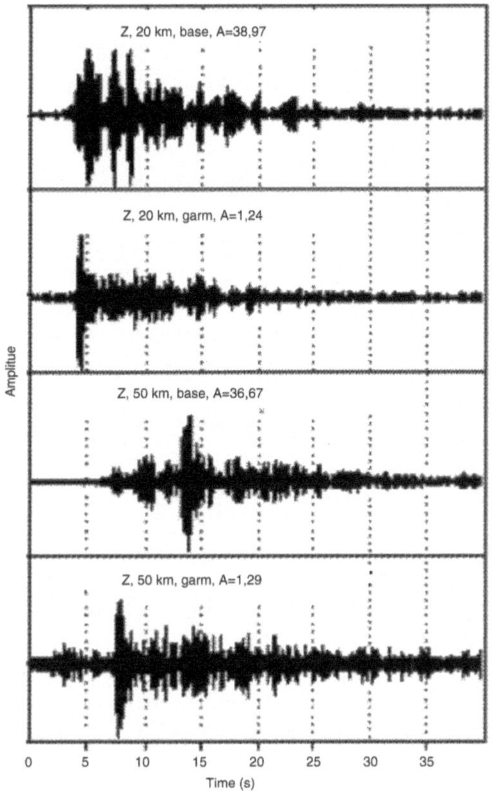

Figure 4 Vibrational seismograms for fundamental and second harmonics at distances of 20 and 50 km. The second harmonic constitutes about 3% at a distance of 20 km and about 17% at a distance of 50 km from the source (Z-component). Greater contrast between the arrivals of primary waves in the second harmonic is observed.

(longitudinal) waves (4.9 s at 20 km, 7.8 s at 50 km), which increase the accuracy of measurement for wave arrival times. From the point of view of seismic physical processes, the higher contrast in the arrival of P-waves results from their saturation at the high frequencies of the second harmonics. The level of nonlinear effects determined by the ratio between the maxima of the secondary and reference seismograms is about 3%.

3. Nonlinear Processing of Vibrational Seismograms

In deep vibroseismic sounding (DVS) of the Earth, there are some difficulties in determining the arrival times using vibrational seismograms obtained from a source located long distances away (hundreds of kilometers). The difficulty stems from the limited power and limited range of the vibrator excitation frequencies, compared to powerful explosions. Here, the influence of external noise on the process of wave detection is significant. Some authors (Glinsky et al., 2000) avoid these difficulties using synchronous stacking of vibrograms (vibrational seismograms) along the lineup axes and over a series of repeating sounding sessions. In their case, the dependence of the total wave amplitude on the direction of summation creates the effect of directional sensitivity, which, by analogy with antennas, can be called a directional characteristic (DC). Attempting to increase the accuracy in determining wave arrival time using a limited number of seismic receivers ($n \sim 5$), raises the notion of nonlinear processing, which forms the basis of the multiplicative antenna. In this case, in contrast to antennas with linear processing, we can achieve more accurate resolution in the direction of wave propagation with nonlinear processing, using the same number of antenna elements. In the case of n sensors, the output signal for nonlinear processing has the form:

$$u_{\Pi} = \overline{u_1 \cdot u_2 \cdot \ldots \cdot u_n} \approx F_E^n \cdot \Psi_1(k, \theta, d), \tag{1}$$

whereas, for linear processing, it has the form:

$$u_{\Sigma} = \overline{u_1 + u_2 + \cdots + u_n} \approx n \cdot F_E \cdot \Psi_2(k, \theta, d), \tag{2}$$

where F_E is the envelope of the seismogram at the output of one sensor, θ is the angle between the direction of wave arrival perpendicular to the line of seismic sensor locations, $k = 2\pi/\lambda$ is the wave number, and d is the distance to the next sensor. Amplitude variations of the functions ψ_1 and ψ_2 are limited, since they are formed from trigonometric functions. Taking this into account, functions F_E are determined in Eqs (1) and (2).

Vibrational seismograms obtained as a result of the correlation (Eq. (1)) correspond to a zero phase signal emitted by the source, and hence its arrival time is determined by the maximum of the wave amplitude. The variations in the wave arrival estimate in the direction of propagation θ were determined by the second derivative (Tikhonov, 1983) in the form $D_\theta = -1/(\mathrm{d}^2 u(\theta)/\mathrm{d}\theta^2)|_{\theta_0}$, where θ_0 corresponds to the best sensitivity of the receivers achieved in the direction of wave-front propagation. The search algorithm for this direction is based on the calculation of $u_{\Pi}^* = \max u_{\Pi}$ for a set of linear hodographs whose angular location is within the angular window chosen initially, and using an expected wave velocity. In this case, the calculation is made by using an exhaustive method. For the particular case at $n \approx 1$, for nonlinear processing $D_{\theta 1}$ has the order $\sim 1/n^2$, and for linear processing has the order $D_{\theta 2} \approx 1/n$. Thus, the error in determining the wave propagation direction, determined by the hodograph's direction of

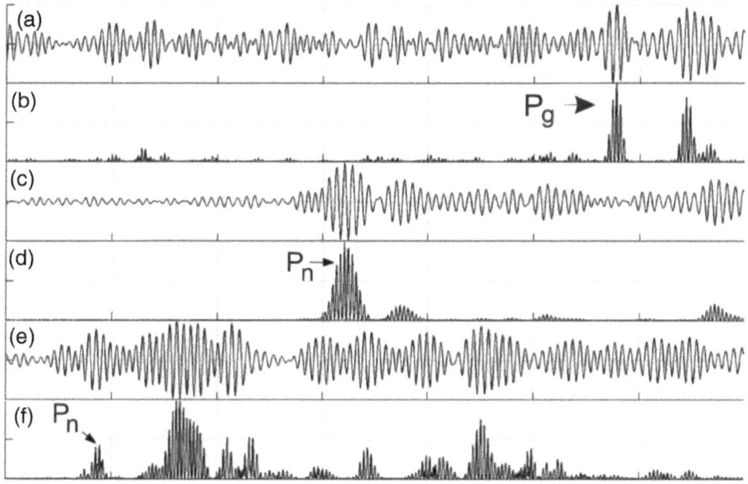

Figure 5 Seismograms with linear processing (a, b, c) and nonlinear processing (d, e, f). Nonlinear processing produces more distinguishable arrivals of primary waves and a higher accuracy in determining their arrival times.

seismic traces, is for nonlinear processing (Eq. (1)) by a factor of n smaller than for linear processing (Eq. (2)).

The high resolution of nonlinear processing is illustrated for the vibrational seismograms obtained from the CV-100 vibrator at DVS in the Degelen-Bystrovka direction (Figure 5). The recording points were located at distances of 304, 342, and 371 km from the source. The sounding was made by sweep-signals in the 5.85-8.0 Hz frequency range, with a frequency sweep duration of 31 min, 29 s.

The top of Figure 5 contains seismograms at (a) 371 km, (b) 342 km, and (c) 304 km, obtained by linear summation of seismic traces along theoretical travel-times (hodographs) corresponding to the waves refracted from the Moho boundary ($V_P = 8.1$ km/s). In Figure 5 (bottom), the seismograms are for: (d) 371 km), (e) 342 km, and (f) 304 km, obtained by nonlinear processing of the same data as for linear processing. Comparing the results of both types of processing shows that after the nonlinear processing, the contrast in P_n-wave arrivals and the sharpness of their maxima are greater than after the linear processing. For these distances, the arrival times t_P are 56, 52.5, and 47.6 s, respectively.

4. Nonlinear Phenomena in Seismic Monitoring

Modern concepts of earthquake source development, based on the kinetic theory of rock failure by Zhurkov (1968) and the multidisciplinary approach by Alekseev et al. (1998), consider the processes of intensive fracturing involved in potential earthquake zones. Following their approach, we use crack density function $\theta(x, y, z, t)$ for spatial coordinates and time to characterize the medium.

For seismic methods, we use the parameters (such as amplitude, frequency, and phase) of observed longitudinal and transverse waves to obtain more detailed information about the changes in fractured zones. Alekseev et al. (2004) suggest that the dynamic characteristics of the wave field are most sensitive to changes in the geological medium's elasticity. Also, the medium's fracturing is a likely physical basis for development of the seismic event. This suggests that the nonlinear components (such as multiple harmonics) in the recorded wave fields should be taken into account (Glinsky et al., 2006). Thus, it is important to find the relationship between the wave field nonlinearity and the parameters of a medium's fracturing.

Such a relationship was obtained by Verbitsky (1987) for a model of a fractured zone in a homogeneous isotropic medium with the elasticity modulii K_1, μ_1 and density ρ_1. Uniformly scattered and randomly-oriented spheroidal voids were taken as the initial model of fracturing. The shape of the voids is determined by the parameter α, which is equal to the ratio between the rotation axis length of a spheroid and the length of its second axis. The distribution of the relative volume of voids between its minimal value $\alpha_{\min}$ and maximal value $\alpha_{\max}$ is described by the function $\varphi(\alpha)$. It is assumed that the length of an elastic wave with the highest frequency propagating in the medium being modeled is much greater than the linear dimensions of the largest voids.

The following relationships are used for the effective elasticity module of a medium with spheroidal voids:

$$K_{(1)} \approx K_0 \left[1 - \varphi_{\mathrm{rel}} a \frac{K_0^2}{p_0 K_1} f(\alpha_0) u_{ll} \right],$$

$$\mu_{(1)} \approx \mu_0 \left[1 - \varphi_{\mathrm{rel}} b \frac{K_0 \mu_0}{p_0 \mu_1} f(\alpha_0) u_{ll} \right].$$

Here $K_0 \approx K_1 (1 + \varphi_{\mathrm{rel}} a F)^{-1}$, $\mu_0 \approx \mu_1 (1 + \varphi_{\mathrm{rel}} b F)^{-1}$,

$$a = \frac{4(1 - \nu_1^2)}{3\pi(1 - 2\nu_1)}, \quad b = \frac{8(1 - \nu_1) \cdot (5 - \nu_1)}{15\pi(2 - \nu_1)}, \quad f(\alpha_0) = \frac{\varphi(\alpha)}{\varphi_{\mathrm{rel}}},$$

$\varphi_{\mathrm{rel}} = \int_{\alpha_0}^{\alpha_{\max}} \varphi(\alpha) d\alpha$ is the initial value of the medium's fracture porosity; $F = \int_{\alpha_0}^{\alpha_{\max}} f(\alpha)/\alpha d\alpha$; ν_1 is Poisson's coefficient, K_0 is the effective compression modulus of the micro fractured medium, and u_{ll} is the sum of diagonal components for the dynamic deformation tensor.

The equation of propagation for plane monochromatic elastic waves along the OX-axis in the medium when only longitudinal motions are present in this medium ($u_x \neq 0, u_z = u_y = 0$) has the form

$$\rho_0 \frac{\partial^2 u_x}{\partial t^2} - M_0 \frac{\partial^2 u_x}{\partial x^2} = B \frac{\partial u_x}{\partial x} \cdot \frac{\partial^2 u_x}{\partial x^2}. \tag{3}$$

At the boundary condition $u_x(0, t) = U_x \sin \omega t$, the solution to the equation in a second approximation has the following form (Verbitsky, 1987):

$$u_x = U_x \sin \omega \left(t \mp \frac{x}{c_P} \right) - \left(\frac{U_x}{2} \right)^2 \frac{B}{M_0} k_P^2 x \cos 2(\omega t \mp k_P x), \tag{4}$$

where $k_P = \omega / c_P$, $B = -3\varphi_{\text{rel}} \frac{k_0}{p_0} (a \frac{k_0^2}{k_1} + \frac{4}{3} b \frac{\mu_0^2}{\mu_1}) f(\alpha_0)$, $M_0 = k_0 + \frac{4}{3}\mu_0$, x—wave travel length.

It follows from Eq. (4) that in a fractured medium, there appears a harmonic with doubled frequency. The harmonic's amplitude is determined by the coefficient B, which depends on the medium's fracturing parameters, the Mach number $M = U_x \omega / c_P$, and the wave travel length x, increasing proportionally with x. This phenomenon was noted earlier as an accumulating nonlinearity in a nonlinearly elastic medium (Polyakova, 1964). Taking into account Eq. (4), the coefficient of nonlinearity for the monochromatic wave is determined by the ratio between the amplitudes of the fundamental and second harmonics:

$$\frac{u_2}{u_1} = \frac{1}{8} \frac{U_x B k_P^2 x}{M_0}. \tag{5}$$

Equation (5) relates the parameters of wave-field nonlinearity to the medium's fracturing (determined by the parameter B), which depends on the sizes of fractures and their distribution density, as well as on the medium's elasticity modulus. Taking into account this dependence and the aforementioned role of fractures in earthquake development, it is likely that the dynamic parameters of wave-field nonlinearity can be successfully used as an important parameter for monitoring.

We analyzed the nonlinearity coefficient from Eq. (5) as a function of the medium's fracturing, the amplitude of particle velocity U_x, and the distance x. Water-saturated fractured granite was chosen as the medium, with the following elasticity parameters: Young's modulus $E = 2.216 \cdot 10^9$ Pa, Poisson's coefficient $\nu = 0.44296$, static pressure $p_0 = 10^3$ Pa, frequency $f = 10$ Hz, and propagation velocity of the P-wave in granite

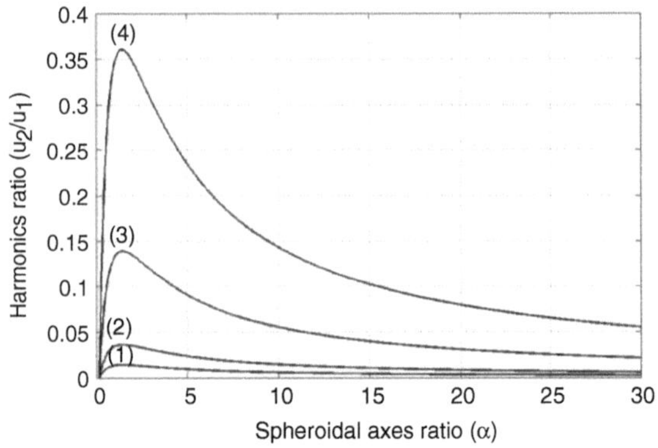

Figure 6 The ratio between the second and fundamental harmonics versus the geometrical parameters of the medium's inhomogeneity due to fracturing. Young's modulus $E = 2.216 \times 10^9$ Pa, Poisson's coefficient $\nu = 0.44296$, static pressure $p_0 = 10^3$ Pa, frequency $f = 10$ Hz, propagation velocity of the P-wave in granite $C_P = 5000$ m/s. Wave travel path: Curves (1) and (2), $x = 10$ km; Curves (3) and (4), $x = 100$ km.

$C_P = 5000$ m/s. The curves in Figure 6 show the nonlinearity coefficient of the monochromatic wave for these parameters versus the ratio between the ellipsoid axes describing an elementary fracture. Curves 1 and 2 were computed for the wave propagation length $x = 10$ km; Curves 3 and 4 for 100 km. The medium's particle velocity $U_x = 2.7 \times 10^{-8}$ m/s corresponds to Curves 1 and 3, the velocity of 70×10^{-8} m/s corresponds to Curves 2 and 4.

The presence of maxima in the plots indicates that the nonlinear effects of wave propagation predominate in inhomogeneous media with limited sizes of inhomogeneities.

The dependence of the second harmonic's amplitude on the propagation distance in an attenuating medium can be described by the following relationship (Zarembo and Krasilnikov, 1966):

$$a_2 = \frac{K_c x \omega^2 a_1^2}{8C_{P,S}^2}. \tag{6}$$

Here, x is the wave travel length, K_c is the nonlinearity coefficient of the medium determined by the expression $K_c \approx \rho\nu(\Delta\nu/\Delta p) \approx (\Delta\nu/\nu)\Delta\theta$, in which ρ is the density, $\Delta\theta$ is the volume deformation variation, Δp is the pressure variation, a_1 is the amplitude of the first harmonic, and $C_{P,S}$ is the velocity of the P- or S-wave. The value of this coefficient estimated by some authors is $K_c \approx 10^3$. As an illustration, Figure 7 presents curves for the nonlinear effect "accumulation" versus distance at given seismic velocities, in a "source-receiver" distance range of 0.3-355 km. Curve 1 here corresponds

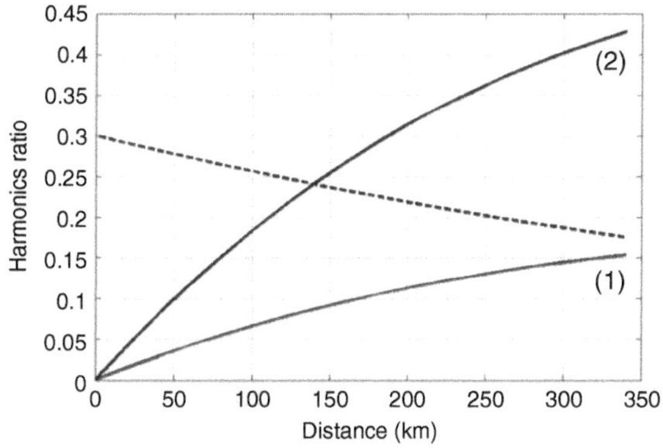

Figure 7 Coefficient of nonlinear distortions versus distance at different seismic velocities: (1) $C_P = 5000$ m/s, (2) $C_S = 3000$ m/s. Attenuation of this coefficient due to wave energy absorption in the medium is shown by the dashed line. The data are for the radiation regime of harmonic oscillations at a basic frequency of 6.3 Hz.

to the velocity $C_P = 5000$ m/s, while curve 2 corresponds to the velocity $C_S = 3000$ m/s. The nonlinearity coefficient is obviously larger for smaller seismic velocities.

It follows from an analysis of Figures 6 and 7 that the ratio between the second and the fundamental (first) harmonics can increase by several times as the wave travel path increases. The presence of maxima in the plots (Figure 6) indicates that the nonlinear effects of wave propagation predominate in inhomogeneous fractured media.

The nonlinearity parameters of the wave field at large distances also carry the influence of the vibrator nonlinearity (Glinsky et al., 2002). Changes in the initial ratios between the second and fundamental harmonics a_{02}/a_{01} vary (depending on the distance) in accordance with the following approximate law, which is valid for long observation distances (hundreds or more kilometers) and low frequencies:

$$\frac{a_{f_2}(x)}{a_{f_1}(x)} = \frac{a_{02}}{a_{01}} \exp[(\alpha_2 - \alpha_1)(x - x_0)], \tag{7}$$

where $\alpha_{1,2} \approx 2.5 \cdot 10^{-4} f_{1,2}$ (1/km) is the medium's attenuation coefficient, and x_0 is the "source-receiver" distance near the nonlinear volume's center.

As an example, the dashed line in Figure 7 shows the attenuation curve (Eq. (7)) for the cases in which $f_1 = 6.3$ Hz and $f_2 = 12.6$ Hz. Therefore, the effect of propagation distance on nonlinear parameters is driven by a resonance linear increase (Eq. (6)) and by an exponential decrease due to attenuation (Eq. (7)). The combined effect of these two factors defines the natural spatial bounds of the nonlinear effect.

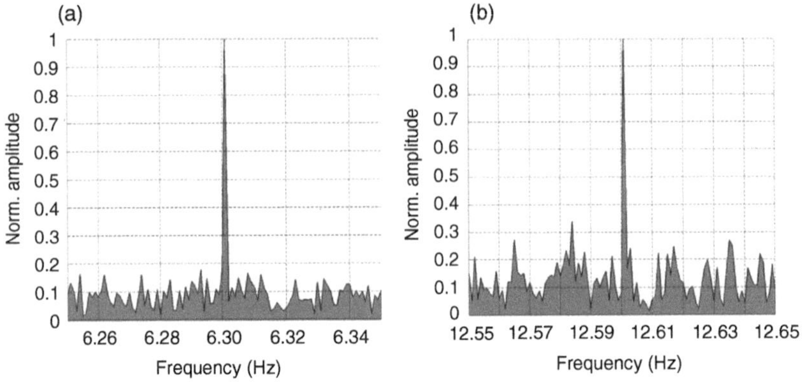

Figure 8 Spectrograms of the fundamental (a) and second (b) harmonics at a distance of 355 km. Here, f is the basic frequency radiated by the source, ρ is the signal/noise ratio, and m is the amplitude of the harmonic at ADC levels.

5. Experimental Results

The nonlinearity effect was estimated using monochromatic signals, based on the amplitude ratios of the second and fundamental harmonics. The spectrograms for the sums of monochromatic signals and noise confirm the possibility of detecting both types of harmonics at a "source-receiver" distance of 355 km—in this case, the signal of the fundamental harmonic had the frequency of 6.3 Hz, whereas the frequency of the second harmonic was 12.6 Hz (as presented in Figures 8(a) and 8(b), respectively). The parameters of the spectrograms are shown in the upper part of the figures: m is the amplitude of the spectral peak at the frequency f, and ρ is the ratio between the amplitude m and the root-mean-square value of noise. It follows from the spectrograms that this ratio for both signals is 0.48. At the same time, the calculated (using Eq. (7)) ratio at this distance and frequency is 0.14. We attribute this difference to the contribution towards the propagation nonlinearity originating in the medium.

It seems proper to take into account the information coming from higher harmonics in seismically active zones, as well as in studying geodynamic processes. We illustrate this below by monitoring changes in the Earth's crust during lunar-solar tides. The amplitude ratios for the second and fundamental harmonics were recorded during periodically repeating (over a 3-hour period) sounding sessions using mono-frequencies generated by a powerful CV-100 seismic vibrator. Each session consisted of sequentially radiated 20-minute-long oscillations at frequencies of 6.3 and 7.0 Hz. These sessions were continuously repeated over a 4-day period.

The main objective of the experiment is described in detail in Glinsky et al. (2000). This paper investigates the possibility of selecting daily and

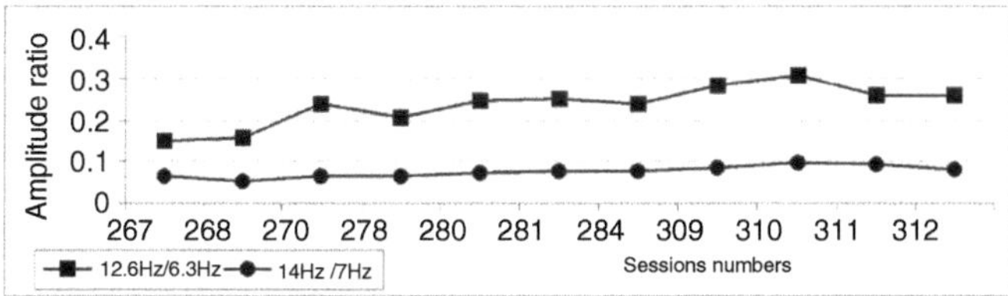

Figure 9 Ratio between the second and fundamental harmonics at the source: Sessions 267-312.

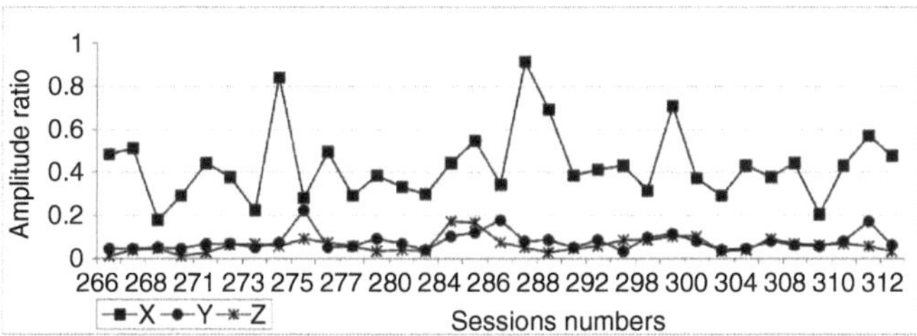

Figure 10 Ratio between the second (12.6 Hz) and fundamental (6.3 Hz) harmonics at the 5th gauge: Sessions 266-312.

half-daily periodicities for varying the parameters of the seismic field for lunar-solar tides. We recorded all three (X, Y, and Z) components. The plots in Figure 9 reflect the ratios between the second and the fundamental harmonics at frequencies of 6.3 and 7 Hz. These data correspond to the parameters of the radiated seismic field in immediate proximity to the source (at a distance of 30 m). It is clear that the greatest nonlinear effect manifests itself at a frequency of 6.3 Hz. Here, the ratio of harmonics varies within 15-30%; at the frequency of 7.0 Hz, this ratio is within 5-10%.

Figure 10 shows similar ratios obtained at a distance of 355 km from the source, corresponding to the components X, Y, and Z, and the source frequency of 6.3 Hz. These data show that the largest values of the ratios are at component X oriented to the vibrator. The ratios of harmonics at the considered distance are, on average, higher than those observed near the source. In particular, we have 50% against 25% at 6.3 Hz. External noise causes fluctuations in the obtained ratios (from session to session).

We relate the observed increase in the nonlinear component at long observational distances to the contribution of propagation nonlinear effects. It follows from this assumption that changes in deformation-related nonlinear effects caused by lunar-solar tides can be detectable. It is known (Melchior, 1966) that many geophysical fields are characterized by daily and half-daily periodicity. Latent periodicity in the series was selected with the use of the discrete Fourier transform, employing the weight function to smooth the edges of the observation series. Figure 11 shows the results from

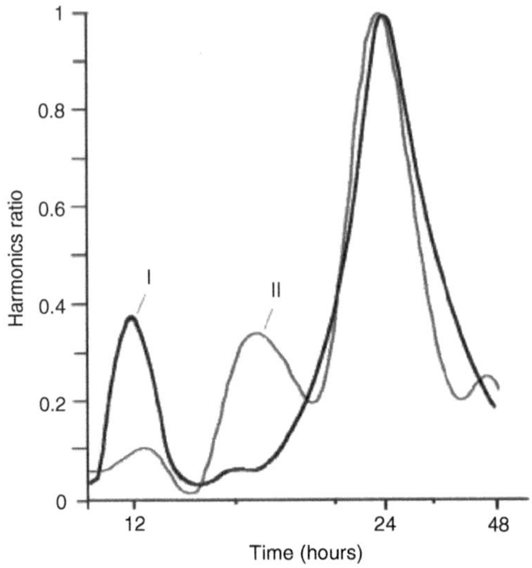

Figure 11 Amplitude spectrum of time variations caused by Earth tides: Curve I—Gravity variations; Curve II—Variations in the relationships between the second and base harmonics during the monitoring of a 355 km long Earth's crust zone, with a sounding frequency of 6.3 Hz.

this selection for the observation series corresponding to the ratios between the amplitudes of the second and first (fundamental) harmonics with sounding frequency 6.3 Hz (Curve II). The curve represents the observation series corresponding to Component Z. For comparison, the plot shows the amplitude spectrum of gravitational variations recorded for the same period (Curve I). It follows that daily periodicity is confidently distinguished within the time series $a_{12.6}(355)/a_{6.3}(355)$. Half-daily periodicity is expressed weakly.

One advantage of the monitoring method proposed here is its stability with respect to fluctuations in the amplitude of radiated waves, owing to the seasonal and instrumental variations. This was confirmed experimentally from the comparative analysis of the test recordings near the source, and of the ratios between the amplitudes of secondary and fundamental waves at a distance of 355 km.

The graphs representing the amplitudes of the fundamental harmonics close to the source depending on the session number (at a frequency of 6.3 Hz and the components X- [dotted line], Y- [solid line], Z- [dashed line]), are presented in Figure 12. This figure shows that some of the sounding sessions had sharp decreases in amplitude levels, caused by the "artificial" decrease in the power of the original CV-100 vibrator. Despite this fact, the stability in detecting half-daily and daily periodicities resulting from the lunar tides, at the remote recording point, was retained. However, the above periodicities are not visible in the amplitude graphs for just the

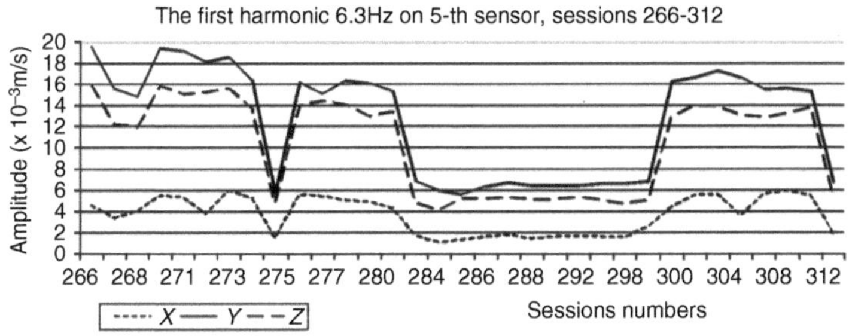

Figure 12 Levels of radiated signals in the source zone for X-, Y- and Z-components.

fundamental harmonic at a frequency of 6.3 Hz. The likely reason for this is that the fluctuations in the radiation oscillation levels bring about corresponding fluctuations in the signal levels at the receiving point. At the same time, when the harmonic ratio is used, the correspondent fluctuations cancel (Figure 9), and it becomes possible to detect daily and half-daily periodicities.

6. Discussion

Active seismic monitoring with the use of powerful seismic vibrators is accompanied by nonlinear wave effects developing during both the radiation and propagation of seismic waves. These effects occur because of imperfections in the source-ground coupling, as well as because of the soil nonlinearity under the vibrator.

Nonlinear wave-propagation effects occur in soil and rocks, which attenuate seismic waves. Accounting for both nonlinearity and attenuation contributes to an increase in the accuracy and sensitivity of active monitoring technology, mostly as a result of the following factors:

- Seismograms of the second harmonic are characterized by sharp arrivals of the first (P-) waves, which increase the accuracy of measurement for the arrival times of waves and their time resolution. The sharpness of the P- wave arrivals is caused by the high frequencies of the second harmonics. This is confirmed by a comparison of P-wave (Figure 13(a)) and S-wave (Figure 13(b)) spectrograms recorded at a distance of 355 km from the vibrator. The frequency band of the required signals lies within 5.5-8.5 Hz. It follows from Figure 13 that the main peak of the spectrogram for the P-wave is closer to the upper edge of the frequency band of the signal, and the main peak for the S-wave is closer to the lower edge. Additional use of the frequency band of second harmonics still further saturates the spectrum of P-waves with higher frequencies,

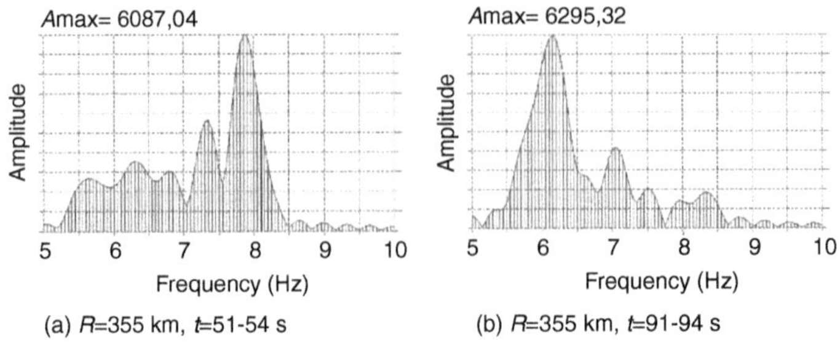

Figure 13 (a) P-waves spectrograms and (b) S-waves spectrograms, recorded at a distance of 355 km from the vibrator.

which contributes to an increase in the contrast of wave arrivals. This is confirmed by detection and measurement of the arrival times of P-waves at distances up to 50 km (Figure 4).

- Nonlinear radiation effects are associated with the choice of the sounding regime: They are maximal in the harmonic regime and can reach 100% and more, because of the development of resonance phenomena at near-resonant frequencies during a long sounding. In the regime of sounding by wideband signals (sweeps), these phenomena are less pronounced, since they do not have enough time to fully develop because of frequency scanning. Quantitatively, within the broadband sounding regime, the nonlinear effects do not exceed 10%. Given this fact, the phenomenon of radiation nonlinearity can be more effectively used in the harmonic regime of sounding.
- Low-frequency vibrators are characterized by a limited effective frequency band of sounding signals. The ratio between the upper and lower frequencies of the frequency band is about 1.5-1.8. This is the reason for the lower contrast in the arrivals of the main wave types and their visibility on the noise background. Our successful attempt to increase accuracy in determining the wave arrival time within seismograms, obtained for a limited frequency range at basic frequencies and at a limited number of seismic receivers ($n \sim 5$), raises the possibility of using the principle of nonlinear processing. It forms the basis of the multiplicative antenna. In this case, in contrast to antennas with linear processing, one can achieve more accurate resolution in the direction of wave propagation with the same number of antenna elements.
- In conducting active vibroseismic monitoring, we must account for the influence of seasonal and instrumental variations in the parameters of the radiation field, since they limit the sensitivity of the vibroseismic method. To eliminate source-related variations, we have proposed a statistical algorithm for processing observational data, based on the measurement of the amplitudes for the second and fundamental harmonics at the background of noise–and the calculation of their ratios. The stability of

the algorithm from the variations in the force characteristics of radiation, and its high sensitivity to variations in the medium's rigidity parameters caused by geodynamic processes, are proved by detection of daily and half-daily periodicities. These were obtained from seismic-observation data from the periods of the Earth's tides at a source-receiver distance of 355 km. The obtained results correlate well with the results from processing of the gravimetric observation data.
- Taking into account the above, we find very likely that high harmonics carry useful information; information that can help in monitoring geodynamic processes in seismically active zones.

7. Conclusions

In this work, we studied the problem of increasing the informational content from the processing of seismic monitoring data, taking into account the nonlinear effects of wave fields at the stages of radiation and propagation of seismic oscillations in elastic media. We estimated the quantitative nonlinear characteristics related to the radiation and propagation of seismic wave fields generated by powerful ground-based vibrators. We developed a statistical algorithm for the processing of recorded seismic data, based on the amplitude ratios of the second and fundamental harmonics. We found that the proposed algorithm is robust, even when variations in the source are present, and that it is sensitive to variations caused by geodynamic processes. This is especially important for long-term monitoring of the processes occurring in seismic-prone regions.

Acknowledgments

This research was supported by the Russian Foundation for Basic Research, Grants No. 07-07-00214-a, 07-05-00858-a, 08-07-10000-k.

References

Alekseev, A.S., Chichinin, I.S., Korneev, V.A., 2005. Powerful low-frequency vibrators for active seismology. Bulletin of the Seismological Society of America 95, 1–17.

Alekseev, A.S., Geza, N.I., Glinsky, B.M., Emanov, A.F., Kashun, V.N., Kovalevsky, V.V., Manshtein, A.K., Mikhailenko, B.G., Seleznev, V.S., Serdukov, S.V., Sobisevitch, A.L., Sobisevitch, L.E., Soloviev, V.M., Khairetdinov, M.S., Chichinin, I.S., Yushin, V.I., 2004. Active Seismology with Powerful Vibrosources. ICM&MG SB RAS, Filial Geo publishing house of the SB RAS, Novosibirsk (in Russian).

Alekseev, A.S., Belonosov, A.S., Petrenko, V.E., 1998. Determination of an integral earthquake precursor with the use of a multidisciplinary model and active vibroseismic monitoring. Bulletin of the Novosibirsk Computing Center, Series: Mathematical Modeling in Geophysics 7, 3–36.

Brekhovskich, L.M., Goncharov, W.W., 1982. Introduction to a Mechanics of the Continuous Medium. Nauka, Moscow (in Russian).

Glinsky, B.M., Khairetdinov, M.S., Sedukhina, G.F., 2006. Geophysical data processing with for nonlinear effects of radiation and propagation of elastic waves. Optoelectronics, Instrumentation and Data Processing 42 (5), 37–44.

Glinsky, B.M., Khairetdinov, M.S., Kovalevsky, V.V., 2002. Nonlinear wave effects in the powerful seismic vibrators zone. In: Nonlinear Acoustics at the Beginning of the 21st Century, vol. 1. pp. 327–330.

Glinsky, B.M., Kovalevsky, V.V., Khairetdinov, M.S., 2000. Vibroseismic monitoring of earthquake prone areas. Volcanology and Seismology 21, 723–730.

Khairetdinov, M.S., 2003. The nonlinear wave effects in vibroseismic. Bulletin of the Novosibirsk Computing Center, Series: Mathematical Modeling in Geophysics 8, 59–68.

Korneev, V.A., Nihei, K.T., Myer, L.R., 1998. Nonlinear Interaction of Plane Elastic Waves: Scientific Report, LBNL-41914. Lawrence Berkeley National Laboratory, Berkeley, California 94720.

Melchior, P., 1966. The Earth Tides. Pergamon Press, New York, NY.

Nikolaev, A.V., 1987. Problems of nonlinear seismics. In: Collection of articles Problems of Nonlinear Seismics. Nauka, Moscow, pp. 5–20.

Rudenko, O.V., Soluyan, S.I., 1975. Theoretical Fundamentals of Nonlinear Acoustics. Nauka, Moscow (in Russian).

Polyakova, A.L., 1964. Nonlinear effects in a solid. Soviet Physics, Solid State 6 (1), 50–54.

Tikhonov, V.I., 1983. Optimal Signal Reception. Radio I Svyaz', Moscow (in Russian).

Verbitsky, T.Z., 1987. Peculiarities of elastic wave propagation in nonlinearly elastic porous media. In: Problems of Nonlinear Seismics. Nauka, Moscow, pp. 94–103.

Zarembo, L.K., Krasilnikov, W.A., 1966. Introduction to Nonlinear Acoustics. Nauka, Moscow (in Russian).

Zhurkov, S.N., 1968. Kinetic concept of strength of solid bodies. Vestnik Akademii. Nauk SSSR 3, 46–52.

Theory of Data Analysis and Interpretation

CHAPTER 17

Maxwell's Equations and Numerical Electromagnetic Modeling in the Context of the Theory of Differential Forms

Michael S. Zhdanov[a]

Contents

[a] Department of Geology and Geophysics, University of Utah

E-mail address: mzhdanov@mines.utah.edu.

Handbook of Geophysical Exploration: Seismic Exploration, Volume 40
ISSN 0950-1401, DOI: 10.1016/S0950-1401(10)04022-X

1. Introduction

The principal physical laws characterizing the behavior and interaction of electric and magnetic fields were unified in the comprehensive theory of electromagnetic fields by James Maxwell in his *Treatise on Electromagnetism*, first published in 1873 (Maxwell, 1954). The formulation of this theory represents one of the most important events in physics since Newton's time. In fact, Maxwell was the first to introduce the mathematical equations and physical laws which govern the electromagnetic field. Any effort to use electromagnetic fields to explore the earth must be firmly based on these physical laws and their mathematical consequences.

The fundamental system of electromagnetic (EM) field equations, Maxwell's equations, was developed by generalization of the basic laws of electromagnetism established in the first half of the 19th century. In the framework of classical theory, the EM field is described by the electric and magnetic vector fields, and Maxwell's equations represent a system of differential equations with respect to these vector fields.

During recent decades, an alternative approach was developed to the formulation of Maxwell's equations. This approach is based on the algebraic theory of differential forms and results in a very compact and symmetric system of differential form equations.

The differential forms were originally introduced in differential geometry to study the properties of the lines and surfaces in multidimensional mathematical spaces. However, it was realized not so long ago that these forms provide a very elegant and powerful tool to study the physical fields as well. We can treat the differential forms as another mathematical language which, similar to the vectorial language, can be used to describe the physical fields. In fact, in a four-dimensional space-time continuum, the differential forms can be treated as linear combinations of the differentials of the flux, the work, and/or the source of the vector fields. Therefore, Maxwell's equations for the differential forms contain the differentials of the flux and work of the electric and magnetic fields. This property of differential forms indicates that it is more suitable to consider the electric and/or magnetic flux and work as major characteristics of the EM field, instead of using the conventional vectorial representations. This approach seems to be quite reasonable from a physical point of view as well, because in physical experiments we, as a rule, measure the flux and the work (or voltage) of the electric and magnetic fields.

Moreover, the remarkable fact is that, based on the fundamental differential equations for the forms in a four-dimensional space, we can demonstrate that any pair of the arbitrary vector fields, $\mathbf{H}(\mathbf{r}, t)$ and $\mathbf{D}(\mathbf{r}, t)$, satisfies a system of differential equations which is similar to Maxwell's classical equations for electromagnetic fields! In other words, we can show that if a 2-form in the four-dimensional space-time E_4 is comprised of two arbitrary vector fields, $\mathbf{H}(\mathbf{r}, t)$ and $\mathbf{D}(\mathbf{r}, t)$, then these fields must automatically satisfy a system of Maxwell's equations.

The goal of this paper is to show that Maxwell's equations appear naturally from the basic equations of field theory for the differential forms. There are no other equations for a pair of nonstationary vector fields but equations of the Maxwell type. The basic laws of electromagnetism are actually imprinted in the fundamental differential relationships between the vector fields and differential forms. This new approach to the formulation and understanding of the basic properties of the laws of electromagnetism has the strong potential to stimulate future development in electromagnetic geophysics.

2. Differential Forms in Vector Field Theory

2.1. Concept of the differential form

In general, we can introduce the differential forms as expressions on which integration operates. There exist differential forms of different degrees depending on the dimensions of the domain of integration. In particular, a differential form of degree p, or a p-form, is an integrand of an integral over a domain of dimension p. We shall start our discussion by introducing the basic concept of differential forms in three-dimensional Euclidean space, where the degree of forms p varies from 0 to 3. A 0-form is a scalar function which is "integrated" over a region of zero dimension. In three-dimensional space the differential forms are closely related to the vector fields, and the algebraic and differential operations on the forms can be defined by conventional vector algebra and calculus, which makes it easier to understand for the reader familiar with conventional vector field theory.

In particular, a differential expression $\mathbf{B} \cdot \mathbf{dl}$, which is integrated over a curve, represents the elementary work, $\mathrm{d}W$, of the vector field $\mathbf{B}$ along an infinitesimally small vector element of curve $\mathbf{dl}$. It is called a *differential 1-form* $\underset{(1)}{\varphi}$:

$$\underset{(1)}{\varphi} = \mathbf{B} \cdot \mathbf{dl} = \mathrm{d}W. \tag{1}$$

A differential expression $\mathbf{B} \cdot \mathbf{ds}$, which is integrated over a surface, describes an elementary scalar flux, $\mathrm{d}F_{\mathbf{B}}^{\mathrm{d}s}$, of the vector field $\mathbf{B}$ through an infinitesimally small vector element of surface **ds**. It is called a *differential 2-form* $\underset{(2)}{\psi}$:

$$\underset{(2)}{\psi} = \mathbf{B} \cdot \mathrm{d}\mathbf{s} = \mathrm{d}F_{\mathbf{B}}^{\mathrm{ds}}. \tag{2}$$

Finally, a differential expression $div\,\mathbf{B}\mathrm{d}v$, which is integrated over a volume, is equal to an elementary source, $\mathrm{d}Q$, of the vector field **B** within an infinitesimally small element of volume $\mathrm{d}v$. We call this expression a *differential 3-form* $\underset{(3)}{\theta}$:

$$\underset{(3)}{\theta} = q\mathrm{d}v = \mathrm{d}Q, \tag{3}$$

where

$$q = div\,\mathbf{B}.$$

It is known that the divergence of the vector field **B** can be treated as a source of this field. Thus, all three forms represent the scalar values of the work, flux, and source of the vector field, respectively.

2.2. Exterior (wedge) product of the differential forms

It can be shown that the conventional differential forms of vector calculus (the expressions that are integrated over a line, surface, or volume) are described by antisymmetric linear functions of one, two, or three vector arguments (Zhdanov, 2009). These linear functions (differential forms) represent new mathematical objects which are very useful in a description of the electromagnetic field equations.

The differential forms provide the most natural and elegant mathematical tool for a description of electromagnetic fields (Lindell, 2004; Zhdanov, 2009). In order to be able to apply these functions to electromagnetic theory, we should introduce the mathematical rules of operation on the differential forms, which define the *algebra* of the differential forms.

The simplest operation is addition. It is obvious that the addition of two antisymmetric linear forms is determined as a conventional summation of two functions. This operation satisfies the traditional commutative and associative laws, and also the distributive laws with respect to multiplication by a scalar.

The multiplication of the differential forms requires the introduction of a special algebraic operation, an exterior product, which we will discuss in detail below.

In the case of vectors, we can use different multiplication operations, e.g., dot and cross products of the vectors. In principle, it is easy to consider a product, f, of two forms, φ and ψ, as a product of two linear functions of vector arguments. For example, if $\varphi = \varphi(\mathbf{dl})$ and $\psi = \psi(\mathbf{da}, \mathbf{db})$, we have

$$f = f\,(\mathbf{dl}, \mathbf{da}, \mathbf{db}) = \varphi(\mathbf{dl})\psi\,(\mathbf{da}, \mathbf{db})\,, \tag{4}$$

where multiplication on the right-hand side of Eq. (4) is conducted in the conventional way as a product of two scalar values, ϕ and ψ. The only problem with this definition is that the product of the two antisymmetric linear functions, ϕ and ψ is no longer an antisymmetric function! Indeed, one can see that

$$f(\mathbf{dl}, \mathbf{da}, \mathbf{db}) \neq -f(\mathbf{da}, \mathbf{dl}, \mathbf{db}).$$

In order to keep the result of multiplication within the class of antisymmetric linear functions, we should apply an antisymmetrization operation to the conventional product (4). As a result, we arrive at the following multiplication operation, which is called the exterior (or wedge) product, denoted by the symbol $\wedge$, and defined as

$$\begin{aligned}\Omega &= \varphi \wedge \psi \\ &= \{\varphi\,(\mathbf{dl})\,\psi\,(\mathbf{da}, \mathbf{db}) + \varphi\,(\mathbf{da})\,\psi\,(\mathbf{db}, \mathbf{dl}) + \varphi\,(\mathbf{db})\,\psi\,(\mathbf{dl}, \mathbf{da}) \\ &\quad - \varphi\,(\mathbf{dl})\,\psi\,(\mathbf{db}, \mathbf{da}) - \varphi\,(\mathbf{da})\,\psi\,(\mathbf{dl}, \mathbf{db}) - \varphi\,(\mathbf{db})\,\psi\,(\mathbf{da}, \mathbf{dl})\}\,. \end{aligned} \tag{5}$$

It is easy to verify that the new linear function, $\Omega = \Omega\,(\mathbf{dl}, \mathbf{da}, \mathbf{db})$, is indeed an antisymmetric function.

In the case of the product of two 1-forms, $\phi(\mathbf{da})$ and $\chi\,(\mathbf{db})$, the exterior product is given by the following formula:

$$\Phi\,(\mathbf{da}, \mathbf{db}) = \varphi\,(\mathbf{da}) \wedge \chi\,(\mathbf{db}) = \varphi\,(\mathbf{da})\,\chi\,(\mathbf{db}) - \varphi\,(\mathbf{db})\,\chi\,(\mathbf{da})\,, \tag{6}$$

which is again an antisymmetric function:

$$\Phi\,(\mathbf{da}, \mathbf{db}) = -\,[\varphi\,(\mathbf{db})\,\chi\,(\mathbf{da}) - \varphi\,(\mathbf{da})\,\chi\,(\mathbf{db})] = -\Phi\,(\mathbf{db}, \mathbf{da})\,. \tag{7}$$

In summary, we can see that the addition and exterior multiplication operations of 1-forms satisfy the following laws:

(1) the commutative and associative laws of addition:

$$\varphi + \chi = \chi + \varphi \quad \text{and} \quad \varphi + (\chi + \eta) = (\varphi + \chi) + \eta; \tag{8}$$

(2) the anticommutative law of exterior multiplication:

$$\varphi \wedge \chi = -\chi \wedge \varphi; \tag{9}$$

(3) the distributive law of multiplication over addition:

$$\varphi \wedge (\chi + \eta) = \varphi \wedge \chi + \varphi \wedge \eta \quad \text{and} \\ (\varphi + \chi) \wedge \eta = \varphi \wedge \eta + \chi \wedge \eta. \tag{10}$$

It comes immediately from the anticommutative law (9) that

$$\varphi \wedge \varphi = 0. \tag{11}$$

We can see now that the exterior product of two 1-forms generates a 2-form, while the exterior product of three 1-forms produces a 3-form.

2.3. Canonical representations of the differential forms in three-dimensional Euclidean space

The important fact of the differential form theory is that any 2-form and any 3-form can be expressed as the exterior products of two 1-forms and three 1-forms respectively. These representations are called the *canonical representations* for the differential forms (Zhdanov, 2009). The following table presents a summary of these canonical representations for differential forms in three-dimensional Euclidean space E_3:

$$\text{0-form}: \underset{(0)}{\varphi} = f, \tag{12}$$

$$\text{1-form}: \underset{(1)}{\varphi} = \sum_{\alpha=x,y,z} \varphi_\alpha \mathrm{d}\alpha = \varphi_x \mathrm{d}x + \varphi_y \mathrm{d}y + \varphi_z \mathrm{d}z = \boldsymbol{\varphi} \cdot \mathbf{dr}, \tag{13}$$

$$\text{2-form}: \underset{(2)}{\psi} = \sum_{\alpha,\beta=x,y,z} \boldsymbol{\psi}_{\alpha\beta} \mathrm{d}\alpha \wedge \mathrm{d}\beta = \psi_{yz} \mathrm{d}y \wedge \mathrm{d}z \\ + \psi_{zx} \mathrm{d}z \wedge \mathrm{d}x + \psi_{xy} \mathrm{d}x \wedge \mathrm{d}y = \boldsymbol{\psi} \cdot \mathbf{d\Sigma}, \tag{14}$$

$$\text{3-form}: \underset{(3)}{\theta} = \theta \mathrm{d}x \wedge \mathrm{d}y \wedge \mathrm{d}z = \theta \mathrm{d}v. \tag{15}$$

In these last formulas we used the following notations:

$$\boldsymbol{\varphi} = \left(\varphi_x, \varphi_y, \varphi_z\right), \quad \mathbf{dr} = (\mathrm{d}x, \mathrm{d}y, \mathrm{d}z), \\ \boldsymbol{\psi} = \psi_{yz} \mathbf{d}_x + \psi_{zx} \mathbf{d}_y + \psi_{xy} \mathbf{d}_z \\ \mathbf{d\Sigma} = \mathrm{d}s_x \mathbf{d}_x + \mathrm{d}s_y \mathbf{d}_y + \mathrm{d}s_z \mathbf{d}_z,$$

where $\mathrm{d}s_x$, $\mathrm{d}s_y$, and $\mathrm{d}s_z$ are the combinations of the exterior products of differentials:

$$\mathrm{d}s_x = \mathrm{d}y \wedge \mathrm{d}z, \quad \mathrm{d}s_y = \mathrm{d}z \wedge \mathrm{d}x, \quad \mathrm{d}s_z = \mathrm{d}x \wedge \mathrm{d}y.$$

2.4. The exterior derivative operation

The calculus of differential forms is based on a special differential operation called the *exterior derivative*. This operation can be treated as a generalization of conventional vector differential operations. In fact, all three different vector differential operations (gradient, divergence, and curl) can be represented as a single exterior differential operator.

0-forms

In the case of the simplest 0-form described by a function $f(\mathbf{r})$, the exterior differential is equivalent to the full differential of the function:

$$\mathrm{d}f = \frac{\partial f}{\partial x}\mathrm{d}x + \frac{\partial f}{\partial y}\mathrm{d}y + \frac{\partial f}{\partial z}\mathrm{d}z = grad\, f(\mathbf{r}) \cdot \mathbf{dr}. \tag{16}$$

We can see that in this case the exterior differential operator d can be treated as a counterpart of the vector *del* operator ∇. We can introduce a symbolic differential 1-form d_{E_3} as follows:

$$\mathrm{d}_{E_3} = \mathrm{d}x\frac{\partial}{\partial x} + \mathrm{d}y\frac{\partial}{\partial y} + \mathrm{d}z\frac{\partial}{\partial z} = \mathbf{dr} \cdot \nabla. \tag{17}$$

Then Eq. (16) can be written as

$$\mathrm{d}f = \mathrm{d}E_3 f = \mathbf{dr} \cdot \nabla f = \nabla f \cdot \mathbf{dr}.$$

Therefore, the exterior derivative of the 0-form is equivalent to the gradient of the corresponding scalar field.

1-forms

The exterior differential of the 1-form field $\phi(\mathbf{r})$ is calculated as an exterior product of the differential operator 1-form d_{E_3} and a given 1-form ϕ:

$$\begin{aligned} \mathrm{d}\varphi &= \mathrm{d}_{E_3} \wedge \varphi = \mathrm{d}_{E_3} \wedge (\varphi \cdot \mathbf{dr}) \\ &= curl\, \varphi \cdot \mathbf{d\Sigma} = [\nabla \times \varphi] \cdot \mathbf{d\Sigma} \end{aligned} \tag{18}$$

where we took into account representations (13) and (14) for the differential forms. Therefore, the exterior derivative of the 1-form is equivalent to the curl of the corresponding vector field.

2-forms

The exterior differential of the 2-form field $\psi(\mathbf{r})$ is equal to:

$$\begin{aligned} \mathrm{d}\boldsymbol{\psi} &= \mathrm{d}_{E_3} \wedge \psi = \mathrm{d}_{E_3} \wedge (\boldsymbol{\psi} \cdot \mathbf{d\Sigma}) \\ &= (div\, \boldsymbol{\psi})\, \mathrm{d}v = (\nabla \cdot \boldsymbol{\psi})\, \mathrm{d}v. \end{aligned} \tag{19}$$

According to the canonical representation (14), every 2-from can be described by a vector field $\boldsymbol{\psi}$. Therefore, the exterior derivative of the 2-form is equivalent to the divergence of the corresponding vector field.

3-forms

The exterior differential of the 3-form $\underset{(3)}{\theta} = \theta\, \mathrm{d}x \wedge \mathrm{d}y \wedge \mathrm{d}z$ can be calculated as follows:

$$\mathrm{d}_{E_3} \wedge \underset{(3)}{\theta} = \mathrm{d}\theta \wedge \mathrm{d}x \wedge \mathrm{d}y \wedge \mathrm{d}z.$$

This last expression is equal to zero according to the anticommutative law of the wedge product:

$$\mathrm{d}_{E_3} \wedge \underset{(3)}{\theta} = \left(\frac{\partial \theta}{\partial x}\mathrm{d}x + \frac{\partial \theta}{\partial y}\mathrm{d}y + \frac{\partial \theta}{\partial z}\mathrm{d}z\right) \wedge \mathrm{d}x \wedge \mathrm{d}y \wedge \mathrm{d}z = 0. \quad (20)$$

Thus we can see that, the exterior differential operation in three-dimensional space corresponds to either the gradient, curl, or divergence of conventional vector calculus:

$$\text{0-forms}: \ \mathrm{d}_{E_3} \underset{(0)}{\varphi} = \mathrm{d}_{E_3} f = grad\, f\,(\mathbf{r}) \cdot \mathbf{dr} = \nabla f\,(\mathbf{r}) \cdot \mathbf{dr}, \quad (21)$$

$$\text{1-forms}: \ \mathrm{d}_{E_3} \underset{(1)}{\varphi} = \mathrm{d}_{E_3} \wedge (\varphi \cdot \mathbf{dr}) = curl\, \varphi \cdot \mathrm{d}\boldsymbol{\Sigma} = [\nabla \times \varphi] \cdot \mathrm{d}\boldsymbol{\Sigma}, \quad (22)$$

$$\text{2-forms}: \ \mathrm{d}_{E_3} \underset{(2)}{\varphi} = \mathrm{d}_{E_3} \wedge (\boldsymbol{\psi} \cdot \mathrm{d}\boldsymbol{\Sigma}) = (div\, \boldsymbol{\psi})\, \mathrm{d}v = (\nabla \cdot \boldsymbol{\psi})\, \mathrm{d}v, \quad (23)$$

$$\text{3-forms}: \ \mathrm{d}_{E_3} \underset{(3)}{\varphi} = \mathrm{d}_{E_3} \wedge (\theta\, \mathrm{d}v) = 0. \quad (24)$$

The beauty of the exterior differentiation operator is that it includes all varieties of vector differential operations. In addition, this operator allows us to extend naturally the differentiation operation for multidimensional spaces.

3. Nonstationary Field Equations and Differential Forms

Geophysical methods are based on studying the propagation of the different physical fields within the earth's interior. Two of the most widely used fields in geophysics are seismic and electromagnetic fields, which are typical nonstationary fields. These fields can be represented as vector fields in four-dimensional Euclidean space-time E_4. In this paper we will derive the general system of differential equations for nonstationary vector fields. We will demonstrate that these equations are nothing else but Maxwell-type equations.

3.1. Nonstationary vector fields and differential forms in four-dimensional space E_4

The most effective way of presenting the theory of nonstationary (time-dependent) fields is based on the theory of differential forms. We introduce the four-dimensional Euclidean space E_4, which has three conventional spatial coordinates, $x_1 = x$, $x_2 = y$ and $x_3 = z$, and the fourth coordinate, x_4, equal to time: $x_4 = t$.

Any scalar, U, or vector, $\mathbf{A}$, functions of the space coordinates (x_1, x_2, x_3) and time coordinate t can be treated as functions defined in the four-dimensional space E_4. We can also introduce arbitrary vector fields $\mathbf{H}(\mathbf{r}, t)$, $\mathbf{D}(\mathbf{r}, t)$, and $\mathbf{j}(\mathbf{r}, t)$, all of which are nonstationary (time-dependent) vector functions in three-dimensional space, and a scalar function $q(\mathbf{r}, t)$. The remarkable fact is that any pair of nonstationary fields, $\mathbf{H}(\mathbf{r}, t)$ and $\mathbf{D}(\mathbf{r}, t)$, satisfies a set of differential equations which have exactly the same structure as Maxwell's equation of electromagnetic theory! We will derive these equations using the differential form theory.

One can show that, using the vector and scalar fields $\mathbf{A}, \mathbf{H}, \mathbf{D}, \mathbf{j}$ and q, we can define differential forms of five different orders in the four-dimensional space E_4. These forms can be expressed using the three-dimensional vector notations as follows (Zhdanov, 2009):

$$\text{0-forms}: \underset{(0)}{\Omega} = U, \tag{25}$$

$$\text{1-forms}: \underset{(1)}{\Omega} = \alpha = \mathbf{A} \cdot \mathrm{d}\mathbf{r} - U\mathrm{d}t, \tag{26}$$

$$\text{2-forms}: \underset{(2)}{\Omega} = \psi = \mathbf{D} \cdot \mathrm{d}\boldsymbol{\Sigma} - (\mathbf{H} \cdot \mathrm{d}\mathbf{r}) \wedge \mathrm{d}t, \tag{27}$$

$$\text{3-forms}: \underset{(3)}{\Omega} = \gamma = q\mathrm{d}v - (\mathbf{j} \cdot \mathrm{d}\boldsymbol{\Sigma}) \wedge \mathrm{d}t, \tag{28}$$

$$\text{4-forms}: \underset{(4)}{\Omega} = \theta = q\mathrm{d}v \wedge \mathrm{d}t. \tag{29}$$

3.2. Differential form equations

It is known that any p-form in four-dimensional space E_4 can be split into two terms which are called its *spatial*, Ω_s, *and temporal*, Ω_τ, *components* (Lindell, 2004; Fecko, 2006):

$$\underset{(p)}{\Omega} = \Omega_s + \Omega_t \wedge \mathrm{d}t.$$

Note that the time coordinate and the spatial coordinates are mutually orthogonal in E_4. Therefore, any differential form equation in the space E_4, e.g.,

$$\underset{(p)}{\Omega} = 0,$$

can be split into separate equations for the spatial and temporal components:

$$\Omega_s = 0 \quad \text{and} \quad \Omega_\tau = 0.$$

We will summarize below the basic differential equations for the differential forms in the four-dimensional space E_4.

3.3. Exterior derivative of a scalar field and a generalized source 1-form

We begin with the exterior derivative of the 0-form:

$$\mathrm{d}\underset{(0)}{\Omega} = \mathrm{d}U = grad\, U \cdot \mathbf{dr} + \frac{\partial}{\partial t} U \mathrm{d}t = \underset{(1)}{g}\,. \tag{30}$$

According to Eq. (26), the 1-form $\underset{(1)}{g}$ can be written as:

$$\underset{(1)}{g} = g_{1s} + g_{1\tau}\mathrm{d}t = \mathbf{g}_1 \cdot \mathbf{dr} + g_{1\tau}\mathrm{d}t. \tag{31}$$

Splitting Eq. (30) into its spatial and temporal parts, we find:

$$g_{1s} = \mathbf{g}_1 \cdot \mathbf{dr} = grad\, U \cdot \mathbf{dr} \quad \text{and} \quad g_{1\tau} = \frac{\partial}{\partial t} U. \tag{32}$$

The 1-form $\underset{(1)}{g}$ is called *a generalized source form of the 0-form field* $\underset{(0)}{\Omega}$. Its spatial component is equal to the work of the gradient of a scalar field U, along a vector element $\mathbf{dr}$, while its temporal component is equal to the time derivative of the scalar field U.

3.4. Exterior derivative of a four-potential and a generalized source 2-form

It can be demonstrated that the exterior derivative of the 1-form α (so-called four-potential) is given by the following expression (Zhdanov, 2009):

$$\mathrm{d}\underset{(1)}{\Omega} = \mathrm{d} \wedge \alpha = curl\, \mathbf{A} \cdot \mathbf{d\Sigma} - \left(grad\, U + \frac{\partial \mathbf{A}}{\partial t} \right) \cdot \mathbf{dr} \wedge \mathrm{d}t = \underset{(2)}{g}, \tag{33}$$

where $\underset{(2)}{g}$ is a generalized source 2-form:

$$\underset{(2)}{g} = g_{2s} + g_{2\tau}\mathrm{d}t = \mathbf{j}_A \cdot \mathrm{d}\boldsymbol{\Sigma} - \left(grad\, U + \frac{\partial \mathbf{A}}{\partial t} \right) \cdot \mathbf{dr} \wedge \mathrm{d}t, \qquad (34)$$

and $\mathbf{j}_A = curl\,\mathbf{A}$.

Splitting Eq. (34) into its spatial and temporal equations, we find:

$$g_{2s} = \mathbf{j}_A \cdot \mathrm{d}\boldsymbol{\Sigma} \quad \text{and} \quad g_{2\tau} = -\left(grad\, U + \frac{\partial \mathbf{A}}{\partial t} \right) \cdot \mathbf{dr}. \qquad (35)$$

3.5. Exterior derivative of a 2-form and a four-current

Similarly, we can find the exterior derivative of the 2-form:

$$\mathrm{d}\underset{(2)}{\Omega} = \mathrm{d} \wedge \psi = (div\,\mathbf{D})\,\mathrm{d}v + \left(\frac{\partial}{\partial t}\mathbf{D} - curl\,\mathbf{H} \right) \cdot \mathrm{d}\boldsymbol{\Sigma} \wedge \mathrm{d}t = \underset{(3)}{g}, \qquad (36)$$

where $\underset{(3)}{g}$ is a generalized source 3-form or so-called four-current γ^{ψ}:

$$\underset{(3)}{g} = g_{3s} + g_{3\tau}\mathrm{d}t = q^{\psi}\mathrm{d}v - \left(\mathbf{j}^{\psi} \cdot \mathrm{d}\boldsymbol{\Sigma}\right) \wedge \mathrm{d}t = \gamma^{\psi}, \qquad (37)$$

and

$$\mathbf{j}^{\psi} = curl\,\mathbf{H} - \frac{\partial}{\partial t}\mathbf{D} \quad \text{and} \quad q^{\psi} = div\,\mathbf{D}. \qquad (38)$$

From Eqs (37) and (38) we find the spatial and temporal components of the generalized source 3-form (four-current γ^{ψ}):

$$g_{3s} = (div\,\mathbf{D})\,\mathrm{d}v, \quad g_{3\tau} = \left(\frac{\partial}{\partial t}\mathbf{D} - curl\,\mathbf{H} \right) \cdot \mathrm{d}\boldsymbol{\Sigma}. \qquad (39)$$

3.6. Exterior derivatives of a 3-form and a 4-form

Finally, the exterior derivative of a 3-form is equal to:

$$\mathrm{d}\underset{(3)}{\Omega} = \mathrm{d} \wedge \gamma = -\left(div\,\mathbf{j} + \frac{\partial}{\partial t}q \right) \mathrm{d}v \wedge \mathrm{d}t = \underset{(4)}{g}. \qquad (40)$$

The generalized source 4-form $\underset{(4)}{g}$, according to Eq. (29), can be written as:

$$\underset{(4)}{g} = g_{3\tau}\mathrm{d}t = q^{\gamma}\mathrm{d}v \wedge \mathrm{d}t, \tag{41}$$

where

$$q^{\gamma} = -div\,\mathbf{j} - \frac{\partial}{\partial t}q.$$

The exterior derivative of the 4-form θ in a four-dimensional space is always equal to zero:

$$\mathrm{d} \wedge \theta = 0.$$

We can summarize all these results as:

$$\text{0-forms}: \ \mathrm{d}U = grad\,U \cdot \mathrm{d}\mathbf{r} + \frac{\partial}{\partial t}U\mathrm{d}t, \tag{42}$$

$$\text{1-forms}: \ \mathrm{d} \wedge \alpha = curl\,\mathbf{A} \cdot \mathrm{d}\boldsymbol{\Sigma} - \left(grad\,U + \frac{\partial \mathbf{A}}{\partial t}\right) \cdot \mathrm{d}\mathbf{r} \wedge \mathrm{d}t, \tag{43}$$

$$\text{2-forms}: \ \mathrm{d} \wedge \psi = (div\,\mathbf{D})\,\mathrm{d}v + \left(\frac{\partial}{\partial t}\mathbf{D} - curl\,\mathbf{H}\right) \cdot \mathrm{d}\boldsymbol{\Sigma} \wedge \mathrm{d}t, \tag{44}$$

$$\text{3-forms}: \ \mathrm{d} \wedge \gamma = -\left(div\,\mathbf{j} + \frac{\partial}{\partial \tau}q\right)\mathrm{d}v \wedge \mathrm{d}t. \tag{45}$$

$$\text{4-forms}: \ \mathrm{d} \wedge \theta = 0. \tag{46}$$

4. Ampere-Type Differential Forms and a Continuity Equation

According to Eqs (36) and (37), we have the following differential equation for any 2-form $\psi = \mathbf{D} \cdot \mathrm{d}\boldsymbol{\Sigma} - (\mathbf{H} \cdot \mathrm{d}\mathbf{r}) \wedge \mathrm{d}\tau$:

$$\mathrm{d} \wedge \psi = \gamma^{\psi}, \tag{47}$$

where the corresponding four-current 3-form γ^{ψ} is equal to:

$$\gamma^{\psi} = q^{\psi}\mathrm{d}v - \left(\mathbf{j}^{\psi} \cdot \mathrm{d}\boldsymbol{\Sigma}\right) \wedge \mathrm{d}t, \tag{48}$$

and

$$j^{\psi} = curl\,\mathbf{H} - \frac{\partial}{\partial t}\mathbf{D} \quad \text{and} \quad q^{\psi} = div\,\mathbf{D}. \tag{49}$$

Note that, according to the basic property of the exterior derivative operator, the double application of the external differential is identically equal to zero. Therefore, the four-current γ^{ψ} satisfies the following equation:

$$d \wedge \gamma^{\psi} = d \wedge (d \wedge \psi) = 0. \tag{50}$$

This last equation can be written, according to formula (40), as:

$$div\, \mathbf{j}^{\psi} + \frac{\partial}{\partial t} q^{\psi} = 0. \tag{51}$$

We can see that Eq. (49) represents Maxwell's equations exactly if vector fields **D** and **H** are treated as the electric and magnetic fields, respectively. Correspondingly, Eq. (51) represents a conservation law for the four-current and is called *the continuity equation* because it has the form of the continuity equation of electromagnetic theory.

Equation (47) is called *a fundamental differential equation for 2-forms*, because any 2-form in the four-dimensional space E_4 must satisfy this equation. At the same time, Eq. (49) are nothing else but Maxwell's first and fourth equations for the electric field **D** and magnetic field **H**, which describe Ampere's law of electromagnetism with Maxwell's displacement current $\partial \mathbf{D}/\partial t$. Thus, Maxwell's equations appear naturally from the general theory of differential forms.

The 2-form ψ, which satisfies the fundamental equation (47), is called *an Ampere-type differential form*. We should note, however, that actually every 2-form in the four-dimensional Euclidean space E_4 is an Ampere-type form. Its spatial component, ψ_s, is called an *electric induction* 2-form D, while its temporal component, ψ_τ, is called a magnetic 1-form H:

$$D = \psi_s = \mathbf{D} \cdot d\mathbf{\Sigma}, \quad H = \psi_\tau = \mathbf{H} \cdot d\mathbf{r}. \tag{52}$$

Thus, the Ampere-type differential form can be written as:

$$\psi = D - H \wedge dt. \tag{53}$$

5. Faraday-type Differential Forms and Four-potential

Let us consider now a special class of 2-forms, which satisfies Eq. (47) with a zero right-hand part:

$$d \wedge \phi = 0. \tag{54}$$

In other words, we assume now that the corresponding four-current is equal to zero, $\gamma_{\phi} = 0$, and the 2-form ϕ is an exact form.

In this case, according to de Rham's theorem (Lindell, 2004), there exists a 1-form (four-potential) α,

$$\alpha = \mathbf{A} \cdot \mathrm{d}\mathbf{r} - U \mathrm{d}t,$$

such that

$$\phi(\mathbf{r}) = \mathrm{d} \wedge \alpha(\mathbf{r}). \tag{55}$$

Equation (55) can be written in an equivalent way as:

$$\begin{aligned} \phi &= \mathbf{B} \cdot \mathrm{d}\boldsymbol{\Sigma} + (\mathbf{E} \cdot \mathrm{d}\mathbf{r}) \wedge \mathrm{d}t = \mathrm{d} \wedge \alpha(\mathbf{r}) \\ &= \mathit{curl}\, \mathbf{A} \cdot \mathrm{d}\boldsymbol{\Sigma} - \left(\mathit{grad}\, U + \frac{\partial \mathbf{A}}{\partial t} \right) \cdot \mathrm{d}r \wedge \mathrm{d}t, \end{aligned} \tag{56}$$

where $\mathbf{B}$ and $\mathbf{E}$ are some conventional nonstationary (time-dependent) vector functions in three-dimensional space. These functions, according to formulas (54) and (38), satisfy the following equations:

$$\mathit{curl}\, \mathbf{E} = -\frac{\partial \mathbf{B}}{\partial t} \quad \text{and} \quad \mathit{div}\, \mathbf{B} = 0. \tag{57}$$

Remarkably, Eqs (57) are nothing else but Maxwell's second and third equations for electric field $\mathbf{E}$ and magnetic field $\mathbf{B}$, which describe the Faraday law of electromagnetic induction and the fact of an absence of magnetic charges. That is why the 2-form ϕ, which satisfies Eq. (54), is called *a Faraday-type differential form*. Its spatial component, ϕ_s, is called a *magnetic induction* 2-form $\mathbf{B}$, while its temporal component, ϕ_τ, is called an electric 1-form E:

$$B = \phi_s = \mathbf{B} \cdot \mathrm{d}\boldsymbol{\Sigma}, \qquad E = \phi_\tau = \mathbf{E} \cdot \mathrm{d}\mathbf{r}. \tag{58}$$

Thus, the Faraday-type differential form can be written as:

$$\psi = B + E \wedge \mathrm{d}t. \tag{59}$$

Equation (56) shows that every Faraday-type form can be expressed by the corresponding four-potential α. Splitting Eq. (56) into its spatial and temporal equations, we find that:

$$B = \mathbf{B} \cdot \mathrm{d}\boldsymbol{\Sigma} = curl\,\mathbf{A} \cdot \mathrm{d}\boldsymbol{\Sigma}, \quad E = \mathbf{E} \cdot \mathbf{dr} = -\left(grad\, U + \frac{\partial \mathbf{A}}{\partial t}\right) \cdot \mathbf{dr}. \tag{60}$$

Finally we arrive at the conventional representation for the vector fields **B** and **E** by the vector **A** and scalar U potentials:

$$\mathbf{B} = curl\,\mathbf{A} \quad \text{and} \quad \mathbf{E} = -\left(grad\, U + \frac{\partial \mathbf{A}}{\partial t}\right). \tag{61}$$

6. MAXWELL'S EQUATIONS

6.1. Basic equations in the theory of electromagnetic fields

Maxwell's equations consist of the two vector equations and two scalar equations shown below:

$$\nabla \times \mathbf{H} = \mathbf{j} + \frac{\partial \mathbf{D}}{\partial t} = \mathbf{c}, \tag{62}$$

$$\nabla \times \mathbf{E} = -\frac{\partial \mathbf{B}}{\partial t}, \tag{63}$$

$$\nabla \cdot \mathbf{B} = 0, \tag{64}$$

$$\nabla \cdot \mathbf{D} = q. \tag{65}$$

Here, **H** and **B** are the vector magnetic and induction fields, respectively; **E** and **D** are the vector electric and displacement fields, respectively; q is the electrical charge density; **j** is the conduction current density; and **c** is the total current density (the sum of conduction and displacement currents). The pairs of fields, **E** and **D**, **H** and **B**, are related by the following expressions, known as the constitutive equations:

$$\mathbf{D} = \varepsilon\,\mathbf{E}, \tag{66}$$

$$\mathbf{B} = \mu\,\mathbf{H}, \tag{67}$$

where ε and μ are dielectric and magnetic permeabilities of the medium, respectively.

Maxwell's equations were introduced by generalization of the basic laws of electromagnetism established in the first half of the 19th century. It is interesting to note that in fact all these equations can be derived directly from the basic differential equations of field theory, formulated above. Indeed, let us introduce two electromagnetic differential 2-forms, M and F, according to the following formulae:

$$M = \mathbf{D} \cdot \mathrm{d}\mathbf{\Sigma} - (\mathbf{H} \cdot \mathbf{dr}) \wedge \mathrm{d}t = D - H \wedge \mathrm{d}t, \tag{68}$$

$$F = \mathbf{B} \cdot \mathrm{d}\mathbf{\Sigma} + (\mathbf{E} \cdot \mathbf{dr}) \wedge \mathrm{d}t = B + E \wedge \mathrm{d}t, \tag{69}$$

where $D = \mathbf{D} \cdot \mathrm{d}\mathbf{\Sigma}$, $H = \mathbf{H} \cdot \mathbf{dr}$, $B = \mathbf{B} \cdot \mathrm{d}\mathbf{\Sigma}$, and $E = \mathbf{E} \cdot \mathbf{dr}$.

Following Misner et al. (1973) and Deschamps (1981), we will call these forms Maxwell's field, M, and force field, F, respectively. Using the basic properties of differential 2-forms, discussed above, we can write the following differential equations for these forms:

$$\mathrm{d} \wedge M = \gamma^e \tag{70}$$

$$\mathrm{d} \wedge F = \gamma^m, \tag{71}$$

where the corresponding electric, γ^e, and magnetic, γ^m, four-currents are equal to:

$$\gamma^e = q\mathrm{d}v - (\mathbf{j} \cdot \mathrm{d}\mathbf{\Sigma}) \wedge \mathrm{d}t, \tag{72}$$

$$\gamma^m = q^m \mathrm{d}v - \left(\mathbf{j}^m \cdot \mathrm{d}\mathbf{\Sigma}\right) \wedge \mathrm{d}t. \tag{73}$$

Here, the functions q^m and $\mathbf{j}^m$ are the magnetic charge density and the magnetic current density, respectively.

According to formulae (47) and (49), from the differential equation (70) for Maxwell's field M we obtain immediately Maxwell's first and fourth equations (62) and (65):

$$\mathit{curl}\,\mathbf{H} = \mathbf{j} + \frac{\partial}{\partial t}\mathbf{D} \quad \text{and} \quad \mathit{div}\,\mathbf{D} = q. \tag{74}$$

Taking into account that the external differential of the four-current γ^e, according to (70), is equal to zero,

$$\mathrm{d} \wedge \gamma^e = \mathrm{d} \wedge \mathrm{d} \wedge M = 0,$$

and considering formula (40), we arrive at the *continuity equation* for electric current density $\mathbf{j}$ and the charge density q:

$$\nabla \cdot \mathbf{j} = -\frac{\partial q}{\partial t}. \tag{75}$$

In a similar way, from the differential equation (71) for the force field F we obtain immediately a generalization of Maxwell's second and third equations (63) and (64), which allows the existence of the magnetic charges:

$$curl\,\mathbf{E} = -\mathbf{j}^m - \frac{\partial}{\partial t}\mathbf{B} \quad \text{and} \quad div\,\mathbf{B} = q^m. \tag{76}$$

Note that the magnetic four-current γ^m satisfies the same differential equation as the electric four-current:

$$\mathrm{d} \wedge \gamma^m = \mathrm{d} \wedge \mathrm{d} \wedge F = 0.$$

Therefore, the magnetic charges and current, in general cases, are related by the continuity equation as well:

$$\nabla \cdot \mathbf{j} = -\frac{\partial q^m}{\partial t}. \tag{77}$$

Introducing the magnetic charges makes Maxwell's equation symmetrical. However, in the real world we do not observe the magnetic charges, which results in a Faraday-type equation for the force field:

$$\mathrm{d} \wedge F = 0. \tag{78}$$

This equation, written in vectorial notation, brings us to Maxwell's original second and third equations (63) and (64):

$$curl\,\mathbf{E} = -\frac{\partial}{\partial t}\mathbf{B} \quad \text{and} \quad div\,\mathbf{B} = 0.$$

Thus, the whole system of Maxwell's equations automatically appears from the general theory of nonstationary fields. This remarkable fact demonstrates the power of mathematical theory. We can see now that the basic laws of electromagnetism are actually hidden in the fundamental differential relationships between the vector fields and the differential forms.

In summary, we can represent now the full system of Maxwell's equations in a very compact and elegant form as follows:

$$\mathrm{d} \wedge M = \gamma^e, \tag{79}$$

$$\mathrm{d} \wedge F = 0, \tag{80}$$

where

$$M = \mathbf{D} \cdot \mathrm{d}\mathbf{\Sigma} - (\mathbf{H} \cdot \mathrm{d}\mathbf{r}) \wedge \mathrm{d}t = D - H \wedge \mathrm{d}t, \tag{81}$$

and

$$F = \mathbf{B} \cdot \mathrm{d}\boldsymbol{\Sigma} + (\mathbf{E} \cdot \mathrm{d}\mathbf{r}) \wedge \mathrm{d}t = B + E \wedge \mathrm{d}t. \tag{82}$$

It was demonstrated above that any Faraday-type form can be expressed by the corresponding four-potential α. Taking into account that the force field F is a Faraday-type form, we can use Eq. (60), according to which,

$$B = \mathbf{B} \cdot \mathrm{d}\boldsymbol{\Sigma} = curl\,\mathbf{A} \cdot \mathrm{d}\boldsymbol{\Sigma}, \quad E = \mathbf{E} \cdot \mathrm{d}\mathbf{r} = -\left(grad\,U + \frac{\partial \mathbf{A}}{\partial t} \right) \cdot \mathrm{d}\mathbf{r}. \tag{83}$$

From the last formula we obtain a classical representation of the magnetic and electric fields using *electrodynamic potentials*, $\mathbf{A}$and U:

$$\mathbf{B} = curl\,\mathbf{A} \quad \text{and} \quad \mathbf{E} = -\left(grad\,U + \frac{\partial \mathbf{A}}{\partial t} \right). \tag{84}$$

7. Integral Formulations of the Differential Form Equations for Maxwell's Field and Force Field

Equations (79) and (80) represent a differential (localized) formulation of the laws of electromagnetism. In applications, it is useful to have the integral formulations of the same laws. For example, we will demonstrate below that the integral formulations are useful in numerical modeling of EM fields.

7.1. Faraday's electromagnetic induction law

Let us consider Maxwell's equation for the force field first:

$$\mathrm{d} \wedge F = 0. \tag{85}$$

We can integrate this equation over a geometrical element C_p ($p = 1, 2, 3, 4$) from the four-dimensional Euclidean space E_4:

$$\int_{C_p} \mathrm{d} \wedge F = 0, \tag{86}$$

where the geometrical element C_p can be treated as a p-dimensional domain in a multidimensional space. For example, we can consider a geometrical element C_3 in a form of the "cylinder" over some surface S in three-dimensional Euclidean space E_3, with the conventional spatial coordinates,

$x_1 = x$, $x_2 = y$, and $x_3 = z$, and a time interval $T = (t_0 \leq t \leq t_0 + \Delta t)$. Using standard mathematical notations, we can present the geometrical element C_3 in the form:

$$C_3 = T \times S.$$

According to a general Stokes's theorem (Fecko, 2006), the integral of the exact p-form $\underset{(p)}{\Omega} = \mathrm{d} \wedge \underset{(p-1)}{\Omega}$ over a geometrical element C_p is equal to the integral of the (p-1)-form $\underset{(p-1)}{\Omega}$ over the boundary, ∂C_p, of the geometrical element C_p:

$$\int_{C_p} \underset{(p)}{\Omega} = \int_{C_p} \mathrm{d} \wedge \underset{(p-1)}{\Omega} = \int_{\partial C_p} \underset{(p-1)}{\Omega}, \quad p = 1, 2, 3, 4. \tag{87}$$

According to Maxwell's equation (85), the force field F is an exact 2-form. Integrating both sides of Eq. (85) over the geometrical element $C_3 = T \times S$ and taking into account the general Stokes's theorem (87), we obtain

$$\int_{C_3} \mathrm{d} \wedge F = \int_{\partial C_3} F = 0. \tag{88}$$

Substituting expression (85) for the force field in Eq. (88), we have:

$$\begin{aligned} \int_{\partial C_3} F &= \int_{\partial C_3} \mathbf{B} \cdot \mathrm{d}\boldsymbol{\Sigma} + \int_{\partial C_3} (\mathbf{E} \cdot \mathrm{d}\mathbf{r}) \wedge \mathrm{d}t \\ &= \int_{\partial C_3} B + \int_{\partial C_3} E \wedge \mathrm{d}t = 0. \end{aligned} \tag{89}$$

A simple geometrical consideration shows that

$$\partial C_3 = \partial T \times S - T \times \partial S = [t_0 + \Delta t] \times S - [t_0] \times S - T \times \partial S. \tag{90}$$

Taking into account the geometrical structure (90) of the boundary ∂C_3, we can calculate the integrals in Eq. (89):

$$\int_{\partial C_3} B = \int_{\partial C_3} B(\mathbf{r}, t) = \int_{\partial C_3} B(\mathbf{r}, t_0 + \Delta t) - \int_{\partial C_3} B(\mathbf{r}, t_0), \tag{91}$$

and

$$\int_{\partial C_3} E \wedge \mathrm{d}t = \int_{\partial S} \int_T E(\mathbf{r}, t)\, \mathrm{d}t. \tag{92}$$

Thus, Eq. (89) is reduced to

$$\int_{\partial C_3} B + \int_{\partial C_3} E \wedge \mathrm{d}t = \int_S [B(\mathbf{r}, t_0 + \Delta t) - B(\mathbf{r}, t_0)] + \int_{\partial S} \int_T E(\mathbf{r}, t) \wedge \mathrm{d}t = 0. \tag{93}$$

Let us assume that we have an infinitesimal time interval $\Delta t \to 0$. In this case Eq. (93) can be simplified as

$$\int_S [B(\mathbf{r}, t_0 + \Delta t) - B(\mathbf{r}, t_0)] = -\Delta t \int_{\partial S} E(\mathbf{r}, t_0),$$

which leads to the final equation:

$$\frac{\partial}{\partial t} \int_S B(\mathbf{r}, t)|_{t_0} = -\int_{\partial S} E(\mathbf{r}, t_0). \tag{94}$$

We can recall now that $B = \mathbf{B} \cdot \mathrm{d}\boldsymbol{\Sigma}$ is a magnetic induction flux 2-form, and $E = \mathbf{E} \cdot \mathbf{dr}$ is an electric voltage 1-form. Therefore, Eq. (94) represents a conventional Faraday's law for electromagnetic induction:

$$\int_{\partial S} \mathbf{E}(\mathbf{r}, t_0) \cdot \mathbf{dr} = -\frac{\partial}{\partial t} \int_S \mathbf{B}(\mathbf{r}, t) \cdot \mathrm{d}\boldsymbol{\Sigma}|_{t_0}. \tag{95}$$

7.2. Integral formulation of Ampere's law

In a similar way we can demonstrate that the differential form equation for Maxwell's field (70) results in Ampere's electromagnetic law. Indeed, let us reproduce Eq. (70) for convenience:

$$\mathrm{d} \wedge M = \gamma^e. \tag{96}$$

Integrating both sides of this equation over the geometrical element $C_3 = T \times S$ and taking into account the general Stokes's theorem (87), we obtain:

$$\int_{C_3} \mathrm{d} \wedge M = \int_{\partial C_3} M = \int_{\partial C_3} \gamma^e. \tag{97}$$

Substituting expressions (68) and (72) for Maxwell's field M and four-current γ^e in Eq. (97), we have:

$$\int_{\partial C_3} (D - H \wedge \mathrm{d}t) = \int_{C_3} [q\mathrm{d}v - j \wedge \mathrm{d}t], \tag{98}$$

where

$$j = \mathbf{j} \cdot \mathrm{d}\boldsymbol{\Sigma} \tag{99}$$

is an electric current 2-form.

Assuming that we have an infinitesimal time interval $\Delta t \to 0$, and proceeding in analogy with the force field equations above, we can calculate the integrals in Eq. (88) as

$$\begin{aligned} \int_{\partial C_3} (D - H \wedge \mathrm{d}t) = & \int_S [D(\mathbf{r}, t_0 + \Delta t) - D(\mathbf{r}, t_0)] \\ & - \Delta t \int_{\partial S} H(\mathbf{r}, t_0) \end{aligned} \tag{100}$$

and

$$\int_{T \times S} q\mathrm{d}v = 0, \quad \int_{T \times S} j \wedge \mathrm{d}t = -\Delta t \int_S j(\mathbf{r}, t_0). \tag{101}$$

Substituting expressions (100) and (101) back into Eq. (98), we arrive at the final integral form of the first Maxwell's equation:

$$\frac{\partial}{\partial t} \int_S D(\mathbf{r}, t)|_{t_0} - \int_{\partial C} H(\mathbf{r}, t_0) = - \int_S j(\mathbf{r}, t_0). \tag{102}$$

Taking into account that $D = \mathbf{D} \cdot \mathrm{d}\boldsymbol{\Sigma}$ is an electric displacement flux 2-form, $H = \mathbf{H} \cdot \mathrm{d}\mathbf{r}$ is magnetic work 1-form, and $j = \mathbf{j} \cdot \mathrm{d}\boldsymbol{\Sigma}$ is an electric current 2-form, we obtain the classical integral formulation of Ampere's law:

$$\int_{\partial S} \mathbf{H} \cdot \mathrm{d}\mathbf{r} = \int_S (\mathbf{j} \cdot \mathrm{d}\boldsymbol{\Sigma}) + \frac{\partial}{\partial t} \int_S \mathbf{D} \cdot \mathrm{d}\boldsymbol{\Sigma}. \tag{103}$$

7.3. Integral equations for Maxwell's field and force field in the frequency domain

We can obtain the integral equations for Maxwell's field and force field in the frequency domain by introducing frequency domain differential forms:

$$\{E(\mathbf{r},\omega), D(\mathbf{r},\omega), H(\mathbf{r},\omega), B(\mathbf{r},\omega), j(\mathbf{r},\omega)\}$$
$$= \frac{1}{2\pi}\int_{-\infty}^{\infty}\{E(\mathbf{r},\omega), D(\mathbf{r},\omega), H(\mathbf{r},\omega), B(\mathbf{r},\omega), j(\mathbf{r},\omega)\}\mathrm{e}^{-\mathrm{i}\omega t}\mathrm{d}t. \quad (104)$$

Applying the Fourier transform (104) to Eqs (94) and (102), we obtain:

$$\int_{\partial S} E(\mathbf{r},\omega) = \mathrm{i}\omega\int_{S} B(\mathbf{r},\omega) \quad (105)$$

and

$$\int_{\partial S} H(\mathbf{r},\omega) = \int_{S} j(\mathbf{r},\omega) - \mathrm{i}\omega\int_{S} D(r,\omega). \quad (106)$$

Equations (105) and (106) provide integral representations of Maxwell's differential form equations in the frequency domain.

8. Numerical Modeling Using Differential Forms

There are several ways to obtain discrete analogs of Maxwell's equations. In the vast majority of numerical algorithms, the model region is discretized into a number of prisms as shown in Figure 1 (Zhdanov, 2002).

A Cartesian coordinate system is defined with the z axis directed downward, and the x axis directed to the right. The indices i, k, and l are used to number the grid points in the x, y, and z directions, respectively. The electromagnetic parameters, σ, μ, and ε are assumed to be constant within each elementary prism. We denote this grid by Σ:

$$\Sigma = \left\{ (x_i, y_k, z_l) \begin{array}{ll} x_1 = x', \; x_{N_I} = x'' & x_{i+1} = x_i + \Delta x_i \\ & i = 1, 2, \ldots, N_I \\ y_1 = y', \; y_{N_K} = y'' & y_{k+1} = y_k + \Delta y_k \\ & k = 1, 2, \ldots, N_K \\ z_1 = z', \; z_{N_L} = z'' & z_{l+1} = z_l + \Delta z_l \\ & l = 1, 2, \ldots, N_L \end{array} \right\}.$$

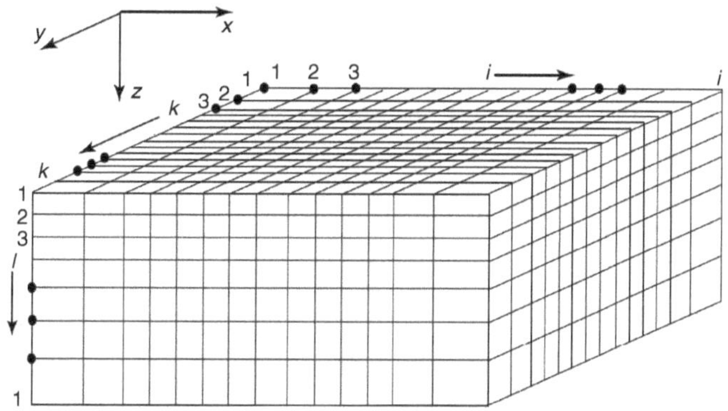

Figure 1 The model region is discretized into a number of prisms. The indices i, k, and l are used to number the grid points in the x, y, and z directions, respectively. The electromagnetic parameters, σ, μ, and ε are assumed to be constant within each elementary prism.

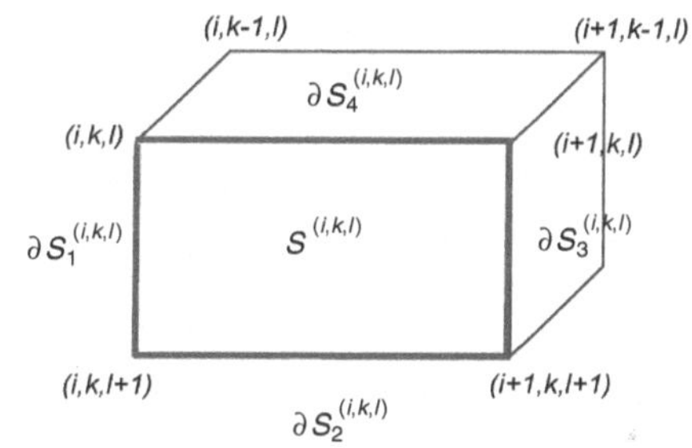

Figure 2 Discretization of the electromagnetic differential forms on a rectangular grid.

Let us consider one prismatic cell of the grid (Figure 2). We denote by S a face of the prism, bounded by a rectangular contour ∂S. We consider the integral Eq. (105) for EM differential forms written for a face of the prism.

We can evaluate the contour integral in Eq. (105) as a sum of four integrals over the edges of the face of the prism:

$$\int_{\partial S_1} E(\mathbf{r},\omega) + \int_{\partial S_2} E(\mathbf{r},\omega) + \int_{\partial S_3} E(\mathbf{r},\omega) + \int_{\partial S_4} E(\mathbf{r},\omega)$$
$$= \mathrm{i}\omega \int_S B(\mathbf{r},\omega). \tag{107}$$

For example, at the prism face $S^{(i,k,l)}$ parallel to the vertical plane xz (see Figure 2) and having the upper left corner at the node (i, k, l) the integral

Eq. (107) can be written as follows:

$$\int_{\partial S_1^{(i,k,l)}} E(\mathbf{r},\omega) + \int_{\partial S_2^{(i,k,l)}} E(\mathbf{r},\omega) + \int_{\partial S_3^{(i,k,l)}} E(\mathbf{r},\omega) + \int_{\partial S_4^{(i,k,l)}} E(\mathbf{r},\omega) = \mathrm{i}\omega \int_{S^{(i,k,l)}} B(\mathbf{r},\omega). \tag{108}$$

We denote by $E_m(i,k,l)$ a voltage along the edge $\partial S_m^{(i,k,l)}$, $m = 1, 2, 3, 4$,

$$E_m(i,k,l) = \int_{\partial S_m^{(i,k,l)}} E(\mathbf{r},\omega),$$

and by $B(i,k,l)$ a magnetic induction flux through the prism's face $S^{(i,k,l)}$,

$$B(i,k,l) = \int_{S^{(i,k,l)}} B(\mathbf{r},\omega).$$

Using these notations, we can write Eq. (108) in the form:

$$\sum_{m=1}^{4} E_m(i,k,l) = B(i,k,l). \tag{109}$$

Similar algebraic expressions can be obtained for other faces of an elementary prism. We derive analogous formulae using Eq. (106) for EM differential forms written for a face of the prism. Combining all these algebraic equations together, we arrive at a full system of linear equations for the discretized values of the flux and voltage (work) of an electromagnetic field on a rectangular grid.

It is important to emphasize that the system of algebraic equations for the fluxes and voltages (work) of an electromagnetic field, derived above, provides an exact representation of the original system of Maxwell's equations for the differential forms. At the same time, any discretization of the classical system of Maxwell's equations for the vector fields based on finite difference or finite element methods results in some approximate representation of the vector fields. This property of the numerical methods based on differential form equations opens a possibility for developing a very accurate technique for electromagnetic modeling, especially in the case of a high conductivity contrast.

Another advantage of modeling based on the differential form equations is that the corresponding fluxes and voltages (work) of EM fields are continuous on the faces and edges, respectively, of homogeneous prisms.

9. CONCLUSIONS

We have demonstrated in this paper that the major differential equations for nonstationary vector fields can be expressed in a very simple form (47). The simplicity and symmetry of this equation indicates that the differential forms provide a natural representation for vector fields in four-dimensional space, E_4. Note that these forms are introduced as a linear combination of the flux of the vector field $\mathbf{D}$ through a vector element of the surface, $D = \mathbf{D} \cdot \mathbf{d\Sigma}$, and a time differential multiplied by the work of the vector field $\mathbf{H}$ along a vector element of a line, $H = \mathbf{H} \cdot \mathbf{dr}$. The differential equations for a pair of arbitrary vector fields in a four-dimensional Euclidean space have a structure identical to Maxwell's system of equations. Therefore, the basic laws of the classical theory of electromagnetic fields are encoded in the mathematical structure of the differential forms.

An important feature of the differential form of Maxwell's equations (79) and (80) is that they describe the relationships between the elementary fluxes and the work of the different EM field components, while the original Maxwell's equations (62)–(65) deal with the vectors of the EM fields themselves. Thus, the new mathematical form of Maxwell's equations emphasizes the importance of the fluxes and the work of the EM field. We should conclude that, the flux of the field through a given surface and the work of the field along a given path indeed represent the most important physical entities which are studied and measured in geophysical experiments. That is why the new form of Maxwell's equations (79) and (80) appears to be extremely well suited for the description of geophysical EM phenomena.

ACKNOWLEDGMENTS

The author acknowledges the support of the University of Utah Consortium for Electromagnetic Modeling and Inversion (CEMI), which includes BAE Systems, Baker Atlas Logging Services, BGP China National Petroleum Corporation, BHP Billiton World Exploration Inc., BP, EMGS, ENI S.p.A., ExxonMobil Upstream Research Company, Fugro, Halliburton Energy Services, Information Systems Laboratories, Newmont Mining Co., OHM, Petrobras, PGS, Rio Tinto - Kennecott, Rocksource, Russian Research Center Kurchatov Institute, Schlumberger, Science Applications International Co., Shell International Exploration and Production Inc., StatoilHydro, Sumitomo Metal Mining Co., Woodside Energy, and Zonge Engineering and Research Organization.

REFERENCES

Deschamps, G.A., 1981. Electromagnetics and differential forms. Proc. IEEE 69 (6), 676–696.
Fecko, M., 2006. Differential Geometry and Lie Groups for Physicists. Cambridge University Press, 697 pp.
Lindell, I.V., 2004. Differential Forms in Electromagnetics. IEEE Press, John Wiley & Sons, 253 pp.
Maxwell, J.C., 1954. A Treatise on Electricity and Magnetism. Dover Publications, New York (unabridged and unaltered republication of the third edition of 1891), 506 pp.
Misner, C.W., Thorne, K.S., Wheeler, J.A., 1973. Gravitation. Freeman, San Francisco, 1279 pp.
Zhdanov, M.S., 2002. Geophysical Inverse Theory and Regularization Problems. Elsevier, 609 pp.
Zhdanov, M.S., 2009. Geophysical Electromagnetic Theory and Methods. Elsevier, 856 pp.

CHAPTER 18

3D Electromagnetic Holographic Imaging in Active Monitoring of Sea-Bottom Geoelectrical Structures

Michael S. Zhdanov[a], Martin Cuma[b], *and* Takumi Ueda[c]

Contents

1. Introduction

During the last few years marine controlled-source electromagnetic (MCSEM) surveys have become widely used for offshore petroleum exploration. The main target of this survey is the sea-bottom petroleum reservoir, which is usually characterized by a low electrical conductivity anomaly within relatively conductive sea-bottom sediments. There is growing interest in the interpretation of MCSEM data based on 3D geoelectrical models. The conventional approach based on standard 3D forward modeling and inversion meets significant difficulties because of

[a] Department of Geology and Geophysics, University of Utah, USA
[b] Center for High Performance Computing, University of Utah, USA
[c] Geological Survey of Japan, Japan

E-mail address: mzhdanov@mines.utah.edu (M.S. Zhdanov).

Handbook of Geophysical Exploration: Seismic Exploration, Volume 40
ISSN 0950-1401, DOI: 10.1016/S0950-1401(10)04023-1

the enormous amount of computations required in the case of the multitransmitter and multireceiver data acquisition systems typical for marine CSEM surveys.

There is an alternative approach to the solution of this problem, which is based on the ideas of electromagnetic holography and/or migration (Zhdanov, 1981; Zhdanov and Frenkel, 1983a,b; Zhdanov, 1988; Zhdanov and Keller, 1994; Zhdanov et al., 1996; Zhdanov and Traynin, 1997; Zhdanov, 1999, 2001, 2002; Tompkins, 2004; Mittet et al., 2005; Wan and Zhdanov, 2005a,b; Zhdanov et al., 2006).

The physical principles of EM holography parallel those underlying optical holography and seismic migration. The recorded amplitudes and phases of an EM field scattered by an object form a broadband EM hologram. As in optical and radiowave holography, we can reconstruct the volume image of the object by "illuminating" the broadband EM hologram by a reference signal. While in the optical case this can be performed optically, yielding a visible image, in the case of a broadband EM field the reconstruction is done numerically, using computer transformation. In fact EM holography/migration, similar to seismic migration, is based on a special form of downward continuation of the observed field, which can be computed as a solution of the boundary value problem for the adjoint Maxwell's equations.

In the current paper we consider an application of this approach to the interpretation of a typical sea-bed logging (SBL) survey which consists of a set of sea-bottom receivers and a moving electrical bipole transmitter. The receivers record the magnitude and the phase of the frequency domain (FD) electromagnetic field generated by the moving transmitter and scattered back by sea-bottom geoelectrical structures. The combined electromagnetic signal in the receivers forms a broad-band EM hologram of the sea-bottom geological target (e.g., a petroleum reservoir). In order to reconstruct the geoelectrical image of the target, we replace a set of receivers with a set of auxiliary transmitters located in the positions of the receivers. The strength and the phase of the signal transmitted by these auxiliary transmitters are determined according to the parameters of the observed field in the receivers. These transmitters generate an EM field, which is called a backscattering or migration field. The vector cross-power spectrum of the background field (the field generated by the original transmitter in a medium without a target) and the backscattering field produces a numerical reconstruction of a volume image of the conductivity distribution (Zhdanov, 2001).

We should note, however, that the frequency of the EM signal used in a marine EM is very low, about 1 Hz. In this low frequency range, the EM field propagates in sea-bottom formations according to the diffusion equation (Zhdanov and Keller, 1994), which results in a relatively low resolution of the geoelectrical image obtained by the numerical algorithm described above. In order to improve the resolution of the EM holographic imaging, we should apply the migration iteratively. The development of the corresponding method of iterative migration with application to MCSEM data constitutes the main subject of the present paper.

2. Marine Controlled-Source Electromagnetic (MCSEM) Method

The MCSEM method for hydrocarbon exploration has a long history that stretches back to the beginning of the 20th century (Berdichevsky et al., 1989; Srnka et al., 2006). One can look back to the beginning of marine CSEM research more than 80 years ago with direct current (DC) surveys off the Cornish coast of England. Significant progress, however, was only made in the 1960s and 1970s through efforts by university and government groups, especially for pure geophysical, earth science research, based mainly on the natural marine magnetotelluric (MMT) method (e.g., Novysh and Fonarev, 1966; Trophimov and Fonarev, 1972). There have been several EM methods for mapping and imaging sub-sea-bottom geoelectrical structures (e.g., Sinha et al., 1990; Chave et al., 1991), especially for application to the study of the oceanic lithosphere and actively spreading mid-ocean ridges (Young and Cox, 1981; Evans et al., 1994; Constable and Cox, 1996; MacGregor and Sinha, 2000; MacGregor et al., 2001). Marine geoelectrical investigations have been intensively developed in the former Soviet Union. A review of this research can be found in Berdichevsky et al. (1989).

The turning point for the marine CSEM method came almost 25 years ago. According to Srnka et al. (2006), at the beginning of the 1980s major western oil companies began investigating marine CSEM theory and computer modeling. The companies found results that looked promising for deep-water applications (i.e., the deeper the water, the better), and field tests were proposed over the known reservoirs. The conclusion of the companies at that time was, however, that the time for marine CSEM in the petroleum industry had not yet arrived; this was mainly due to the lack of suitable acquisition equipment, inadequate computational capability, a rapidly growing emphasis on the new, powerful 3D marine seismic technology, and limited deep-water business opportunities. As a result, marine CSEM research was deferred. For about the last 20 years the 3D seismic survey has been a de facto standard and priority choice for offshore hydrocarbon exploration with the cooperation of gravity/magnetic and well logging.

We should note, however, that Russian geophysicists had used marine EM for HC exploration since the early 1970s. In the latter half of the 1990s, western oil and gas companies found that, in some potential hydrocarbon reservoir areas, the seismic method had failed to detect the hydrocarbons, or did not have enough resolution because of a complicated

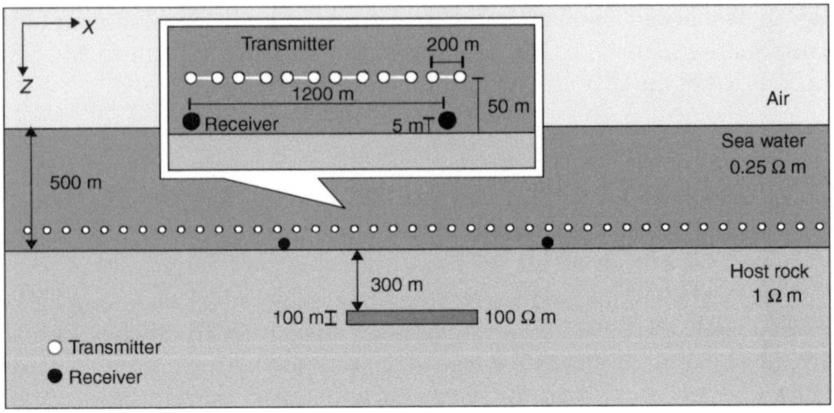

Figure 1 Sketch of the SBL survey and model design.

subsurface structure. Therefore, several research projects were initiated to find appropriate exploration techniques in such areas, including re-examination of marine EM methods around the world—not only by the academic community, but also by several major western petroleum companies. During the past two decades, great advances in both equipment and computational resources have created a favorable wind for EM methods, especially the MCSEM method for petroleum exploration.

3. Frequency Domain Electromagnetic (FDEM) Migration of MCSEM Data

Let us consider a typical MCSEM survey consisting of a set of electric field receivers located at the sea bottom and an electric bipole transmitter moving at some elevation above the sea bottom, as shown in Figure 1. We assume that the electrical conductivity in the model can be represented as the sum of the background conductivity $\sigma = \sigma_b$ and an anomalous conductivity $\Delta\sigma$ distributed within some local inhomogeneity D associated with the location of the petroleum reservoir.

The background conductivity is formed by a horizontally layered model consisting of nonconductive air, a conductive seawater layer, and a horizontally homogeneous (layered) section of a sea-bottom formation. The receivers are located at points with radius-vector r_j $(j = 1, 2, 3, \ldots, J)$ in some Cartesian coordinate system. Every receiver R_j records electric and magnetic field components of the field generated by an electric bipole transmitter moving above the receivers. We denote this field as $\mathbf{E}_i(r_j)$, $\mathbf{H}_i(r_j)$, where i is the index of the corresponding transmitter, T_i, located at the point $\mathbf{r}_i$ $(i = 1, 2, 3, \ldots, I)$.

Let us consider the data observed by one receiver, R_j. According to the reciprocity principle, the electric field component excited at $\mathbf{r}_j$ in the direction of $\mathbf{b}$ by an electric current element $\mathbf{a}$ at $\mathbf{r}_i$ is identical to the electric field component excited at $\mathbf{r}_i$ in the direction of $\mathbf{a}$ by an electric current element $\mathbf{b}$ at $\mathbf{r}_j$ (Zhdanov, 2002, p. 226):

$$\mathbf{E}_i(\mathbf{r}_j) \cdot \mathbf{b} = \mathbf{E}_j^E(\mathbf{r}_i) \cdot \mathbf{a}. \tag{1}$$

Similarly, the magnetic field component excited at $\mathbf{r}_j$ in the direction of $\mathbf{b}$ by an electric current element $\mathbf{a}$ at $\mathbf{r}_i$ is equal to the electric field component (multiplied by the minus sign) excited at $\mathbf{r}_i$ in the direction of $\mathbf{a}$ by a magnetic current element $\mathbf{b}$ at $\mathbf{r}_j$:

$$\mathbf{H}_i(\mathbf{r}_j) \cdot \mathbf{b} = -\mathbf{E}_j^H(\mathbf{r}_i) \cdot \mathbf{a}. \tag{2}$$

Therefore, one can substitute a reciprocal survey configuration for the original survey, assuming that we have electric, T_j^E, and magnetic, T_j^H, dipole transmitters located in the position of the receiver, R_j, and a set of receivers measuring the reciprocal electric fields, $\mathbf{E}_j^E(\mathbf{r}_i)$ and $\mathbf{E}_j^H(\mathbf{r}_i)$ in the positions of the original transmitters, T_i.

We can now calculate the backscattering (or migration) field for the data collected by one fixed sea-bottom receiver, R_j. Consider, for example, the reciprocal electric fields $\mathbf{E}_j^E(\mathbf{r}_i)$. This field can be represented as the sum of the background and anomalous parts:

$$\mathbf{E}_j^E(\mathbf{r}_i) = \mathbf{E}_j^b(\mathbf{r}_i) + \mathbf{E}_j^{E_a}(\mathbf{r}_i), \tag{3}$$

where the background electric field, $\mathbf{E}_j^b(\mathbf{r}_i)$ is generated by the electric dipole transmitter T_j^E in a model with a given background conductivity σ_b. The residual electric field, $\mathbf{R}_{E_j}(\mathbf{r}_i)$ is equal to the difference between the background and "observed" reciprocal field:

$$\mathbf{R}_{E_j}(\mathbf{r}_i) = \mathbf{E}_j^b(\mathbf{r}_i) - \mathbf{E}_j^E(\mathbf{r}_i) = -\mathbf{E}_j^{E_a}(\mathbf{r}_i). \tag{4}$$

According to the definition (Zhdanov, 2002), the backscattering (migrated) residual field is a field generated in the background medium by a combination of all electric dipole transmitters located at points $\mathbf{r}_i$ with the current moments determined by the complex conjugate residual field

$\mathbf{R}_{E_j}^*(\mathbf{r}_i)$ according to the following formula:

$$\mathbf{E}_j^m(\mathbf{r}) = \mathbf{E}_j^m(\mathbf{r}; \mathbf{R}_{E_j}^*) = \sum_{i=1}^{I} \mathbf{G}_E(\mathbf{r}|\mathbf{r}_i)\mathbf{R}_{E_j}^*(\mathbf{r}_i), \tag{5}$$

where the lower subscript j shows that we migrate the field observed by the receiver R_j, and $\mathbf{G}_E$ is the electric Green's tensor for the layered (background) conductivity model σ_b. Therefore, the migration field can be computed as a superposition of 1D weighted (by the corresponding receiver residual) responses generated by electric dipoles with the unit moments located at every transmitter position in the model with the background conductivity σ_b. This 1D electric dipole modeling is a very fast process, which results in the fast migration algorithm.

Formula (5) allows us to reconstruct the migration field everywhere in the medium under investigation. It can be shown that this transformation is stable with respect to the noise in the observed data. At the same time the spatial distribution of the migration field is closely related to the conductivity distribution in the medium. However, one needs to apply the corresponding imaging conditions to enhance the conductivity image produced by the EM migration. We will discuss this problem in the following sections of the paper.

In the general case of multiple receivers, the migration field is generated in the background medium by all electric dipole transmitters located above all receivers, R_j, having the current moments determined by the complex conjugate residual field $\mathbf{R}_{E_j}^*(\mathbf{r}_i)$:

$$\mathbf{E}^m(\mathbf{r}) = \sum_{j=1}^{J}\sum_{i=1}^{I} \mathbf{G}_E(\mathbf{r}|\mathbf{r}_i)\mathbf{R}_{E_j}^*(\mathbf{r}_i). \tag{6}$$

According to formula (5), we have:

$$\mathbf{E}^m(\mathbf{r}) = \sum_{j=1}^{J} \mathbf{E}_j^m(\mathbf{r}). \tag{7}$$

Therefore, the total migration field for all receivers can be obtained by summation of the corresponding migration field computed for every individual receiver.

A remarkable fact is that the migration of both electric and magnetic field data is actually reduced to the same forward problem for the electric field generated by the electric dipole transmitters. The only difference is that, in the case of the electric field receivers, we use the electric observed data to determine the electric current moment in the reciprocal transmitters. In the case of the magnetic receivers, the observed magnetic data are used to determine electric current moments of the receivers.

4. Electromagnetic Imaging Using Joint Migration of Electric and Magnetic Fields

The principles of EM holography/migration imaging are very similar to those of optical holography (Zhdanov, 2001). They can be summarized as follows.

(1) We "illuminate" the background media by a reciprocal electric dipole (in the case of electric observations) located in the actual receivers' positions to generate the "electric mode" background EM field $\{\tilde{\mathbf{E}}^{bE}, \tilde{\mathbf{H}}^{bE}\}$. Alternatively, we "illuminate" the background media by a reciprocal magnetic dipole (in the case of the magnetic field observations) located in the actual receivers' positions to generate the "magnetic mode" background EM field $\{\tilde{\mathbf{E}}^{bE}, \tilde{\mathbf{H}}^{bE}\}$.

(2) We "illuminate" the background media by artificial transmitters located in the positions of the true transmitters and represented by equivalent (fictitious) electric current dipoles. In the case of electric observations, the current moments are determined by the complex conjugate anomalous electric field observed in the true receiver for the given transmitter position. The electromagnetic field produced by this system of artificial electric dipoles generates the "electric mode" migration (backscattering) anomalous field $\{\tilde{\mathbf{E}}^{mE}, \tilde{\mathbf{H}}^{mE}\}$. In the case of magnetic observations, the current moments are determined by the complex conjugate anomalous electric field multiplied by the factor $(-i\omega\mu)$. The electromagnetic field produced by this system of artificial transmitters generates the "magnetic mode" migration (backscattering) anomalous field $\{\tilde{\mathbf{E}}^{mE}, \tilde{\mathbf{H}}^{mE}\}$.

(3) In the case of electric field observations, the geoelectrical image of the sea-bottom inhomogeneities, $\mathbf{I}_0^E$, is formed by summation of the cross-power spectrum of the "electric mode" background and migration fields:

$$\mathbf{I}_0^E = \mathrm{Re}\,\Sigma_{\omega_n}\left(\tilde{\mathbf{E}}^{bE}\cdot\tilde{\mathbf{E}}^{mE}\right), \tag{8}$$

where summation is done over all frequencies ω_n of the recorded fields.

(4) In the case of magnetic field observations, the geoelectrical image of the sea-bottom inhomogeneities, $\mathbf{I}_0^H$ is formed by calculating the cross-power spectrum of the "magnetic mode" background and migration

fields:

$$\mathbf{I}_0^H = \text{Re}\,\Sigma_{\omega n}\left(\tilde{\mathbf{E}}^{bH}\cdot\tilde{\mathbf{E}}^{mH}\right). \tag{9}$$

(5) In the case of joint migration of the electric and magnetic observed data, the geoelectrical image of the sea-bottom inhomogeneities, $\mathbf{I}_0^{EH}$, is formed by summation of the "electric mode" and "magnetic mode" images:

$$\mathbf{I}_0^{EH} = \text{Re}\,\Sigma_{\omega n}\left[\left(\tilde{\mathbf{E}}^{bE}\cdot\tilde{\mathbf{E}}^{mE}\right)+\left(\tilde{\mathbf{E}}^{bH}\cdot\tilde{\mathbf{E}}^{mH}\right)\right]. \tag{10}$$

Note that, in the case of multireceiver observations, the final image is produced by summation of all migration images generated for each receiver.

The images generated by formulas (8) to (10) are usually slightly distorted due to the different sensitivities of the observed data to the geoelectrical anomalies located at different depths and at different horizontal positions. To account for different sensitivities of the data to the conductivity distribution, we use an additional weighting function, W_m:

$$\sigma \approx -k(\mathbf{W}_m^*\mathbf{W})^{-1}\mathbf{I}_0, \tag{11}$$

where stands for any of the migration images introduced above, and k is some scaling coefficient.

The model parameter weighting matrix W_m is computed using the integrated sensitivity $\mathbf{S}$ as follows:

$$\mathbf{W}_m = \mathbf{S}^{\frac{1}{2}}, \tag{12}$$

where the integrated sensitivity is determined using the following formula (Zhdanov, 2002, p. 80):

$$\mathbf{S} = \text{diag}(\mathbf{F}^*\mathbf{F})^{\frac{1}{2}}. \tag{13}$$

In the last formula, $\mathbf{F}$ stands for the Fréchet derivative matrix of the corresponding EM forward modeling operator. By weighting the migration image $\mathbf{I}_0^{EH}$ with the integrated sensitivity, we assure that the observed data are equally sensitive to the conductivity variations within every part of the domain of investigation. As a result, we generate the electrical conductivity image which correctly reflects the volume distribution of the anomalous conductivity.

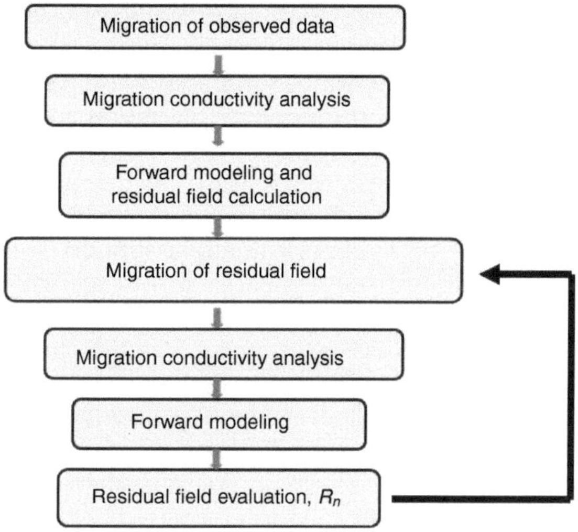

Figure 2 A scheme of the computational flow for iterative migration.

5. Regularized Iterative Migration

We have demonstrated above that migration imaging can be treated as the first iteration in the solution of the electromagnetic inverse problem. Obviously, we can obtain better imaging results if we repeat the iterations. Following Zhdanov (2002), we can describe the method of iterative migration as follows. On every iteration we calculate the theoretical electromagnetic response $\tilde{\mathbf{E}}_n$ for the given geoelectrical model σ_n, obtained in the previous step, we calculate the residual field between this response and the observed field, $\tilde{\mathbf{R}}_E^n$ and then we migrate the residual field. The geoelectrical image is computed, according to Eq. (8), as a sum over the frequencies of the dot product of the migrated residual field and the theoretical response $\tilde{\mathbf{E}}^n$. This image is corrected by the integrated sensitivity $\mathbf{S}$ to calculate the new geoelectrical model σ_n on the basis of expression (11). Iterative migration is terminated when the residual field becomes smaller than the required accuracy level of the data fitting. Figure 2 shows a scheme of the computational flow for iterative migration.

In fact, iterative migration results in rigorous inversion. It was demonstrated in Portniaguine and Zhdanov (1999) and Zhdanov (2002) that images with sharp boundaries can be recovered by regularized inversion algorithms based on a special family of stabilizing functionals. Particularly, the minimum support (MS) functional was found to be useful in the solution of this problem. It selects the inverse model within the class of models with a minimum volume of a domain with anomalous parameter distribution. This class of models describes the compact objects which are typical targets, for example, in mineral and hydrocarbon exploration. A similar approach can be applied in the case of the iterative migration transformation. We call this technique focusing iterative migration. Numerical implementation of focusing migration is similar to focusing inversion (Zhdanov, 2002). Particularly, images with sharp boundaries can be recovered using the minimum support (MS) or minimum gradient support (MGS) stabilizing functionals. This technique is implemented in our algorithm of focusing migration of electric and magnetic field data.

Another advantage of iterative migration is based on the fact that it allows us to include an a priori model of the target in the iterative process in a similar way as in the case of conventional inversion. The details of this technique can be found in (Zhdanov, 2002).

Note that every iteration of the migration algorithm requires two forward modeling computations: one to compute the migration field, and another one to compute the predicted data in the receivers. In this work, we use recently developed migration code that is parallelized over the Z dimension of the migration domain. For calculation of the migration and predicted fields we use an integral equation (IE) based on a parallel computer program. This enables us to considerably reduce the computation time and also model larger problems by increasing the migration domain size and the number of the cells used for the migration domain discretization.

6. Migration of Synthetic MCSEM Data

We have analyzed the principles of the iterative EM migration outlined above using as an example synthetic MCSEM data, computed for the models shown below.

6.1. Model 1

In the first set of numerical experiments, we assume that a synthetic MCSEM survey is conducted in relatively shallow water with a sea depth of 300 m. The survey consists of eleven sea-bottom receivers and an electric dipole transmitter moving along a line passing directly above the receivers at an elevation 50 m above the seafloor. The transmitter generates a frequency domain EM field every 200 m along the towing line, which is extended from −3000 m to 3000 m. Eleven seafloor electric receivers are located 5 m above the sea bottom along the x coordinates from $x = -2500$ m to $x = 2500$ m with a 500 m spacing. The background layered geoelectrical model consists of a seawater layer with a thickness of 300 m, a resistivity of 0.25 Ohm m, and homogeneous sea-bottom sediments with a resistivity of 1 Ohm m. There is

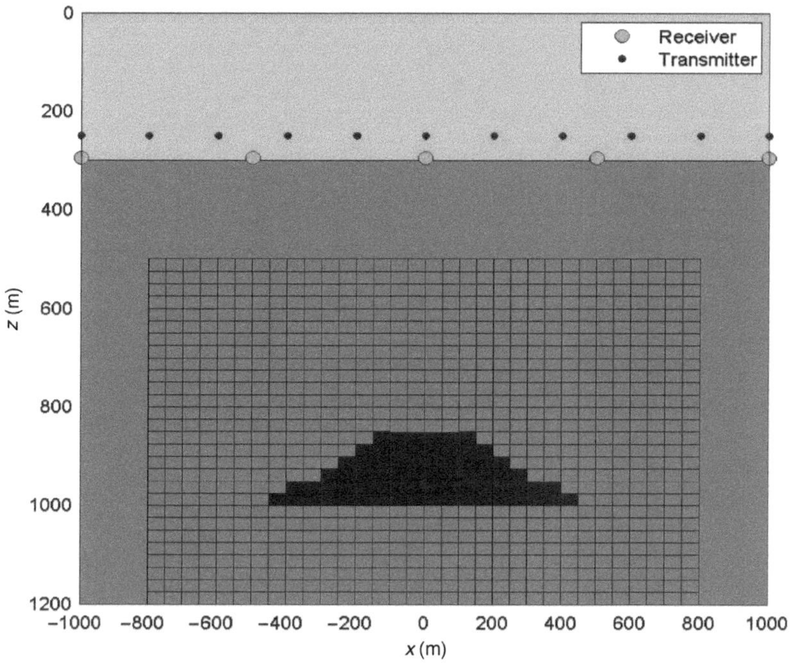

Figure 3 A vertical section of the true resistivity model along the survey line. The reservoir resistivity (blue) is 100 Ohm m, the background resistivity (red) is 1 Ohm m.

an anticlinal oil reservoir located in the seafloor sediments at a depth between 850 m and 1000 m below the sea level with a resistivity of 100 Ohm m and a maximum horizontal diameter of 900 m. A vertical section of the true resistivity model along the survey line is shown in Figure 3. The migration domain is discretized with a cell size of $25 \times 25 \times 25$ m^3 resulting in 28,672 cells.

The transmitter generates an EM field at frequencies of 0.25 and 0.75 Hz. The receivers measure the in-line component of the electric fields, E_x, and the cross-line component of the magnetic fields, H_y, simultaneously. The synthetic MCSEM data for this model were calculated using the parallel integral equation forward modeling code PIE3D developed by Yoshioka and Zhdanov (2005).

The observed data for this model can be represented as the plots of the total electric field E_x and magnetic field H_y recorded in the receivers, the plots of the same fields normalized by the absolute values of the background electric and magnetic fields respectively, and the plots of the phase difference between the observed and background electric fields (Figures 4 through 7). We show magnitude versus offset (MVO) and phase versus offset (PVO) plots of the total and normalized fields in these figures. Note that we have

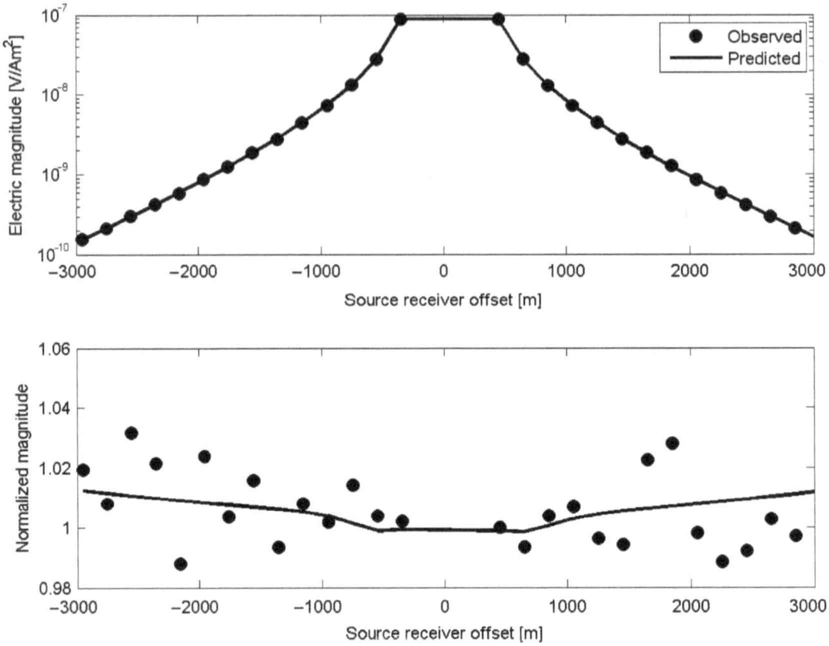

Figure 4 The top panel shows magnitude versus offset (MVO) plots of the total electric field, while the bottom panel presents the MVO plot of the same field normalized by the absolute values of the background electric fields. The observed data contaminated by noise are shown by the dots. The solid line corresponds to the data predicted for the migration resistivity model.

contaminated the synthetic observed data with random Gaussian noise. The noise level increases linearly from 1% at zero offset up to 7% at 10,000 m offset to simulate the typical noise behavior in the field MCSEM data.

We consider a 3D migration of the electric and magnetic field data for Model 1. We can realize a full 3D migration for the data observed by a few receivers located along a profile, because we generate a 3D migration field by a set of reciprocal transmitters. The migration field generated by these transmitters propagates within the medium in all directions, creating a 3D image of the target.

We perform two iterative migration calculations. The first calculation uses no a priori model. We run 35 iterations of the migration. The location and shape of the reservoir are resolved fairly well (Figure 8). The holographic image is slightly too deep and the smaller upper part is not visible.

In the second calculation, we have used an anticlinal a priori model with resistivity 3 Ohm m that encloses the true reservoir. We again ran 35 iterations. However, the a priori model was used only during the 30 initial iterations. The reservoir location is now resolved much clearer (Figure 9). Figures 4 through 7 show the predicted data computed for the migration

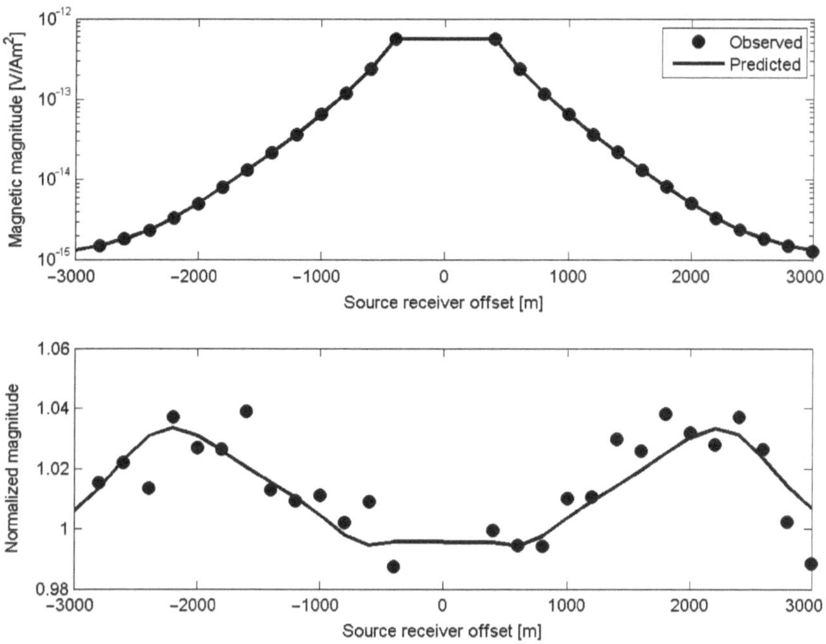

Figure 5 The top panel shows magnitude versus offset (MVO) plots of the total magnetic field, while the bottom panel presents the MVO plot of the same field normalized by the absolute values of the background magnetic fields. The observed data contaminated by noise are shown by the dots. The solid line corresponds to the data predicted for the migration resistivity model.

model shown in Figure 9. The predicted data show a good fit to the noisy observed data.

6.2. Model 2

In the second test of the numerical experiments, we consider a geoelectrical model of the sea-bottom formation shown in Figure 10. The MCSEM survey consists of two parallel transmitter-receiver profiles in the x direction. Each line consists of five sea-bottom receivers and an electric dipole transmitter moving along a line passing directly above the receivers at an elevation 50 m above the seafloor. In order to reduce the calculation cost, we assume that the transmitter generates a frequency domain EM field every 500 m along the towing line, which is extended from -4000 m to 4000 m. A total of ten seafloor electric receivers are located 5 m above the sea bottom along the x coordinates from $x = -2000$ to $x = 2000$ m with a 1000 m spacing at $y = -500$ m and $y = +500$ m. The background layered geoelectrical model consists of a seawater layer with a thickness of 300 m, a resistivity of 0.25 Ohm m, and homogeneous sea-bottom sediments with a

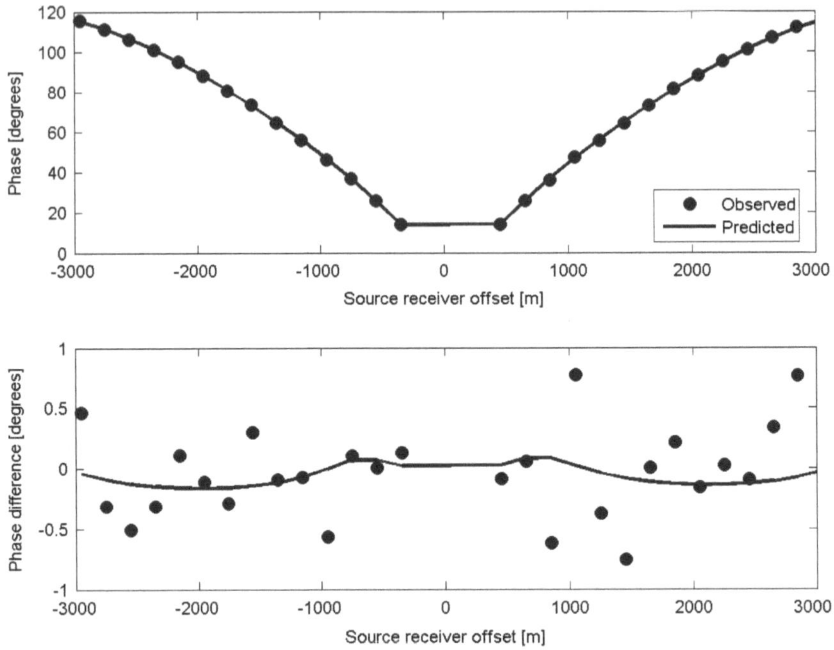

Figure 6 The top panel shows phase versus offset (PVO) plots of the total electric field, while the bottom panel presents the PVO plot of the phase difference between the observed and background electric fields. The observed data contaminated by the noise are shown by the dots. The solid line corresponds to the data predicted for the migration resistivity model.

resistivity of 1 Ohm m. There is an L-shaped reservoir located in the seafloor sediments at a depth between 800 m and 1000 m below sea level with a resistivity of 100 Ohm m, and a horizontal size of 2000 m by 2000 m. A 3D sketch of the true resistivity model and transmitter-receiver locations is shown in Figure 10. The size of the cells in the migration domain is $100 \times 100 \times 25$ m^3 resulting in 44,800 cells.

The transmitter generates an EM field at frequencies of 0.125, 0.25 and 0.75 Hz. The receivers measure the in-line component of the electric fields, E_x, and the cross-line component of the magnetic fields, H_y, simultaneously.

We present in Figures 11 and 12 the observed electric data for this model, as an example. We show magnitude versus offset (MVO) and phase versus offset (PVO) plots of the total and normalized fields in these figures. Note that we have contaminated the synthetic observed data with random Gaussian noise. The noise level increases linearly from 1% at zero offset up to 7% at 10,000 m offset to simulate the typical noise behavior in field MCSEM data.

We apply the joint 3D migration to the electric and magnetic data with an L-shaped a priori model that encloses the true reservoir and has a

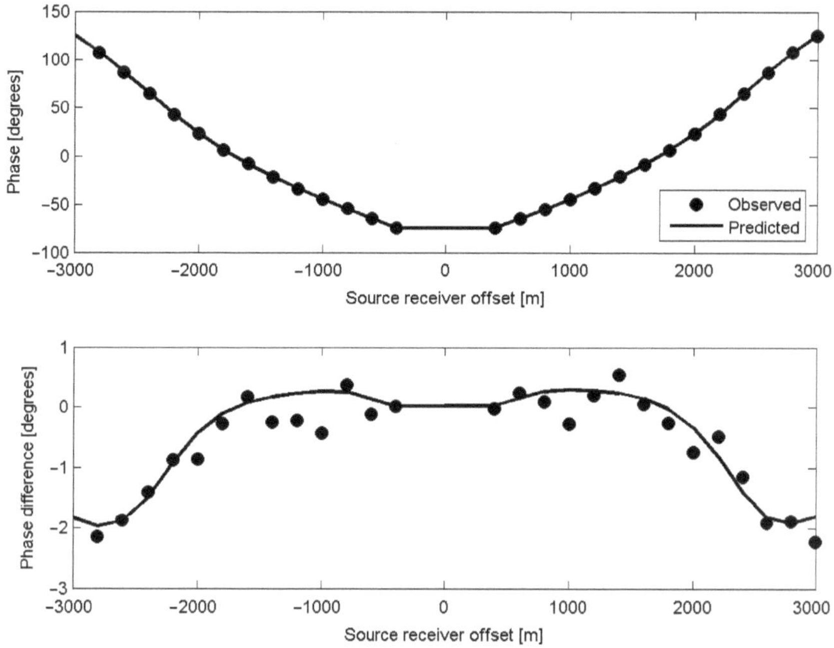

Figure 7 The top panel shows phase versus offset (PVO) plots of the total magnetic field, while the bottom panel presents the PVO plot of the phase difference between the observed and background magnetic fields. The observed data contaminated by the noise are shown by the red dots. The blue solid line corresponds to the data predicted for the migration resistivity model.

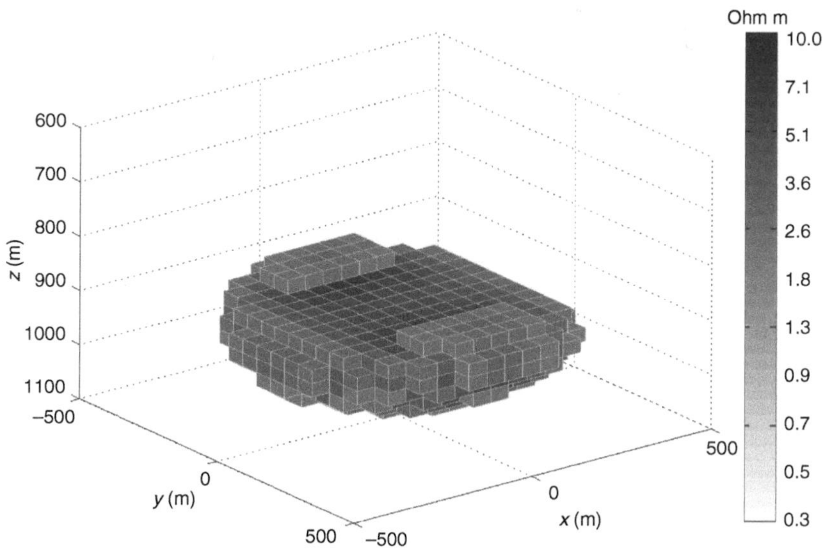

Figure 8 Final 3D holographic image of Model 1 without using an a priori model.

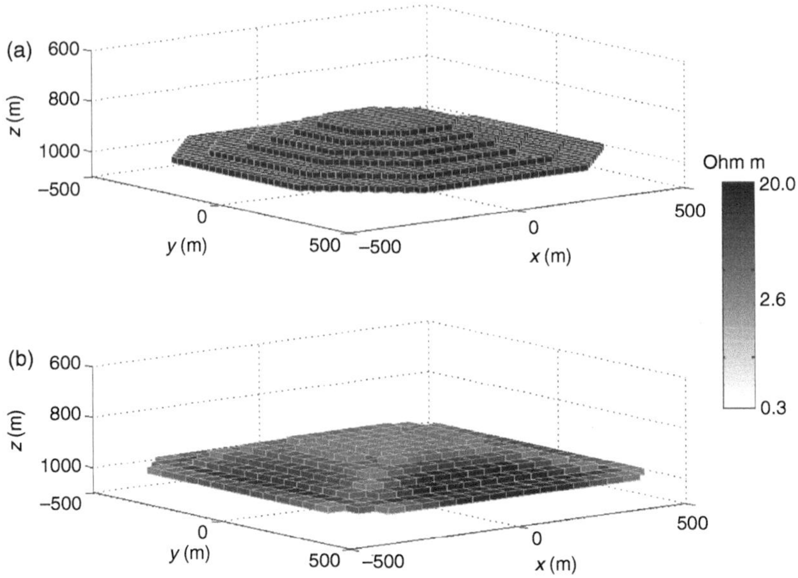

Figure 9 The figure shows (a) the true model and (b) the final 3D holographic image obtained by joint iterative migration of the EM data for Model 1.

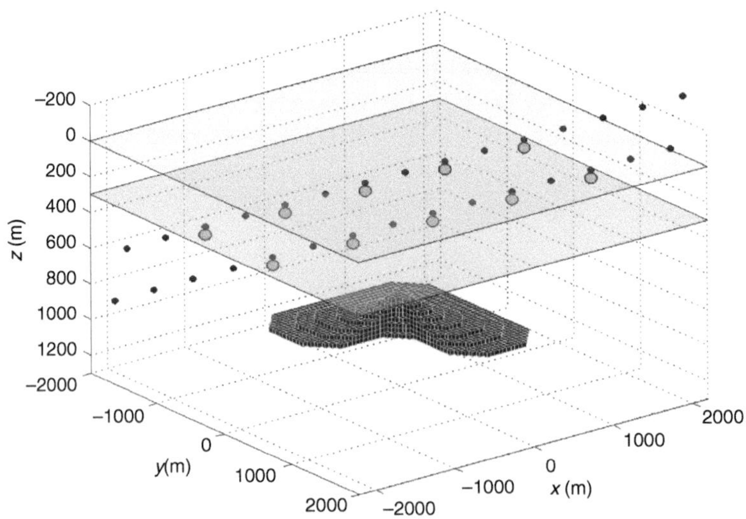

Figure 10 A 3D sketch of Model 2. The blue dots show the transmitter towing profiles and the green dots show the location of a sea-bottom receiver. The host rock has a resistivity of 1 Ohm m, while the resistivity of the L-shaped reservoir is 100 Ohm m.

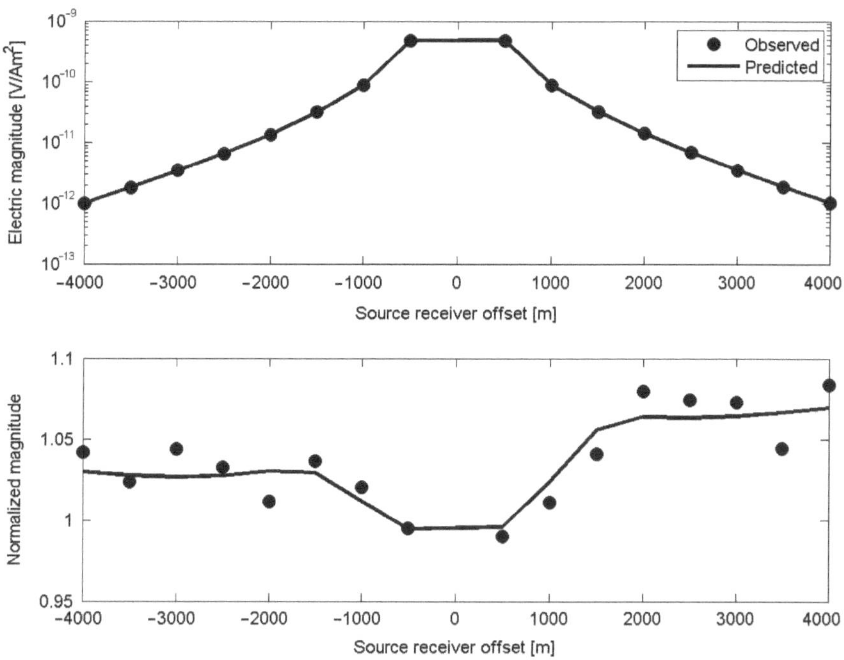

Figure 11 The top panel shows magnitude versus offset (MVO) plots of the total electric field, while the bottom panel presents an MVO plot of the same field normalized by the absolute values of the background magnetic field. The observed data contaminated by noise are shown by the dots. The solid line corresponds to the data predicted for the migration resistivity model.

resistivity of 3 Ohm m. We ran 35 iterations of focusing migration. The a priori model was applied only for the 30 initial iterations. The normalized residual reaches 8% at iteration #35. The corresponding holographic image is shown in Figure 13 as an X-Y plan view at a depth of $z = 962.5$ m and in Figure 14 as a 3D rendering of the true body and migration result. The shape of the recovered reservoir is close to the a priori model but we notice at the corners of the reservoir smaller resistivity values, which are rounded in the true model. Also, the smaller top of the reservoir is not recovered as well as the thicker bottom. We have also plotted in Figures 11 and 12 the predicted data computed for the migration model shown in Figure 14.

7. Inversion of Troll Gas Province MCSEM Data

We have applied the 3D EM migration techniques, including both fast migration imaging and iterative migration, to the interpretation of marine

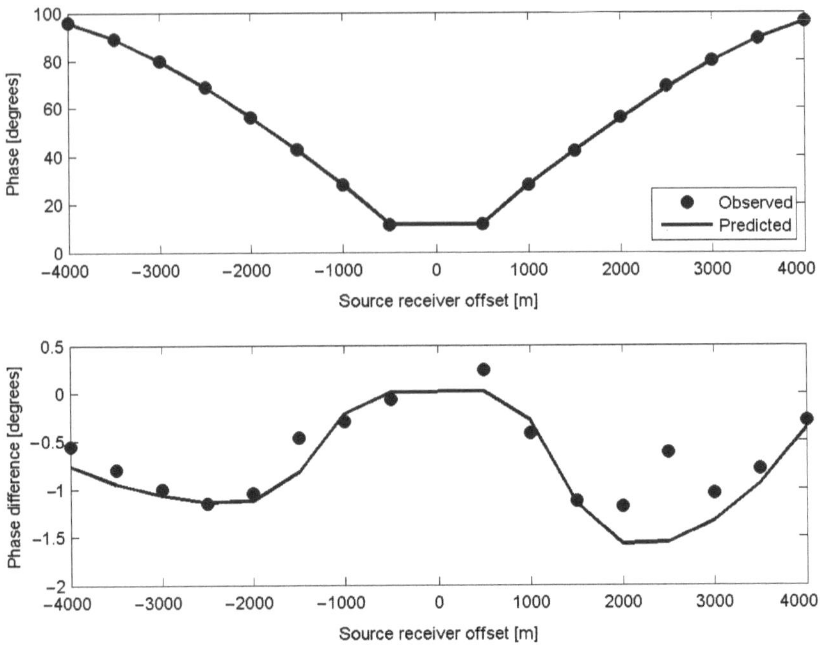

Figure 12 The top panel shows phase versus offset (PVO) plots of the total electric field, while the bottom panel presents a PVO plot of the phase difference between the observed and background magnetic fields. The observed data contaminated by noise are shown by the dots. The solid line corresponds to the data predicted for the migration resistivity model.

EM data collected by EMGS and Statoil at the Troll West Gas Province (TWGP), offshore Norway (Johansen et al., 2005).

The Troll province is located in the northeastern part of the North Sea (Figure 15). This reservoir area is the largest gas discovery on the Norwegian Continental Shelf. The Troll province is separated into three subregions such as Troll East, Troll West Gas Province (TWGP) and Troll West Oil Province (TWOP), as shown in Figure 16. The Seabed Logging (SBL - a trademark of the MCSEM survey) was conducted over the smaller section of TWGP shown in Figure 17. The hydrocarbon reservoir is formed by Jurassic sandstones with a gas layer of up to 160 m. The sandstone saturated with hydrocarbon sands is characterized by a high average resistivity in the range of 200-500 Ohm m. This highly resistive layer is located at a depth of 1400 m below sea level (1100 m below the seafloor). The host sand sediments are filled with saline water and have relatively low resistivity in the range of 0.5-2 Ohm m. Geoelectrical structures such as this highly resistive hydrocarbon target within conductive background sediments makes TWGP very well suited for the MCSEM field test.

A marine CSEM survey was conducted using 24 receivers, deployed at the sea bottom along a line crossing the Western Gas Province. The

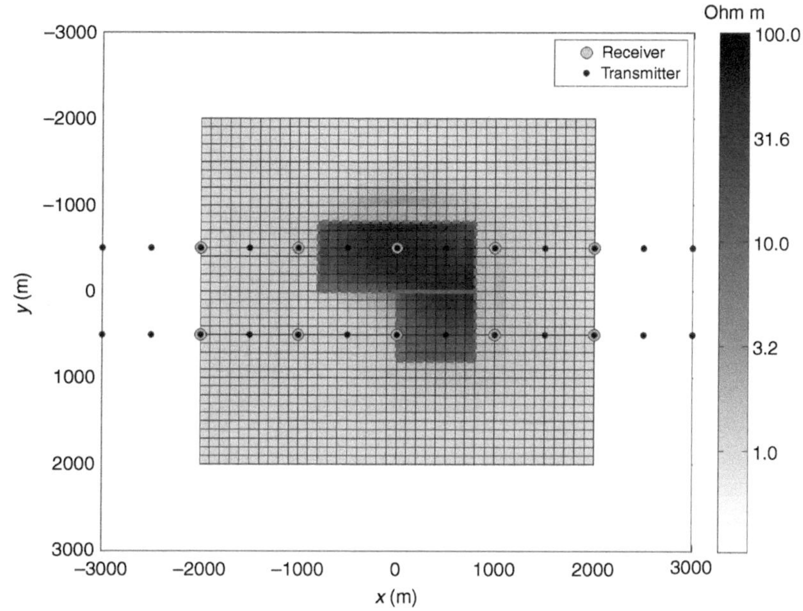

Figure 13 2D *X-Y* plan view images of the final joint iterative migration result for Model 2 obtained by 3D migration of the EM fields.

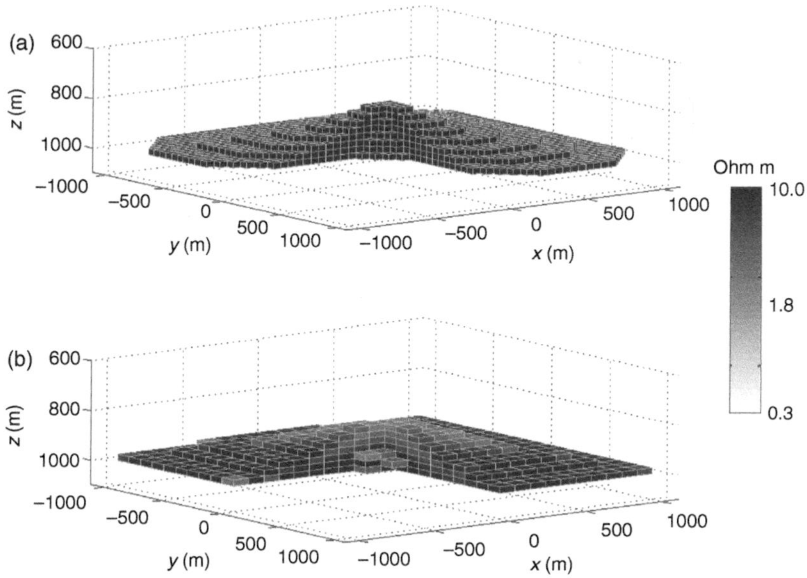

Figure 14 3D images of (a) the true model and (b) the final joint iterative migration result for Model 2 obtained by 3D migration of the EM fields.

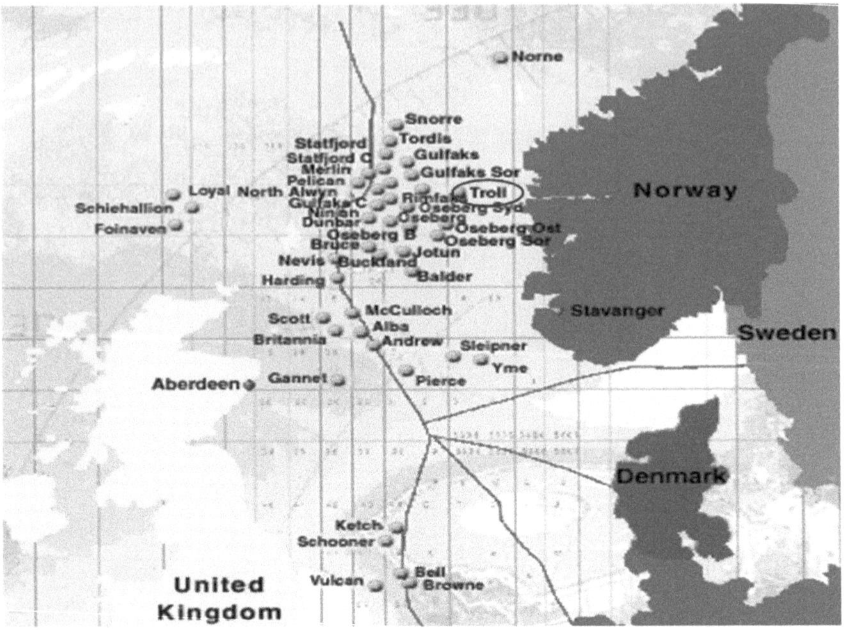

Figure 15 A map of the Troll West Gas Province (TWGP) location.

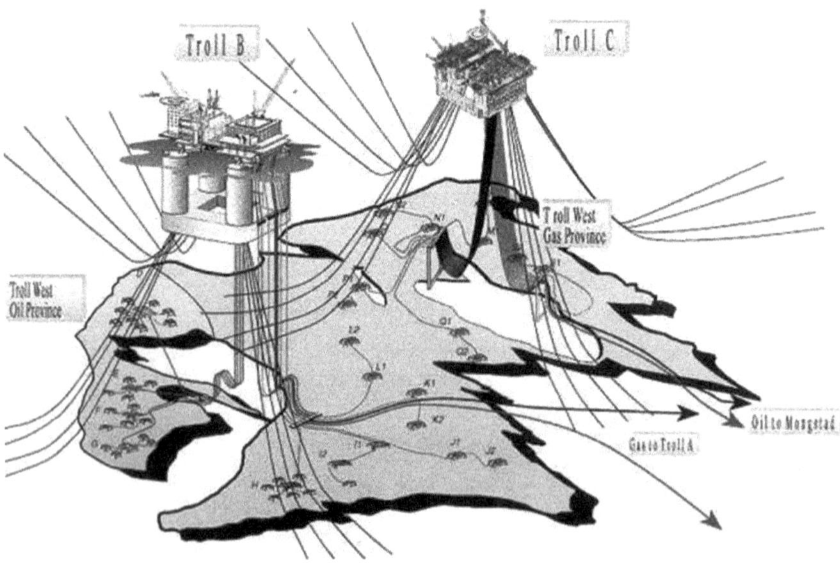

Figure 16 A schematic view of the Troll West Gas Province (TWGP) (after http://www.offshore-technology.com/projects/troll/).

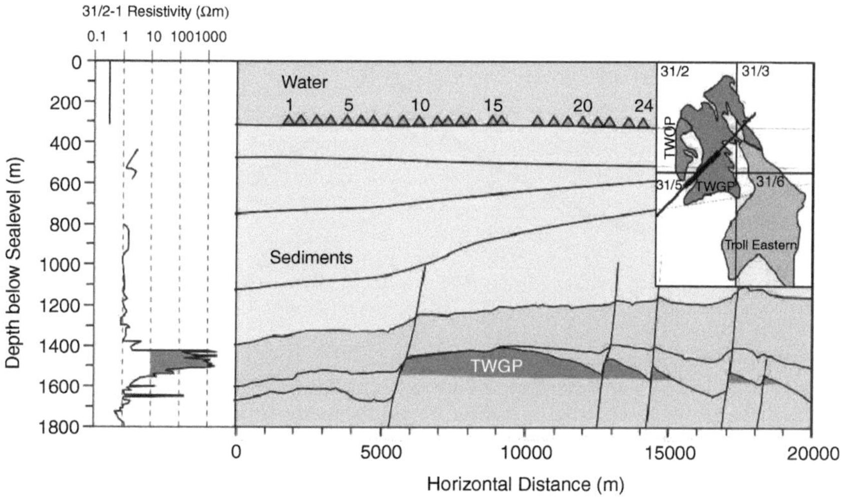

Figure 17 A simplified geological model along the MCSEM survey line in the Troll West Gas Province (TWGP), offshore Norway (after Johansen et al. (2005)).

transmitter was a horizontal electric bipole (HEB) with a length of 230 m towed by the survey vessel. The transmitting bipole generated a sine wave signal with a base frequency 0.25 Hz. The peak-to-peak current varied from zero up to 1000 amperes. The height of the towed transmitter was kept at a level of 40 m above the seafloor by an umbilical cable connected to the survey vessel. The observed EM data were measured in the time domain. These data were numerically transformed into the frequency domain and recorded as amplitudes and phases versus source-receiver offset curves. According to Johansen et al. (2005), the data quality is reliable with the source-receiver offset up to 8 km. Figure 17 shows a simplified geological model along the MCSEM survey line in the Troll West Gas Province (TWGP), offshore Norway (Johansen et al., 2005).

We should note that there are several publications dedicated to the inversion of the Troll MCSEM data (Chen et al., 2004; Hoversten et al., 2006; Gribenko and Zhdanov, 2007, etc.). In the paper by Hoversten et al. (2006), the finite difference-based inversion was used, while Gribenko and Zhdanov (2007), applied a rigorous inversion based on integral equation (IE) forward modeling.

In order to apply the migration algorithm developed in this paper to the Troll MCSEM data, we have selected a 1D layered background structure based on the 1D inversion (with a known water depth equal to 338 m). As a result of 1D inversion, the following parameters of the horizontally layered geoelectrical structure were obtained: the water conductivity is equal to 3 S/m and the conductivity of the sea-bottom sediment is 0.42 S/m.

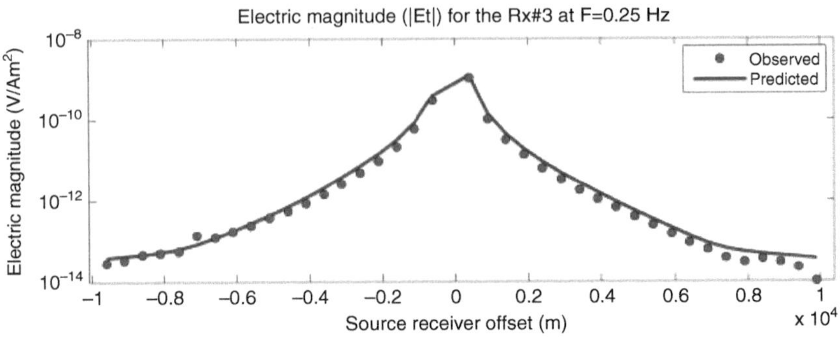

Figure 18 Amplitude versus offset (AVO) plots of the Troll Field data at a frequency of 0.25 Hz. The red dots show the observed in-line electric field data, while the predicted data for a model obtained by iterative migration are plotted by the solid blue line.

We have used this two-layer model as a background geoelectrical structure in our imaging and regularized iterative migration.

We have selected a domain of migration 21 km and 9 km long in the x- and y-directions, respectively, and 1.5 km deep in the vertical (z) direction from 400 m down to 1900 m below sea level, where the x axis of the Cartesian coordinates is oriented along the MCSEM profile, and the z axis is directed downward. This migration domain is discretized in $84 \times 18 \times 60 = 90,720$ cells, with the cell sizes 250, 500, and 25 m in the x, y and z directions, respectively, as shown in Figure 17. With previous knowledge of the seismic profile we designed an a priori model with the top layer in the x-direction being the seismic lines that mark the top of the reservoir in Figure 17 and the bottom at 1575 m with a resistivity of 3 Ohm m. We ran 70 iterations of migration. Note that the a priori model was applied only on the first 10 iterations and was switched off during the subsequent focusing.

The transmitters are located every 500 m along the transmitter line with a maximum offset equal to 10 km. We have used the data with the minimum offset of 500 m. Altogether, we used 24 transmitter locations for every receiver. In order to minimize the computational costs only the lower frequency of 0.25 Hz was selected. At the same time the EM field at this frequency penetrates deep enough in the conductive sediments to provide a clear response for the resistive reservoir. Amplitude versus offset (AVO) plots of the Troll Field data at a frequency 0.25 Hz are shown in Figure 18. The red dots show the observed in-line electric field data, while the predicted data for a model obtained by iterative migration are plotted by the solid blue line.

Figure 19 represents a 3D holographic image of the geoelectrical model obtained by migration imaging. Figure 20 represents overlapping of migration results with the geological interpreted section. One can see good agreement between the migration results and the geological section. The

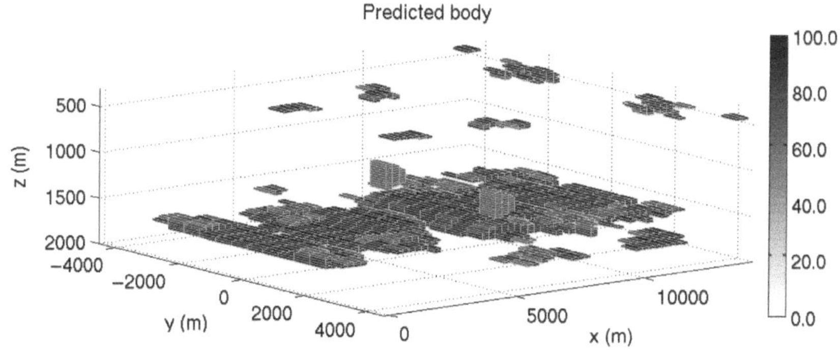

Figure 19 A 3D holographic image of Troll West Gas Province (TWGP), North Sea, obtained by iterative migration. The domains colored in blue show the migration image with the predicted resistivity above 50 Ohm m.

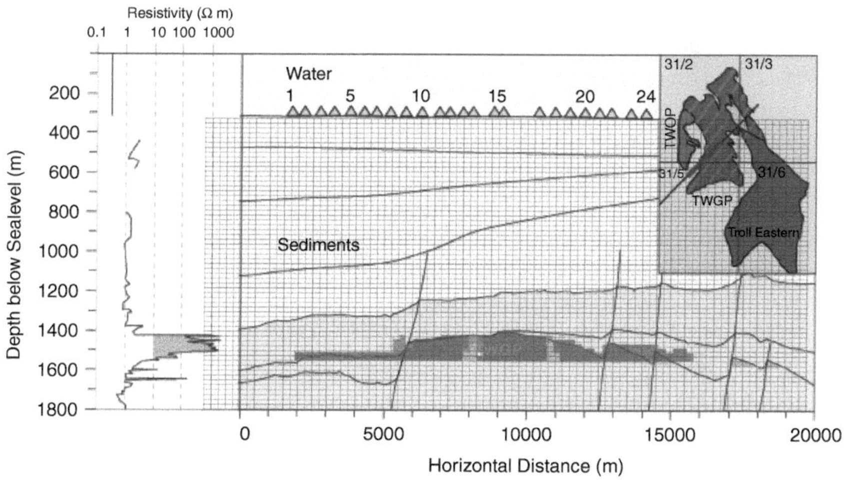

Figure 20 A cross section of the result of iterative migration overlapping on the geological cross section. The blue area overlapped with the geological section shows the migration image with the predicted resistivity above 20 Ohm m.

holographic/migration imaging method, which has been developed in this paper, has the ability to detect a strong resistivity anomaly in the area of the Jurassic sandstone reservoir.

8. Conclusion

Electromagnetic holography/migration was originally introduced for interpretation of land EM data. However, this technique is most effective in the case of relatively dense EM surveys, which are difficult to implement on land. The MCSEM surveys with their dense system of transmitters and receivers happen to be extremely well suited to application of the EM holography/migration technique. In this paper we illustrate all the basic principles of EM holography/migration in application to MCSEM data interpretation. We show that, by using the reciprocity principle, we can represent the system of moving transmitters and fixed sea-bottom receivers by a survey with fixed sea-bottom transmitters and multiple seawater receivers. The migration (backscattering) field is produced by a combination of all electric dipole transmitters operating simultaneously according to the recorded signal in the receivers. The cross-power spectra of the migration and background electric fields generate a volume image of the anomalous conductivity distribution in sea-bottom formations.

In order to improve the resolution and quality of the migration image, we applied an iterative migration by repetitive backscattering of the residual field within the background medium. The backscattered field was computed using an IE based parallel computer code.

The basic principles of the migration imaging formulated in this paper were implemented in the computer code and are tested on a typical model of a sea-bottom petroleum reservoir. The holographic/migration imaging method was applied to interpretation of the practical MCSEM data acquired at Troll West Gas Province by Statoil and EMGS. The interpretation results show that migration can be treated as a prospective method of MCSEM data interpretation.

ACKNOWLEDGMENTS

The authors acknowledge the support of the University of Utah Consortium for Electromagnetic Modeling and Inversion (CEMI), which includes BAE Systems, Baker Atlas Logging Services, BGP China National Petroleum Corporation, BHP Billiton World Exploration Inc., British Petroleum, Centre for Integrated Petroleum Research, EMGS, ENI S.p.A., ExxonMobil Upstream Research Company, INCO Exploration, Information Systems Laboratories, MTEM, Newmont Mining Co., Norsk Hydro, OHM, Petrobras, Rio Tinto - Kennecott, Rocksource, Russian Research Center Kurchatov Institute, Schlumberger, Shell International Exploration and Production Inc., Statoil, Sumitomo Metal Mining Co., and Zonge Engineering and Research Organization.

We are thankful to Dr. Jonny Hesthammer and Rocksource for stimulating this research and fruitful discussions.

The authors thank Dr. Jens Danielsen of EMGS and Dr. Tage Røsten of StatoilHydro, and the Troll license group consisting of StatoilHydro, Petoro,

Norske Shell, Total, and ConocoPhillips, for providing the TWGP SBL data and permission to publish the result.

REFERENCES

Berdichevsky, M.N., Zhdanova, O.N., Zhdanov, M.S., 1989. Marine Deep Geoelectricity. Nauka, Moscow, 90 pp.

Chave, A.D., Constable, S.C., Edwards, R.N., 1991. Electrical exploration methods for the seafloor. In: Nabighian, M.N., Corbett, J.D. (Eds.), Electromagnetic methods in applied geophysics. In: Applications: Society of Exploration Geophysicists, vol. 2. Tulsa, Oklahoma, pp. 931–966.

Chen, J., Hoversten, G.M., Vasco, D.W., Rubin, Y., Hou, Z., 2004. Joint inversion of seismic AVO and EM data for gas saturation estimation using a sampling-based stochastic mode: Presented at the 74th Annual International Meeting, SEG.

Constable, S., Cox, C., 1996. Marine controlled source electromagnetic sounding - ii. The PEGASUS experiment: Journal of Geophysical Research 97, 5519–5530.

Evans, R.L., Sinha, M.C., Constable, S., Unsworth, M.J., 1994. On the electrical nature of the axial melt zone at 13 on the East Pacific Rise. Journal of Geophysical Research 99, 577–588.

Gribenko, A., Zhdanov, M.S., 2007. Rigorous 3D inversion of marine CSEM data based on the integral equation method. Geophysics 72 (2), WA73–WA84.

Johansen, S.E., Amundsen, H.E.F., Rosten, T., Ellingsrud, S., Eidesmo, T., Bhuyian, A.H., 2005. Subsurface hydrocarbons detected by electromagnetic sounding. First Break 23 (3), 31–36.

Hoversten, G.M., Cassassuce, F., Gasperikova, E., Newman, G.A., Chen, J., Rubin, Y., Hou, Z., Vasco, D., 2006. Direct reservoir parameter estimation using joint inversion of marine seismic and CSEM data. Geophysics 71, C1–C13.

MacGregor, L.M., Sinha, M.C., 2000. Use of marine controlled source electromagnetic sounding for sub-basalt exploration. Geophysical Prospecting 48, 1091–1106.

MacGregor, L.M., Sinha, M.C., Constable, S., 2001. Electrical resistivity structure of the Valu Fa Ridge, Lau basin, from marine controlled source electromagnetic sounding. Geophysical Journal International 146, 217–236.

Mittet, R., Maao, F., Aakervik, O.M., Ellingsrud, S., 2005. A two-step approach to depth migration of low frequency electromagnetic data: Presented at the 75th Annual International Meeting, SEG.

Novysh, V.V., Fonarev, G.A., 1966. Some results of the electrimagnetic study in the Arctic ocean. Geomagnetism and Aeronomy 6 (2), 406–409.

Portniaguine, O., Zhdanov, M.S., 1999. Focusing geophysical inversion images. Geophysics 64, 874–887.

Sinha, M.C., Patel, P.D., Unsworth, M.J., Owen, T.R.E., MacCormack, M.G.R., 1990. An active source electromagnetic sounding system for marine use. Marine Geophysical Research 12, 29–68.

Srnka, L.J., Carazzone, J.J., Ephron, M.S., Eriksen, E.A., 2006. Remote reservoir resistivity mapping. The Leading Edge 25, 972–975.

Trophimov, I.L., Fonarev, G.A., 1972. Some results of the magnetotelluric profiling in the Arctic ocean. Izvestia USSR Academy of Sciences 2, 81–92.

Tompkins, M.J., 2004. Marine controlled-source electromagnetic imaging for hydrocarbon exploration: interpreting subsurface electrical properties. First Break 22 (8), 27–33.

Wan, L., Zhdanov, M.S., 2005a. Rapid seabed imaging by frequency domain electromagnetic migration: Proceedings of Annual Meeting of the Consortium for Electromagnetic Modeling and Inversion, 169–186.

Wan, L., Zhdanov, M.S., 2005b. Rapid seabed imaging by frequency domain electromagnetic migration. Presented at the 75th Annual International Meeting, SEG.
Yoshioka, K., Zhdanov, M.S., 2005. Electromagnetic forward modeling based on the integral equation method using parallel computers: 74th Annual International Meeting, SEG, Expanded Abstracts.
Young, P.D., Cox, C.S., 1981. Electromagnetic active source sounding near the East Pacific Rise. Geophysical Research Letters 8, 1043–1046.
Zhdanov, M.S., 1981. Continuation of nonstationary electromagnetic fields in geoelectrical problems. Izvestia Akademii Nauk SSSR, Fizika Zemly 12, 60–69.
Zhdanov, M.S., 1988. Integral Transforms in Geophysics. Springer-Verlag.
Zhdanov, M.S., Frenkel, M.A., 1983a. The solution of the inverse problems on the basis of the analytical continuation of the transient electromagnetic field in reverse time. Journal of Geomagnetism and Geoelectricity 35, 747–765.
Zhdanov, M.S., Frenkel, M.A., 1983b. Electromagnetic migration. In: Hjelt, S.E. (Ed.), The Development of the Deep Geoelectric Model of the Baltic Shield, Part 2. Univ. of Oulu, Oulu, pp. 37–58.
Zhdanov, M.S., Keller, G., 1994. The Geoelectrical Methods in Geophysical Exploration. Elsevier, p. 873.
Zhdanov, M.S., Traynin, P., Booker, J., 1996. Underground imaging by frequency domain electromagnetic migration. Geophysics 61, 666–682.
Zhdanov, M.S., Traynin, P., 1997. Migration versus inversion in electromagnetic imaging technique. Journal of Geomagnetism and Geoelectricity 49, 1415–1437.
Zhdanov, M.S., 1999. Electromagnetic migration. In: Deep Electromagnetic Exploration. Springer-Verlag, Narosa Publishing House, New Delhi, pp. 283–298.
Zhdanov, M.S., 2001. Method of broad band electromagnetic holographic imaging: US Patent # 6,253,100 B1.
Zhdanov, M.S., 2002. Geophysical Inverse Theory and Regularization Problems. Elsevier, p. 609.
Zhdanov, M.S., Ueda, T., Gribenko, A., 2006. Iterative migration in marine CSEM data interpretation. In: Proceedings of Annual Meeting of the Consortium for Electromagnetic Modeling and Inversion, pp. 181–202.

CHAPTER 19

Foundations of the Method of EM Field Separation into Upgoing and Downgoing Parts and its Application to MCSEM Data

Michael S. Zhdanov[a] *and* Shuming Wang[a]

Contents

1. Introduction

The problem of electromagnetic field separation into upgoing and downgoing parts is one of the oldest problems of geophysics. It was addressed

[a] Department of Geology and Geophysics, University of Utah, USA

E-mail address: mzhdanov@mines.utah.edu (M.S. Zhdanov).

Handbook of Geophysical Exploration: Seismic Exploration, Volume 40
ISSN 0950-1401, DOI: 10.1016/S0950-1401(10)04024-3

first by the great German mathematician Carl Friedrich Gauss in the middle of the 19th century in his study of the nature of the geomagnetic field and its separation into external and internal parts (Chapman and Bartels, 1940). The Gaussian method was originally applied to models with a spherical or plane earth surface. This restriction was removed by Kertz (1954) and Siebert (1962), by using the theory of analytical functions of complex variables. Both Gaussian and Kertz-Siebert methods were developed for potential magnetic field separation in nonconductive media only. In the pioneering paper by Zhdanov (1980) a generalized method of electromagnetic (EM) field separation into upgoing and downgoing parts was introduced. Later on this technique was extended to field separation into parts related to sources located in different regions of the space, including external and internal, normal and anomalous parts of a transient EM field. It was also extended to the separation of EM anomalies into the surface and deep parts and to the determination of the main part of a deep anomaly (Berdichevsky and Zhdanov, 1984; Zhdanov, 1988; Zhdanov et al., 1996). This technique was also extended to the separation of EM fields measured at the sea bottom (Zhdanova and Zhdanov, 1999). All of these publications provide a solid foundation for the electromagnetic field separation and its application in EM geophysics.

Recently, we have observed a renewed interest in the problem of upgoing/downgoing EM field separation, generated by increased research and development of marine controlled source electromagnetic (MCSEM) methods. A problem that arises in the MCSEM method is that EM energy may travel from the source to the receiver along many paths. For example, seafloor receivers of EM data measure not only a response from the sea-bottom geoelectrical formations (the "upgoing field") but they also measure a direct part of the primary field from the source, the field travelling from the source to the sea surface and reflected back to the sea bottom (a so-called airwave), and the natural magnetotelluric field. The latter parts of the total EM field form the "downgoing field" because the sources of these fields are located above the receivers (above the seafloor). The methods of EM field decomposition into upgoing and downgoing components in application to MCSEM data are discussed in the recent paper by Amundsen et al. (2006). They presented a constructive method of the decomposition of the EM field given on a horizontal plane, and provided practical examples of its application for 1D and 2D models.

The main goal of our paper is to review the basic principles of EM field separation. The most general approach to the solution of this problem is based on the formalism of the Stratton-Chu type integrals, introduced by Zhdanov (1980, 1988). We begin our paper with a brief discussion of the theory of the Stratton-Chu type integrals and its application to the field separation. This theory is used for decomposition of the EM field measured on an arbitrary surface, e.g., on a seafloor with a variable bathymetry.

In a case where the data are measured on a horizontal plane, the most effective technique for field separation can be developed using a spatial Fourier transform in the (k, ω) domain (Berdichevsky and Zhdanov, 1984; Zhdanova and Zhdanov, 1999). Finally, we introduce a new method of field separation based on using horizontal gradients of the observed EM fields. This method allows us to develop a fast and accurate separation of 3D EM fields into upgoing and downgoing fields, which can be effectively used for interpretation of MCSEM field data.

2. Integral Transforms of Electromagnetic Fields Using Stratton-Chu Type Integrals

A general integral method of EM field decomposition into upgoing and downgoing components has been developed by Zhdanov (1980, 1988). We will discuss below the basic ideas of this method as applied to MCSEM data. Our approach is based on using Stratton-Chu Type integrals. A summary of the basic definitions and properties of these integrals is provided in Appendices A and B, for convenience. We will demonstrate below that this theory provides the foundations for the solution of the problem of EM field separation into upgoing and downgoing parts.

2.1. Basic equations of upgoing and downgoing fields

Let us consider a typical MCSEM survey conducted in an area with rough bathymetry. The conductivity of seawater is known and it is equal to σ_w. Air is characterized by a complex conductivity σ_0. In fact, for the typically low frequencies used in the MCSEM method, we can neglect the displacement currents in the air and assume that $\sigma_0 = 0$, however, for completeness, we will keep this term in the equations. Sea-bottom formations are characterized by an arbitrary 3D distribution of the conductivity, $\sigma(\mathbf{r})$, where $\mathbf{r}$ is a radius vector of the given point in some Cartesian system of coordinates (x, y, z). The frequency domain EM field is generated by an electric bipole transmitter Tx located at some depth within the sea-water layer. This field is measured by a system of receivers located at surface S at a small elevation (usually a few meters) above the sea bottom. We assume that S is a smooth surface extending to infinity in the horizontal directions. The magnetic permeability is everywhere equal to the free space constant μ_0.

The electromagnetic field in this model satisfies the following equations:

$$\nabla \times \mathbf{H} = \begin{cases} \sigma_0 \mathbf{E}, & \mathbf{r} \in A, \\ \sigma_w \mathbf{E} + \mathbf{j}^e, & \mathbf{r} \in W, \\ \sigma(\mathbf{r})\mathbf{E}, & \mathbf{r} \in B, \end{cases} \qquad \nabla \times \mathbf{E} = i\omega\mu_0 \mathbf{H}, \tag{1}$$

where A denotes the upper half-space occupied by the atmosphere, W represents the sea-water layer, B denotes the lower half-space formed by the sea-bottom geoelectrical formations with the conductivity $\sigma(\mathbf{r})$, and $\mathbf{j}^e$ is an extraneous current in the transmitter. Note that all components of the electromagnetic field are continuous at surface S, because it is located within the homogeneous seawater layer.

We can represent the electromagnetic field $\mathbf{E}$, $\mathbf{H}$ as a sum of the upgoing $\mathbf{E}^u$, $\mathbf{H}^u$ (internal) and downgoing $\mathbf{E}^d$, H^d (external) components:

$$\mathbf{E} = \mathbf{E}^u + \mathbf{E}^d, \quad \mathbf{H} = \mathbf{H}^u + \mathbf{H}^d. \tag{2}$$

The upgoing field $\mathbf{E}^u$, $\mathbf{H}^u$ is due to the excess electric currents, $\mathbf{j}^B$, induced within the conductive sea bottom, in the model where the conductivity of the atmosphere, the upper half-space A, is equal to the electric conductivity of the seawater. It satisfies the equations:

$$\nabla \times \mathbf{H}^u = \begin{cases} \sigma_w \mathbf{E}^u, & \mathbf{r} \in A, \\ \sigma_w \mathbf{E}^u + \mathbf{j}^e, & \mathbf{r} \in W, \\ \sigma_w \mathbf{E}^u + \mathbf{j}^B, & \mathbf{r} \in B, \end{cases} \qquad \nabla \times \mathbf{E}^u = i\omega\mu_0 \mathbf{H}^u, \tag{3}$$

where

$$\mathbf{j}^B = [\sigma(\mathbf{r}) - \sigma_w]\mathbf{E}. \tag{4}$$

The downgoing field $\mathbf{E}^d$, $\mathbf{H}^d$ is produced by the current in the transmitter and the "excess" electric currents, $\mathbf{j}^A$, distributed within the atmosphere, in the model where the conductivity both in the sea-bottom formations (the lower half-space B), and in the atmosphere (the upper half-space A), is equal to the electric conductivity of the seawater. It satisfies the equations:

$$\nabla \times \mathbf{H}^d = \begin{cases} \sigma_w \mathbf{E}^d + \mathbf{j}^A, & \mathbf{r} \in A, \\ \sigma_w \mathbf{E}^d + \mathbf{j}^e, & \mathbf{r} \in W, \\ \sigma_w \mathbf{E}^d, & \mathbf{r} \in B, \end{cases} \qquad \nabla \times \mathbf{E}^d = i\omega\mu_0 \mathbf{H}^d, \tag{5}$$

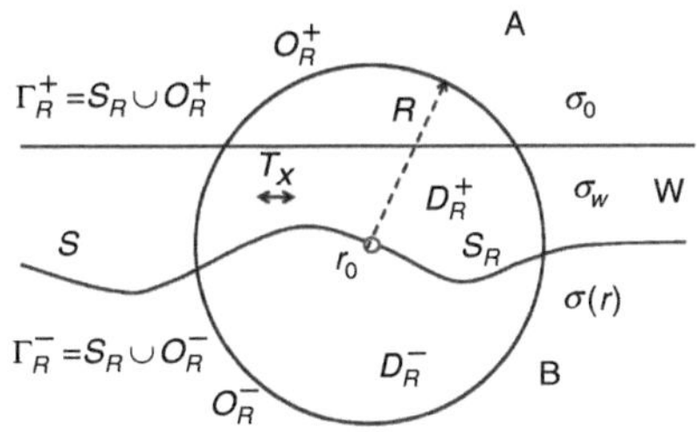

Figure 1 A scheme illustrating the application of the Stratton-Chu type integrals to the separation of the upgoing and downgoing fields.

where

$$\mathbf{j}^A = [\sigma_0 - \sigma_w]\mathbf{E}. \tag{6}$$

It is easy to verify that, by adding the left- and right-hand sides of Eqs (5) and (3), we obtain the original system of Maxwell's equations for the total field (1).

2.2. Application of the Stratton-Chu type integrals for field separation

The problem of field decomposition can be formulated as follows: find the upgoing and downgoing components at the surface of observation S from the total EM field $\mathbf{E}$, $\mathbf{H}$, observed also at S.

We will apply the theory of the Stratton-Chu type integrals to solve this problem. Let us take a point at the surface of observation, $\mathbf{r}_0 \in S$, and draw a sphere O_R of radius R with the center at $\mathbf{r}_0$ (see Figure 1). We select the radius R so large that it completely encloses the electric bipole transmitter T_x. We denote by O_R^- the part of the sphere lying below the observational surface S, and by O_R^+ the part of the sphere lying above the observational surface S within the water layer W and within the upper half-space A of the atmosphere. S_R is the part of the surface of observation found inside the sphere. We denote by Γ_R^+ a piece-wise smooth close surface formed by the semisphere O_R^+ and S_R, $\Gamma_R^+ = O_R^+ \cup S_R$ (see Figure 1). This surface Γ_R^+ is a closed boundary of the upper part, D_R^+, of the ball bounded by a sphere O_R, located above surface S. In a similar way, we denote by Γ_R^- a piece-wise smooth closed surface formed by the lower semisphere O_R^- and S_R: $\Gamma_R^- = O_R^- \cup S_R$. This surface Γ_R^- is a closed boundary of the lower part, D_R^-, of the ball located below surface S.

We can introduce now the Stratton-Chu integrals over the piece-wise smooth closed surface Γ_R^+:

$$\mathbf{C}_{\Gamma_R^+}^{E}(\mathbf{r}') = \iint_{\Gamma_R^+} [(\mathbf{n} \cdot \mathbf{E})\nabla G_w + (\mathbf{n} \times \mathbf{E}) \times \nabla G_w + \mathrm{i}\omega\mu(\mathbf{n} \times \mathbf{H})G_w]\mathrm{d}s, \tag{7}$$

$$\mathbf{C}_{\Gamma_R^+}^{H}(\mathbf{r}') = \iint_{\Gamma_R^+} [(\mathbf{n} \cdot \mathbf{H})\nabla G_w + (\mathbf{n} \times \mathbf{H}) \times \nabla G_w + \sigma_w(\mathbf{n} \times \mathbf{E})G_w]\mathrm{d}s, \tag{8}$$

where $\mathbf{r}' \in D_R^+$, $\mathbf{n}$ is the unit vector of an inward-pointing normal to D_R^+, and G_w is the fundamental Green's function for the Helmholtz equation in a homogeneous full space with the seawater conductivity σ_w:

$$G_w(\mathbf{r}'|\mathbf{r}) = -\frac{1}{4\pi\,|\mathbf{r}-\mathbf{r}'|}\exp(\mathrm{i}k_w|\mathbf{r}-\mathbf{r}'|), \quad k_w = \sqrt{\mathrm{i}\omega\mu_0\sigma_w}, \mathrm{Re}\, k_w > 0. \tag{9}$$

The Stratton-Chu type integrals are linear operators with respect to the electric and magnetic fields, $\mathbf{E}$ and $\mathbf{H}$. Therefore, according to equations (2), integrals (7) and (8) can be represented as the sums of similar Stratton-Chu type integrals, calculated for the values of the upgoing and downgoing fields, respectively:

$$\mathbf{C}_{\Gamma_R^+}^{E}(\mathbf{r}') = \mathbf{C}_{\Gamma_R^+}^{E^u}(\mathbf{r}') + \mathbf{C}_{\Gamma_R^+}^{E^d}(\mathbf{r}'), \quad \mathbf{C}_{\Gamma_R^+}^{H}(\mathbf{r}') = \mathbf{C}_{\Gamma_R^+}^{H^u}(\mathbf{r}') + \mathbf{C}_{\Gamma_R^+}^{H^d}(\mathbf{r}'). \tag{10}$$

Note that the upgoing EM field, $\mathbf{E}^u$, $\mathbf{H}^u$, inside domain D_R^+ satisfies the equations:

$$\nabla \times \mathbf{H}^u = \sigma_w\mathbf{E}^u, \quad \nabla \times \mathbf{E}^u = \mathrm{i}\omega\mu_0\mathbf{H}^u, \tag{11}$$

and vanishes at infinity. Hence, according to the property (2) of the Stratton-Chu type integrals, we have the following:

$$\mathbf{C}_{\Gamma_R^+}^{E^u}(\mathbf{r}') = \begin{cases} \mathbf{E}^u(\mathbf{r}'), & \mathbf{r}' \in D_R^+ \\ 0, & \mathbf{r}' \in C\overline{D}_R^+, \end{cases} \quad \mathbf{C}_{\Gamma_R^+}^{H^u}(\mathbf{r}') = \begin{cases} \mathbf{H}^u(\mathbf{r}'), & \mathbf{r}' \in D_R^+ \\ 0, & \mathbf{r}' \in C\overline{D}_R^+, \end{cases} \tag{12}$$

where D_R^+ and $C\overline{D}_R^+$ are the domains which are internal and external with respect to the closed surface Γ_R^+.

In a similar way, we introduce the Stratton-Chu integrals over a piecewise smooth closed surface Γ_R^-, which can also be decomposed into Stratton-Chu type integrals, calculated for the values of the upgoing and downgoing fields, respectively:

$$\mathbf{C}^E_{\Gamma_R^-}(\mathbf{r}') = \mathbf{C}^{E^u}_{\Gamma_R^-}(\mathbf{r}') + \mathbf{C}^{E^d}_{\Gamma_R^-}(\mathbf{r}'), \quad \mathbf{C}^H_{\Gamma_R^-}(\mathbf{r}') = \mathbf{C}^{H^u}_{\Gamma_R^-}(\mathbf{r}') + \mathbf{C}^{H^d}_{\Gamma_R^-}(\mathbf{r}'). \quad (13)$$

We can recall, according to Eqs (5), that the downgoing EM field, $\mathbf{E}^d$, $\mathbf{H}^d$, inside domain $D_R^- \subset B$ satisfies the equations:

$$\nabla \times \mathbf{H}^d = \sigma_w \mathbf{E}^d, \quad \nabla \times \mathbf{E}^d = i\omega\mu_0 \mathbf{H}^d, \quad (14)$$

and vanishes at infinity. Hence, according to the property (2) of the Stratton-Chu type integrals, we have:

$$\mathbf{C}^{E^d}_{\Gamma_R^-}(\mathbf{r}') = \begin{cases} -\mathbf{E}^d(\mathbf{r}'), & \mathbf{r}' \in D_R^- \\ 0, & \mathbf{r}' \in C\overline{D}_R^-, \end{cases} \quad \mathbf{C}^{H^d}_{\Gamma_R^-}(\mathbf{r}') = \begin{cases} -\mathbf{H}^d(\mathbf{r}'), & \mathbf{r}' \in D_R^- \\ 0, & \mathbf{r}' \in C\overline{D}_R^-, \end{cases} \quad (15)$$

where D_R^- and $C\overline{D}_R^-$ are the domains which are internal and external with respect to the closed surface Γ_R^-, and the minus sign on the right side of Eqs (15) is due to the fact that the normal vector n is directed outward to domain D_R^- (see Figure 1).

Summing up Eqs (12) and (15), we have:

$$\mathbf{C}^{E^u}_{\Gamma_R^+}(\mathbf{r}') + \mathbf{C}^{E^d}_{\Gamma_R^-}(\mathbf{r}') = \begin{cases} \mathbf{E}^u(\mathbf{r}'), & \mathbf{r}' \in D_R^+ \\ -\mathbf{E}^d(\mathbf{r}'), & \mathbf{r}' \in D_R^-, \end{cases}$$

$$\mathbf{C}^{H^u}_{\Gamma_R^+}(\mathbf{r}') + \mathbf{C}^{H^d}_{\Gamma_R^-}(\mathbf{r}') = \begin{cases} \mathbf{H}^u(\mathbf{r}'), & \mathbf{r}' \in D_R^+ \\ -\mathbf{H}^d(\mathbf{r}'), & \mathbf{r}' \in D_R^-. \end{cases} \quad (16)$$

One can readily obtain the limit values of the last equations at the surface S_R thanks to the continuity of the electric and magnetic fields at this surface:

$$\lim_{r' \to r_0^+} [\mathbf{C}^{E^u}_{\Gamma_R^-}(\mathbf{r}') + \mathbf{C}^{E^d}_{\Gamma_R^+}(\mathbf{r}')] = \mathbf{E}^u(\mathbf{r}_0), \quad \mathbf{r}_0 \in S_R, \quad \mathbf{r}_0^+ \in \overline{D}_R^+,$$

$$\lim_{r' \to r_0^+} [\mathbf{C}^{H^u}_{\Gamma_R^-}(\mathbf{r}') + \mathbf{C}^{H^d}_{\Gamma_R^+}(\mathbf{r}')] = \mathbf{H}^u(\mathbf{r}_0), \quad \mathbf{r}_0 \in S_R, \quad \mathbf{r}_0^+ \in \overline{D}_R^+, \quad (17)$$

and

$$\lim_{r'\to r_0^-}[\mathbf{C}^{E^u}_{\Gamma_R^-}(\mathbf{r}')+\mathbf{C}^{E^d}_{\Gamma_R^+}(\mathbf{r}')]=-\mathbf{E}^d(\mathbf{r}_0),\quad \mathbf{r}_0\in S_R,\quad \mathbf{r}_0^-\in\overline{D}_R^-,$$

$$\lim_{r'\to r_0^-}[\mathbf{C}^{H^u}_{\Gamma_R^-}(\mathbf{r}')+\mathbf{C}^{H^d}_{\Gamma_R^+}(\mathbf{r}')]=-\mathbf{H}^d(\mathbf{r}_0),\quad \mathbf{r}_0\in S_R,\quad \mathbf{r}_0^-\in\overline{D}_R^-. \quad (18)$$

The limit values of the Stratton-Chu type integrals can also be defined by Eqs (88) and (89), for example:

$$\lim_{r'\to r_0^+}[\mathbf{C}^{E^u}_{\Gamma_R^-}(\mathbf{r}')+\mathbf{C}^{E^d}_{\Gamma_R^+}(\mathbf{r}')]=\mathbf{C}^{E^u}_{O_R^-}(\mathbf{r}_0)+\mathbf{C}^{E^u}_{S_R}(\mathbf{r}_0)+\frac{1}{2}\mathbf{E}^u(\mathbf{r}_0)$$
$$+\mathbf{C}^{E^d}_{O_R^+}(\mathbf{r}_0)+\mathbf{C}^{E^d}_{S_R}(\mathbf{r}_0)+\frac{1}{2}\mathbf{E}^d(\mathbf{r}_0),$$

$$\lim_{r'\to r_0^-}[\mathbf{C}^{E^u}_{\Gamma_R^-}(\mathbf{r}')+\mathbf{C}^{E^d}_{\Gamma_R^+}(\mathbf{r}')]=\mathbf{C}^{E^u}_{O_R^-}(\mathbf{r}_0)+\mathbf{C}^{E^u}_{S_R}(\mathbf{r}_0)-\frac{1}{2}\mathbf{E}^u(\mathbf{r}_0)$$
$$+\mathbf{C}^{E^d}_{O_R^+}(\mathbf{r}_0)+\mathbf{C}^{E^d}_{S_R}(\mathbf{r}_0)-\frac{1}{2}\mathbf{E}^d(\mathbf{r}_0), \quad (19)$$

where the singular integrals $\mathbf{C}^{E^u}_{S_R}(\mathbf{r}_0)$ and $\mathbf{C}^{E^d}_{S_R}(\mathbf{r}_0)$ are determined in terms of the Cauchy principal value. Similar equations can be written for magnetic fields as well.

Proceeding to the limit for $R\to\infty$, and equating the Eqs (17) and (19), we obtain:

$$\mathbf{E}^u(\mathbf{r}_0)=\frac{1}{2}[\mathbf{E}^u(\mathbf{r}_0)+\mathbf{E}^d(\mathbf{r}_0)]+\mathbf{C}^{E^u}_S(\mathbf{r}_0)+\mathbf{C}^{E^d}_S(\mathbf{r}_0), \quad (20)$$

and

$$\mathbf{E}^d(\mathbf{r}_0)=\frac{1}{2}[\mathbf{E}^u(\mathbf{r}_0)+\mathbf{E}^d(\mathbf{r}_0)]-\mathbf{C}^{E^u}_S(\mathbf{r}_0)-\mathbf{C}^{E^d}_S(\mathbf{r}_0), \quad (21)$$

where we take into account that, according to the radiation conditions, integrals $\mathbf{C}^{E^u}_{O_R^-}(\mathbf{r}_0)$ and $\mathbf{C}^{E^d}_{O_R^+}(\mathbf{r}_0)$ vanish when $R\to\infty$.

Finally, according to formulas (2), we have

$$\mathbf{E}^u(\mathbf{r}_0)=\frac{1}{2}\mathbf{E}(\mathbf{r}_0)+\mathbf{C}^E_S(\mathbf{r}_0), \quad (22)$$

and

$$\mathbf{E}^{d}(\mathbf{r}_0) = \frac{1}{2}\mathbf{E}(\mathbf{r}_0) - \mathbf{C}_S^E(\mathbf{r}_0). \tag{23}$$

Similar expressions can be derived for magnetic field components as well:

$$\mathbf{H}^{u}(\mathbf{r}_0) = \frac{1}{2}\mathbf{H}(\mathbf{r}_0) + \mathbf{C}_S^H(\mathbf{r}_0), \tag{24}$$

and

$$\mathbf{H}^{d}(\mathbf{r}_0) = \frac{1}{2}\mathbf{H}(\mathbf{r}_0) - \mathbf{C}_S^H(\mathbf{r}_0). \tag{25}$$

In formulas (22) through (25), expressions $\mathbf{C}_S^E(\mathbf{r}_0)$ and $\mathbf{C}_S^H(\mathbf{r}_0)$ stand for the Stratton-Chu type integrals over the entire observation surface S, determined at a singular point $\mathbf{r}_0 \in S$ in terms of the Cauchy principal value:

$$\mathbf{C}_S^E(\mathbf{r}_0) = \iint_S [(\mathbf{n}\cdot\mathbf{E})\nabla G(\mathbf{r}_0|\mathbf{r}) + (\mathbf{n}\times\mathbf{E})\times\nabla G(\mathbf{r}_0|\mathbf{r}) + i\omega\mu_0(\mathbf{n}\times\mathbf{H})G(\mathbf{r}_0|\mathbf{r})]ds, \tag{26}$$

$$\mathbf{C}_S^H(\mathbf{r}_0) = \iint_S [(\mathbf{n}\cdot\mathbf{H})\nabla G(\mathbf{r}_0|\mathbf{r}) + (\mathbf{n}\times\mathbf{H})\times\nabla G(\mathbf{r}_0|\mathbf{r}) + \sigma_w(\mathbf{n}\times\mathbf{E})G(\mathbf{r}_0|\mathbf{r})]ds, \tag{27}$$

and $G_w(\mathbf{r}_0|\mathbf{r})$ is the fundamental Green's function for the Helmholtz equation in a homogeneous full space with the seawater conductivity σ_w, defined by Eq. (9). Equations (22) through (25) describe a general integral transformation of the total EM field, observed on the arbitrary surface S within the water layer, into upgoing and downgoing parts. These equations serve as a theoretical foundation of the method of up/down decomposition of the EM field.

3. Spatial Fourier Transform Method of EM Field Separation into Upgoing and Downgoing Parts

Consider again a 3D geoelectrical model of a marine CSEM survey. We assume now that the background conductivity in this model is formed by a horizontally layered model consisting of nonconductive air, a conductive seawater layer with the conductivity σ_w, and the homogeneous part of a sea-bottom formation. The sea-bottom inhomogeneities are located within domain V with the conductivity $\sigma = \sigma_b + \Delta\sigma$, varying spatially. Note that the frequencies of the EM field used in the marine EM are very low (usually less than 10 Hz). Therefore we can ignore the displacement currents. In the marine CSEM method, measurements are conducted at the sea bottom, while the electric dipole transmitter is located above the sea bottom. We introduce a Cartesian coordinate system with the origin at sea level and axis z directed downward. We assume also that the seafloor is flat, and the receivers are located at the horizontal plane $z = z_0$. It was demonstrated by Berdichevsky and Zhdanov (1984) that in this situation the field decomposition can be easily conducted in the spatial frequency (k, ω) domain. A similar approach was considered recently in Amundsen et al. (2006) as well. We will show below that these spectral based decomposition formulas are equivalent to the general Stratton-Chu decomposition transforms in the case of a flat surface of observation.

3.1. Electromagnetic field in the (k, ω) domain

We begin with Maxwell's equations (78), written for low frequency (less than 10 Hz) EM fields within a homogeneous seawater layer with the conductivity σ_w:

$$\begin{aligned} &\nabla \times \mathbf{H} = \sigma_w \mathbf{E}, \quad (a) \\ &\nabla \times \mathbf{E} = \mathrm{i}\omega\mu\mathbf{H}, \quad (b) \\ &\nabla \cdot \mathbf{H} = 0, \quad (c) \\ &\nabla \cdot \mathbf{E} = 0. \quad (d) \end{aligned} \tag{28}$$

Separating Maxwell's equations (28), we arrive at the following Helmholtz equations:

$$\begin{aligned} &\nabla^2\mathbf{E} + k_w^2\mathbf{E} = 0, \\ &\nabla^2\mathbf{H} + k_w^2\mathbf{H} = 0, \end{aligned} \tag{29}$$

where $k_w = \sqrt{\mathrm{i}\omega\mu_0\sigma_w}$.

Let us define spatial Fourier spectrums of the electric and magnetic fields using the following expressions:

$$\mathbf{e}(k_x, k_y, z) = \int_{-\infty}^{\infty}\int \mathbf{E}(x, y, z)\exp[\mathrm{i}(k_x x + k_y y)]\mathrm{d}x\mathrm{d}y, \tag{30}$$

$$\mathbf{h}(k_x, k_y, z) = \int_{-\infty}^{\infty}\int \mathbf{H}(x, y, z)\exp[\mathrm{i}(k_x x + k_y y)]\mathrm{d}x\mathrm{d}y, \tag{31}$$

where k_x and k_y are spatial frequencies.

The corresponding inverse Fourier transforms are as follows:

$$\mathbf{E}(x, y, z) = \frac{1}{4\pi^2} \int_{-\infty}^{\infty} \int \mathbf{e}(k_x, k_y, z) \exp[-\mathrm{i}(k_x x + k_y y)] \mathrm{d}k_x \mathrm{d}k_y, \quad (32)$$

$$\mathbf{H}(x, y, z) = \frac{1}{4\pi^2} \int_{-\infty}^{\infty} \int \mathbf{h}(k_x, k_y, z) \exp[-\mathrm{i}(k_x x + k_y y)] \mathrm{d}k_x \mathrm{d}k_y. \quad (33)$$

Applying a Fourier transform to equations (29), we obtain:

$$\frac{\partial^2}{\partial z^2}\mathbf{e} = \nu^2 \mathbf{e}, \quad \text{and} \quad \frac{\partial^2}{\partial z^2}\mathbf{h} = \nu^2 \mathbf{h}, \quad z_b \geq z \geq 0, \quad (34)$$

where $\nu = (k_x^2 + k_y^2 - \mathrm{i}\omega\mu\sigma_w)^{1/2}$, $\mathrm{Re}\nu \geq 0$, is the wavenumber in the $(\mathbf{k}, \omega)$ domain, and z_b is the depth of the sea bottom.

The general solutions of the last equations have the following form:

$$\mathbf{e}(k_x, k_y, z) = \mathbf{e}^u \exp[\nu(z - z_0)] + \mathbf{e}^d \exp[-\nu(z - z_0)], \quad z_b \geq z \geq 0, \quad (35)$$

$$\mathbf{h}(k_x, k_y, z) = \mathbf{h}^u \exp[\nu(z - z_0)] + \mathbf{h}^d \exp[-\nu(z - z_0)], \quad z_b \geq z \geq 0, \quad (36)$$

where z_0 is the depth of the location of the receivers, and $\mathbf{e}^{u,d}$ and $\mathbf{h}^{u,d}$ are the spectrums of the upgoing and downgoing components of the electromagnetic field to be determined. The goal is to find the upgoing and downgoing components of the field.

3.2. Separation of the observed EM field into upgoing and downgoing components

Following Berdichevsky and Zhdanov (1984), we can develop a technique for separation of the observed EM field into upgoing and downgoing components based on Eqs (35) and (36). Indeed, direct analysis of Eqs (35) and (36) shows that this problem can be solved by differentiation of both sides of these equations with respect to the vertical coordinate z:

$$\frac{\partial}{\partial z}\mathbf{e}(k_x, k_y, z) = \nu\mathbf{e}^u \exp[\nu(z - z_0)] - \nu\mathbf{e}^d \exp[-\nu(z - z_0)], \quad z_b \geq z \geq 0, \quad (37)$$

$$\frac{\partial}{\partial z}\mathbf{h}(k_x, k_y, z) = \nu\mathbf{h}^u \exp[\nu(z - z_0)] - \nu\mathbf{h}^d \exp[-\nu(z - z_0)], \quad z_b \geq z \geq 0. \quad (38)$$

Multiplying both sides of Eq. (35) by ν and adding Eq. (37), we have

$$\nu \mathbf{e}(k_x, k_y, z) + \frac{\partial}{\partial z}\mathbf{e}(k_x, k_y, z) = 2\nu \mathbf{e}^u \exp[\nu(z - z_0)]. \quad (39)$$

A similar combination of Eqs (36) and (38) produces:

$$\nu \mathbf{h}(k_x, k_y, z) + \frac{\partial}{\partial z}\mathbf{h}(k_x, k_y, z) = 2\nu \mathbf{h}^u \exp[\nu(z - z_0)]. \quad (40)$$

From the last two expressions we find immediately that:

$$\mathbf{e}^u = \frac{1}{2}\left[\mathbf{e}(k_x, k_y, z) + \frac{1}{\nu}\frac{\partial}{\partial z}\mathbf{e}(k_x, k_y, z)\right]\exp[-\nu(z - z_0)] \quad (41)$$

and

$$\mathbf{h}^u = \frac{1}{2}\left[\mathbf{h}(k_x, k_y, z) + \frac{1}{\nu}\frac{\partial}{\partial z}\mathbf{h}(k_x, k_y, z)\right]\exp[-\nu(z - z_0)]. \quad (42)$$

Thus, the problem of field separation is reduced to computations of the vertical derivatives of the electric and magnetic fields. The last problem can be solved using the spectral form of Maxwell's equations (28).

Let us apply the Fourier transform (30) and (31) to both sides of Eqs (28):

$$\begin{aligned} &\delta_z \times \mathbf{h} = \sigma_w \mathbf{e}, \quad &(a)\\ &\delta_z \times \mathbf{e} = i\omega\mu \mathbf{h}, \quad &(b)\\ &\delta_z \cdot \mathbf{h} = 0, \quad &(c)\\ &\delta_z \cdot \mathbf{e} = 0, \quad &(d) \end{aligned} \quad (43)$$

where δ_z stands for a symbolic vector $(-ik_x, -ik_y, \partial/\partial z)$.

It follows directly from the last equations that the spectrums of the vertical derivatives of the different components of the electromagnetic field can be found as a linear combination of the spectrum of the fields themselves.

For example, we can find from Eqs (43)c and (43)d that

$$\frac{\partial}{\partial z}h_z = ik_x h_x + ik_y h_y, \quad (44)$$

and

$$\frac{\partial}{\partial z} e_z = \mathrm{i}k_x e_x + \mathrm{i}k_y e_y. \tag{45}$$

In a similar way, we determine from Eqs (43)a and (43)b that

$$\frac{\partial}{\partial z} h_x = -\mathrm{i}k_x h_z + \sigma_w e_y,$$
$$\frac{\partial}{\partial z} h_y = -\mathrm{i}k_y h_z - \sigma_w e_x,$$
$$\frac{\partial}{\partial z} e_x = -\mathrm{i}k_x e_z + \mathrm{i}\omega\mu h_y,$$
$$\frac{\partial}{\partial z} e_y = -\mathrm{i}k_y e_z - \mathrm{i}\omega\mu h_x.$$

Combining these formulas together, we have:

$$\frac{\partial}{\partial z} \mathbf{h} = \hat{\mathbf{K}}_{xy} \mathbf{h} + \sigma_w \hat{\boldsymbol{\eta}} \mathbf{e}, \tag{46}$$

$$\frac{\partial}{\partial z} \mathbf{e} = \mathrm{i}\omega\mu \hat{\boldsymbol{\eta}} \mathbf{h} + \hat{\mathbf{K}}_{xy} \mathbf{e}, \tag{47}$$

where we use the following notations:

$$\hat{\mathbf{K}}_{xy} = \begin{bmatrix} 0 & 0 & -\mathrm{i}k_x \\ 0 & 0 & -\mathrm{i}k_y \\ \mathrm{i}k_x & \mathrm{i}k_y & 0 \end{bmatrix}, \quad \hat{\boldsymbol{\eta}} = \begin{bmatrix} 0 & 1 & 0 \\ -1 & 0 & 0 \\ 0 & 0 & 0 \end{bmatrix}.$$

Substituting Eqs (46) and (47) into (41) and (42), we finally find that:

$$\mathbf{e}^u = \frac{1}{2} \left[\left(\hat{\mathbf{I}} + \frac{1}{\nu} \hat{\mathbf{K}}_{xy} \right) \cdot \mathbf{e} + \frac{\mathrm{i}\omega\mu}{\nu} \hat{\boldsymbol{\eta}} \cdot \mathbf{h} \right] \exp[-\nu(z - z_0)] \tag{48}$$

and

$$\mathbf{h}^u = \frac{1}{2} \left[\left(\hat{\mathbf{I}} + \frac{1}{\nu} \hat{\mathbf{K}}_{xy} \right) \cdot \mathbf{h} + \frac{\sigma_w}{\nu} \hat{\boldsymbol{\eta}} \cdot \mathbf{e} \right] \exp[-\nu(z - z_0)]. \tag{49}$$

We can write the last formulas using scalar notations, assuming $z = z_0$:

$$e_x^u = \frac{1}{2} e_x - \frac{\mathrm{i}k_x}{2\nu} e_z + \frac{\mathrm{i}\omega\mu}{2\nu} h_y, \tag{50}$$

$$e_y^u = \frac{1}{2}e_y - \frac{\mathrm{i}k_y}{2\nu}e_z - \frac{\mathrm{i}\omega\mu}{2\nu}h_x, \tag{51}$$

$$e_z^u = \frac{1}{2}e_z + \frac{\mathrm{i}k_x}{2\nu}e_x + \frac{\mathrm{i}k_y}{2\nu}e_y, \tag{52}$$

$$h_x^u = \frac{1}{2}h_x - \frac{\mathrm{i}k_x}{2\nu}h_z + \frac{\sigma_w}{2\nu}e_y, \tag{53}$$

$$h_y^u = \frac{1}{2}h_y - \frac{\mathrm{i}k_y}{2\nu}h_z - \frac{\sigma_w}{2\nu}e_x, \tag{54}$$

$$h_z^u = \frac{1}{2}h_z + \frac{\mathrm{i}k_x}{2\nu}h_x + \frac{\mathrm{i}k_y}{2\nu}h_y. \tag{55}$$

Note that, according to Eqs (43)a and (43)b, the vertical components of the spectrum of the electric and magnetic fields are equal to:

$$e_z = \frac{\mathrm{i}k_y h_x - \mathrm{i}k_x h_y}{\sigma_w}, \quad h_z = \frac{\mathrm{i}k_y e_x - \mathrm{i}k_x e_y}{\mathrm{i}\omega\mu}. \tag{56}$$

3.3. Convolution form of decomposition operators

Consider, as an example, the x component of the electric field. Applying the inverse Fourier transforms to Eq. (50), we have:

$$\begin{aligned} E_x^u = {} & \frac{1}{2}E_x - \frac{1}{8\pi^2}\int_{-\infty}^{\infty}\int \frac{\mathrm{i}k_x}{\nu}e_z(k_x, k_y, z)\exp[-\mathrm{i}(k_x x + k_y y)]\mathrm{d}k_x\mathrm{d}k_y \\ & + \frac{\mathrm{i}\omega\mu}{2}\frac{1}{4\pi^2}\int_{-\infty}^{\infty}\int \frac{1}{\nu}h_y(k_x, k_y, z)\exp[-\mathrm{i}(k_x x + k_y y)]\mathrm{d}k_x\mathrm{d}k_y. \end{aligned} \tag{57}$$

Using the convolution theorem (Arfken and Weber, 1995, pp. 863–865), we can write Eq. (57) in the form:

$$\begin{aligned} E_x^u(x', y', z) = {} & \frac{1}{2}E_x(x', y', z) \\ & + \frac{1}{2}\int_{-\infty}^{\infty}\int \frac{\partial}{\partial x'}K(x' - x, y' - y)E_z(x, y, z)\mathrm{d}x\mathrm{d}y \\ & + \frac{\mathrm{i}\omega\mu}{2}\int_{-\infty}^{\infty}\int K(x' - x, y' - y)H_y(x, y, z)\mathrm{d}x\mathrm{d}y, \end{aligned} \tag{58}$$

where $K(x', y')$ is the convolution kernel function for a separation operator given by the formula:

$$K(x', y') = \frac{1}{4\pi^2}\int_{-\infty}^{\infty}\int \frac{1}{\nu}\exp[-\mathrm{i}(k_x x' + k_y y')]\mathrm{d}k_x\mathrm{d}k_y. \tag{59}$$

Note that both integrals in Eq. (58) are determined in the sense of the Cauchy principal value, which means that these integrals are equal to the limits of the following integrals:

$$\int_{-\infty}^{\infty}\int \frac{\partial}{\partial x'}K(x'-x,y'-y)E_z(x,y,z)\mathrm{d}x\mathrm{d}y$$
$$=\lim_{\rho\to 0}\iint_{S_\rho}\frac{\partial}{\partial x'}K(x'-x,y'-y)E_z(x,y,z)\mathrm{d}x\mathrm{d}y, \tag{60}$$

$$\int_{-\infty}^{\infty}\int K(x'-x,y'-y)H_y(x,y,z)\mathrm{d}x\mathrm{d}y$$
$$=\lim_{\rho\to 0}\iint_{S_\rho}K(x'-x,y'-y)H_y(x,y,z)\mathrm{d}x\mathrm{d}y, \tag{61}$$

where S_ρ is the entire horizontal plane XY with the exclusion of a circle $(x'-x)^2+(y'-y)^2\leqslant\rho$.

The integral in Eq. (59) can be expressed using the tabulated integral $J_1(x',y',z')$ (Zhdanov and Keller, 1994):

$$J_1(x',y',z')=\frac{1}{2\pi}\int_{-\infty}^{\infty}\int\frac{\exp[-\nu(z'-z)]}{\nu}\exp[-\mathrm{i}(k_xx'+k_yy')]\mathrm{d}k_x\mathrm{d}k_y$$
$$=\frac{\exp(\mathrm{i}k_wr)}{r}, \tag{62}$$

where $r=\sqrt{x'^2+y'^2+(z'-z)}$ and $k_w=\sqrt{\mathrm{i}\omega\mu\sigma_w}$.

Comparing Eqs (59) and (62), we can see that

$$K(x',y')=\frac{1}{2\pi}J_1(x',y',z')_{z'=z_b}. \tag{63}$$

Note that function J_1 is nothing else but the fundamental Green's function (9) for the Helmholtz equation in a homogeneous full space with the sea-water conductivity σ_w, multiplied by (-4π):

$$J_1(x',y',z_b)=-4\pi G_w(\mathbf{r}'|\mathbf{r}_0), \tag{64}$$

where $\mathbf{r}'=(x',y',z_b)$ and $\mathbf{r}_0=(0,0,z)$. Thus we have:

$$K(x',y')=-2G_w(\mathbf{r}'|\mathbf{r}_0). \tag{65}$$

Therefore, integral transformation (58) can be written as:

$$E_x^u(x', y', z) = \frac{1}{2}E_x(x', y', z) - \int_{-\infty}^{\infty}\int \frac{\partial}{\partial x'}G_w(\mathbf{r}'|\mathbf{r}_0)E_z(x, y, z)\mathrm{d}x\mathrm{d}y$$
$$- \mathrm{i}\omega\mu \int_{-\infty}^{\infty}\int G_w(\mathbf{r}'|\mathbf{r}_0)H_y(x, y, z)\mathrm{d}x\mathrm{d}y. \quad (66)$$

We can derive analogous convolution-type expressions for all other upgoing/downgoing components of the electromagnetic field observed on a horizontal plane.

Direct algebraic calculations (which we do not present here because of their length) show that similar convolution forms of the decomposition operators can be obtained directly from the Stratton-Chu decomposition transformations (22)–(25).

4. Electromagnetic Field Separation into Upgoing and Downgoing Parts Using Horizontal Gradients

We should note that the methods of EM field decomposition based on Stratton-Chu type transforms and on the Fourier transform in the (k, ω) domain require measurements of the total EM field on the entire surface of observation S. In practice, however, we measure these fields at a discrete number of observation points located within a limited part of S. That is why it is important to develop simplified methods of field decomposition which would require measurements at just a few points. The simplest way to solve this problem is based on an assumption that the observed data can be approximated by a vertically propagated plane EM wave. In this case, we can use only one spatial frequency, $k_x = k_y = 0$, and the spectral decomposition formulas (50)–(55) are reduced to the simple algebraic formulas for field separation at a given point. This approach, for example, was discussed in (Amundsen et al., 2006) and is widely used for decomposition of field MCSEM data.

However, one can produce a more accurate but still rapid decomposition method assuming just slow horizontal variations of the EM field observed within the area of observation. In this case, the EM field within the seawater can be expressed approximately as (Zhdanov et al., 1996; Zhdanov, 2002, 2009):

$$\begin{aligned}\mathbf{E}(x, y, z) &= \mathbf{Q}_E^u(x, y, z)\exp[-\mathrm{i}k(z - z_0)] + \mathbf{Q}_E^d(x, y, z)\exp[\mathrm{i}k(z - z_0)],\\ \mathbf{H}(x, y, z) &= \mathbf{Q}_H^u(x, y, z)\exp[-\mathrm{i}k(z - z_0)] + \mathbf{Q}_H^d(x, y, z)\exp[\mathrm{i}k(z - z_0)],\end{aligned} \quad (67)$$

where $z_b \geq z \geq 0$, z_b is the depth of the sea bottom, z_0 is the depth of the location of the receivers, and $k = k_w = \sqrt{\mathrm{i}\omega\mu\sigma_w}$.

We assume that the vector functions $\mathbf{Q}_E^u$, $\mathbf{Q}_E^d$, $\mathbf{Q}_H^u$, and $\mathbf{Q}_H^d$ vary relatively slowly with the depth:

$$|\partial \mathbf{Q}_{E,H}^{u,d}/\partial z| \ll |\partial \exp[\pm ik(z - z_0)]/\partial z|. \tag{68}$$

The goal is to find the upgoing and downgoing components of the EM field. This problem can be solved by differentiation of both sides of Eq. (67) with respect to the vertical coordinate z:

$$\begin{aligned} \partial \mathbf{E}(x, y, z)/\partial z &= ik\mathbf{Q}_E^d \exp[ik(z - z_0)] - ik\mathbf{Q}_E^u \exp[-ik(z - z_0)], \\ \partial \mathbf{H}(x, y, z)/\partial z &= ik\mathbf{Q}_H^d \exp[ik(z - z_0)] - ik\mathbf{Q}_H^u \exp[-ik(z - z_0)]. \end{aligned} \tag{69}$$

Multiplying both sides of Eq. (67) by ik and adding (69), we have:

$$\begin{aligned} ik\mathbf{E}(x, y, z) - \partial \mathbf{E}(x, y, z)/\partial z &= 2ik\mathbf{Q}_E^u \exp[-ik(z - z_0)], \\ ik\mathbf{H}(x, y, z) - \partial \mathbf{H}(x, y, z)/\partial z &= 2ik\mathbf{Q}_H^u \exp[-ik(z - z_0)]. \end{aligned} \tag{70}$$

From the last expression we find immediately that:

$$\begin{aligned} \mathbf{Q}_E^u(x, y, z) &= \frac{1}{2}[\mathbf{E}(x, y, z) - \frac{1}{ik}\partial \mathbf{E}(x, y, z)/\partial z] \exp[ik(z - z_0)], \\ \mathbf{Q}_H^u(x, y, z) &= \frac{1}{2}[\mathbf{H}(x, y, z) - \frac{1}{ik}\partial \mathbf{H}(x, y, z)/\partial z] \exp[ik(z - z_0)]. \end{aligned} \tag{71}$$

Thus the problem of field separation is reduced to computations of the vertical derivatives of the electric and magnetic fields. The last problem can be solved using Maxwell's equations.

Within the homogeneous seawater layer the EM field satisfies Maxwell's equations (28). It follows directly from Eq. (28) that the vertical derivatives of the different components of the electromagnetic field can be found as a linear combination of the fields themselves. For example, we can find from Eqs (28)c and (28)d that

$$\begin{aligned} \frac{\partial}{\partial z} H_z &= -\frac{\partial}{\partial x} H_x - \frac{\partial}{\partial y} H_y, \\ \frac{\partial}{\partial z} E_z &= -\frac{\partial}{\partial x} E_x - \frac{\partial}{\partial y} E_y. \end{aligned} \tag{72}$$

In a similar way, we determine from (28)a and (28)b that

$$\begin{aligned}
\frac{\partial}{\partial z}H_x &= \frac{\partial}{\partial x}H_z + \sigma_w E_y, \\
\frac{\partial}{\partial z}H_y &= \frac{\partial}{\partial y}H_z - \sigma_w E_x, \\
\frac{\partial}{\partial z}E_x &= \frac{\partial}{\partial x}E_z + \mathrm{i}\omega\mu H_y, \\
\frac{\partial}{\partial z}E_y &= \frac{\partial}{\partial y}E_z - \mathrm{i}\omega\mu H_x.
\end{aligned} \tag{73}$$

Substituting Eqs (73) into (71), we can write the last formulas using scalar notations, assuming $z = z_0$:

$$\begin{aligned}
Q^u_{E_x} &= \frac{1}{2}E_x - \frac{1}{2\mathrm{i}k}\frac{\partial}{\partial x}E_z - \frac{\mathrm{i}\omega\mu}{2\mathrm{i}k}H_y, \\
Q^u_{E_y} &= \frac{1}{2}E_y - \frac{1}{2\mathrm{i}k}\frac{\partial}{\partial y}E_z + \frac{\mathrm{i}\omega\mu}{2\mathrm{i}k}H_x, \\
Q^u_{E_z} &= \frac{1}{2}E_z + \frac{1}{2\mathrm{i}k}\frac{\partial}{\partial x}E_x + \frac{1}{2\mathrm{i}k}\frac{\partial}{\partial y}E_y, \\
Q^u_{H_x} &= \frac{1}{2}H_x - \frac{1}{2\mathrm{i}k}\frac{\partial}{\partial x}H_z - \frac{\sigma_w}{2\mathrm{i}k}E_y, \\
Q^u_{H_y} &= \frac{1}{2}H_y - \frac{1}{2\mathrm{i}k}\frac{\partial}{\partial y}H_z + \frac{\sigma_w}{2\mathrm{i}k}E_x, \\
Q^u_{H_z} &= \frac{1}{2}H_z + \frac{1}{2\mathrm{i}k}\frac{\partial}{\partial x}H_x + \frac{1}{2\mathrm{i}k}\frac{\partial}{\partial y}H_y,
\end{aligned} \tag{74}$$

where $Q^u_{E_x}$, $Q^u_{E_y}$, $Q^u_{E_z}$ are the scalar components of vector $\mathbf{Q}^u_E$, and $Q^u_{H_x}$, $Q^u_{H_y}$, $Q^u_{H_z}$ are the scalar components of vector $\mathbf{Q}^u_H$, respectively.

According to Eqs (28)a and (28)b, the vertical components of the electric and magnetic fields are equal to:

$$\begin{aligned}
E_z &= \frac{1}{\sigma_w}\left[\frac{\partial}{\partial x}H_y - \frac{\partial}{\partial y}H_x\right], \\
H_z &= \frac{1}{\mathrm{i}\omega\mu}\left[\frac{\partial}{\partial x}E_y - \frac{\partial}{\partial y}E_x\right].
\end{aligned} \tag{75}$$

Therefore, in principle, in order to solve the upgoing/downgoing field separation problem, one can measure the horizontal components of the EM field only.

After we determine the vector coefficients $\mathbf{Q}_E^u$, $\mathbf{Q}_E^d$, and $\mathbf{Q}_H^u$, $\mathbf{Q}_H^d$, we can find the upgoing and downgoing fields themselves, e.g.:

$$\begin{aligned} \mathbf{E}^u(x, y, z) &= \mathbf{Q}_E^u \exp[-ik(z - z_0)], \\ \mathbf{H}^u(x, y, z) &= \mathbf{Q}_H^u \exp[-ik(z - z_0)]. \end{aligned} \tag{76}$$

Note that the expressions $\frac{\partial}{\partial x}E_x$, $\frac{\partial}{\partial y}E_y$, $\frac{\partial}{\partial x}E_z$, $\frac{\partial}{\partial y}E_z$, etc. represent the horizontal gradients of the electric and magnetic field components. These horizontal gradients can be calculated numerically using finite differences, e.g.:

$$\frac{\partial}{\partial x}E_x \approx \frac{E_x(x + \Delta x, y, z) - E_x(x, y, z)}{\Delta x}. \tag{77}$$

By substituting the finite difference expressions into formulas (74), we arrive at the finite difference expression for the upgoing and downgoing fields.

5. Numerical Examples of Marine EM Data Decomposition

In order to check the effectiveness of the separation technique developed above based on the horizontal gradients, we have conducted a number of numerical experiments with synthetic EM data. We present some of these results below.

5.1. Model 1

We begin our numerical study with Model 1, shown in Figure 2. The geoelectrical section of this model is formed by a seawater layer with a resistivity of 0.4 Ohm m and a thickness of 340 m, and conductive sea-bottom sediments with a resistivity of 1 Ohm m, respectively. A 3D reservoir with 100 Ohm m resistivity is embedded in the sea bottom at a depth of 500 m below the seafloor. The EM field in the model is generated by a horizontal electric bipole transmitter towing behind the ship at an elevation of 50 m above the seafloor. The receivers measuring all six components of the EM field (or four horizontal components only) are located at the seafloor.

As an example, we consider the field generated by one transmitter located at the point whose horizontal coordinates are $x = -10$ km and $y = 0$, where the origin of the coordinates, $x = 0$ and $y = 0$, is at the center of the reservoir. The receivers are located along several survey lines with a separation of 500 m in both x and y directions, as shown in Figure 3.

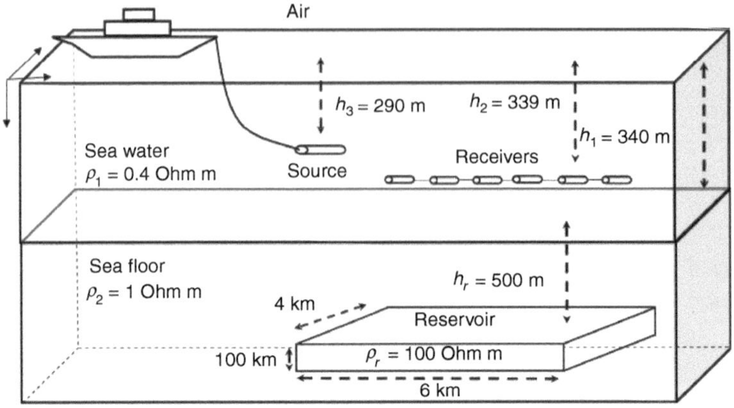

Figure 2 3D view of Model 1. The geoelectrical section of this model is formed by a seawater layer with a resistivity of 0.4 Ohm m and a thickness of 340 m, and conductive sea-bottom sediments with a resistivity of 1 Ohm m, respectively. A 3D reservoir with 100 Ohm m resistivity is embedded in the sea bottom at a depth of 500 m below the seafloor.

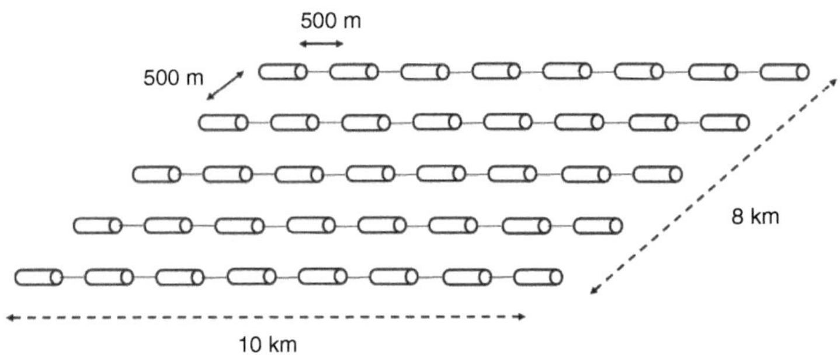

Figure 3 The array of receivers for Model 1. Each spatial step along the line of receivers is 500 meters, and the distance between two neighboring survey lines is also 500 meters.

Figures 4 and 5 show the results of the decomposition of the x component of the electric field, E_x, and the y component of the magnetic field, H_y, into upgoing and downgoing parts along one of the observational profiles. The top panels in these figures show the plots of the amplitude vs. offset curves for the total and upgoing fields, while the bottom panels present the similar plots of the phase vs. offset curves. The calculation of the upgoing field component was done using the method of horizontal gradients introduced above.

We should note that the problem of upgoing/downgoing field separation can be solved using the data along one survey line only. However, the most profound result can be seen in the maps of the upgoing and downgoing parts of the field (see Figures 6 and 7). One can clearly see in these figures

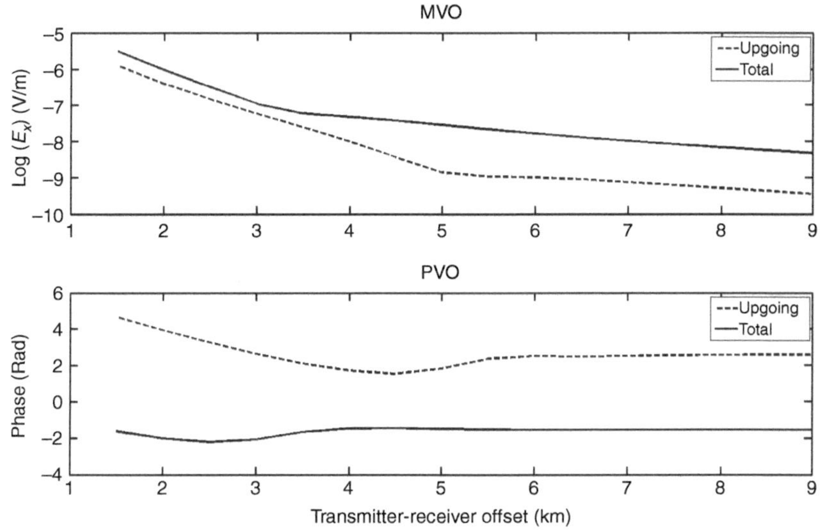

Figure 4 Model 1: Example of the separation of the in-line electric field E_x at a frequency of 0.5 Hz into the upgoing and downgoing parts. The top panel shows the plots of the amplitude vs. offset curves for the total and upgoing fields, while the bottom panel presents similar plots of the phase vs. offset curves.

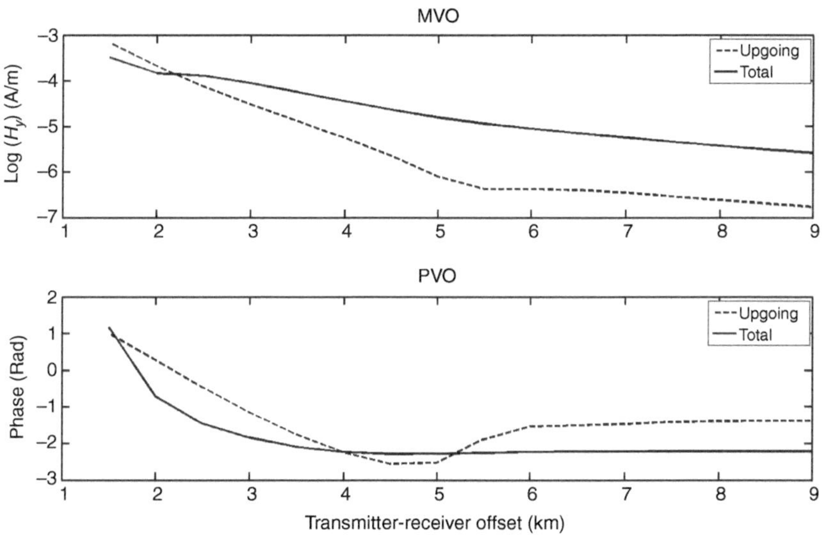

Figure 5 Model 1: Example of the separation of the in-line magnetic field H_y at a frequency of 0.5 Hz into the upgoing and downgoing parts. The top panel shows the plots of the amplitude vs. offset curves for the total and upgoing fields, while the bottom panel presents similar plots of the phase vs. offset curves.

that the separation of the upgoing part of the total electric field results in effective imaging of the horizontal location of the reservoir. A similar effect is observed in the maps of the horizontal component of the magnetic field H_y after a decomposition transformation.

Note that in Figures 6 and 7 a dashed white rectangle outlines the location of a resistive reservoir. From these results, it is clear, the upgoing field (real and/or imaginary parts, amplitude and/or phase) separated using the technique, reflects the location of the target much better than the total EM field data. Numerical data, shown above, were generated for a frequency of 0.5 Hz. We also used other frequencies from 0.1 Hz to 10 Hz and other locations of the transmitting bipole source to test our technique. We have obtained similar results for all these frequencies and different transmitter locations.

5.2. Model 2

We have constructed Model 2 to test the separation technique for a more complicated model representing a geological structure with two reservoirs embedded within the conductive sediments at a depth of 500 m and 600 m below the seafloor, respectively. Figure 8 shows a 3D view of the model, while Figure 9 presents a plan view of the reservoir locations. As an example, we assume that a horizontal electric bipole transmitter is located in the central survey line at a distance of 7 km to the left side of the reservoir 1 shown in Figure 9. The receiver array is the same as in Model 1 with spatial steps, both along the survey line and perpendicular to the survey line, equal to 500 m, as shown in Figure 3.

Figures 10 and 11 indicate the effectiveness of separation for Model 2, in which a dashed white rectangle outlines the location of 3D resistive reservoirs. All these results demonstrate that the developed technique of the total field decomposition into upgoing and downgoing parts can be effectively used for fast analysis of observed sea-bottom EM data. The upgoing field produced using the separation technique reflects well the locations of the sea-bottom resistive structures.

6. Conclusions

The modern technique of EM field decomposition into upgoing and downgoing parts has its roots in the classical methods of geomagnetic field separation into external and internal parts. The most general approach to the solution of this problem is based on the theory of the Stratton-Chu type integrals. This approach makes it possible to separate MCSEM data observed on the seafloor in areas with rough bathymetry. In the case of a flat seafloor, the separation of MCSEM data can be done using the Fourier transform in the (k, ω) domain. However, spatial spectrum transformation, in a general case, requires measurement of the data in a relatively large area

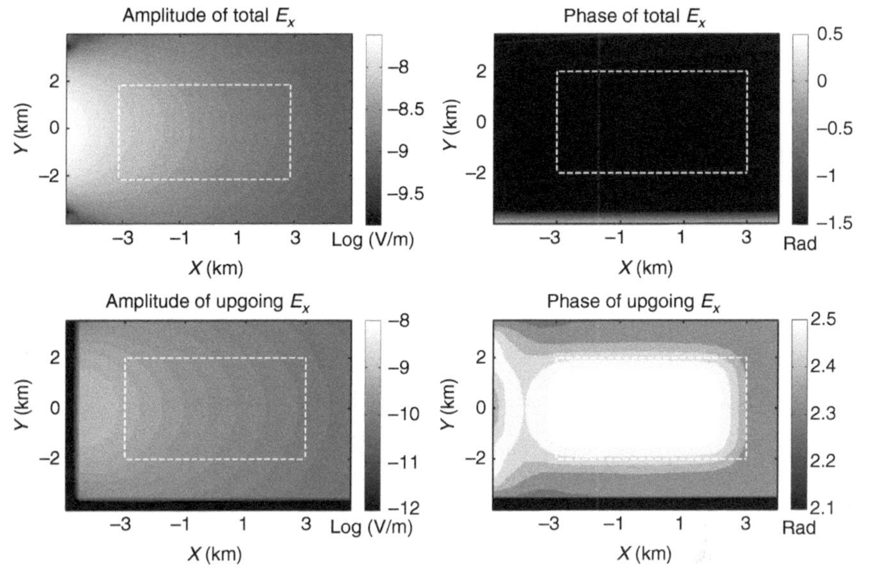

Figure 6 Model 1: The top panels show maps of the amplitude (top left panel) and phase (top right panel) of the total in-line electric field E_x at a frequency of 0.5 Hz. The bottom panels present maps of the amplitude (bottom left panel) and phase (bottom right panel) of the upgoing in-line electric field E_x^u separated using the method of horizontal gradients.

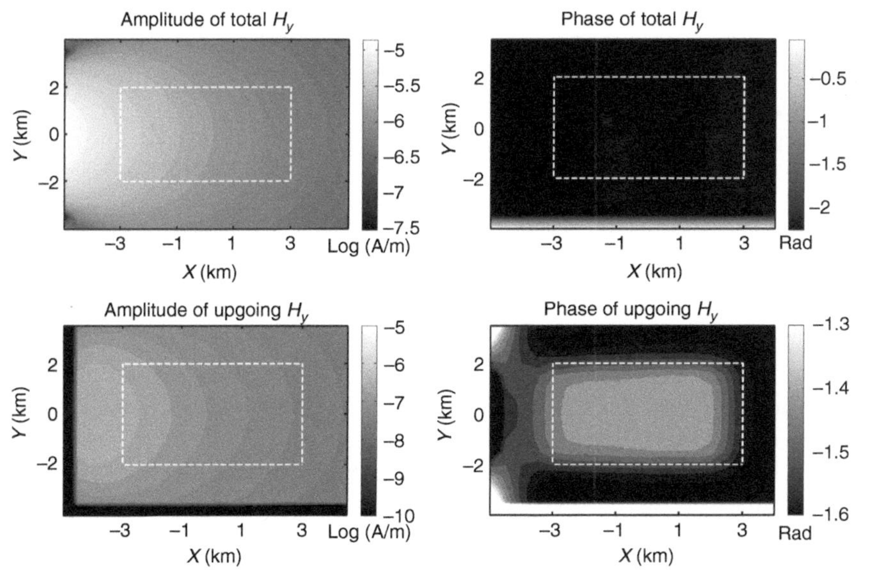

Figure 7 Model 1: The top panels show maps of the amplitude (top left panel) and phase (top right panel) of the total magnetic field H_y at a frequency of 0.5 Hz. The bottom panels present maps of the amplitude (bottom left panel) and phase (bottom right panel) of the upgoing magnetic field H_y^u separated using the method of horizontal gradients.

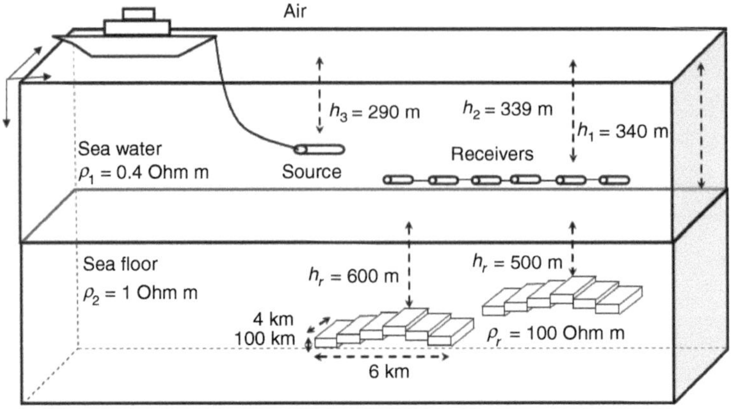

Figure 8 3D view of Model 2. The geoelectrical section of this model is formed by a seawater layer with a resistivity of 0.4 Ohm m and a thickness of 340 m, and conductive sea-bottom sediments with a resistivity of 1 Ohm m, respectively. Two reservoirs with 100 Ohm m resistivity are embedded in the sea bottom at depths of 500 m and 600 m below the seafloor.

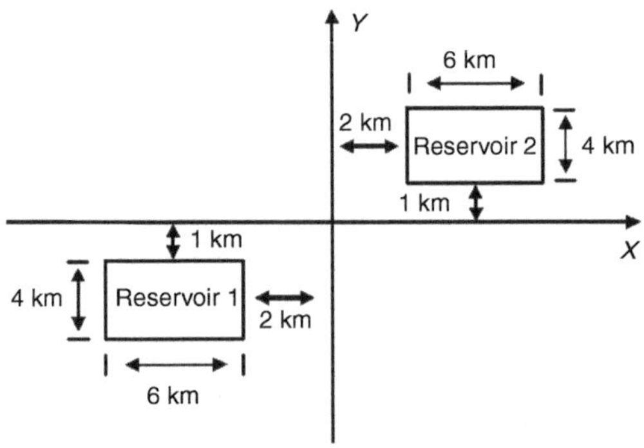

Figure 9 Model 2: Plan view of the resistive reservoir locations.

of observation. This method can be used for a single point separation under the assumption that the field can be approximated by a plane wave, which is valid in the far field of the source only.

We have developed a method of solving this problem which is free of these limitations of the spectral decomposition method. It is based on the calculations of the horizontal gradients of the field, and therefore it can be used for a decomposition of MCSEM data measured at two points only. The numerical study shows that this novel method can be used as a rapid transformation of MCSEM data for a qualitative evaluation of the location of the typical exploration targets, e.g., HC reservoirs.

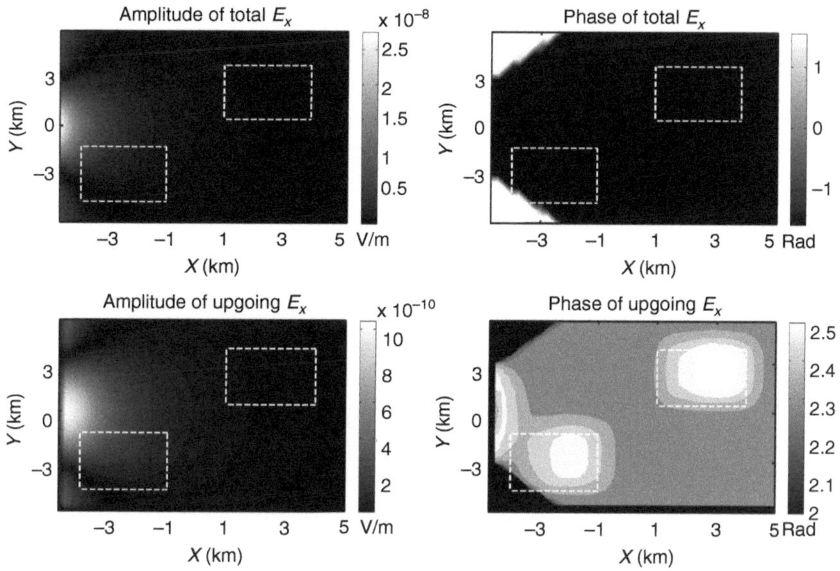

Figure 10 Model 2: The top panels show maps of the amplitude (top left panel) and phase (top right panel) of the total in-line electric field E_X at a frequency of 0.5 Hz. The bottom panels present maps of the amplitude (bottom left panel) and phase (bottom right panel) of the upgoing in-line electric field E_x^u separated using the method of horizontal gradients.

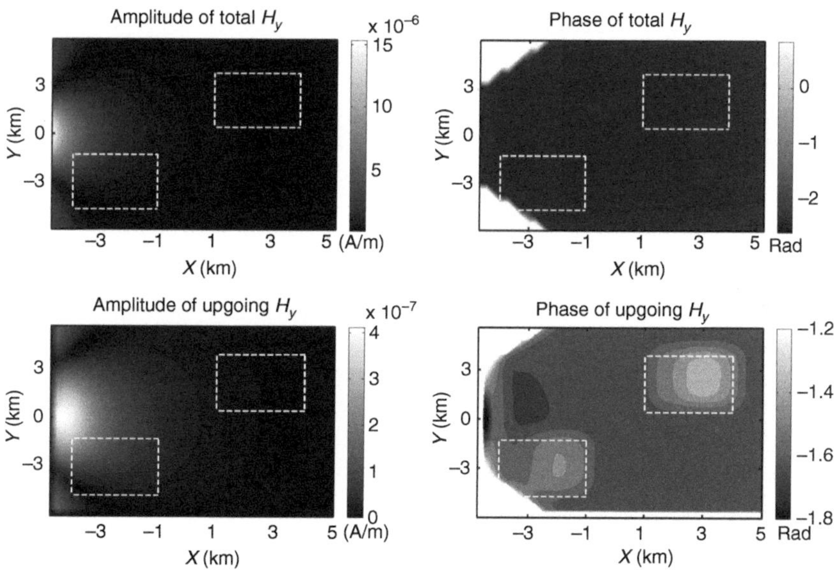

Figure 11 Model 2: The top panels show maps of the amplitude (top left panel) and phase (top right panel) of total magnetic field H_y at a frequency of 0.5 Hz. The bottom panels present maps of the amplitude (bottom left panel) and phase (bottom right panel) of the upgoing magnetic field H_y^u separated using the method of horizontal gradients.

ACKNOWLEDGMENTS

The authors acknowledge the support of the University of Utah Consortium for Electromagnetic Modeling and Inversion (CEMI), which includes BAE Systems, Baker Atlas Logging Services, BGP China National Petroleum Corporation, BHP Billiton World Exploration Inc., BP, EMGS, ENI S.p.A., ExxonMobil Upstream Research Company, Fugro, Halliburton Energy Services, Information Systems Laboratories, NewmontMining Co., OHM, Petrobras, PGS, Rio Tinto - Kennecott, Rocksource, Russian Research Center Kurchatov Institute, Schlumberger, Science Applications International Co., Shell International Exploration and Production Inc., StatoilHydro, Sumitomo Metal Mining Co., Woodside Energy, and Zonge Engineering and Research Organization.

APPENDIX A. STRATTON-CHU INTEGRAL FORMULAS

In these appendices we present a short review of the theory of the Stratton-Chu type integrals, developed by Zhdanov (1988).

Let us consider a 3D domain D bounded by a smooth surface S. We assume that domain D is filled by a homogeneous medium with the conductivity σ, magnetic permeability μ, and dielectric constant ε. The frequency-domain electromagnetic field, $\mathbf{E}$, $\mathbf{H}$, in the model is generated by some sources located outside domain D. The time dependence is expressed by an exponential function $\exp(-i\omega t)$. This field satisfies inside D (up to its boundary S) the following Maxwell's equations:

$$\begin{aligned} \nabla \times \mathbf{H} &= \tilde{\sigma}\mathbf{E}, \\ \nabla \times \mathbf{E} &= i\omega\mu\mathbf{H}, \\ \nabla \cdot \mathbf{H} &= 0, \\ \nabla \cdot \mathbf{E} &= 0, \end{aligned} \tag{78}$$

where $\tilde{\sigma} = \sigma - i\omega\varepsilon$ stands for the complex electric conductivity of the medium.

One can formulate a boundary-value problem for the EM field as follows. We assume that the EM field components are known everywhere on the surface S. The goal is to find the values of the electric, $\mathbf{E}(\mathbf{r}')$, and magnetic, $\mathbf{H}(\mathbf{r}')$, fields inside the domain, $\mathbf{r}' \in \mathrm{D}$. A solution of this problem is provided by the classical Stratton-Chu integral formulas (Stratton, 1941):

$$\iint_S [(\mathbf{n} \cdot \mathbf{E})\nabla G + (\mathbf{n} \times \mathbf{E}) \times \nabla G + i\omega\mu(\mathbf{n} \times \mathbf{H})G]ds = \mathbf{E}(\mathbf{r}'), \quad \mathbf{r}' \in \mathrm{D}, \tag{79}$$

$$\iint_S [(\mathbf{n}\cdot\mathbf{H})\nabla G + (\mathbf{n}\times\mathbf{H})\times\nabla G + \tilde{\sigma}(\mathbf{n}\times\mathbf{E})G]\mathrm{d}s = H(\mathbf{r}'), \quad \mathbf{r}' \in \mathrm{D}, \tag{80}$$

where $\mathbf{n}$ is the unit vector of an inward pointing normal to S; $G = G(\mathbf{r}'|\mathbf{r})$ is the fundamental Green's function for the Helmholtz equation:

$$G(\mathbf{r}'|\mathbf{r}) = -\frac{1}{4\pi\,|\mathbf{r}-\mathbf{r}'|}\exp(\mathrm{i}\tilde{k}|\mathbf{r}-\mathbf{r}'|),$$

and $\tilde{k}$ is the corresponding wave number:

$$\tilde{k} = \sqrt{\mathrm{i}\omega\mu\tilde{\sigma}} = \sqrt{\mathrm{i}\omega\mu\sigma + \omega^2\mu\varepsilon}\ \mathrm{Re}\,\tilde{k} > 0. \tag{81}$$

The Green's function satisfies the equation:

$$\nabla^2 G + \tilde{k}^2 G = \delta(\mathbf{r}-\mathbf{r}'), \tag{82}$$

where $\delta(\mathbf{r}-\mathbf{r}')$ is a delta-function.

APPENDIX B. STRATTON-CHU TYPE INTEGRALS AND THEIR PROPERTIES

The concept of the Stratton-Chu type integrals was introduced by Zhdanov (1980, 1988).

Let some vector fields $\mathcal{E}$ and $\mathcal{H}$ be specified on the inner side of a closed smooth boundary S of a domain D_i and have continuously differentiable tangential components, $\mathcal{E}_\tau$ and $\mathcal{H}_\tau$. We assume that the normal components of these fields, $\mathcal{E}_n$ and $\mathcal{H}_n$, are related to the tangential components by the following expressions:

$$\mathcal{E}_n = -\frac{1}{\tilde{\sigma}}\nabla_s\cdot(\mathrm{n}\times\mathcal{H}_\tau), \quad \mathcal{H}_n = -\frac{1}{\mathrm{i}\omega\mu}\nabla_s\cdot(\mathrm{n}\times\mathcal{E}_\tau). \tag{83}$$

Equations (83) represent a two-dimensional form of Maxwell's equations (Zhdanov, 1988).

We consider the expressions:

$$C_S^E(\mathbf{r}') = \iint_S [(\mathrm{n}\cdot\mathcal{E})\nabla G + (\mathrm{n}\times\mathcal{E})\times\nabla G + \mathrm{i}\omega\mu(\mathrm{n}\times\mathcal{H})G]\mathrm{d}s, \tag{84}$$

and

$$C_S^H(\mathbf{r}') = \iint_S [(\mathrm{n} \cdot \mathcal{H})\nabla G + (\mathrm{n} \times \mathcal{H}) \times \nabla G + \tilde{\sigma}(\mathrm{n} \times \mathcal{E})G]\mathrm{d}s. \quad (85)$$

These expressions are equal to the integrals from formulas (79) and (80), if $\mathcal{E} = \mathbf{E}$ and $\mathcal{H} = \mathbf{H}$. They are called Stratton-Chu type integrals (Zhdanov, 1980, 1988), and functions $\mathcal{E}$ and $\mathcal{H}$ are their densities.

The Stratton-Chu type integrals have several important properties which make them very useful in EM applications.

(1) Everywhere in the whole space outside the surface S, the integrals C_S^E and C_S^H satisfy Maxwell's equations:

$$\begin{aligned} \nabla \times \mathbf{C}_S^H &= \tilde{\sigma}\mathbf{C}_S^E \\ \nabla \times \mathbf{C}_S^E &= \mathrm{i}\omega\mu\mathbf{C}_S^H, \\ \nabla \cdot \mathbf{C}_S^H &= 0, \\ \nabla \cdot \mathbf{C}_S^E &= 0. \end{aligned} \quad (86)$$

(2) If the densities of the Stratton-Chu type integrals are equal to the boundary values of the electromagnetic fields $\mathbf{E}$ and $\mathbf{H}$, satisfying the Maxwell's equations (78) in the domain D right up to the surface S and vanishing at infinity, then we have:

$$\mathbf{C}_S^E(\mathbf{r}') = \begin{cases} \mathbf{E}(\mathbf{r}'), & \mathbf{r}' \in D_i, \\ 0, & \mathbf{r}' \in D_e, \end{cases} \quad \mathbf{C}_S^H(\mathbf{r}') = \begin{cases} \mathbf{H}(\mathbf{r}'), & \mathbf{r}' \in D_i, \\ 0, & \mathbf{r}' \in D_e, \end{cases} \quad (87)$$

where D_i and D_e are the inner and outer domains bounded by S.

(3) At the points on the inner ($\mathbf{r}_0^+$) and outer (r_0^-) sides of the surface S there exist limit values of the Stratton-Chu type integrals equal to:

$$\begin{aligned} \mathbf{C}_i^E(\mathbf{r}_0) &= \lim_{\mathrm{r}' \to \mathbf{r}_0^+} \mathbf{C}_S^E(\mathbf{r}') = \mathbf{C}_S^E(\mathbf{r}_0) + \frac{1}{2}\mathcal{E}(\mathbf{r}_0), \\ \mathbf{C}_i^H(\mathbf{r}_0) &= \lim_{\mathrm{r}' \to \mathrm{r}_0^+} \mathbf{C}_S^H(\mathbf{r}') = \mathbf{C}_S^H(\mathbf{r}_0) + \frac{1}{2}\mathcal{H}(\mathbf{r}_0), \quad \mathbf{r}' \in D_i, \mathbf{r}_0 \in S \end{aligned} \quad (88)$$

and

$$\begin{aligned} \mathbf{C}_e^E(\mathbf{r}_0) &= \lim_{\mathbf{r}' \to \mathbf{r}_0^-} \mathbf{C}_S^E(\mathbf{r}') = \mathbf{C}_S^E(\mathbf{r}_0) - \frac{1}{2}\mathcal{E}(\mathbf{r}_0), \\ \mathbf{C}_e^H(\mathbf{r}_0) &= \lim_{\mathrm{r}' \to \mathbf{r}_0^-} \mathbf{C}_S^H(\mathbf{r}') = \mathbf{C}_S^H(\mathbf{r}_0) - \frac{1}{2}\mathcal{H}(\mathbf{r}_0), \quad \mathbf{r}' \in D_e, \mathbf{r}_0 \in S, \end{aligned} \quad (89)$$

where the singular integrals $\mathbf{C}_S^E(\mathbf{r}_0)$ and $\mathbf{C}_S^H(\mathbf{r}_0)$ are determined in the sense of the Cauchy principal value.

Thus, the Stratton-Chu type integrals have discontinuities on S, which are equal to their densities:

$$\mathbf{C}_i^E(\mathbf{r}_0) - \mathbf{C}_e^E(r_0) = \mathcal{E}(\mathbf{r}_0), \quad \mathbf{C}_i^H(\mathbf{r}_0) - \mathbf{C}_e^H(\mathbf{r}_0) = \mathcal{H}(\mathbf{r}_0). \tag{90}$$

At the same time, the sums of the limit values of the Stratton-Chu type integrals are equal to the double singular integrals:

$$\mathbf{C}_i^E(\mathbf{r}_0) + \mathbf{C}_e^E(\mathbf{r}_0) = 2\mathbf{C}_0^E(\mathbf{r}_0), \quad \mathbf{C}_i^H(\mathbf{r}_0) + \mathbf{C}_e^H(\mathbf{r}_0) = 2\mathbf{C}_0^H(\mathbf{r}_0). \tag{91}$$

The interested reader can find the proof of properties (1) through (3) in the monograph by Zhdanov (1988).

REFERENCES

Arfken, G., Weber, H.J., 1995. Mathematical Methods for Physicists, Fourth Edition. Academic Press.

Amundsen, L., Løseth, L., Mittet, R., Ellingsrud, S., Ursin, B., 2006. Decomposition of electromagnetic fields into upgoing and downgoing components. Geophysics 71, G211–G223.

Berdichevsky, M.N., Zhdanov, M.S., 1984. Advanced Theory of Deep Geomagnetic Sounding. Elsevier.

Chapman, S., Bartels, J., 1940. Geomagnetism, vol. 2. Clarendon Press.

Kertz, W., 1954. Modelle für erdmagnetischen induzierte elektrische Ströme in Untergrund. Nachr. Akad. Wiss. Göttingen Math.-Phys. Klasse, Abt. 11 A, 101–110.

Siebert, M., 1962. Die Zerlegung eines zweidimensionalen Magnetfeldes in äusseren und inneren Antei mit Hilfe der Cauchyschen Integralformed. Z. Geophys. 5, 231–236.

Stratton, J.A., 1941. Electromagnetic Theory. McGraw-Hill Book Company.

Zhdanov, M.S., 1980. Cauchy integral analogs for the separation and continuation of electromagnetic fields within conducting matter. Geophysical Surveys 4, 115–136.

Zhdanov, M.S., 1988. Integral Transforms in Geophysics. Springer-Verlag.

Zhdanov, M.S., 2002. Geophysical Inverse Theory and Regularization Problems. Elsevier.

Zhdanov, M.S., 2009. Geophysical Electromagnetic Theory and Methods. Elsevier.

Zhdanov, M.S., Keller, G., 1994. The Geoelectrical Methods in Geophysical Exploration. Elsevier.

Zhdanov, M.S., Traynin, P., Booker, J., 1996. Underground imaging by frequency domain electromagnetic migration. Geophysics 61, 666–682.

Zhdanova, O.N., Zhdanov, M.S., 1999. Methods for the analysis and interpretation of the sea floor electromagnetic fields. In: Deep Electromagnetic Exploration. Springer-Verlag.

CHAPTER 20

Automatic Travel Time Determination from a Frequency-domain Transfer Function: The Sompi Event Analysis

Yoko Hasada[a,b], Mineo Kumazawa[b,c], Kayoko Tsuruga[b,d], Takahiro Kunitomo[b,c], *and* Junzo Kasahara[b,c]

Contents

1. Introduction

ACROSS (Accurately-Controlled Routinely-Operated Signal System) is a subsurface exploring/monitoring system whose concept of data acquisition is distinctive: data obtained by the system constitute a complex

[a] Graduate School of Environmental Studies, Nagoya University, Japan
[b] Japan Atomic Energy Agency, Japan
[c] Faculty of Science, Shizuoka University, Japan
[d] Geophysical Services Department, JGI Inc., Japan

E-mail address: hasada@nagoya-u.jp (Y. Hasada).

Handbook of Geophysical Exploration: Seismic Exploration, Volume 40
ISSN 0950-1401, DOI: 10.1016/S0950-1401(10)04025-5

discrete frequency series within a limited bandwidth (Kumazawa et al., 2000). To utilize the ACROSS system, we must establish a series of analysis methods, by which the subsurface structure is estimated from the observed frequency-domain data. As the first step, we developed a method to extract "events" localized in a time domain from frequency-domain data; this is called Sompi event analysis (Hasada et al., 2001).

The data observed by ACROSS is a transfer function between the source input and the receiver output sampled at discrete frequencies in a limited bandwidth. It carries information along various paths between the transmitter and the receiver. In order to distinguish the information about each path from other paths, we attempt to translate a transfer function into a pulse sequence in the time domain, which is equivalent to an impulse response function if the frequency range of the transfer function is infinitely wide. The Sompi event analysis method has been developed to extract "events" localized in a time domain from a transfer function in the frequency domain, which is the basic analysis method of ACROSS (Hasada et al., 2001). Although the validity of Sompi event analysis has been confirmed through various numerical tests, especially for analysis of dispersive waves, there are some problems in its practical application. Here, we introduce a revised procedure, by means of weighted least-squares fitting, based on maximum likelihood estimation.

2. Theory

2.1. Concept of the model

In Sompi event analysis, we assume that the complex frequency sequence to be analyzed is a transfer function or Green's function between the source input and the receiver output sampled at discrete sequential frequencies. Through a kind of autoregressive modeling in the frequency domain, we obtain a set of "events" characterized by complex travel time and complex amplitude, in which the former expresses travel time and attenuation, whilst the latter expresses amplitude and phase angle.

2.2. Model of pulse sequence

Using ACROSS, we measure the transfer function at discrete sequential frequencies in a limited band. A transfer function, which indicates how much the sinusoidal wave of each frequency is amplified and delayed in phase during transmission through a linear system, is obtained by dividing the complex frequency spectrum of the output signal by that of the input signal. The inverse Fourier transform of the transfer function yields an impulse response function.

Assuming a certain medium with neither dispersion nor attenuation, we consider that the waveform $f(t)$ observed at a certain distance from the source is represented as a superposition of delayed pulses similar to the source time function $f_0(t)$,

$$f(t) = \sum_{m=1}^{M} A_m f_0(t - \tau_m), \tag{1}$$

where A_m and τ_m are the amplification factor and the delay time of the m-th wave arrival, respectively. In the frequency domain, the Fourier transform of $f(t)$ is described using the source spectrum $F_0(\omega)$ as

$$F(\omega) = \sum_{m=1}^{M} A_m \mathrm{e}^{-\mathrm{i}\omega\tau_m} F_0(\omega). \tag{2}$$

The transfer function between the source spectrum and the observation spectrum is obtained as

$$H(\omega) = \frac{F(\omega)}{F_0(\omega)} = \sum_{m=1}^{M} A_m \mathrm{e}^{-\mathrm{i}\omega\tau_m}, \tag{3}$$

which corresponds to the impulse response function

$$h(t) = \sum_{m=1}^{M} A_m \delta(t - \tau_m). \tag{4}$$

The transfer function in Eq. (3) contains the exponentials of imaginary numbers, whose real and imaginary parts are cosine and sine functions, respectively. Figure 1(a) and 1(b) show impulse response functions composed of one impulse and the corresponding transfer functions. The oscillation "frequency" in the complex transfer function is equal to the delay time in the impulse response function.

To consider attenuation and phase shift, it is useful to introduce complex values for both amplification factor and delay time. We define the complex amplitude as

$$\alpha_m = A_m \mathrm{e}^{\mathrm{i}\phi_m} \tag{5}$$

and the complex travel time as

$$q_m = \tau_m + \mathrm{i}\upsilon_m. \tag{6}$$

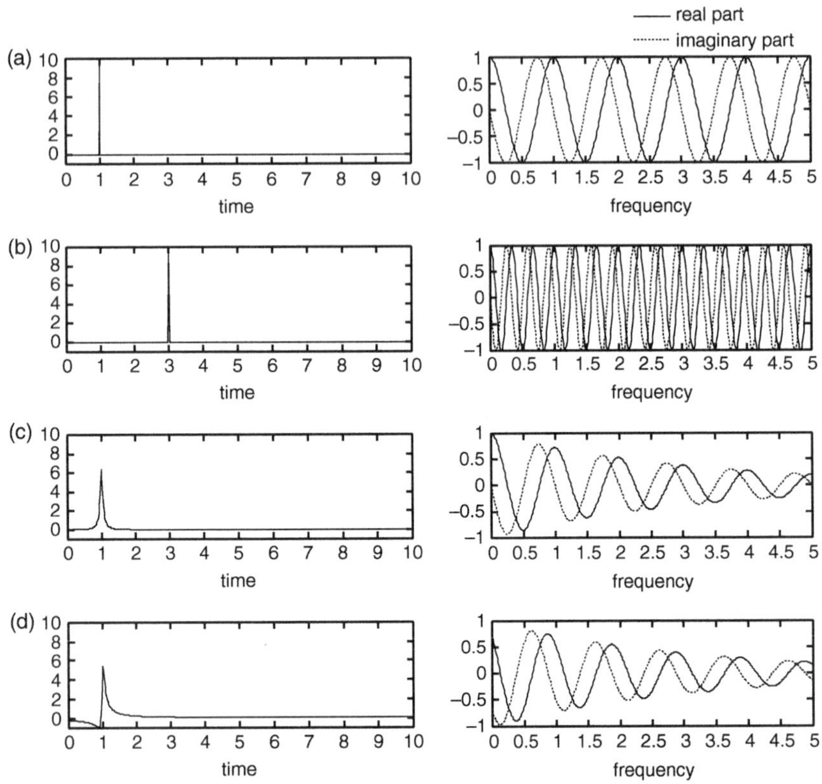

Figure 1 The impulse response functions (left) and the corresponding transfer functions in the frequency domain (right) for a nondispersive medium without attenuation (a and b) and with attenuation (c and d).

Then we have a model for transfer functions as

$$H(\omega) = \sum_{m=1}^{M} \alpha_m e^{-i\omega q_m}, \tag{7}$$

with the corresponding impulse response function being

$$h(t) = \frac{1}{\pi} \sum_{m=1}^{M} \Re\left[\frac{i\alpha_m}{t - q_m}\right], \tag{8}$$

where $\Re$ indicates the real part. Figure 1(c) and (d) show examples for Eq. (7) and (8). The transfer function defined by Eq. (7) is a superposition of decaying or growing sinusoids. Figure 2 shows the schematic image of our model containing plural decaying sinusoids. We refer to each decaying or growing sinusoid as a "wave element" representing a pulse in the time

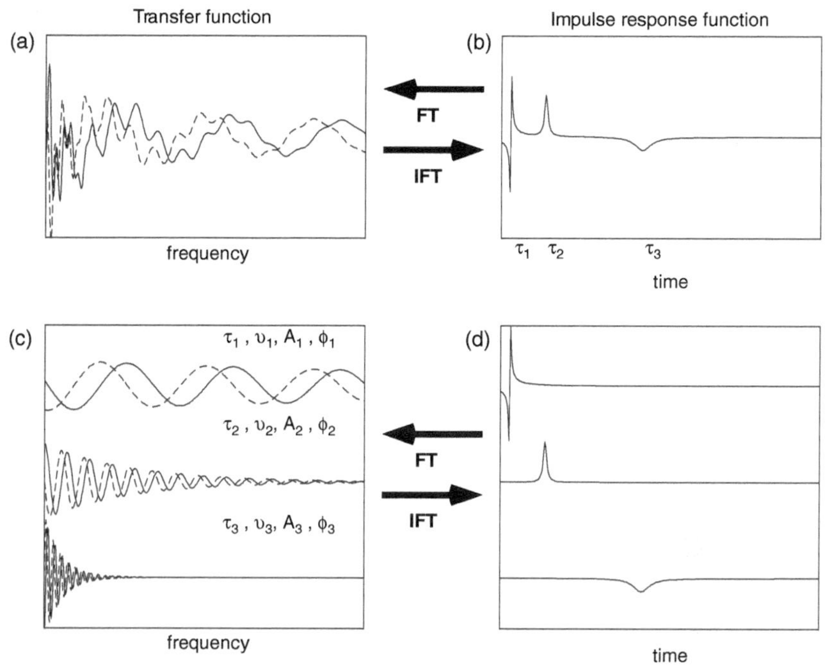

Figure 2 A schematic image of a model in Sompi Event Analysis: (a) an example of a transfer function composed of the three decaying sinusoids shown in (c); (b) the corresponding impulse response function. The three waves in (c) correspond to the three pulses in (d). The impulse response functions in (b) and (d) are calculated by a complex Lorentzian model and are not Fourier transforms of (a) and (c).

domain. From Eq. (7), the growth rate is given by the imaginary part of the complex travel time, υ_m, which is required to be negative for the energy not to diverge to infinity. The constraint for real $h(t)$ is $H(\omega) = H^*(-\omega)$ for negative ω, where $*$ denotes the complex conjugate.

Each pulse in Eq. (8) has the form of a linear combination of the Lorentz function and its Hilbert transform ($\pi/2$ radian phase shift); we refer to the content of the square bracket as a "complex Lorentzian" (Hasada et al., 2001). When α_m is a real number, or ϕ_m in Eq. (5) is zero, the pulse becomes a Lorentz function whose half-value width is $-2\upsilon_m$. This function has long tails on both sides of the peak, which seems to violate causality, since an attenuating medium necessarily involves dispersion, or frequency dependence. To satisfy causality and how to deal with dispersive transfer functions will be shown in the following section.

2.3. Modeling dispersive waves

Here, for convenience, we consider a homogeneous dispersive media. The transfer function is written using a frequency dependent wave number $k(\omega)$

and propagating distance x,

$$H(\omega) = \alpha \mathrm{e}^{-\mathrm{i}k(\omega)x}. \tag{9}$$

Expanding $k(\omega)$ into a Taylor series around ω_0 up to the first order as

$$k(\omega) = k(\omega_0) + \left.\frac{\partial k}{\partial \omega}\right|_{\omega=\omega_0} (\omega - \omega_0) = \frac{\omega_0}{c(\omega_0)} + \frac{\omega - \omega_0}{u(\omega_0)}, \tag{10}$$

we obtain the transfer function

$$H(\omega) = \alpha \mathrm{e}^{-\mathrm{i}\omega_0 \frac{x}{c(\omega_0)}} \mathrm{e}^{-\mathrm{i}(\omega-\omega_0)\frac{x}{u(\omega_0)}} = \tilde{\alpha} \mathrm{e}^{-\mathrm{i}(\omega-\omega_0)\tilde{q}}, \tag{11}$$

where $c(\omega_0)$ and $u(\omega_0)$ denote the phase velocity and group velocity at the frequency ω_0, respectively. Equation (11) is similar to Eq. (7), with the complex amplitude and the complex travel time defined as

$$\tilde{\alpha} = \alpha \mathrm{e}^{-\mathrm{i}\omega_0 \frac{x}{c(\omega_0)}}, \quad \tilde{q} = \frac{x}{u(\omega_0)}. \tag{12}$$

This indicates that the complex travel time obtained by our method is not related to the phase velocity, but to the group velocity.

Thus, our model, as defined by Eq. (7), is approximately adaptable for a dispersive wave if the frequency range of the data to be analyzed is narrow enough. It is effective in dividing the data into segments, with bandwidth suitable for the frequency dependency inherent in the data when we turn to analyzing highly dispersive waves.

2.4. Modeling observed data

We assume that the observed transfer function X_j consists of the signal sequence H_j and the noise sequence E_j.

$$X_j = H_j + E_j \quad (j = 1, \ldots, N), \tag{13}$$

where the real and the imaginary parts of E_j have an independent and identical normal distribution $N(0, \sigma_j^2)$, whose variance depends on the frequency. In the case of ACROSS observation, the errors of the measured transfer function can be evaluated by taking root mean squares of the data at the vicinal frequencies of the transmitted signals. The estimated error of the transfer function ε_j is assumed to be proportional to the noise standard deviation σ_j.

The signal sequence H_j is the superposition of decaying sinusoids, as stated above

$$H_j = \sum_{m=1}^{M} \alpha_m e^{-2\pi i(j-1)\Delta f q_m}. \tag{14}$$

The parameters to be determined are the complex travel times q_m and the complex amplitudes α_m. Because the order of model M is also unknown, we change M in an arbitrary range and attempt to find the optimum model.

2.5. Determination of the complex travel times

Hasada et al. (2001) proposed a procedure to determine complex travel times and complex amplitudes from observed transfer functions by means of autoregressive modeling.

The signal sequence composed of M decaying sinusoids should satisfy the following homogeneous autoregressive (AR) equation of order M.

$$\sum_{l=0}^{M} a_l H_{j-l} = 0 \quad (j = M+1, \ldots, N), \tag{15}$$

where $\boldsymbol{a} = \{a_0, \ldots, a_M\}$ is a complex AR coefficient vector. We determine $\boldsymbol{a}$ by minimizing the fitting error

$$\frac{1}{N-M} \sum_{j=M+1}^{N} \left| \sum_{l=0}^{M} a_l X_{j-l} \right|^2 \rightarrow \min, \tag{16}$$

under a constraint to obtain a non-zero $\boldsymbol{a}$,

$$\boldsymbol{a} \cdot \boldsymbol{a}^* = 1. \tag{17}$$

Minimization of Eq. (16) with the condition Eq. (17), using a Lagrangian undetermined multiplier λ, leads to an eigenvalue problem,

$$\mathbf{P}\boldsymbol{a} = \lambda \boldsymbol{a}. \tag{18}$$

Here $\mathbf{P}$ is a non-Toeplitz Hermitian autocovariance matrix

$$P_{ml} = \frac{1}{N-M} \sum_{j=M+1}^{N} X_{j-l} X_{j-m}^* = P_{lm}^* \quad (l, m = 0, \ldots, M). \tag{19}$$

Solving Eq. (18) yields M eigenvectors. We adopt the eigenvector $\boldsymbol{a}$ corresponding to the minimum eigenvalue, since λ reflects the noise power.

In order to convert the AR coefficient vector $\boldsymbol{a}$ to a set of the complex travel times, we should solve the characteristic equation

$$\sum_{l=0}^{M} a_l z^{-l} = 0, \tag{20}$$

where z is the unit-time-shift operator ($zH_j = H_{j+1}$). The M characteristic solutions $z_1, \ldots, z_M$ of Eq. (20) are related to the complex travel times by

$$z_m = \mathrm{e}^{-2\pi \mathrm{i} \Delta f q_m}, \tag{21}$$

which results in M complex travel times,

$$q_m = \frac{\mathrm{i} \ln z_m}{2\pi \Delta f} \quad (m = 1, \ldots, M). \tag{22}$$

2.6. Determination of the complex amplitudes

The complex amplitudes α_m are determined by least-squares fitting. If we have an estimated error of X_j, weighted least-squares fitting provides a more accurate estimation

$$\sum_{j=1}^{N} w_j \left| X_j - \sum_{m=1}^{M} \alpha_m \mathrm{e}^{-2\pi \mathrm{i}(j-1)\Delta f q_m} \right|^2 \rightarrow \min. \tag{23}$$

The weight w_j is defined in terms of a maximum likelihood procedure, as follows,

$$w_j = \frac{1}{\varepsilon_j^2} \Big/ \sum_{k=1}^{N} \frac{1}{\varepsilon_k^2}, \tag{24}$$

where ε_j is the estimated error of X_j.

2.7. Maximum likelihood approach using linearized model

Hasada et al. (2004a,b) has proposed a procedure to improve estimation by using a maximum likelihood approach through linearization of the model.

Assuming that the estimated complex travel times and complex amplitudes are the initial values of the model parameters,

$$H_j = H_j^0 + \sum_{m=1}^{M} \left(\delta\alpha_m - 2\pi \mathrm{i} \alpha_m^0 \delta q_m \right) \mathrm{e}^{-2\pi \mathrm{i}(j-1)\Delta f q_m^0}, \tag{25}$$

where H_j^0 is the transfer function calculated using the initial values, α_m^0 and q_m^0 are the initial values, and $\delta\alpha_m$ and δq_m are the small variations in parameters, respectively. These small variations are determined by linearized weighted least-squares fitting,

$$\sum_{j=1}^{N} w_j \left| X_j - H_j^0 - \sum_{m=1}^{M} \left(\delta\alpha_m - 2\pi \mathrm{i} \alpha_m^0 \delta q_m \right) \mathrm{e}^{-2\pi \mathrm{i}(j-1)\Delta f q_m^0} \right|^2 \rightarrow \min. \tag{26}$$

Therefore, we have the revised model parameters $q_m = q_m^0 - \delta q_m$ and $\alpha_m = \alpha_m^0 + \delta\alpha_m$.

However, the revised parameters do not always yield a larger likelihood with the original or unlinearized model, particularly when the residual of the original model is large. Therefore we suggest that the revised parameters be adopted only when the AIC (Akaike's Information Criterion) of the new model is less than the original one.

Adding this procedure to every stage of analysis where complex travel times and complex amplitudes have been determined enables a more accurate estimation.

2.8. Determination of the model order

In the original Sompi method for spectral analysis, the model order, or the number of wave elements, is determined by means of a two-parameter AIC proposed by Matsu'ura et al. (1990). One of the two parameters to specify the model is the AR order, the other is the number of wave elements. We describe each model as MODEL(M, M_0), where M is the AR order and M_0 is the number of wave elements, in which the largest M_0 wave elements are adopted from the M candidates in terms of the mean power densities (*MPD*), defined as

$$MPD_m = \frac{1}{N} \sum_{j=1}^{N} \left| \alpha_m \mathrm{e}^{-2\pi i(j-1)\Delta f q_m} \right|^2. \tag{27}$$

The complex amplitudes of MODEL (M, M_0) are recalculated by least-squares fitting for the M_0 wave elements.

The definition of the AIC is

$$AIC = -2l(\hat{\theta}) + 2k, \tag{28}$$

where $l(\hat{\theta})$ is the maximum logarithmic likelihood and k is the number of free parameters. The two-parameter AIC of MODEL (M, M_0) is given by

$$AIC(M, M_0) = 2N \ln 2\pi\hat{\sigma}_0^2 - \sum_{j=1}^{N} \ln w_j + 2N + 8M_0, \tag{29}$$

where $\hat{\sigma}_0^2$ is the maximum likelihood estimate of the noise variance per unit weight calculated by

$$\hat{\sigma}_0^2 = \frac{1}{2N} \sum_{j=1}^{N} w_j \left| X_j - \sum_{m=1}^{M_0} \alpha_m \mathrm{e}^{-2\pi \mathrm{i}(j-1)\Delta f q_m} \right|^2 . \tag{30}$$

We can determine the optimum MODEL (M, M_0) by searching the minimum AIC for a certain range of both M and M_0.

Using the AIC sometimes leads to a model with an improperly high order, in terms of practical application to analysis of observed data, because the parametric model is not fully suitable to a natural system. To avoid false solutions, we suggest screening of the solutions or wave elements by the growth/decay rates and the peak heights. The growth/decay rate of a wave element represents the pulse width in the time domain, it is expressed by the imaginary part of the complex travel time. In the analysis presented here, we adopt the wave elements whose imaginary parts are between $-1/(N\Delta f)$ and $1/(N\Delta f)$. The peak height can be calculated from the growth/decay rates and the complex amplitude as

$$\frac{|\alpha_m|}{N} \frac{1 - \mathrm{e}^{2\pi N \Delta f v_m}}{1 - \mathrm{e}^{2\pi \Delta f v_m}}, \tag{31}$$

where v_m denotes the imaginary part of the complex travel time. We suggest that the threshold of peak height be set at the noise level in the time domain, which can be calculated from errors in the frequency domain or estimated by the time-domain waveform. For both screening procedures reasonable thresholds have to be adjusted according to the purpose.

Figure 3 shows the flow of analysis, including the process by which to search for the optimum model order.

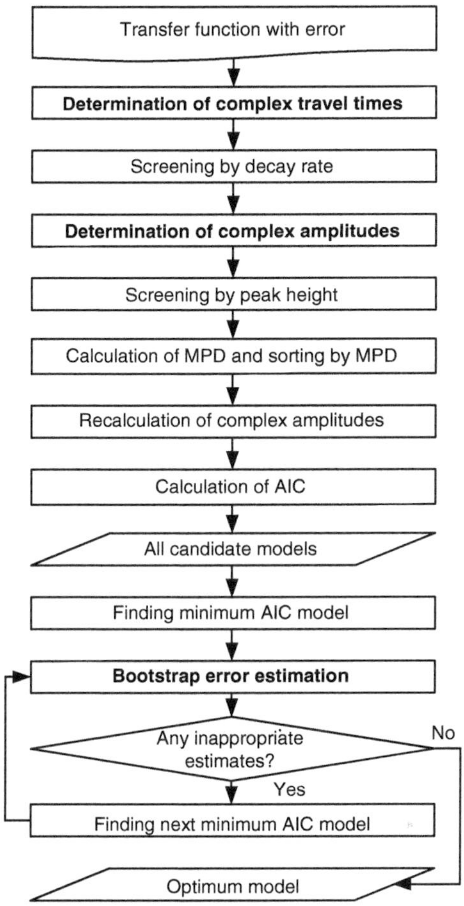

Figure 3 Schematic flow of analysis.

2.9. Error estimation by bootstrap method

A bootstrap method is generally used to evaluate estimation errors. In our analysis, the residuals have to be weighted before resampling if we deal with data whose uncertainties depend on frequency. To identify wave elements obtained from bootstrap replications, we adopt the closeness of the complex travel times. We calculate the distance of all combinations of wave elements from the original dataset and bootstrap replications in the complex plane. Then, in descending order of MPD for the original wave elements, we choose the closest wave element from the bootstrap replication. As with the ordinal bootstrap method, the estimated standard error is given by the standard deviation of all replicated estimates.

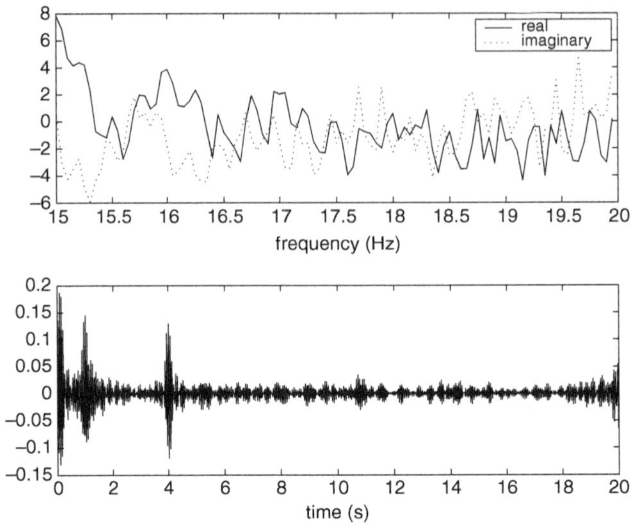

Figure 4 The synthetic transfer function (top) and time-domain waveform (bottom) obtained by Fourier transformation.

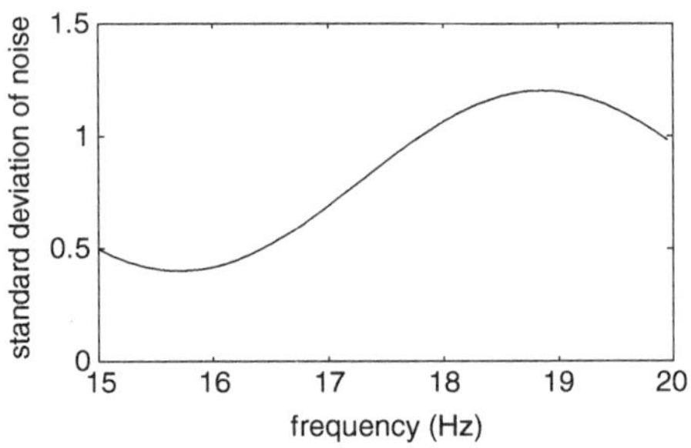

Figure 5 The frequency dependency of noise amplitude.

3. NUMERICAL EXPERIMENT

To demonstrate automatic Sompi event analysis, we analyze a synthetic transfer function consisting of three decaying sinusoids. We synthesize the transfer function by three complex decaying sinusoids and a frequency-dependent random noise sequence. The sampled frequency range is 15-19.95 Hz with an interval of 0.05 Hz. Figure 4 shows the synthetic data, composed of three wave elements at 0.1 s, 1.0 s, and 4.0 s. The frequency dependency of the noise sequence is shown in Figure 5.

We scanned the AR order from 1 to 20, and found the optimum model, MODEL(20, 5). The AIC of all the models are shown in Figure 6. In

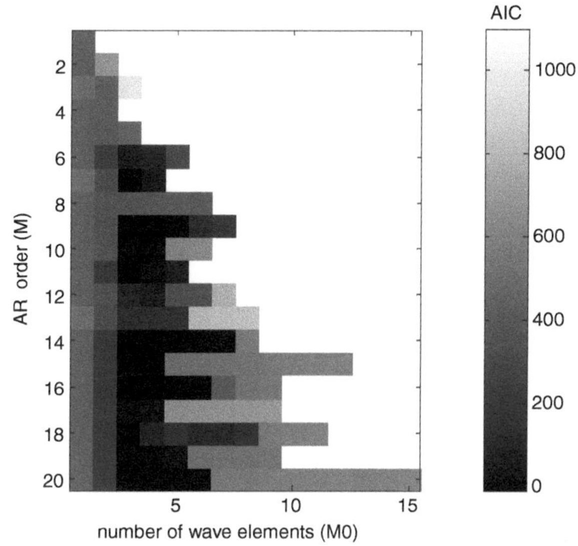

Figure 6 Two-parameter AIC of all models. The minimum AIC model is MODEL(20, 5).

Figure 7, we plot all the resultant wave elements in (a) the time domain and (b) the frequency domain, with Table 1 showing the estimated values of the complex travel times and the complex amplitudes in comparison with the input values. For the three largest wave elements, the estimated values recover the input values within the range of error. The two small redundant wave elements are considered to be created by noise. This result indicates that automatic processing using the Sompi event analysis method provides good estimates of the input complex decaying sinusoids, although it outputs some excessive wave elements, caused by noise.

4. Discussion and Conclusion

We have developed an automatic travel time determination by Sompi event analysis of transfer functions in the frequency domain. In Sompi event analysis, the observed transfer function is modeled as an autoregressive process and decomposed into wave elements. Given that an autoregressive model is based on the solution of a wave equation, the physical significance of the model is clear. However, extended modeling is necessary in cases of multiple reflections or resonance.

Model order determination using a two-parameter AIC has previously been difficult, because this method tends to estimate improperly high order models. To overcome this difficulty and realize automatic processing, we proposed two screening criteria; growth/decay rates and peak heights, which

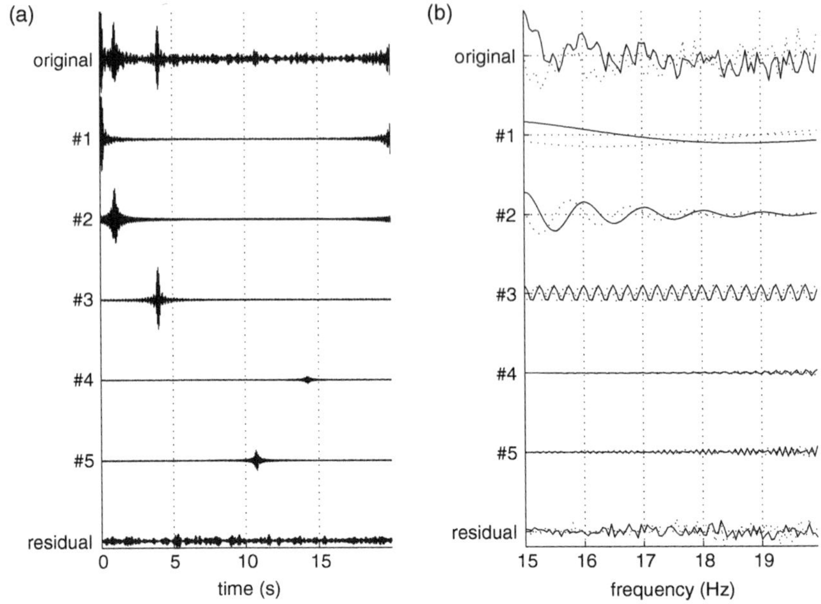

Figure 7 The decomposed wave elements and residual in the (a) time domain and (b) frequency domain, calculated with the parameters by Sompi event analysis. The time-domain waveforms are calculated by Fourier transformation. The largest three wave elements recover the input values well. The smallest two are not input wave elements, but are rather created by noise. The residual has a similar frequency dependency to input noise amplitude.

Table 1 Comparison of the estimated complex travel times and complex amplitudes with the input values. The errors are estimated by the bootstrap method. For the largest three wave elements, the estimated values recover the input values within the range of errors. The two smallest solutions are attributed to noises

	Estimate		Error	Input	
q_1	0.11	−0.02i	0.02	0.10	−0.01i
q_2	1.00	−0.09i	0.03	1.00	−0.10i
q_3	4.01	+0.00i	0.03	4.00	+0.004i
q_4	14.2	+0.1i	0.3		
q_5	10.8	+0.1i	0.1		
α_1	2.3	−1.1i	0.5	2.0	−1.0i
α_2	3.9	+0.7i	0.8	4.0	+1.0i
α_3	1.3	+0.1i	0.6	1.3	+0.0i
α_4	−0.01	−0.00i	0.07		
α_5	−0.03	+0.10i	0.1		

are confirmed, through numerical testing of synthetic data, to be effective in reducing redundant solutions.

Repeated least-squares fitting by means of the linearized model increases the estimation accuracy. For more direct model optimization, some nonlinear optimization techniques should be considered.

REFERENCES

Hasada, Y., Kumagai, H., Kumazawa, M., 2001. Autoregressive modeling of transfer functions in frequency domain to determine complex travel times. Earth Planets Space 53, 3–11.

Hasada, Y., Kumazawa, M., Tsuruga, K., Kunitomo, T., 2004a. Traveltime estimation from the data acquired in frequency domain by ACROSS: Sompi Event Analysis. In: Special issue "Active Monitoring". Chikyu Monthly 61–68 (in Japanese).

Hasada, Y., Kumazawa, M., Tsuruga, K., Kunitomo, T., 2004b. Travel time estimation from a transfer function in frequency domain: The revised Sompi event analysis. In: Fujii, N., Kasahara, J., Higashihara, H., and Ogawa, K. (Eds.), The Proceedings of 1st International Workshop on Active Monitoring in the Solid Earth Geophysics (IWAM04), Task Group for Active Monitoring, Mizunami, Japan. pp. 241–244.

Kumazawa, M., Kunitomo, T., Yokoyama, Y., Nakajima, T., Tsuruga, K., 2000. ACROSS: Theoretical and technical developments and prospect to future applications. Technical report, Japan Nuclear Cycle Development Institute, 9. pp. 115–129 (in Japanese).

Matsu'ura, T., Imanishi, Y., Imanari, M., Kumazawa, M., 1990. Application of a new method of high-resolution spectral analysis, Sompi, for free induction decay of nuclear magnetic resonance. Applied Spectroscopy 44, 618–626.

CHAPTER 21

Effect of Spatial Sampling on Time-lapse Seismic Monitoring in Random Heterogeneous Media

Jun Matsushima[a] *and* Osamu Nishizawa[b]

Contents

1. Introduction

Time-lapse seismic monitoring using active sources is a promising technology to monitor changes in dynamic physical properties as a function of time by analyzing differences between seismic data sets from different epochs. Rapid improvements in oil and gas field monitoring have been made during the last decade, with much attention being focused on time-lapse seismic surveys (e.g., Lumley, 2001). Time-lapse seismic methods have been widely applied to various geophysical surveys such as CO2 sequestration monitoring (Daley et al., 2007; Spetzler et al., 2008), volcanology studies

[a] Graduate School of Engineering, The University of Tokyo, 7-3-1 Hongo, Bunkyo-ku Tokyo 113-8656, Japan
[b] Geological Survey of Japan, AIST, Tsukuba Central 7, 1-1 Higashi, Tsukuba, Ibaraki 305-8567, Japan

E-mail address: jun-matsushima@frcer.t.u-tokyo.ac.jp (J. Matsushima).

Handbook of Geophysical Exploration: Seismic Exploration, Volume 40
ISSN 0950-1401, DOI: 10.1016/S0950-1401(10)04026-7

(Nishimura et al., 2005) and geothermal reservoir monitoring (Matsushima et al., 2004).

Time-lapse analysis assumes implicitly that seismic surveys are repeatable, that is, the seismic response changes in time due only to changes in dynamic physical properties from the subsurface. Although several successful case histories have been reported in the last decade several problems remain in time-lapse analysis, mostly related to approximating the hypotheses mentioned above (Vesnaver et al., 2003). Thus, differences between seismic data sets due to other factors such as variations in acquisition/processing parameters, near-surface conditions, and background noise, must be removed or equalized prior to differencing. The cross-equalization technique (Ross and Altan, 1997) minimizes differences from acquisition and processing changes which increase the repeatability to partially compensate for the imperfect situation.

In most reflection seismic explorations, it is implicitly assumed that the subsurface target heterogeneities are so large and strong that other background heterogeneities cause only small fluctuations in the signals from the target. In this case, a clear distinction can be made between target structures and the small-sale background heterogeneities. However, complicated waveforms appear if the small-scale heterogeneities are significantly strong and are of comparable size to the seismic wavelength. The understanding of seismic wave propagation in random heterogeneous media has been well advanced by many authors on the basis of theoretical studies (Sato and Fehler, 1998), numerical studies (Frankel and Clayton, 1986; Hoshiba, 2000), and experimental studies (Nishizawa et al., 1997; Sivaji et al., 2001).

Wave phenomena in heterogeneous media are important for seismic data processing, but have not been well recognized and investigated in the field of seismic exploration. Numerical studies by Gibson and Levander (1988) indicate that different types of scattered noise can have different effects on the appearance of the final processed section. Gibson and Levander (1990) showed the apparent layering in CMP sections of heterogeneous targets. Emmerich et al. (1993) also concluded that the highly detailed interpretation, which is popular in crustal reflection seismology, is less reliable than believed, as far as the internal structure of scattering zones and scatterer orientations are concerned. Only a few researchers are concerned with how scattered waves affect data processing in seismic reflection studies.

A primary concern of this article is to study the effect of spatial sampling on time-lapse seismic monitoring in random heterogeneous media. According to the Nyquist sampling criterion, it is redundant to deploy sampling intervals smaller than the Nyquist sampling interval. We consider the adoption of highly dense spatial sampling, with intervals smaller than the Nyquist interval, to demonstrate the importance of identical spatial sampling between the base and monitor survey, even when the interval of spatial sampling is smaller than the Nyquist interval. We generate 2-D finite-difference synthetic seismic data as input to this study. Our numerical models have a homogenous structure, upon which randomly distributed heterogeneities are imposed.

2. Proper Spatial Sampling Interval in Seismic Reflection

According to the Nyquist sampling theorem, sampling at two points per wavelength is the minimum requirement for sampling seismic data over the time and space domains; that is, the sampling interval in each domain must be equal to or above twice the highest frequency/wavenumber of the continuous seismic signal being discretized. The phenomenon that occurs as a result of undersampling is known as aliasing. Aliasing occurs when recorded seismic data violate the criterion expressed in Eq. (1).

$$\Delta x \leq \Delta x_N = \frac{v_{\min}}{2f_{\max} \cdot \sin\theta}, \tag{1}$$

where Δx is the spatial sampling interval, which should be equal to or smaller than the spatial Nyquist sampling intervals Δx_N, $v_{\min}$ is the minimum velocity, $f_{\max}$ is the maximum frequency, and θ is the dip angle of the incident plane-wave direction.

On the other hand, in the case of zero-offset, the spatial sample interval should be equal to or smaller than a quarter-wavelength (Grasmueck et al., 2005). Yilmaz (1987) described that aliasing occurs when recorded seismic data violate the criterion expressed in Eq. (2),

$$\Delta x \leq \Delta x_N = \frac{v_{\min}}{4f_{\max} \cdot \sin\theta}. \tag{2}$$

In the presence of structural dips or significant lateral velocity variations, adequate sampling becomes important for both vertical and lateral resolution. For the case of the maximum dip ($\theta = 90$), the spatial Nyquist sampling interval becomes a quarter-wavelength. Thus, quarter-wavelength spatial sampling is a minimum requirement for adequate recording. In practice, however, seismic data are often irregularly and/or sparsely sampled in the space domain because of limitations such as those resulting from difficult topography or a lack of resources. In many cases, proper sampling is outright impossible. In order to avoid aliasing, standard seismic imaging methods discard some of the high frequency components of recorded signals. Valuable image resolution will be lost through processing seismic data (Biondi, 2001). Once seismic data are recorded, it is difficult to suppress aliasing artifacts without resurveying at a finer spatial sampling (Spitz, 1991).

In the case of migration processing, there are three types of aliasing (Biondi, 2001), associated with data, operator, and image spacing. Data space aliasing is the aliasing described above. Operator aliasing, which is common in Kirchhoff migration algorithms, occurs when the migration operator summation trajectory is too steep for a given input seismic trace spacing and frequency content. Image space aliasing occurs when the spatial sampling of the image is too coarse to adequately represent the steeply dipping reflectors that the imaging operator attempts to build during the imaging process. Image space aliasing can be avoided simply by narrowing the image interval.

3. Numerical Simulations

We constructed a synthetic time-lapse seismic data set propagating through two-dimensional random heterogeneous media where a random velocity variation was superimposed on a homogeneous model. The random heterogeneity generates incoherent events by the scattering of waves that contain reflection events. The synthetic data were generated with a second-order finite difference scheme for the constant density two-dimensional acoustic wave equation. In the following, we describe the method for creating models and for calculating the wave field.

To minimize grid dispersion in finite difference modeling, the grid size was set to be about one eighteenth of the shortest wavelength, which was calculated from the minimum velocity of 3600 m/s, the maximum frequency of around 40 Hz ($f_{\max} = 40$), and a 5 m grid spacing. All edges of the finite-difference grid were set to be far from source/receiver locations so that unnecessary events would not disturb the synthetic data. Source/receivers were not located on the edge of the model, but within the model body. In this situation, scattered wave fields generated in the heterogeneous media above the source/receiver locations would be included in the synthetic data. However, this does not affect the conclusions of this article.

3.1. Random heterogeneous model

Random heterogeneous media are generally described by fluctuations of the wave velocity and density, superposed on a homogeneous background. Their properties are given by an autocorrelation function parameterized by the correlation lengths and the standard deviation of the fluctuation. Random media with spatial variations of seismic velocity were generated by the same method as described in Frankel and Clayton (1986). The outline of the scheme is as follows:

(1) Assign a velocity value $v(x, z)$ to each grid point using a random number generator.

(2) Fourier transform the velocity map into the wave number space.
(3) Apply the desired filter in the wavenumber domain.
(4) Fourier transform the filtered data back into the spatial domain.
(5) Normalize the velocities by their standard deviation, centered on the mean velocity.

In this paper, the applied filter (Fourier transform of autocorrelation function, which is equal to the power spectral density function) has a von Karman probability distribution described by Eq. (3):

$$P(k, a) = \frac{4\pi \beta a^2}{\left(1 + k^2 a^2\right)^{\nu+1}}, \tag{3}$$

where k is the wavenumber, β is the Hurst number that controls the components of small scale random heterogeneities, and a is the correlation distance indicating the characteristic heterogeneity size. The wavenumber k we use here is defined by Eq. (4):

$$k = \frac{2\pi}{\lambda}, \tag{4}$$

where λ is the wavelength. We use the above von Karman-type heterogeneous media with $\beta = 0.1$. Saito et al. (2003) described that the value $\beta = 0.1$ is nearly the same as the value for the power spectral density function of velocity fluctuation obtained from well-log data at depths shallower than 10 km (e.g., Shiomi et al., 1997; Goff and Holliger, 2000).

A homogeneous model and source/receiver arrangements are shown in Figure 1(a). Four different source-receiver intervals of width 10, 20, 40, and 80 m were employed; each requiring 201, 101, 51, and 26 sources and receivers, respectively. Random heterogeneities were generated and implemented in the homogeneous model, as shown in Figure 1(b), (the "base" survey). The velocity perturbations shown in Figure 1(b) were normalized to have a standard deviation 3% from the 3800 m/s average and a characteristic heterogeneity size of 50 m ($a = 50$ m). Figure 1(c) (the "monitor" survey) is the same as Figure 1(b) except within the region indicated by the arrow in the figure (the "changing region"), where the average velocity is 4000 m/s and the standard deviation is 3%. The changing region is a square with size of 60 × 60 m.

3.2. Wave field calculation

We employed a 2-D finite difference method for heterogeneous media with random spatial variations in seismic velocity. The source wavelet was the Ricker wavelet with a dominant frequency of 20 Hz. The

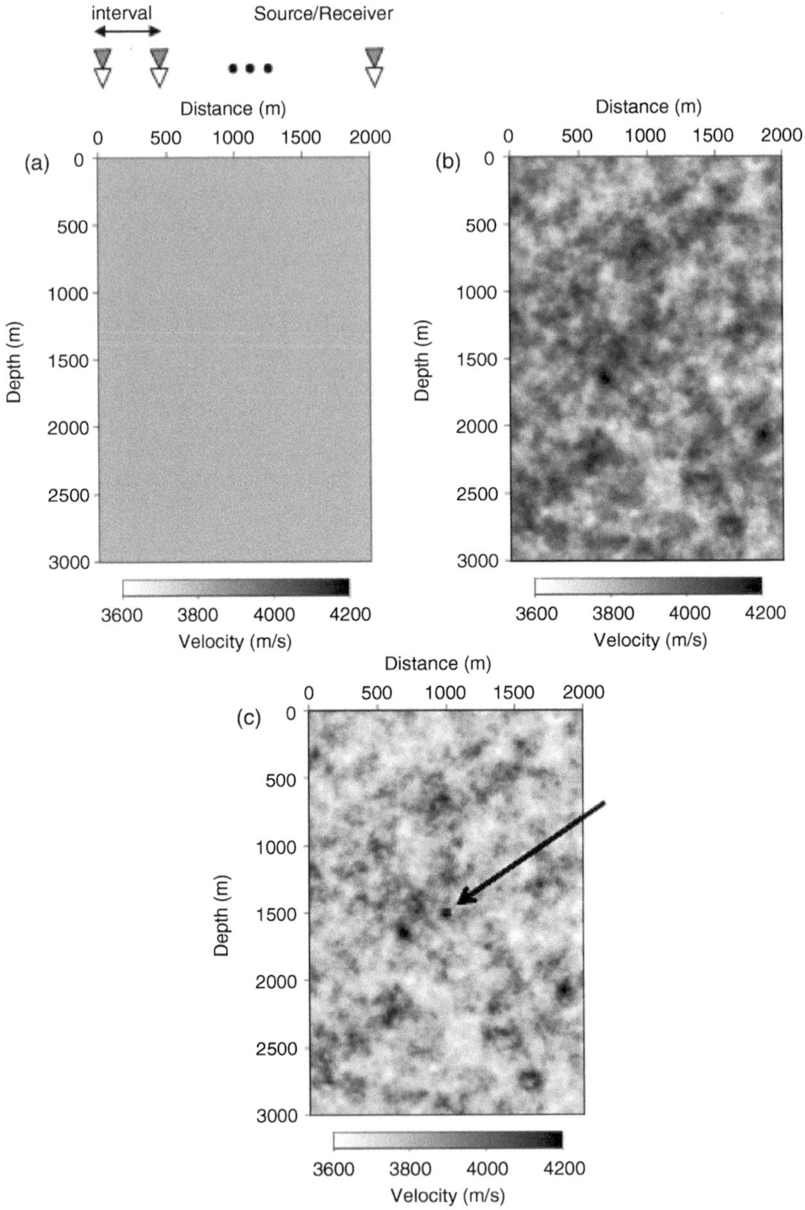

Figure 1 (a). A homogeneous model for numerical simulation examining specifications of data acquisition in time-lapse seismic surveys. (b) The base survey model for a two-dimensional acoustic wave simulation using the finite-difference method. The average velocity is 3800 m/s with a 3% standard deviation and a 50 m correlation distance. (c) The monitor survey model with the same average velocity and standard deviation as for (b), except for the region indicated by the arrow (changing region) within which the average velocity is 4000 m/s, with a 3% standard deviation.

dominant frequency (20 Hz) and the average velocity (3800 m/s) yielded the dominant wavelength (190 m). The product of the wavenumber k and the characteristic length a is used as an index for describing effects of random heterogeneity on seismic waves. In the present cases, the ka values at the dominant wavelengths are about 1.65. Aki and Richards (1980) classified scattering phenomena in terms of two dimensionless numbers ka and kL, where L is the travel distance between the source and receiver. Scattering effects are not so important for very small or very large ka, and these effects become increasingly important with increasing kL. According to this classification, our heterogeneous models are categorized as wave theories for random media where strong scattering may occur and full waveform modeling is required.

In order to remove direct wavelets, the total wave field calculated with the model shown in Figure 1(b), (c) was subtracted from the wave field in a model with a constant velocity of 3800 m/s (Figure 1(a)) to produce the wave field containing the scattered wave field. Figure 2(a) and (b) show an example of the shot gather from the scattered wave field in the case of the base (Figure 1(b)) and monitor surveys (Figure 1(c)) for a source-receiver interval of 10 m, respectively. The shot gathers are full of chaotic diffraction patterns originating from random heterogeneities. In order to examine the differences between Figure 2(a) and (b), the section in Figure 2(b) (monitor data) was subtracted from Figure 2(a) (base data) to produce the difference section shown in Figure 2(c). In Figure 2(c), we can clearly see the diffraction pattern caused by the changing region, and below this event see complicated seismic waves due to scattering generated from the small-scale heterogeneities. The fluctuation depends largely on the characteristic dimension of heterogeneity with respect to the wavelength of the seismic wave and on the intensity of scattering controlled by the impedance contrast (Sivaji et al., 2001).

The frequency-wavenumber (f-k) diagram is helpful for visualizing the sampling of a continuous wave field (Vermeer, 1999). The time window (from 0.65 to 1.05 s) including only scattered wave fields was extracted from each shot gather to calculate an f-k plot. Figure 3(a) through 3(d) show f-k plots of the extracted shot gather from the scattered wave field in the case of base survey (Figure 1(b)) for source-receiver intervals of 10, 20, 40 and 80 m, respectively. According to the spatial Nyquist sampling criterion defined in Eq. (1), Δx_N becomes 45 m ($f_{\max} = 40$, $v_{\min} = 3600$, $\theta = 90$). Thus, spatial sampling less than 45 m is sufficient to prevent spatial aliasing of the scattered wave field. In the case of the 80 m spatial sampling interval of Figure 3(d), the sector of strong amplitudes in the f-k plot would be severely truncated, causing wrap-around effects. On the other hand, in the case of zero-offset defined by Eq. (2), spatial sampling of less than 22.5 m is sufficient to prevent spatial aliasing.

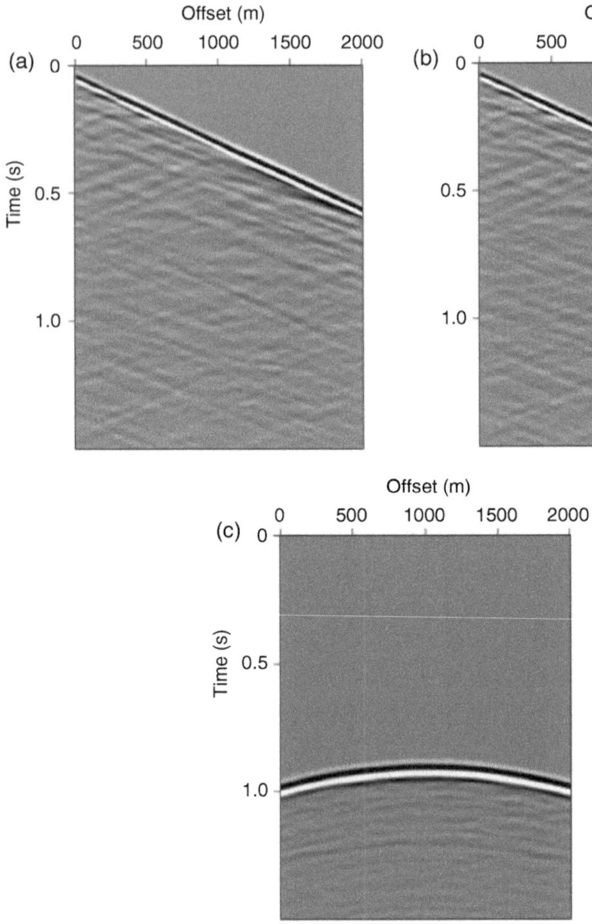

Figure 2 (a) An example of common-shot gather of a scattered wave field calculated for the base survey model (Figure 1(b)). (b) An example of common-shot gather of a scattered wave field calculated for the monitor survey model (Figure 1(c)). (c) Difference image of two common-shot gathers by simple subtraction of Figure 2(b) from Figure 2(a).

4. Results

4.1. Stacked and migrated sections

Two types of data processing, conventional CMP stacking and poststack migration were applied to the present model. For each process, a constant velocity of 3800 m/s was used to obtain the final sections.

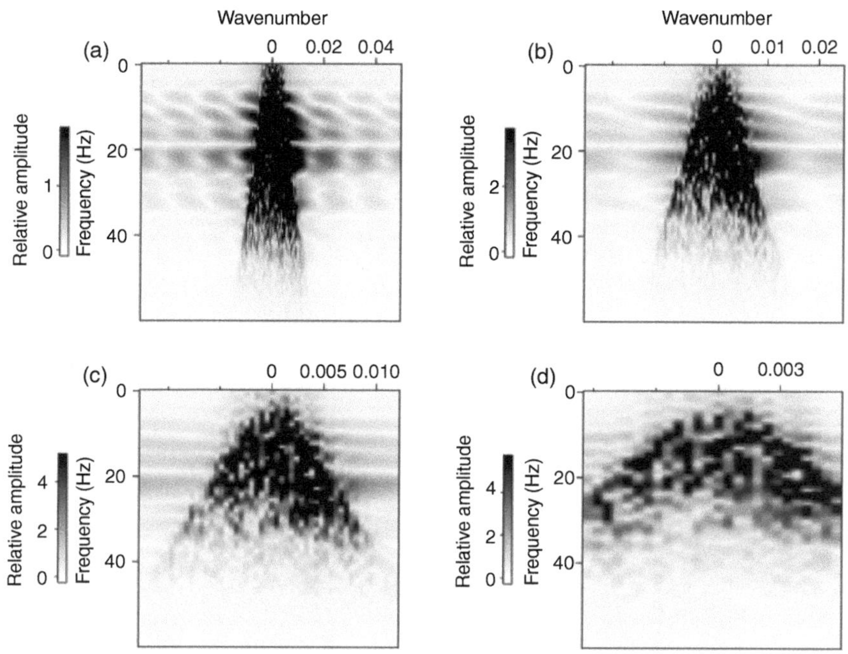

Figure 3 Frequency-wavenumber (f-k) plots of the extracted shot gather of a scattered wave field for spatial sampling intervals of (a) 10 m, (b) 20 m, (c) 40 m, and (d) 80 m.

4.1.1. CMP stacked sections

Figure 4(a) through 4(d) show CMP stacked sections for the monitor survey (Figure 1(c)) with different source/receiver intervals at 10, 20, 40, and 80 m, respectively. The CMP intervals of each model are 5, 10, 20, and 40 m, respectively. Although CMP stacking can act as a powerful mechanism for suppressing multiples and for the attenuation of many types of linear event noises, such as airwaves and ground roll, we can see no significant differences among Figure 4(a) through 4(d). This means that highly dense spatial sampling does not improve the final quality of a section, even when the interval of spatial sampling is smaller than the Nyquist interval. We can hardly identify the diffraction pattern caused by the changing region. Note that image space aliasing occurs in the case of a CMP interval of more than 22.5 m (Figure 4(d)). In each section, we can see many discontinuously subhorizontal and dipping events that do not correlate with velocity heterogeneities of the model. Gibson and Levander (1988) mentioned that the limited bandwidth of the propagating seismic signal, and spatial filtering attributable to CMP stacking, cause these events, bearing no simple relation to the velocity anomalies of the model.

In the same way, we produced CMP stacked sections for the base survey (Figure 1(b)) with different source/receiver intervals at 10, 20, 40, and 80 m,

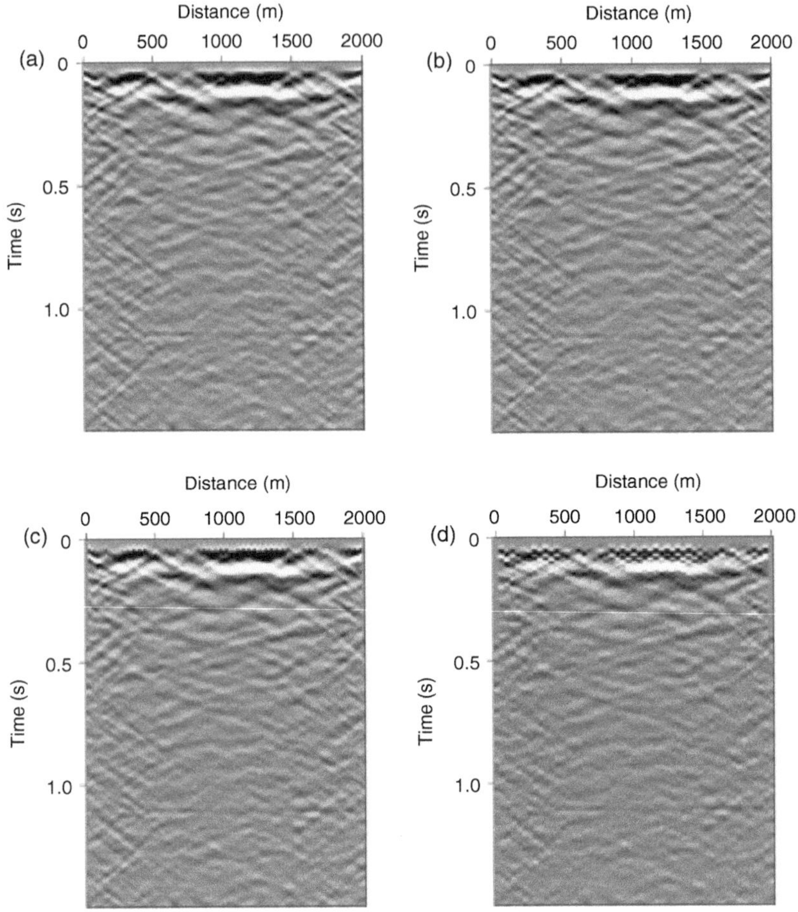

Figure 4 CMP stacked sections for the monitor survey data with different spatial sampling intervals: (a) 10 m, (b) 20 m, (c) 40 m, and (d) 80 m.

respectively. The macroscopic response is the same as for the case of the monitor survey.

4.1.2. Poststack migrated sections

Figure 5(a) through 5(d) show poststack migrated sections using f-k migration (Stolt, 1978) for the monitor survey (Figure 1(c)) with different source/receiver intervals at 10, 20, 40, and 80 m, respectively. We can see no significant differences among Figure 5(a) through 5(d). This means that highly dense spatial sampling does not improve the final quality of a section, even when the interval of spatial sampling is smaller than the Nyquist interval. Although its spatial resolution is improved compared

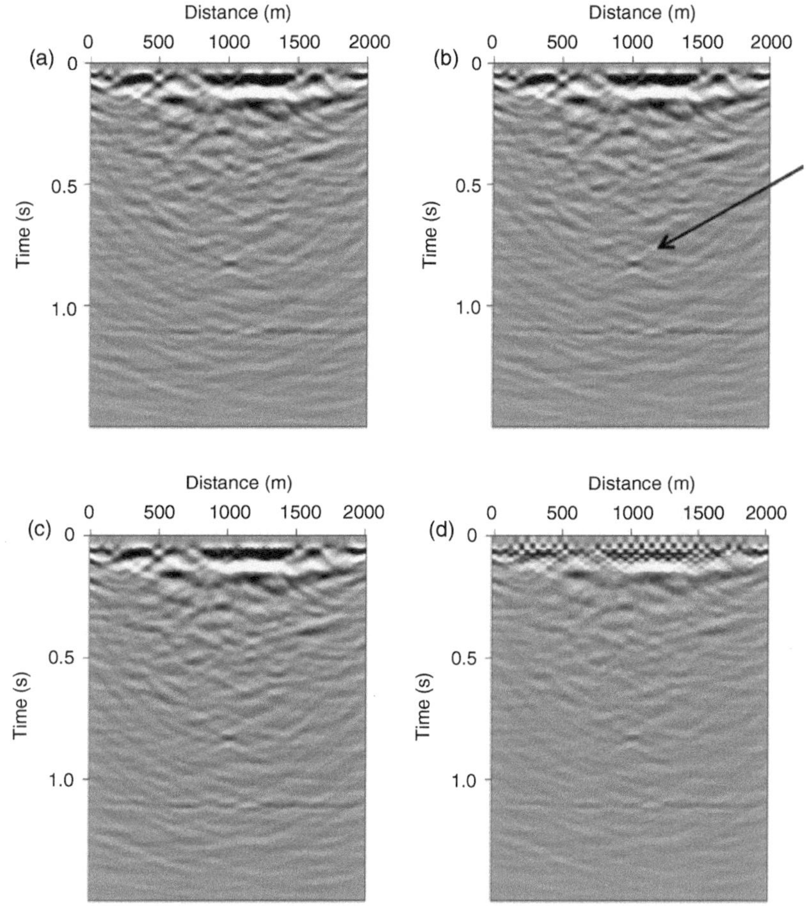

Figure 5 Poststack migrated sections for the monitor survey data with different spatial sampling intervals: (a) 10 m, (b) 20 m, (c) 40 m, and (d) 80 m.

to the CMP stacked image, each poststack migrated section is full of migration noises, producing the appearance that the section is heavily over-migrated, thus reducing the quality of the image. We can slightly identify the changing region indicated by the arrow. In general, migration can improve the lateral resolution by correcting the lateral mispositioning of dipping reflectors or collapsing the diffraction patterns caused by a point scatterer. However, the application of poststack migration here does not improve seismic images in heterogeneous media. It is thought that the reason is that multiple-scattering effects in small-scale heterogeneities do not satisfy the assumption of migration theory based on single scattering. Although migration techniques assume that the seismic data to be migrated consists only of primary reflections and diffractions, these wave fields are attenuated and distorted by heterogeneities and multiple scattered wave fields are generated, producing apparent discontinuities in reflectors or diffractors.

In the same way, we produced the poststack migrated sections for the base survey (Figure 1(b)) with different source/receiver intervals at 10, 20, 40, and 80 m, respectively. The macroscopic response is the same as for the case of the monitor survey except for changing region.

4.2. Difference sections

4.2.1. Differences between base and monitor sections

In order to interpret amplitude and time differences between base and monitor surveys, we apply a simple subtraction of the monitor seismic section from the base seismic section. The CMP monitor sections in Figure 4(a) through 4(d) were subtracted from those of the base survey to produce the difference sections with different source/receiver intervals, as shown in Figure 6(a) through 6(d). Note that the spatial sampling interval of the monitor survey is identical to that of the base survey. We can clearly identify the diffraction pattern due to the changing region, and also beneath this pattern a zone of complicated seismic waves due to seismic wave scattering generated from the small-scale heterogeneities. In Figure 6(a) through 6(d), the entire appearance of the event pattern in difference images is consistent across different sampling intervals, except for Figure 6(d). The section in Figure 6(d) suffers significantly from spatial aliasing effects, causing the different appearance of Figure 6(d) compared to the other difference images.

In the same way, the migrated sections of the monitor survey in Figure 5(a) through 5(d) were subtracted from those of the base survey to produce the difference sections with different source/receiver intervals, as shown in Figure 7(a) through 7(d). Note that the spatial sampling interval of the monitor survey is identical to that of the base survey. We can identify the changing region, and also beneath this region a zone of complicated seismic waves due to seismic wave scattering generated from the small-scale heterogeneities. In Figure 7(a) through 7(d), the entire appearance of the event pattern in difference images is consistent across different sampling intervals, except for Figure 7(d). The section in Figure 7(d) suffers significantly from spatial aliasing effects, causing the different appearance of Figure 7(d) compared to the other difference images.

From above results, it is concluded that although the changing region can be adequately extracted with identical source/receiver intervals between base and monitor surveys, highly dense spatial sampling does not improve the final quality of a difference section even when the interval of spatial sampling is smaller than the Nyquist interval. However, complicated seismic waves below the changing region are due to seismic wave scattering generated from the small-scale heterogeneities, which degrades seismic reflection data,

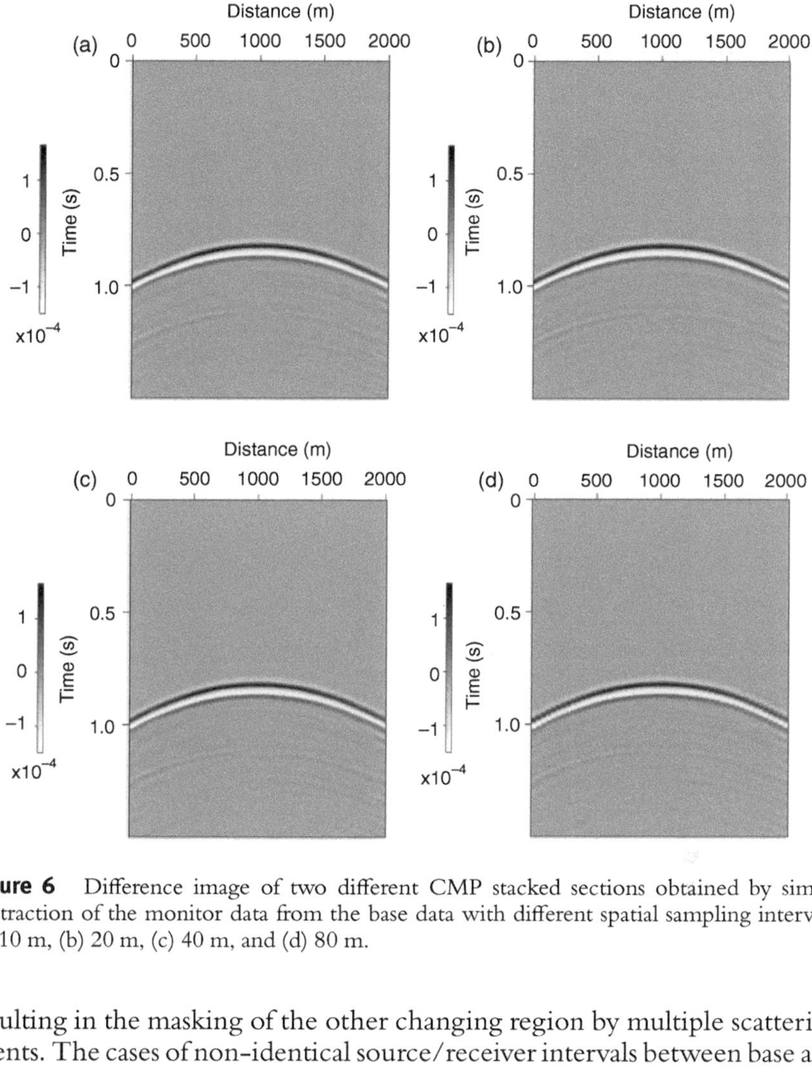

Figure 6 Difference image of two different CMP stacked sections obtained by simple subtraction of the monitor data from the base data with different spatial sampling intervals: (a) 10 m, (b) 20 m, (c) 40 m, and (d) 80 m.

resulting in the masking of the other changing region by multiple scattering events. The cases of non-identical source/receiver intervals between base and monitor surveys are described below.

The CMP section of the monitor survey with a source/receiver interval of 10 m (Figure 4(a)) was subtracted from CMP sections of the base survey with source/receiver intervals at 20, 40, and 80 m to produce the difference sections shown in Figure 8(a) through 8(c). During subtraction, the CMP interval of the monitor section was aligned to those of the base sections, and each section was normalized by its maximum value. In Figure 8(a) through 8(c), the difference images show that significant differences between base and monitor surveys with different sampling intervals are clearly visible; that is, we can clearly identify the diffraction pattern caused by the changing region in Figure 8(a), while we can hardly identify the pattern

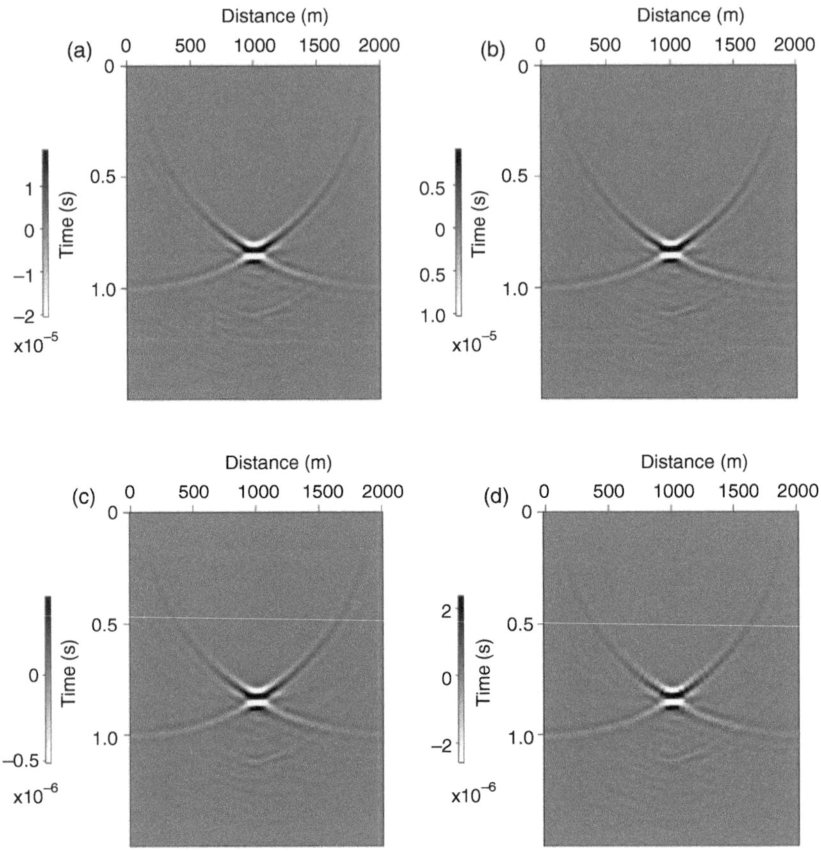

Figure 7 Difference image of two different migrated sections obtained by simple subtraction of the monitor data from the base data with different spatial sampling intervals: (a) 10 m, (b) 20 m, (c) 40 m, and (d) 80 m.

in Figure 8(b) and 8(c). We can see also that the difference sections in Figure 8(a) through 8(c) show numerous dipping events over the whole section. Small values of differences in CMP sections with different sampling intervals are evident as described in next section. The entire appearance of the event pattern of background in difference images is consistent across different sampling intervals. The amplitude of the background in difference images increases with increasing difference of sampling interval. Although small-scale heterogeneities have an influence on waveform with respect to the size of heterogeneities, the scattering effect is dominant at wavelengths comparable to the scale length of heterogeneity (e.g., Nishizawa et al., 1997). Thus, the frequencies of numerous small segments produced from scattered wave fields are attributed to the difference in the size of the heterogeneities.

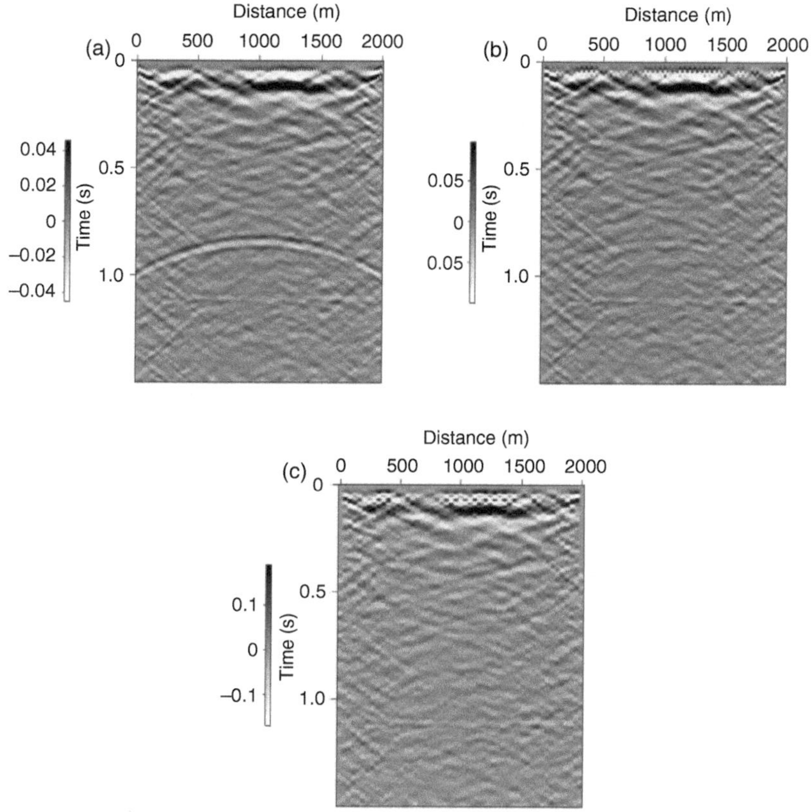

Figure 8 Difference image of two different CMP stacked sections obtained by simple subtraction of the monitor section with a source/receiver interval of 10 m from the base section and different source/receiver intervals: (a) 20 m, (b) 40 m, and (c) 80 m.

In the same way, the migrated section of the monitor survey with a source/receiver interval of 10 m (Figure 5(a)) was subtracted from migrated sections of the base survey with source/receiver intervals at 20, 40, and 80 m to produce the difference sections shown in Figure 9(a) through 9(c). During subtraction, the trace interval of the monitor section was aligned to those of the base sections. Difference images show that significant differences between base and monitor surveys with different sampling intervals are clearly visible; that is, we can clearly identify the diffraction pattern due to the changing region in Figure 9(a) and 9(b) while we can hardly identify the pattern in Figure 9(c). We can see also that the difference sections in Figure 9(a) through 9(c) show numerous migration noises over the whole section. The entire appearance of the event pattern of the background in the difference images is consistent across different sampling intervals. The

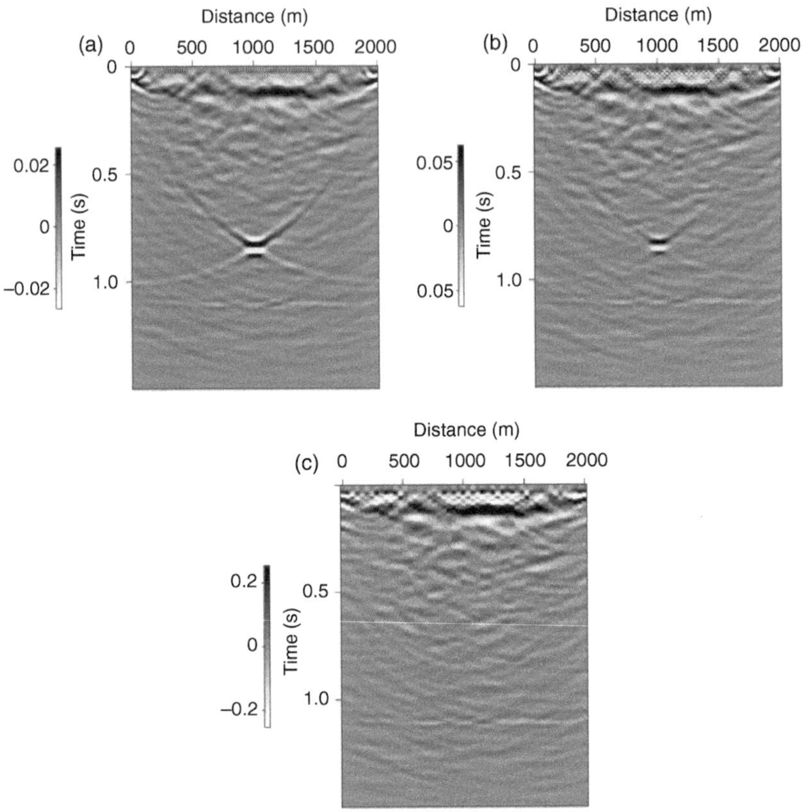

Figure 9 Difference image of two different migrated sections obtained by simple subtraction of the monitor section with a source/receiver interval of 10 m from the base section with different source/receiver intervals: (a) 20 m, (b) 40 m, and (c) 80 m.

amplitude of the background in difference images increases with increasing difference of sampling interval.

4.2.2. Interpretation of difference images

In order to interpret the difference images obtained in our numerical simulations above, we conducted some simple numerical experiments. Figure 10 shows synthetic seismic events consisting of Ricker wavelets with a central frequency of 20 Hz in the t-x plane. In Figure 10, two kinds of events (flat and dipping) exist, with six different spatial sampling intervals (5, 10, 20, 40, 60, 80 m). The dip of the dipping event is 0.3 ms/m. Here a flat event is assumed to be horizontally aligned by a NMO (Normal moveout) correction, while a dipping event is assumed to be an uncorrected reflection from a dipping reflector or diffraction. Figure 11 shows the transformation of the t-x domain shown in Figure 10 into the f-k plane, illustrating that

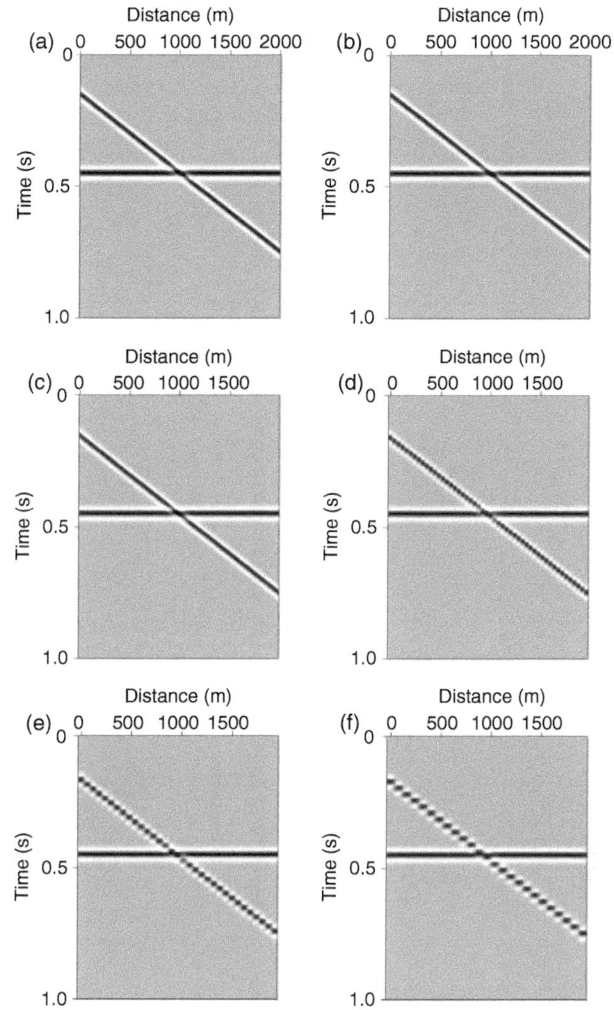

Figure 10 Synthetic seismic gathers consisting of two different events: a flat event (dip = 0) and a dipping event (dip = 0.3 ms/m), in the t-x domain with six different spatial sampling intervals: (a) 5 m, (b) 10 m, (c) 20 m, (d) 40 m, (e) 60 m, and (f) 80 m.

each event in t-x space corresponds to an event with a unique direction from the origin in f-k space. In the case of a spatial sampling interval of more than 40 m, we can clearly see spatial aliasing effects. Each panel of Figure 10 is stacked horizontally to produce a stacked trace (Figure 12). In each trace of Figure 12, we can see a large amplitude event (indicated by Event A) at around 0.5 s., which is generated from a flat event and other events (indicated by Events B and C, respectively) that are generated from a dipping event.

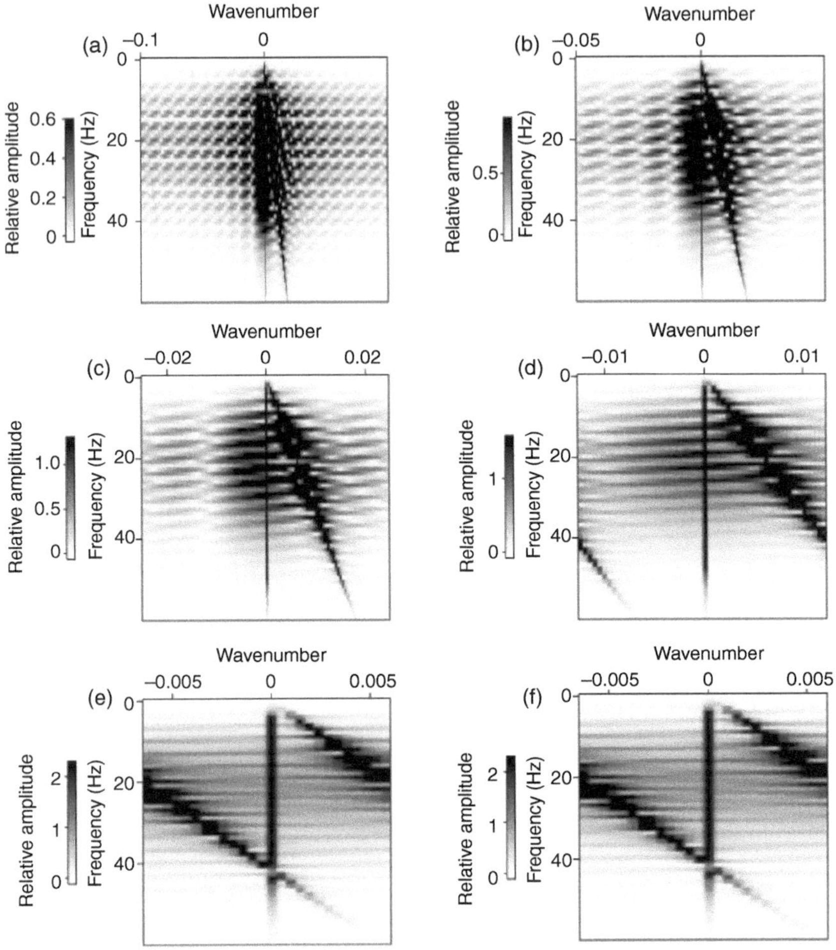

Figure 11 Frequency-wavenumber (f-k) plots of the synthetic gather shown in Figure 10(a) through 10(f).

From another viewpoint, each stacked trace generated from summing the gathers horizontally can be viewed as the interference of Ricker wavelets with different time lags (e.g., Matsushima et al., 1998). When the time lag is zero (the case of a flat event), the wavelets strengthen each other (Event A). When the time lag between wavelets is small, wavelets of a dipping event cancel each other out except at the ends of the dipping event (Event B). As the time lag between wavelets becomes much larger, the interference of the wavelets in a dipping event becomes weaker, leaving uncanceled events (Event C). The existence of Event C is due to data space aliasing, and the existence of Event B is due to a truncation artifact that is related to the finite number of traces. Furthermore, in order to examine the details of

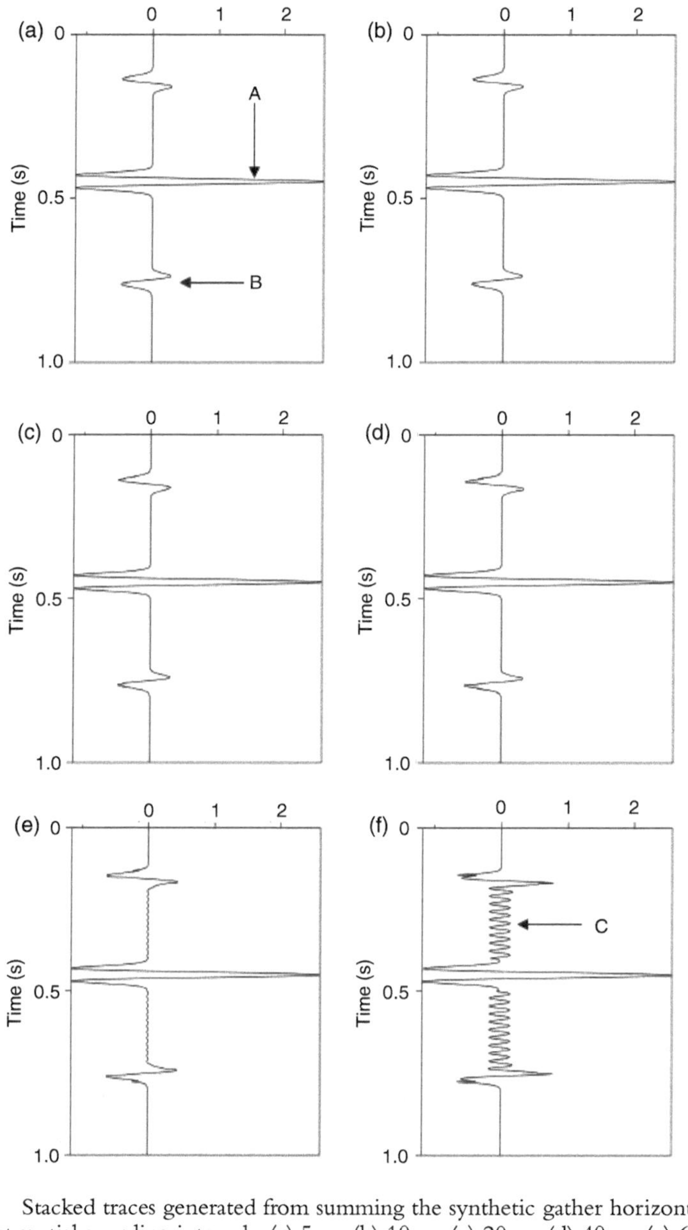

Figure 12 Stacked traces generated from summing the synthetic gather horizontally with six different spatial sampling intervals: (a) 5 m, (b) 10 m, (c) 20 m, (d) 40 m, (e) 60 m, and (f) 80 m.

the difference between traces, the section in Figure 12(a) was subtracted from those in Figure 12(b) through 12(f) to produce the difference sections shown in Figure 13(a) through 13(e). We can see that a large amplitude

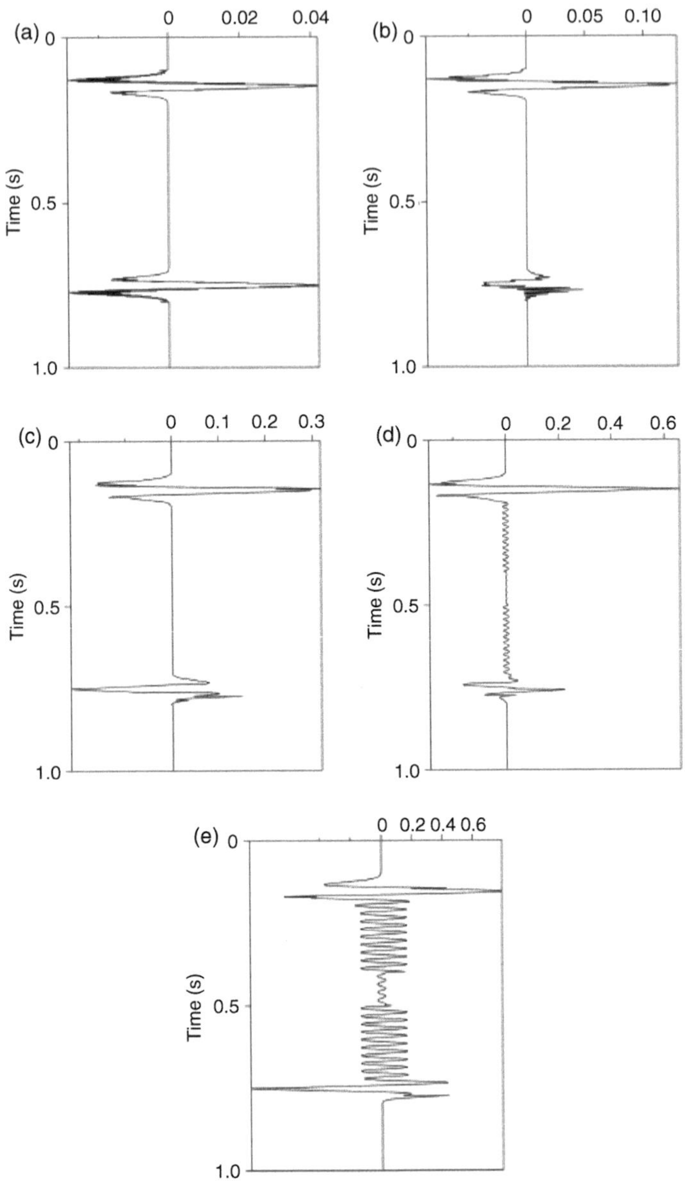

Figure 13 Difference of two different stacked traces obtained by simple subtraction of (a) Figure 12(a) from Figure 12(b), of (b) Figure 12(a) from Figure 12(c), of (c) Figure 12(a) from Figure 12(d), of (d) Figure 12(a) from Figure 12(e) and, of (e) Figure 12(a) from Figure 12(f).

event (Event A) at around 0.5 s. entirely disappeared, and that uncanceled events (Event B and C) remained. We emphasize that difference images of the truncation artifact are not due to seismic modeling or processing artifacts and cannot be removed, even if the spatial sampling is very dense, whereas the artifact due to data space aliasing can be resolved by satisfying the Nyquist sampling criterion.

5. Discussion

5.1. Independent time series noise and induced scattered wave noise

It is important to discriminate between two different types of noise: random noise in time series and a noise-like wave field produced from random heterogeneity. One may regard the scattered waves generated from heterogeneous media as source of random noise in field seismic data. Some authors (e.g., Matsushima et al., 2003) have added random noise to their synthetic data for simulating field seismic data. However, the noise is a consequence of the wave phenomena in heterogeneous media, and is not the same as the noise that randomly appears in the time-series (Levander and Gibson, 1991). Scales and Snieder (1998) concluded that the noise in a seismic wave is not merely a time-series which is independent from the original seismic wave, but is also a signal-induced wave mostly consisting of scattered waves. This is important for seismic data processing but not well recognized in the field of seismic exploration. Thus, this type of random noise is canceled out when CMP stacking is applied. On the other hand, scattering waves generated from random media are now recognized as a mutually dependent noise among the seismic traces, which indicates the interaction between the short-wavelength heterogeneity and the source and reflected wavelet. Although these scattered waves appear as random noise, they are thought to be an accumulation of many scattered waves that in themselves are partially coherent. Thus, this type of scattering noise should be categorized into coherent noise if we classify noise types. In general, coherent noise cannot be reduced merely by increasing the source strength or shortening the sampling interval (higher-fold CMP stacking) after processing the data.

5.2. Spatial sampling interval and the truncation artifact

In random heterogeneous media, two types of data processing, conventional CMP stacking and poststack migration were applied and compared to examine different responses to different sampling intervals. Each data process without data space aliasing achieves very similar final sections for different sampling intervals. Safar (1985) studied the effects of spatial sampling on the lateral resolution of a surface seismic reflection survey when carrying out scatterer point imaging by applying migration, and found almost no effect of spatial sampling on lateral resolution. Safar (1985) also demonstrated the generation of migration noise caused by a large sampling interval. Migration noise is a consequence of spatial aliasing that is related to the frequency, velocity, and dip of a seismic event. A shorter sampling interval cannot improve the spatial resolution very much, even if there is no noise. The same conclusion was obtained by Vermeer (1999). The results we have described above correlate well with those of these previous studies.

However, we found the existence of a small but significant difference between two images generated with adequate sampling. This is due to the truncation artifact that is related to a finite number of traces. Such a truncation error cannot be practically prevented, whereas the discretization error due to coarse spatial sampling can be removed by satisfying the Nyquist sampling criterion. The truncation artifact is dependent on the characteristic heterogeneity size; that is, the frequencies become lower with an increase in the characteristic size of heterogeneity. This is because the scattering effect is dominant at wavelengths comparable to the scale length of heterogeneity. Theoretically, spatial sampling should be sufficiently small to satisfy the Nyquist sampling criterion. However, the present numerical investigations show how the existence of a small but significant difference between two sections affects the final difference images to identify the changing region between base and monitor surveys.

6. Conclusions

We have shown from the viewpoint of spatial sampling how scattered wave fields produced from random isotropic heterogeneity influence the time-lapse seismic monitoring. We used a 2-D finite difference method for numerically modeling acoustic wave propagation. In our numerical experiments, two types of data processing with four different spatial sampling intervals, were applied and compared, which led to the following conclusions:

Highly dense spatial sampling in random heterogeneous media does not seem to improve the final quality of a difference image to identify the changing region between base and monitor surveys. This is because scattering waves generated from random media are not random noises in time series and a noise-like signal-induced coherent noise.

A small but significant difference between two seismic sections generated with different dense samplings exists. This difference, which is attributed to the truncation artifact, causes numerous complicated events over the whole section in heterogeneous media if the spatial sampling interval of base survey is not identical to that of monitor survey. This difference occurs even when the interval of spatial sampling is smaller than the Nyquist interval.

REFERENCES

Aki, K., Richards, P.G., 1980. Quantitative Seismology. W. H. Freeman and Company, San Francisco.

Biondi, B., 2001. Kirchhoff imaging beyond imageing. Geophysics 66, 654–666.

Daley, T.M., Myer, L.R., Peterson, J.E., Majer, E.L., Hoversten, G.M., 2007. Time-lapse crosswell seismic and VSP monitoring of injected CO2 in a brine aquifer. Environmental Geology. doi:10.1007/s00254-007-0943-z.

Emmerich, H., Zwielich, J., Muller, G., 1993. Migration of synthetic seismograms for crustal structures with random heterogeneities. Geophysical Journal of International 113, 225–238.

Frankel, A., Clayton, R., 1986. Finite difference simulations of seismic scattering: Implications for the propagation of short-period seismic waves in the crust and models of crustal heterogeneity. Journal of Geophysical Research 91, 6465–6489.

Gibson, S.B., Levander, R.A., 1988. Modeling and processing of scattered waves in seismic reflection surveys. Geophysics 53, 466–478.

Gibson, S.B., Levander, R.A., 1990. Apparent layering in common-midpoint stacked images of two-dimensionally heterogeneous targets. Geophysics 55, 1466–1477.

Goff, J.A., Holliger, K., 2000. Nature and origin of upper crustal seismic velocity fluctuations and associated scaling properties: Combined stochastic analyses of KTB velocity and lithology logs. Journal of Geophysical Research 104, 13169–13182.

Grasmueck, M., Wegerl, R., Horstmeyer, H., 2005. Full-resolution 3D GPR imaging. Geophysics 70, K12-K19.

Hoshiba, M., 2000. Large fluctuation of wave amplitude produced by small fluctuation of velocity structure. Physics of The Earth and Planetary Interiors 120, 201–217.

Levander, R.A., Gibson, S.B., 1991. Wide-angle seismic reflections from two-dimensional random target zones. Journal of Geophysical Research 96, 10251–10260.

Lumley, D., 2001. Time-lapse seismic reservoir monitoring. Geophysics 66, 50–53.

Matsushima, J., Rokugawa, S., Yokota, T., Miyazaki, T., Kato, Y., 1998. On the relation between the stacking process and the resolution of a stacked section in a crosswell seismic survey. Exploration Geophysics 29, 499–505.

Matsushima, J., Okubo, Y., Rokugawa, S., Yokota, T., Tanaka, K., Tsuchiya, T., Narita, N., 2003. Seismic reflector imaging by prestack time migration in the Kakkonda geothermal field, Japan. Geothermics 32, 79–99.

Matsushima, J., Yokota, T., Okubo, Y., Rokugawa, S., Tanaka, K., Tsuchiya, T., Narita, N., Tani, K., 2004. Repeated seismic reflection measurements in the Kakkonda geothermal field. Journal of Volcanology and Geothermal Research 129, 343–356.

Nishimura, T., Tanaka, S., Yamawaki, T., Yamamoto, H., Sano, T., Sato, M., Nakahara, H., Uchida, N., Hori, S., Sato, H., 2005. Temporal changes in seismic velocity of the crust around Iwate volcano, Japan, as inferred from analyses of repeated active seismic experiment data from 1998 to 2003. Earth Planets Space 57, 491–505.

Nishizawa, O., Sato, T., Lei, X., Kuwahara, Y., 1997. Laboratory studies of seismic wave propagation in inhomogeneous media using a laser Doppler vibrometer. Bulletin of the Seismological Society of America 87, 809–823.

Ross, C.P., Altan, M.S., 1997. Time-lapse seismic monitoring: Some shortcomings in nonuniform processing. The Leading Edge 16, 931–937.

Safar, M.H., 1985. On the lateral resolution achieved by Kirchhoff migration. Geophysics 50, 1091–1099.

Saito, T., Sato, H., Fehler, M., Ohtake, M., 2003. Simulating the envelope of scalar waves in 2D random media having power-law spectra of velocity fluctuation. Bulletin of the Seismological Society of America 93, 240–252.

Sato, H., Fehler, M., 1998. Seismic wave propagation and scattering in the Heterogeneous Earth. Springer-Verlag, New York.

Scales, J.A., Snieder, R., 1998. What is noise ? Geophysics 63, 1122–1124.
Shiomi, K., Sato, H., Ohtake, M., 1997. Broad-band power-law spectra of well-log data in Japan. Geophysical Journal International 130, 57–64.
Sivaji, C., Nishizawa, O., Fukushima, Y., 2001. Relationship between fluctuations of arrival time and energy of seismic waves and scale length of heterogeneity: an inference from experimental study. Bulletin of the Seismological Society of America 91, 292–303.
Spetzler, J., Xue, Z., Saito, H., Nishizawa, O., 2008. Time-lapse seismic crosswell monitoring of CO2 injected in an onshore sandstone aquifer. Geophysical Journal International 172, 214–225.
Spitz, S., 1991. Seismic trace interpolation in the F-X domain. Geophysics 56, 785–794.
Stolt, R., 1978. Migration by Fourier transform. Geophysics 43, 23–48.
Vermeer, G.J.O., 1999. Factors affecting spatial resolution. Geophysics 64, 942–953.
Vesnaver, A.L., Accaino, F., Bohm, G., Madrussani, G., Pajchel, J., Rossi, G., Moro, G.D., 2003. Time-lapse tomography. Geophysics 68, 815–823.
Yilmaz, O., 1987. Seismic Data Processing, Society of Exploration Geophysicists, Tulsa, OK.

CHAPTER 22

Singular Value Decomposition in Active Monitoring Data Analysis

Liao Chengwang[a]

Contents

1. Introduction

The Accurately Controlled Routinely Operated Signal System (ACROSS) intends to monitor very small changes of physical states in the focal region of coming great earthquakes to be predicted through transmitting accurately controlled seismic waves (Kumazawa et al., 2000). What ACROSS observed is a precise transfer function in the frequency domain between the source and the receiver. In a limit band of frequency, the transfer function is modeled as superimposed exponentially damped sinusoids, where each exponentially damped sinusoid indicates a wave component traveling from source to receiver and carries information on the travel time and the attenuation factor of the structure (Hasada et al., 2001). As the first step, the task of data analysis is to retrieve these travel times and attenuation factors from the observed transfer function data. It is essentially a problem of estimating the parameters of exponentially damped sinusoids and is also paid attention to in other areas (Kumaresan and Tufts, 1982; Papadopulos and Nikias, 1991; Kay, 1988; Hua and Sarkar, 1990; Bresler and Makovski, 1986). There are several kinds of method presented to solve

[a] Institute of Seismology, China Earthquake Administration, 40 Hongshan sideways Ave. Wuhan 430071, China

Handbook of Geophysical Exploration: Seismic Exploration, Volume 40

ISSN 0950-1401, DOI: 10.1016/S0950-1401(10)04027-9

such a problem, such as the Maximum-Likelihood method (Parthasarathy and Tufts, 1985), the Pisarenko method (Pisarenko, 1972), the KT method (Tufts and Kumaresan, 1982), the extended Prony method (Kay and Marple, 1981), the ESPRIT method (Roy and Kailath, 1989) and the Subspace fitting method (Viberg and Ottersten, 1991; vander Veen et al., 1993). The method used by (Hasada et al., 2001) is in fact an extended Prony method, which has better performance under a larger dataset and higher than real order. In the case of ACROSS, the transfer function can only be taken as superimposed exponentially damped sinusoids in a limit band of frequency for dispersion. A smaller dataset may lower the performance and a higher order causes the problem of order determination. After examining the above methods, we find that the Subspace fitting method is a better one. It works not only in the case of small dataset but also under a real number of poles. The method is based on singular value decomposition (SVD). So in the next part of this paper, the concept of SVD is introduced and then the algorithm is described. The remainder of the paper tries to validate the algorithm using numerical and field test data.

2. SVD

Singular Value Decomposition (SVD) is a widely used technique to decompose a matrix into several component matrices, exposing many of the useful and interesting properties of the original matrix. Using SVD, we can determine the rank of the matrix, quantify the sensitivity of a linear system to numerical error, or obtain an optimal lower-rank approximation to the matrix.

Let $\boldsymbol{A}$ be a general complex $\boldsymbol{m}$-by-$\boldsymbol{n}$ matrix. The SVD of $\boldsymbol{A}$ is the factorization $\boldsymbol{A} = \boldsymbol{U\Sigma V}^H$, where $\boldsymbol{U}$ and $\boldsymbol{V}$ are unitary, and $\sum = diag(\sigma_1, \ldots, \sigma_r), r = \min(m, n)$, with $\sigma_1 \geq \ldots \geq \sigma_r \geq 0$. The σ_i are called the singular values, the first $\boldsymbol{r}$ columns of $\boldsymbol{V}$ the right singular vectors and the first $\boldsymbol{r}$ columns of $\boldsymbol{U}$ the left singular vectors. If the set rank of A is r, the SVD of $\boldsymbol{A}$ can be simplified as $\boldsymbol{A} = U_r \sum_r V_r^H$, where $U_r$is a m-by-$\boldsymbol{r}$ matrix with the first $\boldsymbol{r}$ columns of $\boldsymbol{U}$ and V_r a n-by-$\boldsymbol{r}$ matrix with the first $\boldsymbol{r}$ columns of $\boldsymbol{V}$ and $\sum$ a r-by-$\boldsymbol{r}$ diagonal matrix with the first r singular values as its elements. If r is substituted by k $(k < r)$, we get a new matrix $\boldsymbol{A}_k$ as $\boldsymbol{A}_k = U_k \sum_k V_k^H$, then $\boldsymbol{A}_k$ is an optimal lower-rank approximation to A and such a SVD is called the truncated SVD (TSVD).

3. Algorithm

As stated above, the transfer function can be modeled as superimposed exponentially damped sinusoids in a limit band of frequency. In the discrete format, the model is generalized as

$$x(n) = \sum_{l=1}^{d} A_l e^{(\alpha_l + i\beta_l)n} = \sum_{l=1}^{d} A_l p_l^n \tag{1}$$

$$y(n) = x(n) + e(n) \quad n = 0, \ldots, N-1 \tag{2}$$

where

$$\alpha_l = \nu_l 2\pi \Delta f \tag{3}$$

$$\beta_l = -\tau_l 2\pi \Delta f \tag{4}$$

$$\tau_l = \frac{x_l}{c} \tag{5}$$

$$\nu_l = -\frac{x_l}{2cQ_l} \tag{6}$$

$$p_l = e^{\alpha_l + i\beta_l} \quad l = 1, \ldots, d \tag{7}$$

x_l is the travel distance of the lth wave component, c is the phase velocity, d is the order and $e(n)$ is the noise. For convenience,we callp_l a pole here.

Firstly, we don't take noise into consideration. Suppose we have obtained N discrete data, denoted as $x(0), x(1), \ldots, x(N-1)$, we construct the $(L+1)$-by-M Hankel matrix as:

$$H = \begin{bmatrix} x(0) & x(1) & \cdots & x(M-1) \\ x(1) & x(2) & \cdots & x(M) \\ \vdots & \vdots & \cdots & \vdots \\ x(L) & x(L+1) & \cdots & x(L+M-1) \end{bmatrix}. \tag{8}$$

We have

$$\begin{aligned} H &= \begin{bmatrix} 1 & \cdots & 1 \\ p_1 & \cdots & p_d \\ p_1^2 & \cdots & p_d^2 \\ \vdots & \cdots & \vdots \\ p_1^L & \cdots & p_d^L \end{bmatrix} \begin{bmatrix} A_1 & & & \\ & A_2 & & \\ & & \ddots & \\ & & & A_d \end{bmatrix} \begin{bmatrix} 1 & p_1 & p_1^2 & \cdots & p_1^{M-1} \\ \vdots & \vdots & \vdots & \cdots & \vdots \\ 1 & p_d & p_d^2 & \cdots & p_d^{M-1} \end{bmatrix} \\ &= AS \end{aligned} \tag{9}$$

where

$$A = \begin{bmatrix} 1 & \cdots & 1 \\ p_1 & \cdots & p_d \\ p_1^2 & \cdots & p_d^2 \\ \vdots & \cdots & \vdots \\ p_1^L & \cdots & p_d^L \end{bmatrix} \tag{10}$$

$$S = \begin{bmatrix} A_1 & & & \\ & A_2 & & \\ & & \ddots & \\ & & & A_d \end{bmatrix} \begin{bmatrix} 1 & p_1 & p_1^2 & \cdots & p_1^{M-1} \\ \vdots & \vdots & \vdots & \cdots & \vdots \\ 1 & p_d & p_d^2 & \cdots & p_d^{M-1} \end{bmatrix}. \tag{11}$$

Denote X as a matrix composed of the first L rows of H and Y the last L rows of H, then

$$X = A_L S \tag{12}$$

$$Y = A_L P S \tag{13}$$

where A_L is the matrix composed of the first L rows of A and

$$P = \begin{bmatrix} p_1 & & & \\ & p_2 & & \\ & & \ddots & \\ & & & p_d \end{bmatrix}. \tag{14}$$

Since

$$Y - \lambda X = A_L P S - \lambda A_L S = A_L(P - \lambda I)S \tag{15}$$

so $p_l, \ldots, p_d$ are general eigenvalues of the matrix pair $\langle X, Y \rangle$.

Suppose we have TSVD of X and Y as

$$X = U_x \Sigma_x V_x^H \tag{16}$$

$$Y = U_y \Sigma_y V_y^H \tag{17}$$

and X and Y share a common row and column space, then we can use U_x and V_x to express Y, $Y = U_x E_y V_x^H$ for example, where $E_y = U_x^H Y V_x$ is a d-by-d matrix. Now

$$Y - \lambda X = U_x(E_y - \lambda \Sigma_x)V_x^H \tag{18}$$

so the general eigenvalues of $Y - \lambda X$ equals those of $\Sigma_x^{-1} E_y$, and since

$$X^+ Y = V_x \Sigma_x^{-1} U_x^H U_x E_y V_x^H = V_x \Sigma_x^{-1} E_y V_x^H \quad (19)$$

we can get general eigenvalues of $Y - \lambda X$ by computing non-zero ones of $X^+ Y$ (or$Y X^+$).

Once we get $p_l, \ldots, p_d, \alpha_l$ and β_l can be computed by

$$\alpha_l = \ln |p_l| \quad (20)$$

$$\beta_l = \arctan [\mathrm{Im}(p_l)/\mathrm{Re}(p_l)] \quad (21)$$

and furthermore τ_l (travel time) and υ_l (relevant with attenuation factor Q) are computed by

$$\tau_l = \frac{-\beta_l}{2\pi \Delta f} \quad (22)$$

$$\upsilon_l = \frac{\alpha_l}{2\pi \Delta f}. \quad (23)$$

To get amplitude the A_l, we can simply use the Least Square method.

When noise is taken into consideration, H should be constructed by $y(n)$ instead of $x(n)$ and X and Y should be projected onto a common row and column space using SVD.

4. Numerical Test

In the numerical test, we apply the algorithm to the structural dispersion analysis of a Rayleigh wave. The model is a simple two-layer one (Figure 1). The P wave velocities of two layers are 4.20 km/s and 6.00 km/s respectively and the S wave velocities 2.43 km/s and 3.45 m/s and the densities 1840 kg/m^3 and 2530 kg/m^3. Thomson-Haskell's method (Zhu, 1988) is used to calculate the phase velocities of up to 3 modes of the Rayleigh wave in the range of 0–400 Hz, with an interval of 0.1 Hz. The group velocities are calculated in the numerical method from the phase velocities data. The group velocity data are divided into 40 consecutive frequency windows with a width of 10 Hz. The algorithm is applied to each frequency window to analyze the travel times, compute the group velocity, taking it as the value at the center frequency of each window (Figure 2). It is seen in Figure 2 that the estimated group velocities are in good agreement with the theoretical ones.

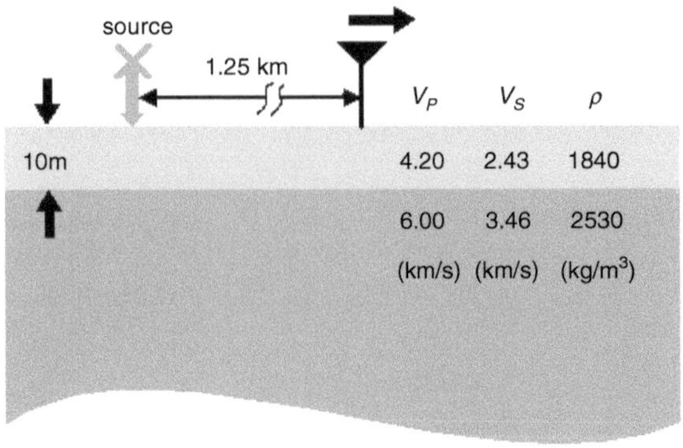

Figure 1 Surface wave model with structural dispersion.

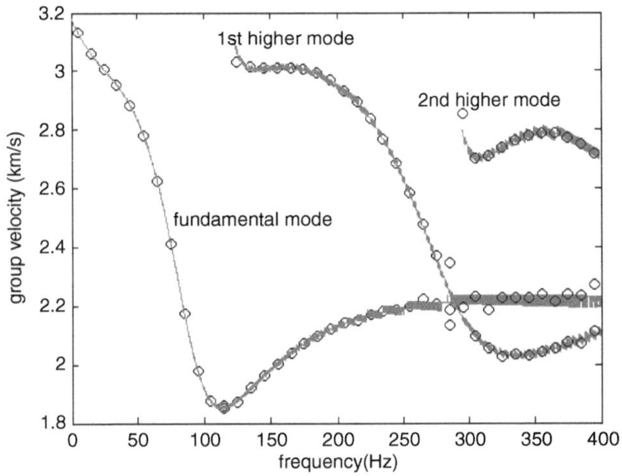

Figure 2 Theoretical (solid lines) and estimated (circles) group velocities of the Rayleigh wave.

5. Field Test

We further examine the algorithm with practical transfer function data observed by our sample ACROSS transmitter. Apparently the best way to prove whether the result is reasonable or not is to associate it directly with an underground structure, but the forward and inverse tools for ACROSS are not ready yet by now. As an alternative, we compare them with those acquired by the multi-filter method from hammer hit data. As we know, the multi-filter method is traditionally used to compute the travel delay of a wave.

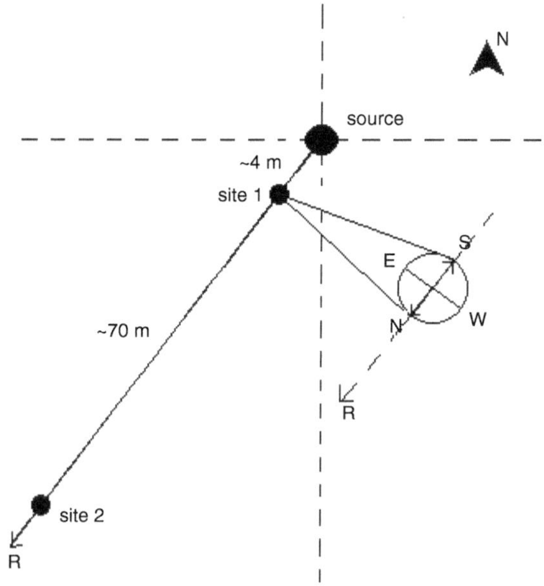

Figure 3 Location map of source and sites.

6. Location of Source and Sites

The ACROSS transmitter is installed on the ground surface, fixed on a concrete base. Two observational sites (site1 and site2) are arranged in a straight line with the source. The distance between site 1 and the source is about 4 m, site1 and site2 are about 70 m apart (Figure 3).

Each site is equipped with a three-component velocity type seismometer (resonant frequency 4.5 Hz) and an Earthquake Data Acquisition System (EDAS). The seismometer is placed with the axis of its NS component along the radical line and the two EDASs' clocks are synchronized with GPS.

7. Experiment and Data Process

Firstly, the transmitter continuously runs from 10 Hz to 25 Hz with a step of 0.1 Hz, lasting 10 s for every frequency. Response data is recorded and its Fourier spectrum is computed for both sites. The transfer functions for the three components are computed by dividing the Fourier spectrum of site2 by those of site1. Our algorithm is then applied to analyze the travel times from site1 to site2 and the attenuation factors of the structure, only the result of the NS component is given (Figure 4).

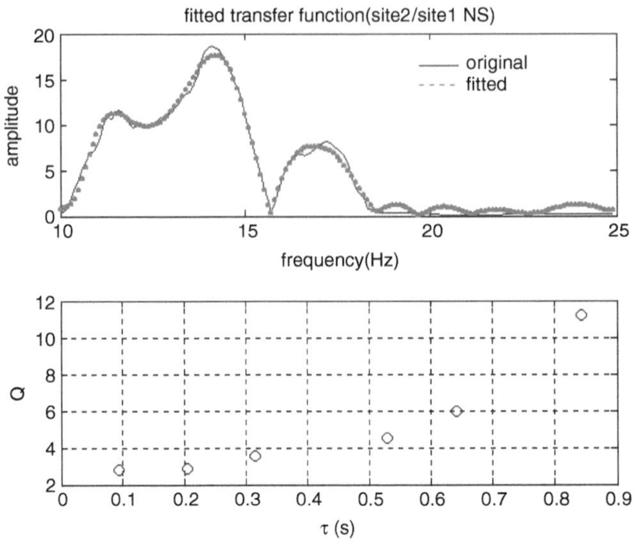

Figure 4 The NS component transfer function between site1 and site2 (above, solid line), the travel times and attenuation factors analyzed (below, circles) and the fitted transfer function using the estimated result (above, dotted line).

As given in Figure 4, the travel times of the wave components from site1 to site2 are estimated to be 0.09 s, 0.20 s, 0.31 s, 0.53 s, 0.64 s and 0.84 s and the transfer function is fitted very well using the estimated 6 poles.

Secondly, we apply hammer hit at the position of the ACROSS transmitter and record response data at both sites. This time the multi-filter method (Zhu, 1988) is used to analyze the travel times of the different wave components from site1 to site2. Figure 5 shows the response data for the hammer hit at both sites and the result analyzed using the multi-filter method (NS component) is given in Figure 6.

As shown in Figure 6, the travel time differences between the two sites are 0.10 s, 0.20 s, 0.37 s, 0.55 s and 0.81 s. They are all almost the same as those obtained by ACROSS. Whether these travel times correspond to real wave components is not clear, and further study is required. We try to validate the algorithm presented in this paper only in this way now.

8. Discussion

An algorithm based on SVD is presented in this paper. The algorithm is validated using numerical and field test data. In the numerical simulation, the algorithm is successfully applied to analyze the structural dispersion of a Rayleigh wave. It should be pointed out that the result is obtained

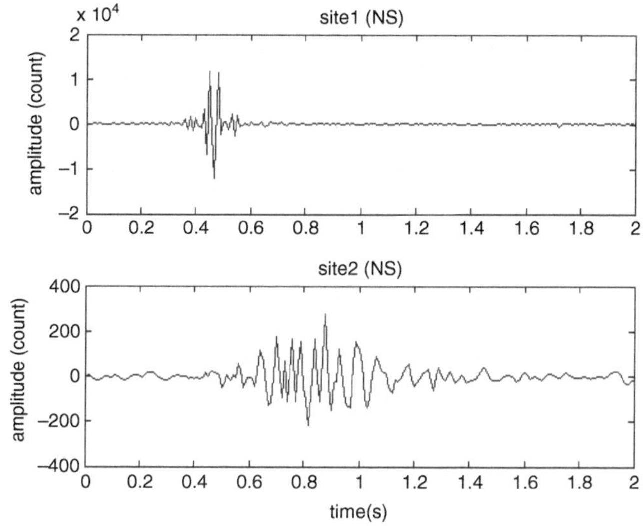

Figure 5 NS component response for hammer hit recorded at site1 (above) and site2 (below).

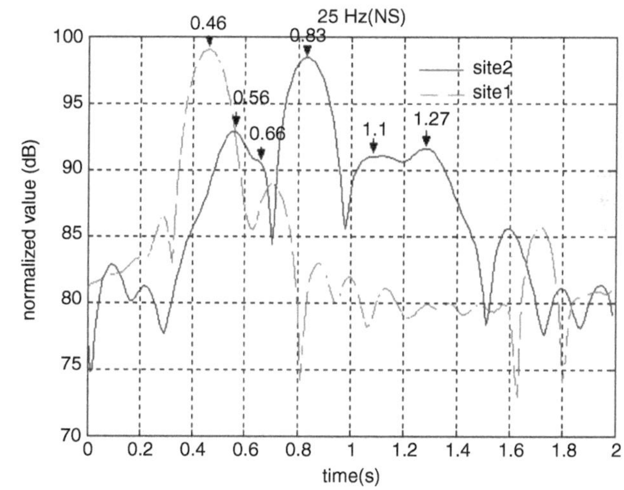

Figure 6 Estimated travel times for site1 (dash line) and site2 (solid line) using the multi-filter method.

under a real number of poles because it is pre-known. It may be not the case in practice. The results may be different if the number of orders is changed. Order determination is a big problem when the algorithm is used to deal with dispersive data, as it is with other methods. In the field test, two experiments were conducted. In both experiments, the two sites are arranged in one straight line with the source. Firstly, the source is ACROSS transmitter and the algorithm is applied to process the transfer function data observed, several travel times are acquired and the transfer function can be fitted well using the estimated poles. Secondly, the hammer hit is used as the source at the position of the ACROSS transmitter and the traditional multi-filter method is used to process the data observed. A few travel times are also obtained. The two experiments' results are almost the same. However more efforts are necessary to check whether these travel times correspond to real wave components. That will be more helpful to validate the algorithm.

REFERENCES

Kumazawa, M., Kunitomo, T., Yokoyama, Y., Nakajima, T., Tsuruga, K., 2000. ACROSS: Theoretical and technical developments and prospect to future applications. Technical report, Japan Nuclear Cycle Development Institute, 9, pp. 115–129.

Hasada, Yoko, Kumagai, Hiroyuki, Kumazawa, Mineo, 2001. Autoregressive modeling of transfer functions in frequency domain to determine complex travel times. Earth Planets Space 53 (1), 3–11.

Kumaresan, R., Tufts, D.W., 1982. Estimating the parameters of exponentially damped sinusoids and pole-zero modeling in noise. IEEE Trans. Acoust. Speech Signal Process. 30 (Dec.), 833–840.

Papadopulos, C.K., Nikias, C.L., 1991. Parameter estimation of exponentially damped sinusoids using higher order statistics. IEEE Trans. Signal Process. 38 (14–37).

Kay, S., 1988. Modern Spectral Estimation. Prentice-Hall, Englwood Cliffs, NJ.

Hua, Y., Sarkar, T.K., 1990. Matrix pencil method for estimating parameters of exponentially damped/undamped sinusoids in noise. IEEE Trans. Acoust. Speech Signal Process. 38 (May), 814–824.

Bresler, Y., Makovski, A., 1986. Exact maximum likelihood parameter estimation of superimposed exponential signals in noise. IEEE Trans. Acoust. Speech Siginal Process. 34 (1081–1089).

Parthasarathy, S., Tufts, D.W., 1985. Maximum-likelihood estimation of parameters of exponentially damped sinusoids. Proc. IEEE 73 (Oct.), 1528–1530.

Pisarenko, V.F., 1972. The retrieval of harmonics from covariance functions. Geophys. J. R. Astron. Soc. 347–366.

Tufts, D.W., Kumaresan, R., 1982. Estimation of frequencies of multiple sinusoids: Making linear prediction perform like maximum likelihood. Proc. IEEE (Sept.), 975–989.

Kay, S.M., Marple, S.L., 1981. Spectrum analysis: A modern perspective. Proc. IEEE 69, 1380–1419.

Roy, R., Kailath, T., 1989. ESPRIT—Estimation of signal parameters via rotational invariance techniques. IEEE Trans. Acoust. Speech Signal Process. ASSP-37, 984–995.

Viberg, M., Ottersten, B., 1991. Sensor array processing based on subspace fitting. IEEE Trans. Signal Process. 39 (May), 1110–1121.

vander Veen, A., Deprettere, E., Swindlehurst, A., 1993. Subspace based signal analysis using singular value decomposition. Proc. IEEE (September).

Zhu, JieShou, 1988. Computation Method in Seismology. Earthquake Press (in Chinese).

PART III

Case Histories

Regional Active Monitoring Experiments

Regional Active Monitoring Experiments

CHAPTER 23

Detection and Continuous Monitoring of Strong Seismic Reflectors at the Subducting Philippine Sea Plate Using ACROSS Methodology

Junzo Kasahara[a,d], Kayoko Tsuruga[a,e], Yoko Hasada[a,b], Takahiro Kunitomo[a,d], Ryoya Ikuta[c,d], Toshiki Watanabe[b], Koshun Yamaoka[b], Naoyuki Fujii[b,d], *and* Mineo Kumazawa[a,d]

Contents

1. Introduction

Continuous efforts to find precursors of catastrophic large earthquakes using seismic, geodetic and geochemical approaches in Japan have almost failed after an earthquake prediction program which has been going for the last 40 years. However, recent studies have revealed that earthquakes do not occur everywhere along the subducting plate boundaries, and the plate boundaries are very heterogeneous from the viewpoint of their frictional state and the strength of the land-ocean plate coupling. It has also been found that earthquake generation is non-steady in time and space, but the same source regions are mostly responsible for repeat large earthquakes (Yamanaka and Kikuchi, 2004). The source region generating repeated earthquakes

[a] Japan Atomic Energy Agency, Japan
[b] Graduate School of Environmental Study, Nagoya University
[c] Earthquake Research Institute, Tokyo University
[d] Present affiliation: Shizuoka University.
[e] JGI, Inc.

Handbook of Geophysical Exploration: Seismic Exploration, Volume 40
ISSN 0950-1401, DOI: 10.1016/S0950-1401(10)04028-0

hereafter will be called an "asperity" and the aseismic slip location will be called a "non-asperity".

Experiments using OBSs (Ocean Bottom Seismometers) and controlled sources in the forearc region of the Japan Trench revealed the presence of strong wide-angle PP reflections from the subducting Pacific slab beneath the north-eastern part of the Japanese Main Island (Fujie et al., 2002; Kasahara et al., 2003a; Mochizuki et al., 2005). The region showing the strong PP reflections has good correlation with an aseismic region since 1926, when the instrumental seismic measurement started. In order to interpret the observation of the strong PP reflections, it is necessary to place a thin (~100 m) and low-velocity Vp (2-4 km/s) layer at the land-ocean plate boundary (Kasahara et al., 2003a; Mochizuki et al., 2005). One of the candidates to create such a low Vp may be fluid, smectite clay, and/or serpentine (Kasahara et al., 2003b).

A similar observation was made by the seismic survey on land. In the land part of the Japanese archipelago, a seismic experiment using chemical explosions also indicated the presence of strong PP reflections along the subducting Philippine Sea Plate beneath the NE of Lake Hamana, in the south-western part of Japan (Matsu'ura et al., 1991; Iidaka et al., 2003; Kodaira et al., 2004). The seismic records at the J5 location (see Figure 1) at the southern end of the NS survey line showed strong reflected arrivals from a depth of 30-40 km on the interplate, which is nearly the same place in the region where an aseismic slip has been observed by the GPS network from August 2000 to February 2006 (Ozawa et al., 2002). The total aseismic slip equals that of an M 7.2 earthquake. Linde and Sacks (1997) suggested that the pre-slip changes before the 1944 Tonankai Earthquake (M 8) (occurred at the subducting Philippine Sea Plate at a depth of 30-40 km beneath Lake Hamana) using the observation of the precursory geodetic change just before the earthquake. The place is in nearly the same location as the one observed by the seismic experiment across the central Japan and the Global Positioning System (GPS) observation. The observations in the NE region of Lake Hamana suggest that the physical and chemical characteristics at the plate boundary beneath Lake Hamana are not in a steady state, but vary with time. In the Tokyo region, Sato et al. (2005) suggested that the region of low PP reflection intensity may relate to the large slip regions during the 1923 Great Kanto Earthquake.

If a rapid temporal change of PP reflection arrivals is caused by the movement of fluid (probably water, at interplates), the change may affect the interplate coupling strength at adjacent large asperities, and may trigger an interplate slip at nearby asperities. These observations may give not only a quantitative estimation for the physical properties and earthquake generation mechanism, but also a qualitative estimation for the precursory slip for the earthquakes. If accelerations of the slip at a non-asperity region are identified by any rapid reflection signature changes, it may be possible to suggest possible triggering of large earthquakes at the adjacent asperities.

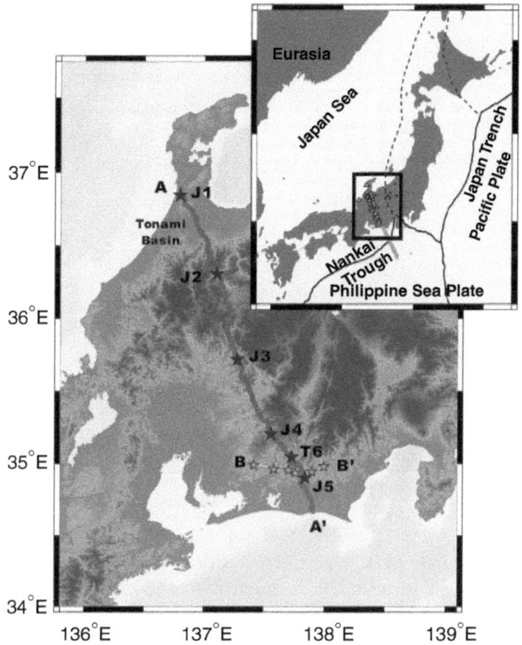

Figure 1 Seismic survey line and location of chemical explosions for the controlled source seismic survey in the NNW-SSE direction in Central Japan (modified after Iidaka et al., 2003). B-BJ5-B' is another seismic experiment in 1990 (Matsu'ura et al., 1991).

This rapid change of reflection signature can be continuously monitored by a combination of an active seismic source, such as Accurately-Controlled Routinely-Operated Signal System (ACROSS) (Kumazawa et al., 2000), and a seismic receiver array located at an appropriate place. The objective of this paper is to illustrate the usefulness of ACROSS for the active monitoring of the aseismic slip acceleration of a PP reflection signature change.

It is very difficult, however, to tell whether this is a PP reflected arrival from the plate boundary or not, if we do not have enough knowledge of the Green's function (transfer function) between the source and the receiver. In order to detect any temporal change of a particular arrival, precise knowledge on each seismic phase is required, because seismic signatures are strongly affected by the details of the crustal structure and ray-paths. Of the good locations to examine the present objective, the interplate region of the NE of Lake Hamana is the best. For this region, we can use the Toki ACROSS seismic source and a receiver array close to the previously surveyed Central Japan seismic line in 2001. The crustal structure along this line is well determined and Toki city, the source location, is close to the survey line (see Figures 1 and 2).

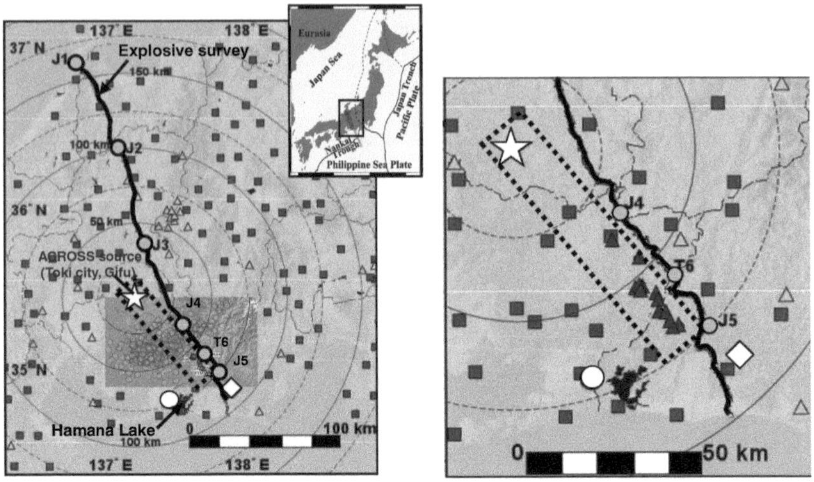

(a) (b)

Figure 2 (a) Location map of the seismic stations used for the ACROSS experiment from November 2004 to September 2005, relative to the controlled source seismic survey performed in 2001. Small open circles show the epicenters of deep low-frequency tremor sources. (b) An enlargement of the ACROSS source shown in Figure 2(a) and the temporal seismic stations. Stars and solid triangles represent the locations of ACROSS sources at Toki city, Gifu, and twenty-two temporal seismic stations, respectively. Solid squares and grey triangles denote permanent seismic stations of the Hi-net seismic network run by NIED (National Research Institute for Earth Sciences and Disaster Prevention) and Nagoya University. The locations of the six chemical explosions and 328 receivers of the seismic experiment in Central Japan in 2001 (Iidaka et al., 2003) are represented by grey circles and thick solid lines, respectively. The open circle and the diamond show newly built ACROSS source locations prepared by Nagoya Univ. and by the Metrological Research Institute, respectively.

2. Evaluation of the Seismic Across for Detection of Reflected Phases

We conducted a seismic experiment using the seismic ACROSS ($F \sim 2.0 \times 10^5 N$) (Kunitomo and Kumazawa, 2004a,b) in Toki city, in conjunction with a temporal receiver array, from December 2004 to September 2005 (Figure 2). FM sweep (Chirp) seismic waves with a 10-20 Hz band with a modulation period of 50 seconds were transmitted from Toki city. The 50 seconds produce a line spectra between 10 and 20 Hz with a 0.02 Hz frequency separation. The rotation speed was controlled by a GPS clock. The rotational directions were reversed every hour. Addition and subtraction of the seismic records from the two transmissions simulated two orthogonal linear vibrations.

Ten temporal seismic stations were installed at an offset distance Δ = 40-75 km from Toki city. Figure 2 shows the locations of the seismic source and the receivers, together with the experimental line and six explosions used during the seismic survey across the Central Japan in 2001 (Iidaka et al., 2003). Each receiver site had one or two sets of tri-axial 4.5 Hz velocity seismometers placed on hard rocks. The observed signals were digitized by 200 or 100 Hz sampling rates using the time-segment-stacking method (Tsuruga et al., 2005) or a continuous recording method on a portable 6-channel A/D recorder Datamark LS-7000.

The sampling clock-rate of every receiver was synchronized to the GPS clock within an accuracy of a few μs. The optimum weighted stacking method (Nagao et al., 2004) was used. The method of data analysis was described in Tsuruga et al. (2005). The transfer functions between Toki and each receiver were obtained by division of the receiving spectra by the source spectra. The details of the method of estimating the S/N using the ratio of averaged non-transmitted frequencies and signal frequencies are described in Tsuruga et al. (2005). Here we analyze the data recorded only in the nighttime (7:00 pm to following 7:00 am) from December 25, 2004 to April 20, 2005. This was done because night-time is much quieter, while daytime data may damage the improvement of S/N with stacking duration. The ACROSS signals exist in the frequency range of 10.25 to 19.45 Hz with a sampling interval of 0.02 Hz.

3. Results

The source spectra at Toki and the recorded spectra at a station OSD located at a 61.2 km offset from Toki as an example is shown in the previous paper (Tsuruga et al., 2005). Details of the data and method of analysis are published in Tsuruga et al. (2005). Dark and gray lines represent the spectrum at each transmitted frequency, and the noise level at a particular spectrum was estimated from the spectral amplitude of nearby spectra (Tsuruga et al., 2005).

The improvement of S/N by increasing the stacking duration is shown in Figure 3(a). The S/N for about 32 days at station OSD located at a 61.2 km offset is up to ten. The width between the two triangles in Figure 3(b) represents twice the error bounds of an estimated standard deviation. Several prominent arrivals are distinguished from the estimation error although waveforms in time domain are obtained using the limited frequency band of 11-19 Hz of the ACROSS transmission (Figure 3(b)). The major arrivals seen in 32-days stacking are clearly identified in 8-days stacking. The S/N for 8-days stacking is approximately equal to five, suggesting the high potential of the ACROSS system for application to the present purpose (Figure 3(a)).

Figure 4 shows the observed record section in time domain between 56 and 74 km offsets obtained by one-month stacking. We can recognize

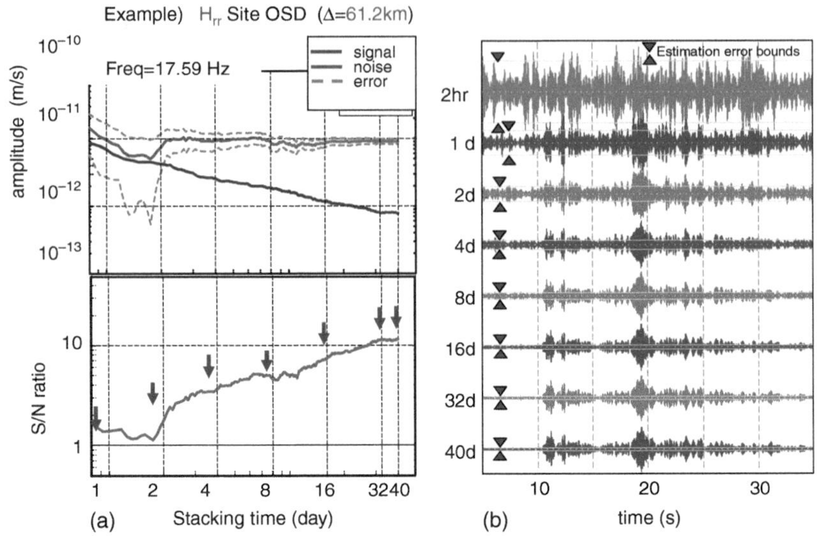

Figure 3 Effect of stacking duration. (a) Improvement of amplitude and S/N by stacking on the Hrr component (radial seismometer component and radial excitation) at the OSD station with an offset distance of 61.2 km. (b) Variation of the observed wave forms at the OSD station with increasing stacking duration. Note the 8-day record exhibits most of the characteristic arrivals seen in the 40-day records. The width shown between the triangles and inverted triangles is the bound of the estimated noise levels.

significant arrivals within the time ranges of 11-16 and of 15-23 seconds. Such phases are also identified, both in the records of the Horai borehole seismic station at $\Delta = 57.4$ km as one of the Hi-net permanent seismic stations by 30-days stacking (Yoshida et al., 2004), and on those of our dense-array sites around $\Delta = 56$-58 km (Ikuta et al., 2005). We also found that the prominent phases appear on not only Hrr, Hzr and Htt but also Htr, Hrt and Htz (Tsuruga et al., 2005). The first subscript of Hij is the receiver component and the second one is the excitation. "r", "t", and "z" are radial, transverse and vertical, respectively. This may suggest that the underground structure is heterogeneous in three dimensions.

4. Discussion

In order to understand the paths of the prominent seismic arrivals around a travel time of 11-16 and 15-23 seconds at $\Delta = 54$-74 km offset distance from Toki, we generated theoretical travel times, ray diagrams and synthetic seismograms. In order to calculate these, we re-interpreted the seismic records obtained by the Research Group for Seismic Expedition in

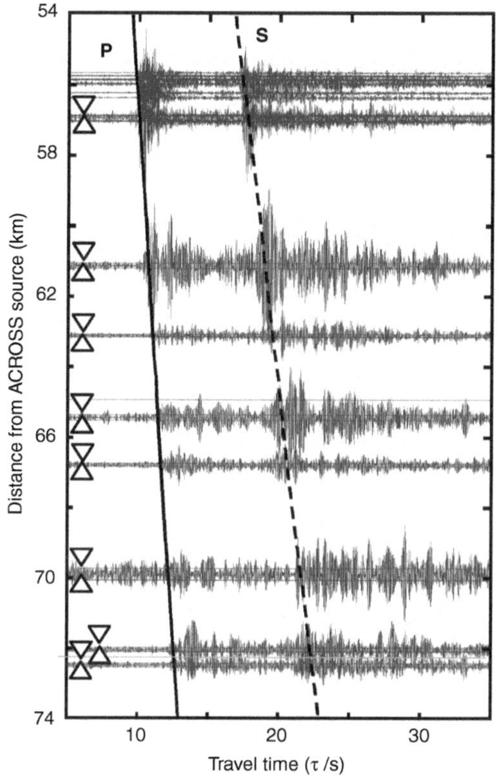

Figure 4 Transfer function Hrr in the time domain for all stations from 54 to 74 km offset distances with a stacking period of 40 days. The thin solid and dotted lines represent the theoretical travel times of refracted P- and S-waves, respectively. Several seismic phases have amplitudes greater than the noise levels (after Tsuruga et al., 2005).

Central Japan (Iidaka et al., 2003; Kodaira et al., 2004) by forward modeling using the "Pasteup" program written by Fujie et al. (2000, 2008). We calculated the theoretical travel times and ray paths by the graph method (Kubota et al., 2009). The theoretical travel times were fitted to the observed seismic records for the six seismic record sections corresponding to the chemical explosions J1-J5. Using this forward modeling, we obtained a crustal structural model shown in Figure 5. This alternative crustal model is basically similar to Iidaka's model, with exception of the presence of a thin low-velocity layer at 20 km depth beneath the central Japan. Similarly to Iidaka's crustal model (Iidaka et al., 2003), we put a thin (~200 m) low-velocity layer (LVL) ($Vp = 3.5$ km/s) at a depth of 25-35 km, just above the subducting slab, to explain the amplitude of the PxP reflected arrivals at the subducting slab. The ray-diagrams of Pg, PxP, PmP and Pl2P for the Toki source are shown in Figure 6, where Pl2P is the phase reflected at

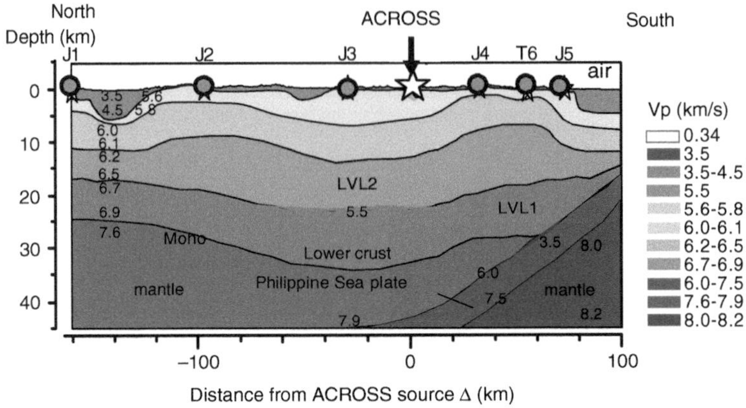

Figure 5 P-wave velocity structure model used in this study. The thickness of the upper crust in the model is approximately 15 to 20 km and the lower crust is 10 to 13 km. This model has two low-velocity layers located at the upper boundary of the Philippine Sea Plate (LVL1) and at the upper boundary of the lower crust (LVL2) (after Tsuruga et al., 2005). The former reflective layer was also suggested by Iidaka et al. (2003) to explain the strong PP reflection phase in the seismic records.

approximately 20 km depth beneath the central Japan. According to the ray diagram, the present receiver positions are not appropriate for observation of the reflected arrivals from the LVL because the PxP arrivals at receivers may come from deeper than the 25-35 km plate boundary on the Philippine Sea Plate if they exist. The most prominent arrival in the current receiver position is likely to be Pg.

The next step for the interpretation of phases (Figure 4) is the interpretation of observed records by means of theoretical seismograms obtained by the above crustal structure (Figure 5). The FDM (Finite Difference Method) (Larsen and Schultz, 1995) was used for computation of the synthetic seismograms. The source function is a horizontal single force with a 5 Hz Ricker wavelet excited at $\Delta = 0$ km. The grid size of the model is 20 m and the time step for computation is 2 ms. Synthetic velocity seismograms (transfer function) in x components with a reduction velocity of 6.0 km/s are shown in Figure 7(a). The synthetic record section indicates that, in the current receiver position, Pg and shallow reflection arrivals are prominent, but the PxP and PmP are not evident.

We compared the observed seismograms and the synthetic ones (Figure 7(b)). The two waveforms fit nicely, proving that the crustal structure seems appropriate and suggesting that the ACROSS source and the current receiver array is good enough to detect the target arrivals.

In addition to the physical property change at deep parts of the Earth's crust and the subducting plate causing temporal changes, there are other important factors which might cause the observed anomalies,

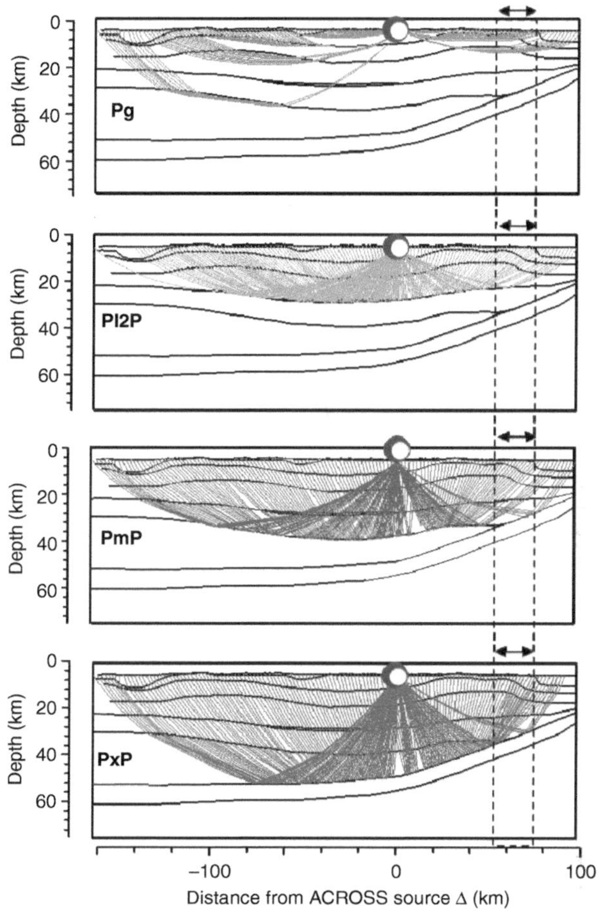

Figure 6 The ray paths (diagrams) from a virtual ACROSS source located at $\Delta = 0$ km of (a) the refraction phases, Pg, through the crust and the reflected phases: (b) Pl2P from the upper boundary of the lower crust, (c) PmP from the bottom of the lower crust, and (d) PxP from the surface of the Philippine Sea Plate. Arrows show the temporary observation zone of this study (after Tsuruga et al., 2005).

such as: (1) source effects, (2) receiver effects, (3) travel path effects, and (4) instrumental effects. We do not precisely know the actual contribution of these effects. The first one can be minimized by monitoring the source signatures near the source. The second one can be minimized by grouping small arrays of nearby receivers. Receivers located in holes may reduce the effects of temperature, humidity, porosity, air pressure, and/or rainfall. It is more difficult to separate temporal changes through ray paths from the reflectivity change at the interplate. These last factors are quite complicated. The characteristics of a seismometer is one of the instrumental effects and

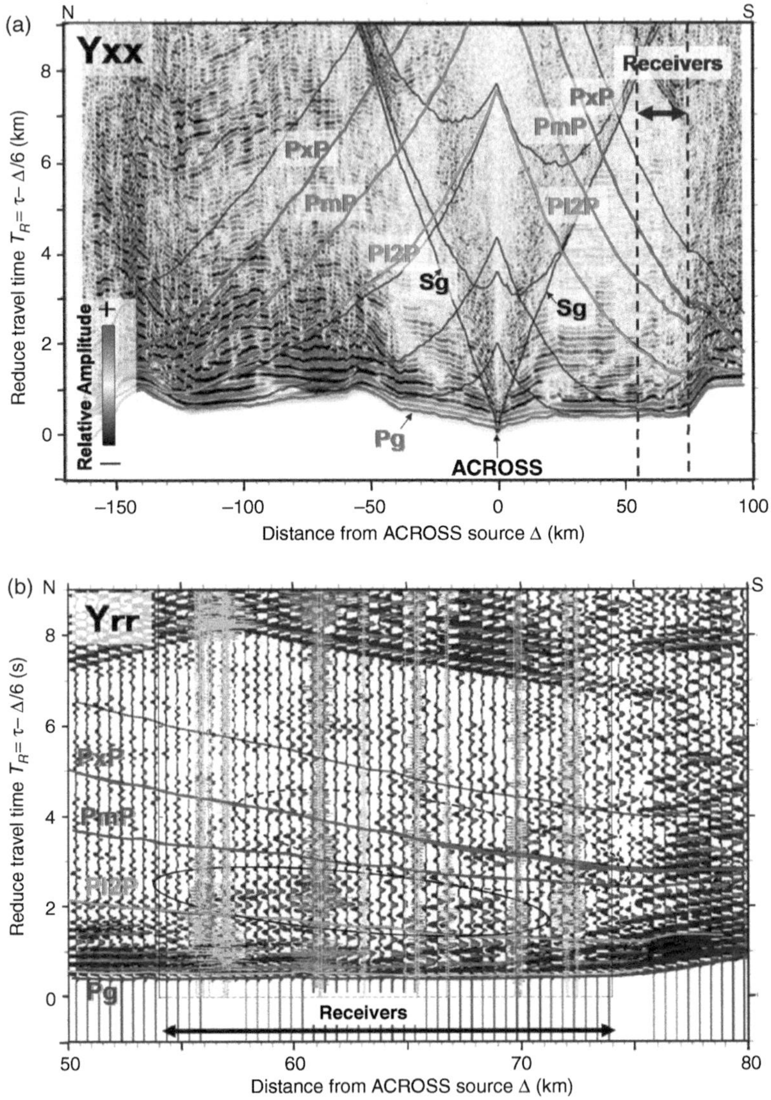

Figure 7 (a) Reduced synthetic velocity seismograms in the horizontal component assuming a horizontal single force at a virtual ACROSS source at $\Delta = 0$ km. A 4-Hz (zero-phase). Ricker wavelet was used as the source time function. (b) Comparison of the transfer function obtained by the ACROSS Tokai experiment with synthetic seismograms at offset distances of 50-80 km (after Tsuruga et al., 2005).

this has been examined (Tsuruga et al., 2010). The effect of seismometer characteristic differences is not large and can be well compensated by the calibration data.

5. Conclusions

In order to examine the validity of ACROSS to detect temporal changes of reflective signatures, we carried out a nine-month experiment from December 2004 to September 2005 using a seismic ACROSS in Toki city and a temporal linear seismic array extending toward the south from Toki city. The reasons behind the choice of this geometrical setting are the known crustal structure along the experimental line and the possibility of the application of the existing high PP reflective zone beneath this line.

Seismic experiments in the Japan Trench in 1996 (Fujie et al., 2002) and 2002 (Mochizuki et al., 2005), and on the central part of Japan in 2001 (Iidaka et al., 2003) revealed strong PP reflections along the subducting plate interfaces of the Pacific Plate and the Philippine Sea Plate. These strong reflective interfaces are considered to be caused by weakly coupled aseismic zones. The accelerated change of reflection signature may be a key to detect rapid changes of the physical properties at the plate boundaries due to migration of weak material, such as fluids. Such a rapid change of the physical properties at aseismic zones may trigger slip at adjacent large asperities and may induce large earthquakes. By continuous active monitoring of such reflections by seismic ACROSS and receiver arrays, it is possible to obtain quantitative information about the plate boundary coupling strength.

The observations show that the 8-days stacking at ~60 km offset distances showed similar prominent arrivals as 40-days stacking. This suggests that ~7- or 8-days stacking can be used to detect some prominent arrivals up to ~60 km offset distances. The seismic characteristics obtained by ACROSS records are equivalent to the down-hole seismometer records given by the High-net seismological network.

By comparison of the observed seismograms and synthetic seismograms, we found a nice fit between synthetic and observed records. It is unfortunate that the reflected arrivals from the subducting plate boundary is absent in the current receiver range. 8-days stacking of received data at the OSD station located in 61.2 km has a sufficient S/N to identify Pg and shallow reflected arrivals.

Although we are aware that the present velocity model may not be the best when considering a three-dimensional heterogeneous structure, the coincidence between the observed seismograms and the synthetic seismograms suggests the applicability of the ACROSS system to detect the reflected arrivals from the lower crust and/or aseismic subducting slab.

To map the distribution of asperities and non-asperities, it is necessary to carry out integrated research, such as natural earthquake observation, mapping of the asperities-non-asperities system by natural and/or controlled sources, continuous monitoring of the asperities-non-asperities system, development of data retrieving systems and studies in rock physics. Kasahara et al. (2006) call this the Exploration of Asperities and Reflectors System (EARS). This paper is one of the EARS components in a feasibility study of active monitoring.

Acknowledgement

This study was carried out in cooperative studies with the Tono Geoscience Center, JAEA and the University of Nagoya. We would like to thank NIED and the University of Nagoya for providing seismic observation data. We also wish to thank the Research Group for Seismic Expedition in Central Japan for permission to use their data.

REFERENCES

Fujie, G., Kasahara, J., Sato, T., Mochizuki, K., 2000. Traveltime and raypath computation: A new method in a heterogeneous media. Geophysical Exploration (Butsuri-Tansa) 53, 1–11.

Fujie, G., Kasahara, J., Hino, R., Sato, T., Shinohara, M., Suyehiro, K., 2002. A significant relation between seismic activities and reflection intensities in the Japan Trench region. Geophys. Res. Lett. 29 (7). doi:10.1029/2001GL013764.

Fujie, G., Kasahara, J., Murase, K., Mochizuki, K., Kaneda, Y., 2008. Interactive analysis tools for the wide-angle seismic data for crustal structure study (Technical Report). Geophysical Exploration (Butsuri-Tansa) 61, 26–33.

Iidaka, T., Iwasaki, T., Moriya, T., Kumakawa, L., Kurashimo, E., Kawamura, T., Tamazaki, F., Koike, K., Aoki, G., 2003. Configuration of subducting Philippine sea plate and crustal structure in the central Japan region. Geophys. Res. Lett. 30. doi:10.1029/20Gl016517.

Ikuta, R., Watanabe, T., Soma, T., Saiga, A., Yamaoka, K., Tsuruga, K., Kunitomo, T., Hasada, Y., Kasahara, J., Satomura, M., Kumazawa, M., Fujii, N., 2005. Monitoring of the state of inter-plate coupling using reflection phase - Preliminary results using a short array. In: Proc. of 2nd International Workshop on Active Geophysical Monitoring of the Earth Lithosphere, Novosibirsk, Russia.

Kasahara, J., Mochizuki, K., Kamimura, A., Nakamura, M., Hino, R., Yamada, T., Sato, T., Moghaddam, P.P., Nishino, M., Nakamura, Y., Kanazawa, T., 2003a. Non-aspereties at subduction zones and estimation of materials at plate boundary. J. Geography 112, 811–827.

Kasahara, J., Toriumi, M., Kawamura, K. (Eds.), 2003b. Role of Water on Earthquake Generation. University Tokyo Press, p. 389 (in Japanese).

Kasahara, J., Tsuruga, K., Mikada, H., Yamaoka, K., Fujii, N., 2006. Time lapse approach in the study of earthquake generation - geophysical exploration of asperities-reflectors system (EARS) for interplate earthquake generation along a subducting oceanic plate. Geophysical Exploration (Butsuri Tansa) 59, 525–538.

Kodaira, S., Iidaka, K., Kato, A., Park, J.-D., Iwasaki, T., Kaneda, Y., 2004. High pore fluid pressure may causes silent slip in the Nankai trough. Science 28 (May), 1295–1298.

Kubota, R., Nishiyama, E., Murase, K., Kasahara, J., 2009. Travel-time estimation of first arrivals and later phases using the modified graph method for a crustal structure analysis. Geophysical Exploration (Butsuri-Tansa) 62, 105–113.

Kumazawa, M., Kunitomo, T., Yokoyama, Y., Nakajima, T., Tsuruga, K., 2000. ACROSS: Theoretical and technical developments and prospect to future applications, Technical report, Japan Nuclear Cycle Development Institute, 9, 115–129.

Kunitomo, T., Kumazawa, M., 2004a. Active Monitoring of the Earth's Structure by the Seismic ACROSS – Development of the Seismic ACROSS and Continuous Transmission at Tono Mine ACROSS Station –, In: Proceedings of "1st International Workshop on Active Monitoring in the Solid Earth Geophysics (IWAM04)", Task Group for Active Monitoring, Mizunami, Japan, pp. 218–222.

Kunitomo, T., Kumazawa, M., 2004b. Active Monitoring of the Earth's Structure by the Seismic ACROSS – Transmitting and Receiving Technologies of the Seismic ACROSS –, In: Proceedings of 1st International Workshop on Active Monitoring in the Solid Earth Geophysics (IWAM04), Task Group for Active Monitoring, Mizunami, Japan, pp. 181–184.

Larsen, S., Schultz, C., 1995. A Technical Report No. UCRL-MA-121792, 18 pp.

Linde, T.A., Sacks, I., 1997. Slow earthquakes and the great Tonankai and Nankaido earthquakes. Eos Trans., AGU 78 (46), Fall Meeting Abst., F156.

Matsu'ura, R.S., Yoshii, T., Moriya, T., Miyamachi, H., Sasaki, Y., Ikami, A., Ishida, M., 1991. Crustal structure of a seismic-reflection profile across the Median and Akaishi Tectonic Lines, central Japan. Bull. Earthq. Res. Inst., Tokyo Univ. 66, 497–516.

Mochizuki, K., Nakamura, M., Kasahara, J., Hino, R., Nakamura, Y., Shinohara, M., Sato, T., Moghadam, P.P., Kanazawa, T., 2005. Intense PP reflection beneath the aseismic forearc slope of the Japan Trench subduction zone and its implication of aseismic slip subduction, 110, B01302. doi:10.1029/2003JB002892.

Nagao, H., Nakajima, T., Kumazawa, M., Kunitomo, T., 2004. Optimum weighted stacking method for acquisition of the ACROSS transfer functions having the maximum signal-to noise ratio. In: Proceedings of "1st International Workshop on Active Monitoring in the Solid Earth Geophysics (IWAM04)", Task Group for Active Monitoring, Mizunami, Japan, pp. 206–270.

Ozawa, S., Murakami, M., Kaidzu, M., 2002. Detection and monitoring of ongoing aseismic slip in the Tokai region, central Japan. Science 298, 1009–1012.

Sato, H., Hirata, N., Koketsu, K., Okaya, D., Abe, S., Kobayashi, R., Matsubara, M., Iwasaki, T., Ito, T., Ikawa, T., Kawanaka, T., Kasahara, K., Harder, S., 2005. Earthquake source faults beneath Tokyo. Science 309, 462–464.

Tsuruga, K., Hasada, Y., Kunitomo, T., Kasahara, J., Ikuta, R., Watanabe, T., Yamaoka, K., Fujii, N., Kumazawa, M., Nagao, H., Nakajima, T., Saiga, A., Satomura, M., 2005. Monitoring of the state of inter-plate coupling using reflection phase – Preliminary results using a long array. In: Proc. of "2nd International Workshop on Active Geophysical Monitoring of the Earth Lithosphere". Novosibirsk, Russia, pp. 23–27.

Tsuruga, K., Kumzawa, M., Kunitomo, T., Shigeta, N., Kasahara, J., 2010. Secondary calibration method of seismometers by utilizing ACROSS signal (in this volume).

Yamanaka, Y., Kikuchi, M., 2004. Asperity map along the subduction zone in northeastern Japan inferred from regional seismic data. J. Geophys. Res. 109, B07307. doi:10:1029/2003JB2683.

Yoshida, Y., Ueno, H., Matsuoka, H., Ishikawa, Y., Kunitomo, T., Kumazawa, M., 2004. The property of ACROSS signals transmitted from the Tono Transmitting Station and observed by JMA network and Hi-net. Chikyu Monthly S47, 124–131.

CHAPTER 24

Active Seismic Monitoring of the San Andreas Fault at Parkfield

Valeri A. Korneev[a] *and* Robert M. Nadeau[b]

Contents

1. Introduction

The town of Parkfield, located on the San Andreas Fault in central California has been the site of intensive multidisciplinary earthquake studies since the 1970s. Moderate-sized earthquakes of about magnitude 6 have occurred on the Parkfield section of the San Andreas Fault at fairly regular intervals—in 1857, 1881, 1901, 1922, 1934, and 1966. The 1857 event was a foreshock of the great Fort Tejon earthquake (magnitude 8(1/4)), which ruptured the fault from Parkfield to the southeast for over 180 miles. Available data suggest that all six moderate-sized Parkfield earthquakes may have been "characteristic," in the sense that they all ruptured at the same area on the fault. The goal of research in the Parkfield area has been to observe the fault and surrounding crust, at close range and at high resolution before, during, and after a large, damaging earthquake, so as to better understand the earthquake process and to provide a scientific basis for earthquake prediction and hazard assessment.

Recognizing this hazard and the regular periodicity of recurring events near Parkfield, the US Geological Survey and the State of California

[a] Lawrence Berkeley National Laboratory, Berkeley, California, USA
[b] University of California at Berkeley, Berkeley, California, USA

Handbook of Geophysical Exploration: Seismic Exploration, Volume 40 2010 Published by Elsevier Ltd.
ISSN 0950-1401, DOI: 10.1016/S0950-1401(10)04029-2

began a comprehensive, long-term Parkfield Earthquake Prediction Project, reported in Bakun and Lindh (1985). More than 100 researchers have been involved in various facets of this project. It was assumed that the next characteristic earthquake would nucleate in the same location as the 1966 event at Middle Mountain, northwest of Parkfield. This area, located at the beginning of the SAF creeping section, has a naturally high seismicity level. Middle Mountain, the central point of the area, is heavily instrumented with various geophysical sensors, including a High-Resolution Seismic Network (HRSN). However, the expected characteristic event occurred in 2004 and was nucleated in the middle of Cholame Valley, 22 km southeast of Parkfield, in the locked section of the SAF, rather far from the instrumented area.

2. HRSN

The HRSN (Figure 1) established at Parkfield, CA in 1987 records exceptionally high-quality data, owing to its 13 closely spaced three-component borehole sensors, its very broadband recordings (0-125 Hz), and its sensitivity (recording events below magnitude 0.5), (Karageorgi et al., 1992). Several aspects of the Parkfield region make it ideal for the study of seismic wave propagation and small earthquakes, and their relationships to tectonic processes. These include the fact that the network spans the expected nucleation region of a repeating Magnitude 6 event and the transition from locked to creeping behavior on the San Andreas fault, the availability of three-dimensional P and S velocity models, a complete seismicity catalogue, a well-defined and simple fault segment, and a homogeneous mode of seismic energy release, as indicated by the earthquake source mechanisms. More than 11,000 earthquakes have been recorded since 1987, in the magnitude range $-1 < M < 5$.

3. Vibroseis Monitoring

A unique data set of seismograms for 720 source-receiver paths was collected as part of a controlled source Vibroseis experiment. In the experiment, seismic waves repeatedly illuminated the epicentral region of the expected M6 event at Parkfield from June 1987 until November 1996. For this effort, a large shear-wave vibrator was interfaced with the three-component (3-C) borehole HRSN, providing precisely timed collection of data for detailed studies of changes in wave propagation associated with stress and strain accumulation in the fault zone. Data collected by the borehole network were examined for evidence of changes associated with the nucleation process of the anticipated M6 earthquake at Parkfield (Karageorgi et al., 1992, 1997). These investigations reported significant

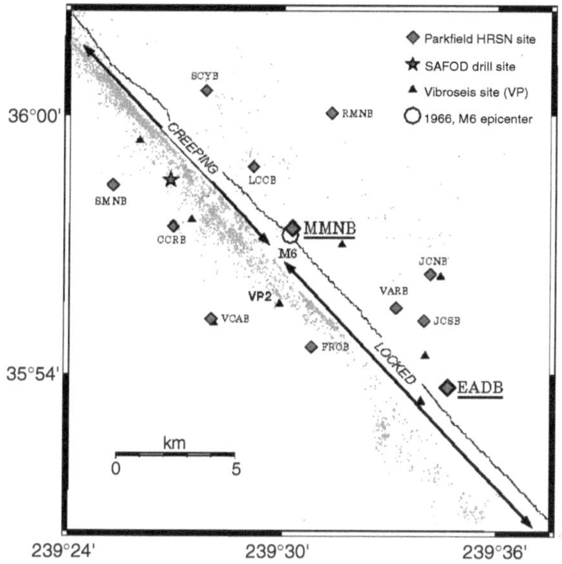

Figure 1 Map showing the San Andreas Fault trace, and the location of the original 10 Parkfield HRSN stations. Relocated seismicity (1987 to mid-1998) is also shown, as are the locations of the 8 source points of the Vibroseis wave propagation monitoring experiment. The epicenter of the 1966 M6 Parkfield main shock and location of the San Andreas Fault Observatory at Depth (SAFOD) drill site are also shown.

travel-time changes in the S- coda for paths crossing the fault zone southeast of the epicenter and above the rupture zone of the 1966 M6 earthquake. Progressively decreasing travel times through the anomalous region changed by over 50 ms by the end of the study. Analysis and modeling of these data and comparison with observed changes in creep, water level, microseismicity, slip-at-depth, and propagation from characteristic repeating microearthquakes showed temporal variations in a variety of wave-propagation attributes—attributes that were synchronous with changes in deformation and local seismicity patterns (Karageorgi et al., 1992; Langbein et al., 1999; Roeloffs, 1998; Nadeau and McEvilly, 1999; Korneev et al., 2000).

The most important result of the Karageorgi et al. (1997, see their Figure 4), studies was the localization of the region of wave-propagation changes to the upper 3-4 km of the San Andreas Fault Zone, just above the locked zone of the expected M6 event. The creeping and seismically active northwestern part of the SAF showed no travel-time changes, while the most substantial changes were observed in the locked southeastern part of the Parkfield region of the SAF. It was hypothesized that these localized changes were in response to changing fluid conditions in the upper crust, induced by tectonic deformation, as well as stress and strain accumulation in the locked section of the fault at depth. Additionally, Vibroseis VSP experiments in the nearby Varian well measured and monitored shear-wave anisotropy (Daley and McEvilly, 1990, 1991).

4. Origin of Changes

At Parkfield, the San Andreas Fault zone is a striking near-vertical low-velocity zone, and it very clearly acts as a waveguide for seismic energy from earthquakes on the fault and from surface sources. Velocity models there show a high Vp/Vs ratio along the fault, both near the surface and at depth, within the fault zone, and a pronounced vertical velocity gradient in the upper 2 km of the section (Michelini and McEvilly, 1991). The geometry of the Vibroseis source and receiver network, the approximate two-dimensionality of the fault zone in the region of the travel-time anomaly, and the existence of detailed P- and S-wave velocity models for the area all combine to provide well-determined constraints for modeling the observations. In Korneev et al. (2000), only data recorded at stations VCAB and JCNB from vibrator site VP2 were considered (Figure 1). At VP2, we have the routine Vibroseis monitoring data from the repeated point source, as well as a cross array of sources with 17 vibrational points at each leg.

The modeling exercise was confined to the VP2 data for VCAB and JCNB stations, for several reasons. Both source-receiver paths are in the anomalous region and reveal substantial travel-time variations. Moreover, the two paths are approximately co-linear and orthogonal to the San Andreas Fault, permitting the use of a two-dimensional formulation in simulating wave propagation. The path sample segments on the two sides of the fault zone have similar length. Finally, the data profile from the closely spaced source array at VP2 defines the spatial coherency of the wavefield, which is helpful in phase identification and interpretation of the recorded wavefield. The velocity model used in numerical simulation incorporates the known properties of the region, where tomographic three-dimensional velocity models have already been determined (Michelini and McEvilly, 1991). A major factor controlling the character of wave propagation at short range from a surface source is the severity of the shallow vertical velocity gradient. Values of velocity gradient in the model were determined by matching the observed and computed direct arrivals in the early part of the seismograms. For the NE side of the fault, the direct arrivals at JCNB could be matched with a velocity profile reduced by a factor 0.76 for that of VCAB, and by 0.5 for the narrow fault zone, modeled as a vertical layer with a thickness of 200 m. Computations were performed using a 2-D elastic finite-difference formulation. Two features dominated the process: energy trapping near the surface by the shallow gradient, and wavefield scattering from the fault zone. Most of the energy is confined to the upper part of the section in multiple reflections at the free surface, producing a complex train of surface-guided waves made up of many arriving phases.

The signatures of the fault zone and shallow gradient on the wavefield are dramatic. In the interval between the first-arriving P and S waves at VCAB, surface-generated multiples and conversions take place. The latter are especially strong for P-waves, e.g., PS and PPS. Strong reflections are also produced by both fault-zone boundaries. At JCNB, the internal fault-zone reflections produce sequences of strong, distinct arrivals following the direct P and S waves. The times in the synthetic seismograms where large travel-time changes were observed in the monitoring project at VCAB and JCNB contain significant energy that has been reflected from the fault zone. This result suggests a ready explanation for the cause of the observed progressively decreasing travel-times: For the path VP2-VCAB, the changes were seen at arrival times after 3.5 s, after the direct waves have passed and the fault-zone reflected waves arrive. On the other hand, the travel-time changes for the VP2-JCNB fault-crossing path begin with the arrival of the direct P wave and occur through the entire seismogram. These results represent strong evidence that the observed variations are most likely caused by changes within the fault zone itself. Observed travel-time changes nicely fit the modeled results for seismograms computed for a 6% velocity increase localized in the narrow fault zone.

Three key conclusions follow from this study: (1) a strong subsurface velocity gradient is a key factor in wave propagation excited by surface-placed sources; (2) use of surface sources allows observations of waves reflected from the SAF and track changes in those waves; (3) observed changes originate in the locked part of the fault core. Unfortunately, the Vibroseis monitoring experiment was terminated in 1996, and the evolution of monitored changes up to the 2004 M6 Parkfield earthquake is unknown.

5. Guided-Wave Imaging of SAF Core

Observations and numerical models testify to the imaging power of fault zone guided waves (FZGW) to characterize, spatially and temporally, the properties and processes within the central cores of major active fault zones. Fault-zone guided waves were identified as such by Aki and co-workers in active-source, surface-to-borehole studies and later in seismograms recorded in or near the fault zone from local earthquakes (Leary et al., 1985; Li et al., 1990). These waves are most visible for sources within a well-developed fault zone and receivers located within the same fault zone segment (Li et al., 1994, 1997) although they appear also to be generated by off-fault surface sources (Korneev et al., 2000). The FZGW appear to be trapped by the presence of material in a fault that has a lower seismic wave velocity than the surrounding, more intact rock, from which it is separated by relatively sharp boundaries—the low-velocity nature of the

San Andreas Fault zone core has long been recognized (Feng and McEvilly, 1983). FZGW are seen in the codas of both the direct P and S waves, but they are usually much stronger in the S-wave coda, with large amplitudes in some cases arriving twice as late as the S-wave travel time (Karageorgi et al., 1997). They usually exhibit lower frequency than the direct P or S waves, and in many cases they appear to be dispersive.

There are compelling reasons to study the FZGW phenomenon. First, it has the potential for defining the structure of the active fault zone at depth. Second, the features that bound rupture extent in large earthquakes (segmentation boundaries, gaps, streaks, or asperities) may be evident in FZGW generation and propagation characteristics. Third and finally, the degree to which processes already under way in the cores of seismogenic fault zones can be detected and monitored is unknown—but is of critical importance in seismology. Detection of transient or systematic changes within the fault core through successful FZGW imaging in four dimensions is a potentially powerful monitoring method (Li et al., 1998; Ben-Zion et al., 2003). This is particularly true at Parkfield, where the San Andreas is a low-velocity zone striking near-vertically (Michelini and McEvilly, 1991) and clearly acting as a waveguide for seismic energy from earthquakes and surface sources (Li et al., 1998; Korneev et al., 2000). Velocity models also show a high Vp/Vs ratio along the fault near the surface and at depth within the fault zone, and a pronounced vertical velocity gradient in the upper 2 km of the section. The geometry of the Vibroseis source and receiver network, the approximate two-dimensionality of the fault zone in the region of the travel-time anomaly, and the existence of detailed P- and S-wave velocity models for the area all combine to provide well-determined constraints in modeling FZGW observations. Numerical modeling and microearthquake data indicate that FZGWs propagate within an SAF zone that is 100 to 200 m wide at seismogenic depths, and has a 20%-40% lower shear-wave velocity than the adjacent unfaulted media. The initial step in our FZGW investigation of the SAF was to look at the nature of FZGWs with respect to the hypocenter source location and receiver position, to map any obvious features in the spatial relationship of source and receiver. Two stations are close enough to the fault zone to provide a conveniently "reversed" profile of sources along the fault (by building "station gathers" of traces).

Results from Korneev et al. (2003) confirm that FZGW at Parkfield are generated within the fault zone and are most prominent late in the coda of S, while also seen in the P coda. Also, FZGW amplitude inversion enables vertical section imaging of the delineation zone between creeping and locked parts of SAF. Most likely, the variation in FZGW attenuation is associated with fracture closure and opening, caused by the evolution of stress-induced changes in loading. Laboratory measurements of time-lapse seismic attenuation in fractured rock under increasing normal load show an initial steady increase in Q (due to closure of preexisting micro-fractures in the rock) that reaches a maximum value plateau, and then decreases rapidly (due to the formation of new fractures) just prior to catastrophic failure of the rock (Sobolev et al., 1996).

The spatial distribution of Q at Parkfield indicates that a similar process may be occurring within the zone of concentrated stress buildup and release associated with magnitude 4 and 5 earthquakes. The process of fracture closure and opening is also expected to involve dewatering and saturation of rocks (respectively), and corresponding changes in water pressure, all of which are expected to result in seismic propagation variations. We interpret the localized zone of FZGW low attenuation as the NW edge of the M6 asperity at Parkfield, at the transition of the SAF locked and creeping behavior at depth, with the high Q, due most likely to dewatering resulting from fracture closure and/or fault-normal compression. A greater understanding of the relationships between FZGW attenuation, and stress-related fracturing processes under conditions of accumulating fault stress, will be needed to obtain information useful for understanding the earthquake nucleation process and for possibly predicting earthquakes.

A particular strength of using FZGWs to study the detailed structure and deformation of an active fault zone is that FZGW propagation is confined to the fault core, making them highly sensitive to any spatial or temporal variations along the fault. In contrast, the propagation paths of direct P- and S-waves from earthquakes occurring within the low-velocity fault zone take place largely outside of the fault core in the higher-velocity country rock. As a result, direct P- and S- arrivals contain little information on the properties of the fault itself. This limits the resolution on fault structure that direct P- and S-waves can provide to about 5 km. In contrast, we have already shown that FZGWs can resolve the details of fault structure on the order of 50 m (Korneev et al., 2000).

By increasing the number of receiver stations located directly on the fault, it should be possible to improve resolution even further. Analysis of seismograms from a cluster of 12 microearthquakes exhibit practically identical waveforms, although the magnitudes of the events belonging to that cluster clearly show strong temporal variations (Figure 2). The ray paths of this cluster intersect the FZ attenuation anomaly, which was found by tomographic inversion. Cross-correlation of the cluster waveforms reveal no travel-time changes for all recorded phases, including direct P, S, and guided waves. In contrast, amplitudes of guided waves show strong changes compared to P- and S-wave amplitudes. These changes correlate spatially and temporally with overall Parkfield seismicity, suggesting their common origin (Korneev and Nadeau, 2004).

The previous studies showed the potential of FZGW for imaging and monitoring the inner part of the FZ. However, only waves generated by natural seismicity were used there, and as a result, the creeping part of the fault comprised most of the fault image. Note that use of active sources would enable seismologists to study the locked part of the SAF, reaching the areas where the actual 2004 M6 Parkfield earthquake originated.

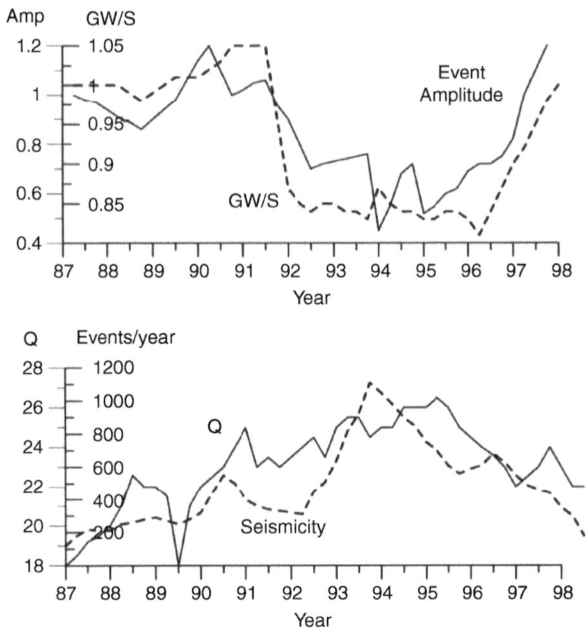

Figure 2 Observed changes during the years 1987-1999 in Parkfield, California. Upper panel shows the amplitude of repeating cluster events (solid line) and amplitude ratio of GW to S waves for the same events (dashed line). Lower panel shows evolution of Q at the cluster hypocenter (solid line) computed using attenuation tomography (Korneev et al., 2003), and year-average seismicity in Parkfield area (dashed line). Guided wave amplitude changes are the most pronounced, whereas no detectable changes were found in body wave travel.

6. Fault Continuity Testing

The continuity of low-velocity layers is a very important problem in geophysics. Promising results from exploration geophysics demonstrate the ability of guided waves to propagate in low-velocity sedimentary layers at distances substantially exceeding those of regular body waves (Korneev and Myer, 2001; Parra et al., 2001).

The detailed studies of seismicity generated by the 1966 Parkfield M6 event revealed very high seismic activity in the Cholame Valley region southeast of Parkfield, as well as ruptures along the Cholame Valley tracing the fault along the surface. In the middle of the valley, the SAF had about a 1.5 km jog to the southwest, after which it continued southeast, staying parallel to the original direction. The jog geometry had no surface evidence and was determined on the basis of epicenter locations for multiple events in this area. Seismic and theoretical studies (Hanna et al., 1972; Eaton et al., 1970) suggest the complex structure of the jog area, with

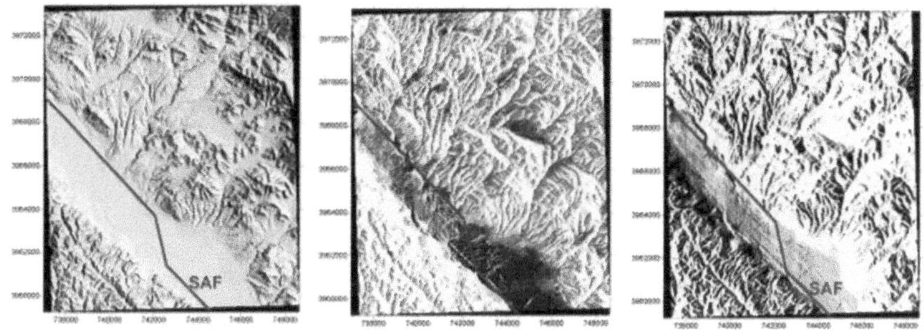

Figure 3 Steep-angle light imaging reveals details of the small-scale topography in the Cholame Valley, which suggests the existence of a fault echelon between SAF strands. Solid line indicates the current mapping of the SAF trace.

interacting segments of two parts of the SAF. Such segments usually represent wide zones, with multiple fractures oriented at 45° with respect to the main fault orientation, and which connect isolated segments of a fault. The corresponding microfracturing was observed in multiple locations of Cholame Valley. The extent and mechanical properties of interacting segments have a strong impact on accumulated strain release.

The importance of this SAF element lies in the fact that this jog was the originating point of the 2004 Parkfield earthquake. The fault-structure studies in Cholame Valley are mostly based on seismic information, because valley sediments cover bedrock and do not allow accurate fault mapping. Nevertheless, we have found that low-amplitude topography features of the valley are apparently affected by bedrock surface geometry. We subsequently analyzed shade relief images by varying color saturations and light-source positions. The shade relief technique has spatial differentiating properties and enables delineation (Figure 3) of both delicate and pronounced topographic details. An existing proposed point of jog connection with the southwest strand of the SAF is observed; the fault can be traced along both sides through this point. Multiple NS-oriented faults that make 45° angles with traces of the SAF are quite visible. Image features also suggest a straight continuation of the NW part of the SAF in the SE direction after passing the jog. At the same time, the NW straight extension of the SE part of the SAF connects it with mapped parts of the South-West Fracture Zone (SWFZ), which extends quasi-parallel to the SAF.

This result suggests that the Cholame Valley fault structure represents two faults separated at some point in time, but which began interacting through the development of shear-faulting in the middle of the valley, leaving some of their old parts temporarily inactive. This hypothesis can be tested using fault-zone guided waves, since inactive faults still should have low-velocity properties, owing to the relatively high concentration of fractures within the zones. A corresponding experiment should include excitation of

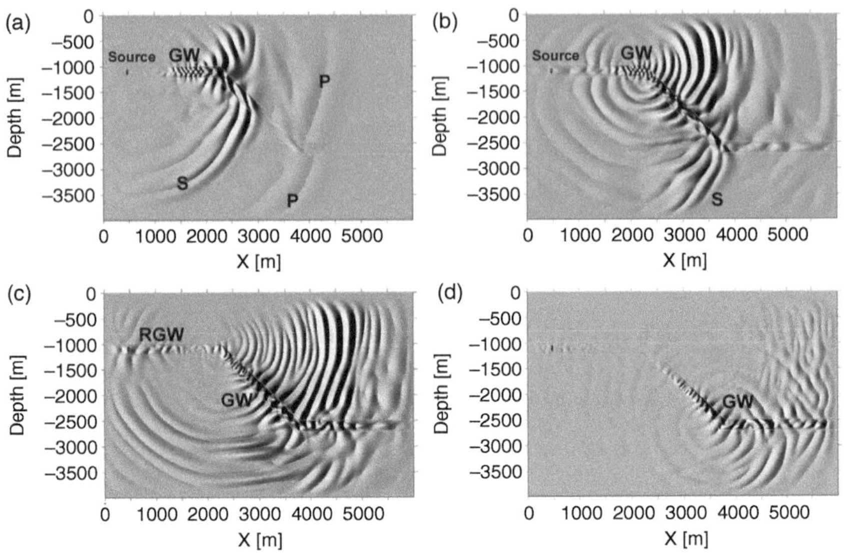

Figure 4 Consecutive snapshots for the SAF jog model. Note the reflected guided waves (RGW) in panel (c), as well as the strong emission of S-wave energy coming from bends. Despite the complex geometry, the GW energy is well trapped in the FZ.

fault-zone guided waves by placing explosive and/or vibroseismic sources on a known existing fault zone (in particular on SWFZ), with surface-seismic-line recording across the anticipated trace of this fault at a 5-10 km distance from the source. Detection of FZGW should indicate the continuity of low-velocity properties along the source-receiver path to support the suggested model of fault structure.

We performed a 2-D numerical experiment to study guided-wave propagation along a jog-shaped waveguide. A finite-difference method with a staggered grid was used for modeling. The embedding medium has wave propagation velocities of Vp = 4000 m/s, Vs = 2.4 m/s, and a density of 2.7 g/cm^3, containing a 200 m wide waveguide with geometry as shown in Figure 4. A vertical-force source was placed in the middle of the waveguide (Figure 4(a)). Panels in Figure 4 represent four consecutive snapshots for field amplitudes as waves propagate in and out of the FZ.

Several interesting effects are shown by this experiment. First, 45° sharp bends keep most of the energy inside of the waveguide. Second, these bends generate visible reflected guided waves (RGW) that propagate backwards, indicating discontinuities in fault properties. Third, the turning direction of guide-wave propagation occurs at the expense of increased S body-wave radiation in the outer medium. The directions of wave propagation are close to the initial propagation direction of the guided waves, as if wave energy exhibits an "inertial" influence (Figure 4(c)). These effects suggest additional potential utilization of guided-wave energy for FZ characterization purposes. For seismology, knowledge of a fault's structure and composition has critical significance for the modeling and prediction of future earthquake scenarios.

7. Discussion and Conclusions

The lack of natural seismic events around the nucleation zone of the 2004 M6 Parkfield earthquake stresses the importance of active controllable seismic monitoring in the zones of potential earthquakes. Monitoring changes associated with processes in the core of the fault zone should provide valuable information about fault conditions and potentially lead to a better understanding of strong earthquakes. Use of a controlled source for guided wave monitoring is most desirable, since with a controlled source, the exact source location and initiation time are known. Explosion sources are not highly repeatable and require special permissions and precautions. A conventional Vibroseis source cannot be used for FZGW monitoring, since its lowest excitation frequency is around 8 Hz, whereas FZGW energy is in the 3-6 Hz band. On the other hand, low-frequency eccentric vibrators (LEV) would be an ideal source for fault zone monitoring using guided waves (Alexeev et al., 2005). With modern modeling techniques, FZGWs can provide detailed images of a fault zone's inner structure and other characteristics. Placing an active, low-frequency source on top of an FZ core and a line of surface receivers intersecting the FZ behind a potential nucleation zone, we can achieve an FZGW monitoring scheme (Figure 5).

In a complementary active monitoring scheme, sources can be placed far outside of the FZ and illuminate a large (10-30 km) FZ area containing a potential nucleation zone. A specially placed line of seismometers can record the waves that intersect the FZ at seismogenic depths. Changes in FZ properties under stress-strain variations will cause corresponding changes in waves transmitted through the FZ and can be interpreted after detection. For this scheme, the use of an ACROSS source (Kunimoto and Kumazawa, 2004) seems the most appropriate, because of its high durability and precision. This scheme is also shown on Figure 5.

Parkfield is an ideal location for active monitoring experiments because of: (1) the number of ongoing important experiments in the area, (2) its high earthquake hazard, (3) its relatively regular earthquake-recurrence interval, and (4) its long-term baseline of monitoring results for various geophysical observables. Understanding the detailed structure, physical properties, and evolution of the fault at depth is critical to advancing our knowledge of earthquake physics, formulating estimates of earthquake hazards, and developing schemes for hazard reduction. FZGW studies show promise for identifying blind faults and fault continuity across

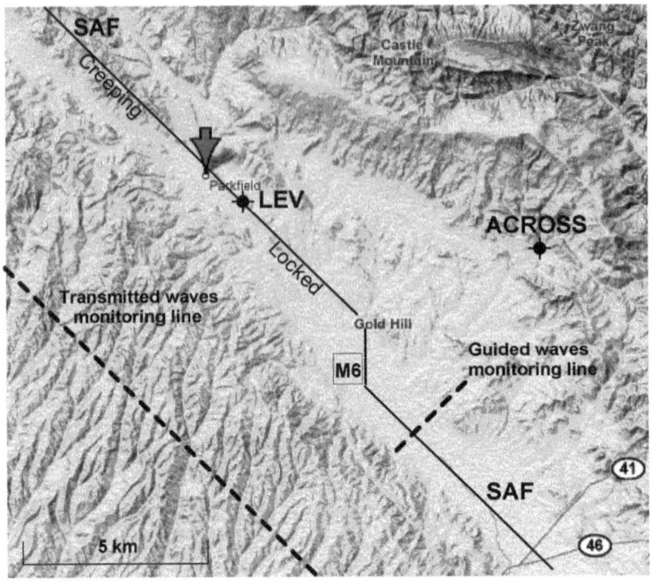

Figure 5 Conceptual SAF active monitoring scheme in the Cholame Valley. The LEV source excites low-frequency GW, which propagate along the SAF to be recorded by the GW monitoring line. Waves excited by the ACROSS source transmit through the SAF at seismogenic depths to be recorded by a transmitted-waves monitoring line. The square indicates the epicenter of the 2004 M6 event.

jogs and at depth. Deep-well seismometers installed during the SAFOD (http://earthquake.usgs.gov/research/parkfield/safod_pbo.php) project also present a major opportunity for controlled source experiments in the Parkfield area.

Acknowledgments

Data processing was done at the Center for Computational Seismology (CCS) at LBNL, which is operated by the University of California for the US Department of Energy (DOE) under Contract No. DE-AC03-76SF00098. The USGS provided financial support for this research through NEHRP Award 1434-95-G-2540, and support for the Berkeley Seismological Laboratory's operation of the HRSN is through Grant No. 07HQAG0014.

REFERENCES

Alexeev, A.S., Chichinin, I.S., Korneev, V.A., 2005. Powerful low-frequency vibrators for active seismology. Bulletin of the Seismological Society of America 95 (1), 1–17.

Bakun, W.H., Lindh, A.G., 1985. The Parkfield, California, prediction experiment. Earthquake Prediction Research 3, 285–304.

Ben-Zion, Y., Peng, Z., Okaya, D., Seeber, L., Armbruster, J.G., Ozer, N., Michael, A.J., Baris, S., Aktar, M., 2003. A shallow fault-zone structure illuminated by trapped waves in the Karadere-Duzce branch of the North Anatolian Fault, western Turkey. Geophysical Journal International 152, 699–717.

Daley, T.M., McEvilly, T.V., 1991. Update of shear-wave anisotropy measurements at the Parkfield Varian well. Seismological Society of America, Annual Meeting.

Daley, T.M., McEvilly, T.V., 1990. Shear-wave anisotropy in the Parkfield Varian Well. Bulletin of the Seismological Society of America 80, 857–869.

Eaton, J.P., O'Neill, M.E., Murdock, J.N., 1970. Aftershocks of 1966 Parkfield-Cholame, California Earthquake: A detailed study. Bulletin of the Seismological Society of America 60 (4), 1151–1197.

Feng, R., McEvilly, T.V., 1983. Interpretation of seismic reflection profiling data for the structure of the San Andreas Fault zone. Bulletin of the Seismological Society of America 73, 1701–1720.

Hanna, W.F., Burch, S.H., Diblee, T.W., 1972. Gravity, magnetics, and geology of the San Andreas fault near Cholame, California, US Geological Survey Professional Paper 646-C.

Karageorgi, E., Clymer, R., McEvilly, T.V., 1992. Seismological studies at Parkfield: II. Search for temporal variations in wave propagation using VibroseisVibroseis. Bulletin of the Seismological Society of America 82, 1388–1415.

Karageorgi, E.D., McEvilly, T.V., Clymer, R.W., 1997. Seismological studies at Parkfield IV: Variations incontrolled-source waveform parameters and their correlation with seismic activity 1987–1994. Bulletin of the Seismological Society of America 87 (1), 39–49.

Korneev, V.A., Myer, L.R., 2001. Testing of sedimentary layer continuity using guided waves. Gas Research Institute Topical Report.

Korneev, V., Nadeau, R., 2004. Vibroseis monitoring of San Andreas Fault in California. In: Proceedings Paper for the International Workshop in Active Monitoring. Mizinami, Japan, pp. 88–95.

Korneev, V.A., Nadeau, R.M., McEvilly, T.V., 2003. Seismological studies at Parkfield IX: Fault zone imaging using guided wave attenuation. Bulletin of the Seismological Society of America 93 (4), 1415–1426.

Korneev, V.A., McEvilly, T.V., Karageorgi, E.D., 2000. Seismological studies at Parkfield VIII: Modeling the observed travel-time changes. Bulletin of the Seismological Society of America 90 (3), 702–708.

Kunimoto, T., Kumazawa, M., 2004. Active monitoring of the earth's structure by seismic ACROSS. In: 1st International Workshop on Active Monitoring in the Solid Earth Geophysics. Mizunami, Japan, pp. 181–184.

Langbein, J., Gwyther, R., Hart, R., Gladwin, M.T., 1999. Slip rate increase at Parkfield in 1993 detected by high-precision EDM and borehole tensor strain meters. Geophysical Research Letters 26, 2529–2532.

Leary, P.C., Li, Y.-G., Aki, K., 1985. Borehole observations of fault zone trapped waves, Oroville, CA. Transaction EOS 66, 976.

Li, Y.-G., Ellsworth, W.L., Thurber, C.H., Malin, P.E., Aki, K., 1997. Fault-zone guided waves from explosions in the San Andreas fault at Parkfield and Cienega Valley, California. Bulletin of the Seismological Society of America 87, 210–221.

Li, Y.-G., Leary, P.C., Aki, K., Malin, P.E., 1990. Seismic trapped modes in the Oroville and San Andreas fault zones. Science 249, 763–766.

Li, Y.-G., Vidale, J.E., Aki, K., Marone, C., Lee, W.H.K., 1994. Fine structure of the Landers fault zone: Segmentation and the rupture process. Science 265, 367–370.

Li, Y.-G., Vidale, J.E., Aki, K., Xu, F., Burdette, T., 1998. Evidence of shallow fault zone strengthening after the 1992 M7.5 Landers, CA, earthquake. Science 279, 217–219.

Michelini, A., McEvilly, T.V., 1991. Seismological studies at Parkfield: I. Simultaneous inversion for velocity structure and hypocenters using B-splines parameterization. Bulletin of the Seismological Society of America 81, 524–552.

Nadeau, R.M., McEvilly, T.V., 1999. Fault slip rates at depth from recurrence intervals of repeating microearthquakes. Science 285, 718–721.

Parra, J.O., Hackert, C., Gorogy, A., Korneev, V.A., 2001. Detection of guided waves between gas wells for reservoir characterization, LBNL-48523. Geophysics 67 (1), 38–49.

Roeloffs, E., 1998. Persistent water level changes in a well near Parkfield, California, due to local and distant earthquakes. Journal of Geophysical Research 103 (B1), 869–889.

Sobolev, G.A., Babichev, O.V., Los, V.F., Kol'tsov, A.V., Ponomaryov, A.V., Ponyatovskaya, V.I., Rozanov, A.S., Stanchits, S.A., Khromov, A.A., Frolov, D.I., Yangquan, L., Jialiu, Z., Jiadong, Q., Shiyu, L., 1996. Precursors of the destruction of water containing blocks of rock. Journal of Earthquake Prediction Research 5.

CHAPTER 25

Characteristics of ACROSS Signals Transmitted from the Tono Transmitting Station and Observed by Hi-net

Yasuhiro Yoshida[a], Hiroshi Ueno[b], Satoshi Takahama[b], Yuzo Ishikawa[c], Sumio Yoshikawa[b], Akio Katsumata[a], Takahiro Kunitomo[d], *and* Mineo Kumazawa[d]

Contents

1. Introduction

The validity of using ACROSS signals for detecting temporal variations in P- and S-wave travel times over short distances (less than several kilometers) has been confirmed by previous studies. Ikuta et al. (2002) detected an S-wave delay of about 1 ms after the 2000 Western Tottori earthquake (Mw 6.6) using ACROSS deployed at Awaji Island. The ACROSS signals transmitted at the surface were observed by seismometers at 800 m and 1700 m depths, with an epicentral distance less than 1 km.

Recent studies (e.g. Kasahara et al., 2001) have demonstrated that the intensity of seismic reflection at the plate boundary is sensitive to the

[a] Meteorological Research Institute, 1-1 Nagamine, Tsukuba, Ibaraki, 305-0031, Japan
[b] Japan Meteorological Agency, 1-3-4 Otemachi, Chiyoda, Tokyo, 100-8122, Japan
[c] Matsushiro Observatory, Japan Meteorological Agency, 3511 Saijo, Nagano, Nagano, 381-1232, Japan
[d] Shizuoka University, 836, Ooya, Sugura, Shizuoka, Shizuoka, 422-8017, Japan

E-mail address: yyoshida@mri-jma.go.jp (Y. Yoshida).

Handbook of Geophysical Exploration: Seismic Exploration, Volume 40
ISSN 0950-1401, DOI: 10.1016/S0950-1401(10)04030-9

coupling strength between the subducting slab and the overriding plate. Using a rate- and state-dependent friction law, a small amount of slow slip occurred before the main rupture (Kato and Hirasawa, 1999), based on computer simulation. This suggests that a temporal variation in the state of the plate boundary is likely to occur before a large earthquake. Active monitoring of the reflectivity at the plate boundary is an important tool for detecting such precursory slow slip and the related change in the physical properties of rocks. In order for investigators to map the plate boundary, either by reflected or refracted waves (e.g., Kasahara et al., 2004), the spatial extent of stations must be comparable in size to the plate boundary's reflective region (in this case, about 50 km). Few studies attempting to monitor the temporal variations in travel time have been conducted for distances greater than 10 km using an active source. This paper analyzes the ACROSS signals transmitted from the Tono station (Kunitomo and Kumazawa, 2004a) by using Hi-net seismometry stations.

The Tokai district where Tono station is located is a good test field site for monitoring the change in state at the plate boundary. In this region, the Philippine Sea plate subducts beneath the Eurasian plate. The assumed future rupture area of the coming Tokai earthquake is beneath the land area, which means that many seismometers must be placed in this region to detect the temporal changes in seismic velocity or the intensity of seismic reflections at the plate boundary.

A seismic experiment was conducted in this region in 2001. Figure 1 includes the profile line of the 2001 experiment crossing central Japan in a NS direction. A clear later phase, explained by the reflected wave at the upper boundary of the subducting Philippine Sea plate (Iidaka et al., 2003), can be detected from the record section of shot J-5. A long-term slow slip (Ozawa et al., 2002) and nonvolcanic low-frequency events (or tremor) occurred in this region (Obara, 2002; Katsumata and Kamaya, 2003). The generation of tremors is considered to be related to the movement of fluid near the subduction zone, suggesting that the seismic velocity or intensity of seismic reflection may change because of fluid movement near the plate boundary.

As part of this Tokai study, the ability to detect seismic ACROSS signals observed at telemetered seismic stations was investigated by Ueno et al. (2003). This study was conducted using ACROSS signals transmitted from the Tono station and observed by JMA and Hi-net stations. The data obtained from the first test broadcast at the Tono seismic ACROSS station were used, as was the source that produced sinusoidal waves of 25 Hz. The time length of one stacking segment is 100 seconds, with stacking taking place about 9000 times over an analysis period of about 11 days. Each segment was weighted by the inverse proportional to the square of the noise level (Nagao et al., 2004). The distance decay of 25 Hz ACROSS signals is proportional to the reciprocal of the epicentral distance. The signal level is sufficiently higher than the noise level up to 100 km. This study sought

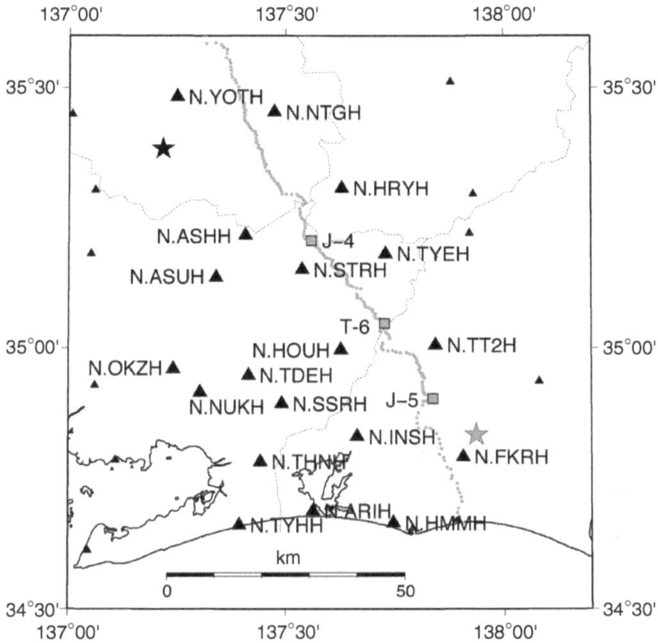

Figure 1 Distribution of stations used in this analysis. The black star denotes the Tono station, and the gray star represents the newly deployed transmitting station. A seismic experiment was conducted in this region in 2001. Shot points are indicated by squares. The profile line is drawn with small gray circles. Large and small triangles identify the stations used for calculating the transfer function and other telemetered stations to JMA.

to demonstrate that ACROSS signal analysis can be done at an epicentral distance up to 100 km.

2. Data and Methods

This study employed Hi-net stations deployed by NIED and equipped with velocity seismometers with a 1 sec eigenperiod in the borehole. All these seismograms are telemetered to JMA in real time. The stations used are SE of the Tono ACROSS transmitting station and cover the Tokai area, with an epicentral distance up to 100 km. Figure 1 plots the distribution of stations used in this study—a total of 19 stations in this analysis.

Data from the fifth transmission experiment at the Tono seismic ACROSS station were analyzed. The experiment lasted nearly 20 months, from February 20, 2004 to October 10, 2005, with interruptions caused by some technical problems accounting for only a few days' disruption within the total time period. It was the longest continuous transmission experiment with the widest FM range ever conducted. The transmission force spectrum for this experiment was virtually flat, between 10.25 to 19.45 Hz, with a corresponding force range of 0.332×10^5 to 1.20×10^5 N. The rotational direction was switched every hour (Kunitomo and Kumazawa, 2004b). The response on excitation by a single force along an arbitrary direction was obtained by the linear combination of the normal and reversed rotational data in the frequency domain. Transfer functions from radial and transverse excitation towards the response of radial, transverse and *z* (plus for downwards) directions were calculated. The six components of the transfer function tensor (*Hrr*, *Hrt*, *Htr*, *Htt*, *Hzr*, *Hzt*) were obtained using normal and reversed rotated data (e.g., *Hzt* denotes the force with a transverse direction and observed in the *z* direction).

The signal spectra interval was 0.02 Hz, with a modulation period of 50 seconds. Data were stacked at intervals of 200 seconds. Signal channels appeared every three noise channels, which evaluated the noise level. The S/N ratio increased with increased total stacking time, and a stacking length of 30 days was used (approximately 12,000 stacking traces), with a sufficient S/N ratio of up to about 60 km.

3. Results

Figure 2 is an example of a transfer function in the frequency domain. $Hzt(\omega)$ for station N.HOUH [epicentral distance: 57 km; azimuth: 139°] is drawn in this figure. Error bars calculated by noise channels are displayed in the amplitude and phase figures. The S/N ratio of more than 2 is sufficient for evaluating the transfer function. Although a large signal-channel phase error occurs where the spectra amplitude is relatively small, the phase error around the spectrum peak region is small. The reliability of all observed data can be evaluated by the error bars in the ACROSS analysis. The spectral energy has two peaks: the first is between 13 and 14 Hz, and the second between 16 and 17 Hz. As plotted in real and imaginary components, the period within the frequency domain of the first peak is about 1/18 Hz and 1/10 Hz in the second. This suggests that the prominent wavelet can be observed at around 18 and 10 seconds in the time domain (Hasada et al., 2001). The 13 to 14 Hz peak corresponds to the S wave; the 16 to 17 Hz peak corresponds to the P wave.

Figure 3 shows the six components of the transfer function tensor in the time domain. The second trace ($Hzt(t)$) is just the inverse Fourier transform of the transfer function in the frequency domain provided in Figure 2 ($Hzt(\omega)$). Large amplitude wavelets can be seen around 10 and 18 seconds, which can be predicted from the frequency-domain transfer functions. These two wavelets correspond to P and S waves. The noise level of the transfer function is small enough to identify these two wavelets. The

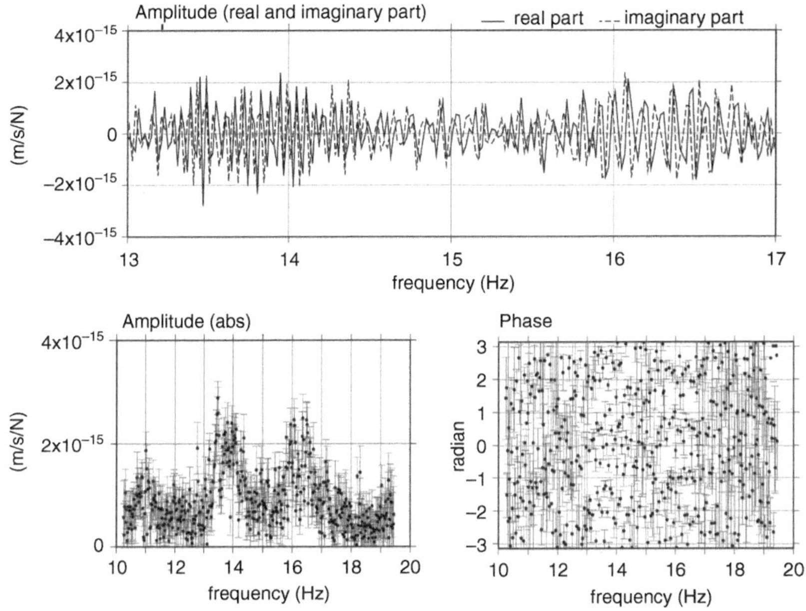

Figure 2 The transfer function in the frequency domain ($Hzt(\omega)$) of station N.HOUH [epicentral distance of 57 km, azimuth of 139°] is shown. The stacking length is 30 days, beginning on February 20, 2004. The upper figure displays the the real and imaginary representation; the lower figure displays the amplitude and phase representation, respectively. Error bars calculated from noise channels are also drawn in the amplitude and phase graphs.

shape of P and S phases for the transfer functions differ from the earthquake waveforms that we often see in seismograms.

The onsets of P and S waves are not clear and sharp, partly because of the limited frequency range (10 to 20 Hz) of the transfer functions. Considering the laterally homogeneous layered media, the amplitudes of *Hzt*, *Hrt*, and *Htr* should be zero. As seen in Figure 3, the amplitude of those components is comparable to that of other components. There are a number of possible reasons for the large amplitude value of those components. One reason is the possible combination of invalid assumptions regarding the horizontally layered media and the related substantial effect of scattering. The other reason is the near-source effect resulting from the heterogeneity of the source region. Further study is necessary to quantitatively evaluate those effects.

Figure 4 illustrates the particle motion of the P and S wave portions at N.HOUH. The principal axis of the P wave particle motion is not parallel to the radial direction for the radial- and transverse-forced cases, but the difference in the angle is not that large (i.e. within 30°), and the linearity is high for all stations. In contrast, the direction of the S wave particle motion is scattered from station to station. The principal axis of S wave particle motion at N.HOUH is almost parallel to the transverse direction, and linearity is

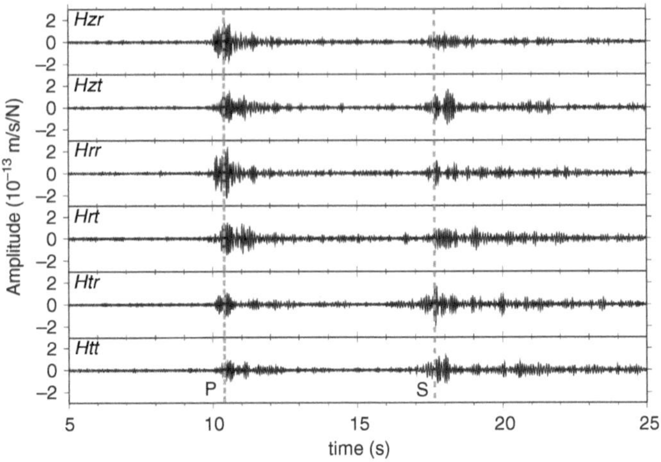

Figure 3 The six components of the transfer function tensor at N.HOUH. The stacking period is the same as in Figure 2. P and S arrivals calculated by the JMA2001 travel-time table (Ueno et al., 2002) are denoted by dashed gray lines. Clear wavelets are observed close to the time of the calculated P and S arrivals. The second trace ($Hzt(t)$) is the inverse Fourier transform of the data presented in Figure 2. The error level of the transfer function is drawn with solid gray lines.

low for both the radial- and transverse-forced sources, suggesting that the effect of the force direction at the source is very small. The direction of wave propagation cannot be detected using only one station. Using seismic array stations or synthetic seismograms is necessary for detecting the arrival direction and identifying seismic phases and scattered waves.

4. Discussion

In the next step, we compared the transfer functions of several stations—record sections for transfer functions of $Hrr(t)$ and $Htt(t)$ are provided in Figure 5. The time axis is reduced by 6 km/s. No significant phases whose amplitudes exceed the noise level can be seen before the P wave arrival time for all components, indicating that the effect of aliasing is not significant when the epicentral distance is less than 60 km. Generally with record sections, the amplitude of the P phase is relatively large in *Hrr* components and the amplitude of the S phase is large in *Htt* components. However, waveform features are very different from station to station. The P and S phases of N.HOUH have a very sharp onset and short duration, and these phases can be identified easily from the transfer function. Conversely, the transfer function at N.ASHH [epicentral distance of 25 km] has an emergent onset and long duration for P and S waves, although the epicentral

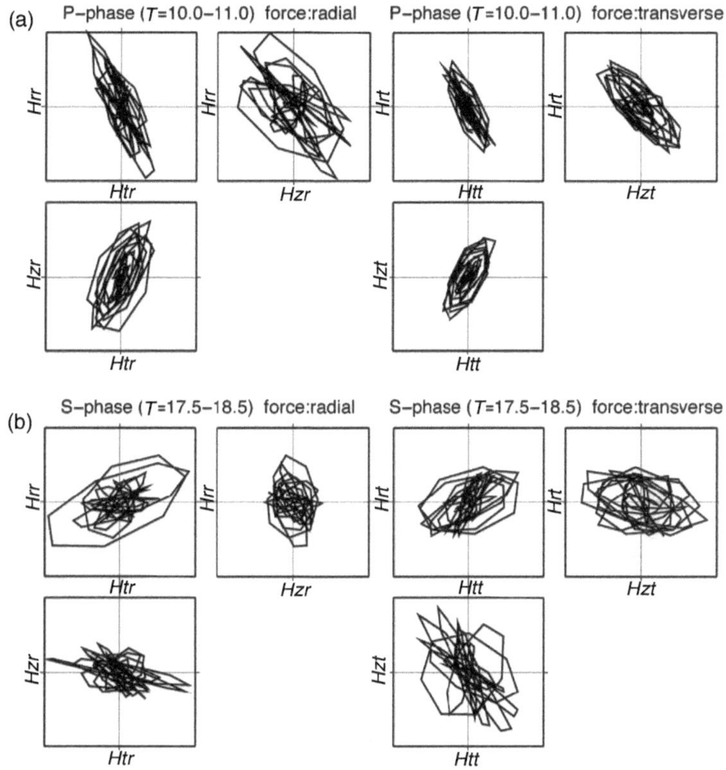

Figure 4 The particle motion of the six components of the transfer function at N.HOUH is displayed: (a) Particle motion of the P wave portion (travel time: 10 to 11 s)—the three-figure group on the left represents the radial forced case, whereas the transverse forced case is depicted on the right; (b) Particle motion of the S wave portion (travel time: 17.5 to 18.5 s)—the left and right three-figures groups indicate the radial and transverse forced case as in (a).

distance is less than that of N.HOUH. This difference suggests that the effect of velocity structure just beneath the observed station is larger than that of the propagation path. (One reason for these phenomena might be the regional differences in scatter distribution.)

Many phases can be seen in this record section. One example is the large amplitude of the later phase observed at N.OKZH [epicentral distance of 47 km] in *Hrr*. This phase can be observed in the large amplitude of the radial-forced case, and its principal axis is 45° from the radial or transverse direction. However, identification of these phases is difficult using only one station; seismic array stations must be deployed to identify them.

Figure 6 shows the temporal variation in the transfer function at N.HOUH for the $Hzt(t)$ component with travel time between 8 and 22 seconds, including the P and S phases. The stacked length of each trace is 10

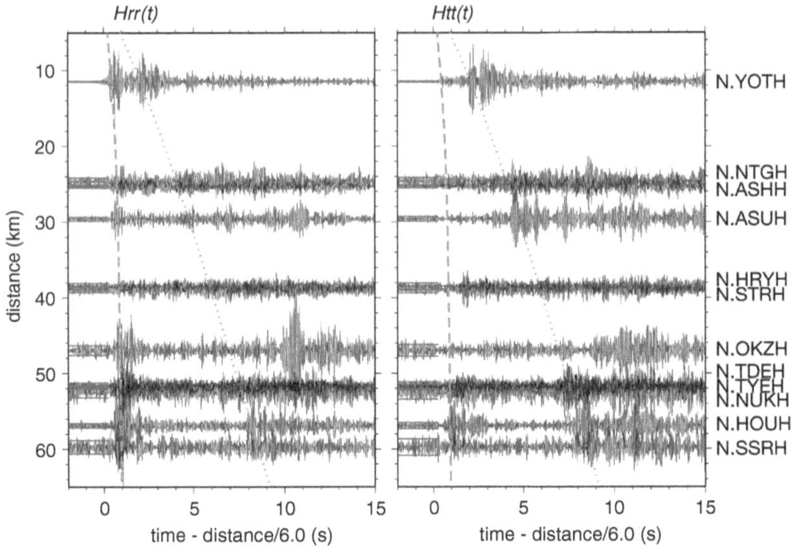

Figure 5 Record section for the transfer functions of $Hrr(t)$ (left) and $Htt(t)$ (right). Twelve Hi-net stations with epicentral distances from 10 to 65 km are included. The legend at the right lists the corresponding station code. The amplitude is normalized by the reciprocal of the epicentral distance. (See Figure 1 for a map view of the station distribution.) The horizontal and vertical axes denote the travel time reduced by 6 km/s and the epicentral distance from the Tono ACROSS transmitting station. The calculated P and S travel times are also shown in dashed and dotted lines.

days. Transfer functions were calculated every 10 days. Over the total analysis interval span of 1 1/2 years, 54 transfer functions from different periods were obtained. The differences in the transfer function, starting with the first trace (from 02/20/2004, for 10 days), are plotted in the upper chart, where the temporal difference with magnitude greater than the noise level at approximately the time of P and S wave arrivals can be seen. The large temporal change corresponds to the large amplitude of the original trace. We can observe an annual change of the transfer function. The amplitude of the temporal annual change is larger in the later parts of the P- and S-wave arrivals. This temporal change might be explained by a change in soil properties or underwater level near the surface, because the later parts contain transmitted or multiple reverberation phases near the earth's surface.

5. Conclusion and Future Plan

A transfer function with adequate S/N ratio was obtained by the Hi-net stations at an epicentral distance of less than 60 km from the Tono ACROSS transmitter. This area coverage is sufficient for mapping temporal

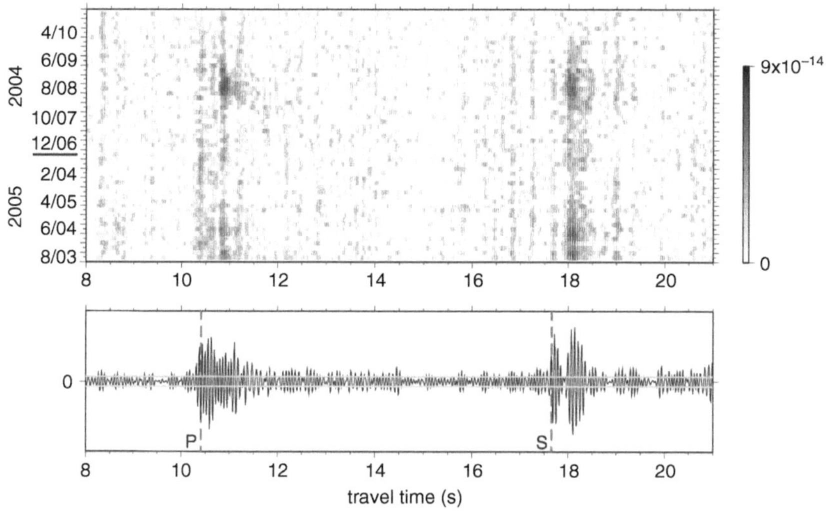

Figure 6 Temporal change of the transfer function ($Hzt(t)$) at N.HOUH for about one and a half years. The first trace (calculated from 02/20/2004) and its noise level (gray lines) can be seen in the lower figure. The calculated P and S arrivals are also shown by vertical dashed lines. Fifty-four transfer functions for every 10 days were calculated. The upper figure plots the difference in the transfer function from the first trace (calculated from 02/20/2004) in grayscale. Only the temporal changes with amplitudes exceeding the noise level are plotted.

changes in physical properties, such as seismic velocity, for a large area within the Tokai region.

The transfer function sequence definitely confirms the presence of temporal variation near the time of P and S wave arrivals, far beyond the uncertainty level specified by S/N. Resolution in time of the detected variation was improved by raising the signal level through the use of a low-frequency signal with less wave attenuation. These improvements were made in 2006 by constructing a new transmitter, deployed in Shizuoka Prefecture under a special research program of the Meteorological Research Institute.

A new seismic ACROSS transmitter was installed at the corner of the Mori town athletic field, located about 10 km SE from the J-5 shot point at which a seismic experiment with explosive sources was conducted in 2001, and found a strong reflection phase from the plate boundary between the Philippine-Sea plate and the Eurasian plate (Iidaka et al., 2003). This is a suitable location for the test site to verify the possibility of monitoring the temporal variation in seismic velocity or the reflection coefficient at the plate boundary. The new ACROSS transmitting station is indicated by a gray star in Figure 1. The primary purpose of this transmitter is to monitor the reflection amplitude of the upper boundary of the Philippine Sea plate (Kasahara et al., 2004).

ACKNOWLEDGMENTS

The waveform data used in this study was obtained from the National Research Institute for Earth Science and Disaster Prevention and the Japan Meteorological Agency.

REFERENCES

Hasada, Y., Kumagai, H., Kumazawa, M., 2001. Autoregressive modeling of transfer functions in frequency domain to determine complex travel times. Earth Planet. Space 53, 3–11.

Iidaka, T., Iwasaki, T., Takeda, T., Moriya, T., Kumakawa, I., Kurashimo, E., Kawamura, T., Yamazaki, F., Koike, K., Aoki, G., 2003. Configuration of subducting Philippine Sea plate and crustal structure in the central Japan region. Geophys. Res. Lett. 30, doi:10.1029/2002GL016517.

Ikuta, R., Yamaoka, K., Miyakawa, K., Kunitomo, T., Kumazawa, M., 2002. Continuous monitoring of propagation velocity of seismic wave using ACROSS. Geophys. Res. Lett. 29, doi:10.1029/2001GL01379.

Kasahara, J., Kamimura, A., Fujie, G., Hino, R., 2001. Influence of water on earthquake generation along subduction zones. Bull. Earthq. Res. Inst. Univ. Tokyo 76, 291–303.

Kasahara, J., Tsuruga, K., Hasada, Y., Yamaoka, K., Fujii, N., Yoshida, Y., Kunitomo, T., Kumazawa, M., 2004. A proposal of imaging of the plate boundary using the active monitoring method. In: Proc. AMSEG2004. pp. 44–48.

Kato, N., Hirasawa, T., 1999. A model for possible crustal deformation prior to a coming large interplate earthquake in the Tokai district, central Japan. Bull. Seism. Soc. Am. 89, 1401–1417.

Katsumata, A., Kamaya, N., 2003. Low-frequency continuous tremor around the Moho discontinuity away from volcanoes in the southwest Japan. Geophys. Res. Lett. 30, doi:10.1029/2002GL015981.

Kunitomo, T., Kumazawa, M., 2004a. Active monitoring of the earth's structure by the seismic ACROSS - Development of the Seismic ACROSS and continuous transmission at Tono Mine ACROSS station. In: Proc. AMSEG2004. pp. 218–222.

Kunitomo, T., Kumazawa, M., 2004b. Active monitoring of the earth's structure by the seismic ACROSS - Transmitting and Receiving Technologies of the Seismic ACROSS. In: Proc. AMSEG2004. pp. 181–184.

Nagao, H., Nakajima, T., Kumazawa, M., Kunitomo, T., 2004. Optimum weighted stacking method for acquisition of the ACROSS transfer functions having the maximum signal-to-noise ratio. In: Proc. AMSEG2004. pp. 266–270.

Obara, K., 2002. Nonvolcanic deep tremor associated with subduction in southwest Japan. Science 296, 1679–1681.

Ozawa, S., Murakami, M., Kaidzu, M., 2002. Detection and monitoring of ongoing aseismic slip in the Tokai region, central Japan. Science 298, 1009–1012.

Ueno, H., Yoshida, Y., Matsuoka, H., Ishikawa, Y., Kunitomo, T., Kumazawa, M., 2003. The observation and analysis of ACROSS signals by seismic networks (Part I). The detection of seismic signals by JMA and Hi-net, Prog. Abst. Joint Meeting.

Ueno, H., Hatakeyama, S., Aketagawa, T., Funasaki, J., Hamada, N., 2002. Improvement of hypocenter determination procedures in the Japan Meteorological Agency. Q. J. Seismol. 65, 123–134 (in Japanese with English abstract).

CHAPTER 26

Active Monitoring Using Submarine Cables—Leveraging Offshore Cabled Observatory for Passive Monitoring

Hitoshi Mikada[a], Junzo Kasahara[b], Naoyuki Fujii[c], *and* Mineo Kumazawa[d]

Contents

1. Introduction

Japan started installing cabled observatories for disaster mitigation purposes in 1978. The Headquarters for Earthquake Research Promotion and the Ministry of Education, Culture, Sports, Science and Technology

[a] Kyoto University, Kyoto-Daigaku Katsura, Nishikyo-ku, Kyoto, 615-8540, Japan

[b] Japan Atomic Energy Agency (Present Affiliation: Shizuoka University, 836 Ohya, Suruga-ku, Shizuoka, 422-8017, Japan)

[c] Nagoya University (Present Address: Midori-Cho, 2-1-504, Ashiya-Shi, Hyogo Pref. 659-0042, Japan)

[d] Atomic Energy Agency (Present Affiliation: Nagoya University, Furo-cho, Chikusa-ku, Nagoya, 464-8601, Japan)

E-mail addresses: mikada@tansa.kumst.kyoto-u.ac.jp (H. Mikada), junz_kshr@globe.ocn.ne.jp (J. Kasahara), snfujii@ipc.shizuoka.ac.jp (N. Fujii), mkumazawa@eps.nagoya-u.ac.jp (M. Kumazawa).

Handbook of Geophysical Exploration: Seismic Exploration, Volume 40

ISSN 0950-1401, DOI: 10.1016/S0950-1401(10)04031-0

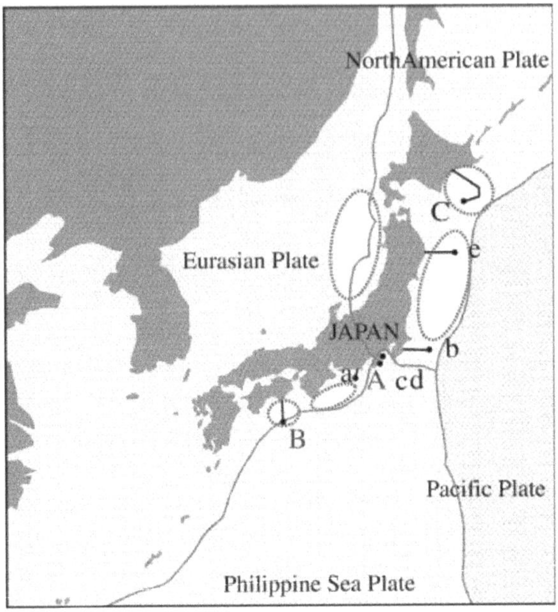

Figure 1 Cable systems, areas of special measurement act, Japanese islands. Eight cabled observatories around Japan for earthquake monitoring and engineering developments (Hirata et al., 2002). They are (a) JMA Off-Suruga, (b) JMA Off-Boso, c) ERI East Off-Izu Peninsula, (d) NIED Sagami-Trough, (e) ERI Off-Sanriku, (A) JAMSTEC Hatsushima (Engineering Development Site), (B) JAMSTEC Off-Muroto, and (C) JAMSTEC Off-Tokachi-Kushiro systems. The circled water areas were advocated by the Headquarters of Earthquake Research Promotion in that the real-time observations are necessary for future potential catastrophic earthquakes.

(MEXT) advocated, in 1996, the installation of cabled observatories for earthquake and tsunami monitoring purposes at least in five sea areas (Figure 1). The advantages of utilizing cabled observatories have been reported as (1) improvements in the detection of micro-earthquakes in seismogenic zones in terms of their magnitudes and locations (Watanabe et al., 2006a), (2) improvements in the real-time detection of tsunamis with a leading time before their arrival to the coast (Matsumoto and Mikada, 2005), (3) detection of long-term seafloor crustal deformations (Mikada et al., 2006), etc. The deployment of earthquake monitoring systems offshore has been proven efficient in the past for studying earthquakes taking place mainly offshore.

Recently, seismological studies of earthquake source mechanisms indicate: (1) a seismic asperity of a fault may break more than once at the same area of the fault in a different earthquake cycles (Nadeau and McEvilly, 1999; Nagai et al., 2001; Hirose and Hirahara, 2002), (2) the rupture process can be complex and in a wide variety of earthquake phenomena is caused not only by an heterogeneity such as asperity, but also from an intrinsic property in the fault system (Hirose and Hirahara, 2002). In the past, passive monitoring, micro-earthquakes and geodetic measurements have been the main target of observations, but the physical properties of fault planes can not be directly estimated through passive observations unless we find a way to monitor these properties directly. Reflection profiling, i.e., an active method, before and after a megathrust earthquake that took place in 2003, showed that there were changes in reflectivity and seismic velocities at the location of its hypocenter (Tsuru et al., 2005). This finding implies that active schemes could be taken to obtain time-variant physical properties of fault planes. Therefore, it is important to resort to active schemes for acquiring physical properties in addition to passive monitoring schemes. The integration of these methods should be the most advantageous way to further understand earthquake phenomena.

We propose (1) passive monitoring of seismogenic zones for precise micro-earthquake locations, source mechanisms, and continuous geodetic monitoring of the seafloor deformation, and (2) active monitoring using multidisciplinary sensors for acquiring changes in the physical properties in the vicinity of the plate interface in seismogenic zones. We would like to demonstrate scientific results from our past studies, both in earthquake and tsunami monitoring, and to show the necessity of further active monitoring schemes. Our discussions will also show the necessity of offshore observations for understanding earthquake generation and for the continuing research and development of multi-purpose observation sensors for future earthquake monitoring systems.

2. Japanese Cabled Observatories

Japanese islands are located at the intersection of four plates (Figure 1). The collision of these plates causes earthquakes at their boundaries every several tens to a few hundred years. Figure 1 represents eight earthquake monitoring cabled observatories, one of which is a test bed for sensor and engineering developments in Japanese waters (Hirata et al., 2002). Since the first cabled observatory was installed in 1978 off the Tokai region (annotated as "a" in Figure 1), the other seven observatories have been successively deployed for monitoring future megathrust earthquakes and associated tsunamis. These systems, mainly equipped with either broadband or highly sensitive seismometers and water pressure gauges, are in continuous operation with all acquired data being telemetered in real-time to the broad community of researchers, the Japan Meteorological Agency (JMA) and other research organizations through the High Sensitivity Seismograph Network (Hi-Net) system of the National Research Institute for Earth Science and Disaster Prevention Center (NIED) (Obara, 2002).

For further research in earthquake seismology or regional tectonic studies, passive monitoring of earthquake activity in the seismogenic zones at plate boundaries is currently the mostly practiced approach. During the decades of system development and accumulated experience of using submarine systems, the number of systems has been insufficient to cover the high-risk region of megathrust earthquakes. This is because the most critical problem in the installation of cabled seafloor observatories is the high cost of deployment (Asakawa et al., 2002).

3. Scientific Advances Brought by the Cabled Observatories

3.1. Earthquake Studies

The installation of seismometers offshore has been well appreciated by seismologists for locating seismic events taking place at the plate boundaries (Hino et al., 1996 for example). It has been shown that earthquakes of much smaller magnitudes were observed after the installation of a cabled observatory in southwest off Hokkaido (Ichiyanagi et al., 2004; Watanabe et al., 2006a; Figure 2). The distribution of micro-earthquake hypocenters has also been well defined for seismogenic zones to give a much finer shape of the zone (Watanabe et al. (2003, 2006b) for example; Figure 3). Figure 3 depicts the distribution of micro-earthquakes that took place off Kushiro-Tokachi after the installation of the cabled observatory. Many events that may not be detectable only by land observations would, therefore, be detected around the area where the cabled observatories have been installed. Ichiyanagi et al. (2004) noted that there was an earthquake swarm which took place about two years before the 2003 Tokachi-oki earthquake (M_{JMA} 8 subduction earthquake) near its hypocenter for two months. Although any causal relationship between the swarm and the mainshock of the M8 subduction earthquake may exist, such a short-duration swarm could not be well located or even detected without real-time telemetered cabled observation. We think that future installation of systems offshore would surely bring much deeper knowledge about seismogenic processes to researchers.

3.2. Tsunami Studies

Pressure gauges attached to cabled observatories are run to detect tsunamis and underwater pressure fluctuations at the locations of the Japanese cabled observatories. Pressure gauges, in general, have the resolution of, depending on the length of averaging window, about sub-millimeter at 1 Hz to lower frequencies. It was shown that the accommodation of tsunami data

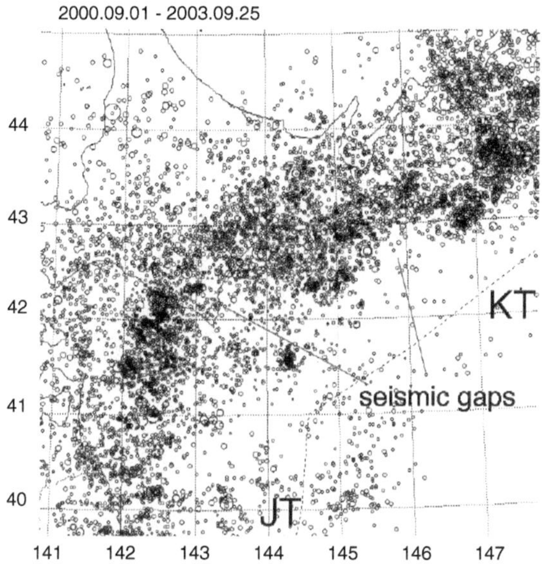

Figure 2 Epicenters of micro-earthquakes after the deployment of the cabled observatory until a day before the 2003 Tokachi-oki earthquake (Watanabe et al., 2006b). Earthquakes with a magnitude greater than four are plotted in this figure. The epicenters of micro-earthquakes in the southwestern part of the Kuril seismogenic zone are almost uniformly distributed except for two seismic gaps off the Nemuro Peninsula (right gap), and off Tokachi (left gap). The seismic gaps correspond to the focal areas of the 1973 Nemuro-Hanto-Oki (M_{JMA} 7.3) and 1952 Tokachi-oki earthquake (M_{JMA} 8.2), respectively. Notations, KT and JT, denote the Kuril and Japan Trenches, respectively.

in the study of an earthquake has enabled us to constrain the depth of earthquakes offshore (Hirata et al., 2003). Hirata et al. (2003) proposed that the observed tsunamis offshore could be compared with numerically simulated tsunamis for the earthquake that caused the observed tsunami. Figure 4 shows the location of the earthquake that they used in their study. Since the source depth of the earthquake, which took place outside the available seismometer network, fluctuated depending on the choice of the seismic velocity structure, it was difficult to estimate the depth with enough accuracy. On the other hand, the observed tsunami time series is a function of the order of the deformation of the seafloor, which strongly depends on the fault parameters including the source depth. The source depth of the earthquake could be estimated through a generalized least square error criterion.

Matsumoto and Mikada (2005) have used an observed tsunami time series to constrain the fault orientation and inclination for an earthquake that took place near the toe of the Nankai trough. The hypocenter and the mechanism of the earthquake were well estimated by land earthquake monitoring, but there was little aftershock after the mainshock. Due to the

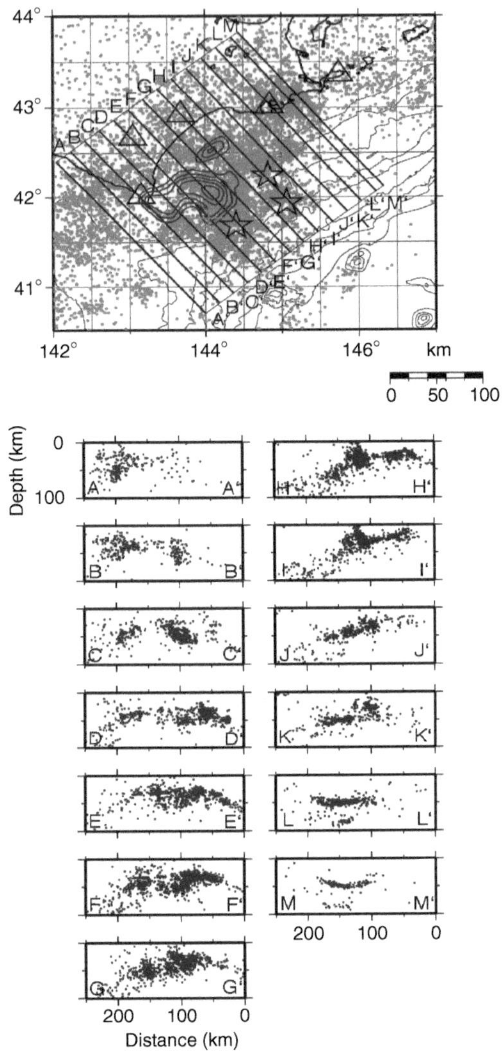

Figure 3 Epicenter and cross sectional hypocenter distributions after the 2003 Tokachi-oki earthquake (Watanabe et al., 2006b). Aftershocks were observed for a period from 26 September 2003 until 31 July 2005. Stars and triangles in the top figure indicate the locations of cabled ocean bottom seismometers (OBS) and land-based stations, respectively. Alphabetically aligned lines from A-A' to M-M' denotes sub-regions for each of which the depth distribution of hypocenters is drawn as shown in the bottom of the figure. Vertical exaggeration is taken as 1:1. Contour lines depict every one-meter displacement, i.e., the asperity map of the 2003 Tokachi-oki earthquake estimated by Yamanaka and Kikuchi (2003), on the plate interface. The maximum displacement is estimated as greater than 5 meters. The red curve between the two lines, H-H' and I-I', indicates the location of the Kushiro Submarine Canyon. A red cross-hair indicates the epicenter of the 2003 Tokachi-oki earthquake. (For interpretation of the references to colour in this figure legend, the reader is referred to the web version of this book.)

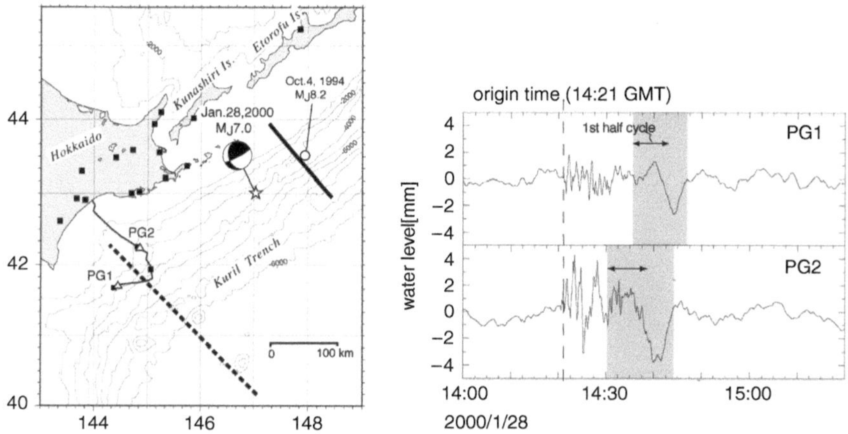

Figure 4 (Left) Location map of an earthquake (January 28, 2000) and the pressure gauges in the study by Hirata et al. (2003). (Right) The observed water pressure at the locations, PG1 and PG2, which are indicated in the figure. The maximum amplitude of the micro-tsunami was about 4 and 6 mm at PG1 and PG2, respectively. They estimated the depth of the earthquake, which is very difficult to determine by land seismic observations, to minimize the discrepancies between the observed and simulated tsunami waveforms within the first cycle of the onset.
Source: Both figures are reproduced from Hirata et al. (2003).

lack of aftershocks, a problem remained in identifying which nodal plane could be the fault that had caused the earthquake. Matsumoto and Mikada (2005) compared two different numerically simulated tsunami time series for each of the fault models with the tsunami observed by a cabled observatory about 100 km away from the shore, and found the difference to conclude which nodal place should be the fault. At the same time, they found that the tsunami arrived about 20 minutes before it hits the shore nearest to the tsunami sensor of the cabled observatory (Figure 5). They claimed that this leading time of 20 minutes could be exploited to enable local inhabitants to evacuate and that it is fundamental to add real-time features to cabled observatories in terms of tsunami disaster mitigation.

Mikada et al. (2004) demonstrated that there was a standing pressure wave generated between the surface and seafloor at the time of the 2003 Tokachi-oki earthquake (Figure 6). When an earthquake deforms the seafloor resulting in a static vertical displacement, a standing wave whose wavelength equals four times the water depth could be generated. While the tsunami height reflects the vertical displacement, the acoustic standing wave has an amplitude almost proportional to the time derivative of the seafloor displacement, i.e., the vertical deformation velocity. At the time of the 2003 Tokachi-oki earthquake, a pressure gauge installed ca. 28 km from the epicenter recorded both tsunami and standing waves with amplitudes of 0.4 and 40 meters, respectively, in equivalent water depths. Mikada et al.

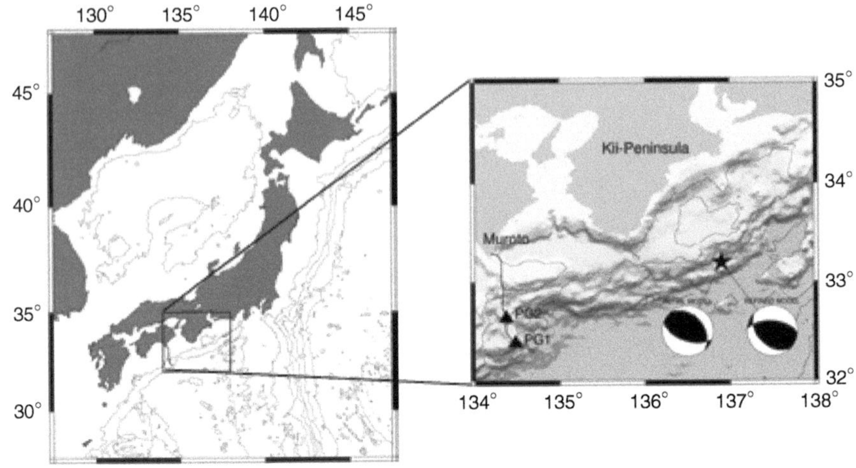

Figure 5 Location of the epicenter of the 2004 off Kii-Peninsula earthquakes (September 5, 2004; Matsumoto and Mikada, 2005). The star indicates the epicenter of both the fore- and the main-shocks. The triangles represent the locations of cabled pressure gauges off Muroto. The two focal mechanisms of the main-shock are estimated in teleseismic waveform modeling by Yamanaka (2004). These two source models are slightly different from each other. The left one indicates a northeast-dipping fault, while the right shows a southwest dipping. Although seismic waveform modeling cannot distinguish which is better, due to a lack of aftershocks, tsunami modeling indicates that the southwest dipping fault model (right source model) is suitable to explain the observed water pressures. Note that the tsunami was observed about 20 minutes before its arrival at the Muroto tidal station, which is indicated in the figure.

(2004) concluded that the seafloor deformation took place in an interval of three seconds at the time of the earthquake and proposed that to deal with seafloor geodetic deformation as a function of time for future tsunami studies.

3.3. Geodetic Studies

The 2003 Tokachi-oki earthquake took place near a cabled observatory. After the main shock, water pressure data from the tidal gauges in the observatory showed a linear trend in the difference between the synthetic and observed water depths. The rates of the linear uplifts at the pressure sensor locations are estimated as 4.1 and 3.9 mm/day respectively at the PG1 and PG2 locations over a period of two months after the mainshock (Mikada et al., 2006). After the main shock, the fluctuation of the difference between the observed and estimated water pressures became unstable and uplift of the seafloor continued.

One of the observatories at the time of the 2003 Tokachi-oki earthquake was the coseismic uplift of the seafloor. If we estimate the distribution of seafloor uplift using one of the rupture models obtained by seismic waveform

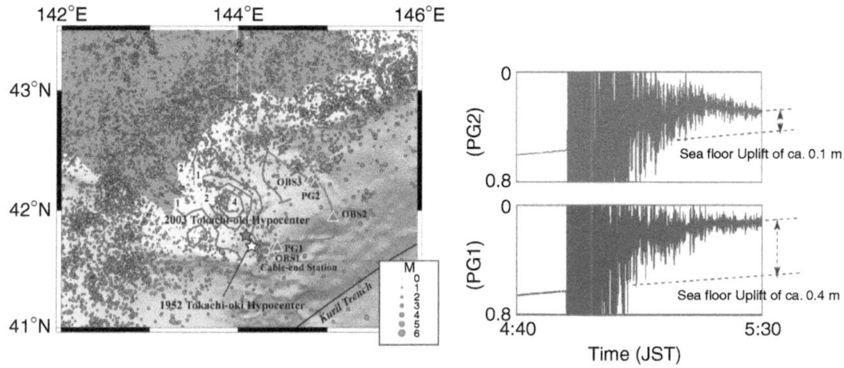

Figure 6 (Left) Locations of sensors in the region of the 2003 Tokachi-oki earthquake. The two open stars depict the hypocenter of the 2003 and 1952 Tokachi-oki earthquakes, respectively (Mikada et al., 2006). The thick line and triangles represent the cable and seismometer locations, respectively. The circle at the end of the thick line and triangles are for a cable-end station and tsunami gauge locations. Contours are drawn every meter for the displacements estimated by a joint inversion of teleseismic and strong ground motion data (Yagi, 2004) for the earthquake. The dark circles are for micro-seismic events from 2000 until August 2003 whose locations were determined by the Japan Meteorological Agency (JMA). Also the magnitudes of the events are indicated by the size of circles. (Right) Observed tidal record acquired by the cabled observatory at the time of the 2003 Tokachi-oki earthquake. Fluctuations in water pressure at the onset of the mainshock have amplitudes about 40 m in equivalent water depth at PG1, while they are about 30 m for PG2. Static discontinuities at the two locations indicate the seafloor uplifts. High amplitude water pressure fluctuations were caused by an abrupt seafloor uplift (Mikada et al., 2004). The tsunami is also included in the tide record whose half amplitudes were about the same as the seafloor uplift values (Mikada et al., 2004). They are all superimposed in the fluctuations.

studies, the recorded static uplift can be well explained as shown in Figure 6. In Figure 7, we could see that the major part of coseismic deformation is located offshore and that there is a tendency in the distribution of aftershocks to be located beneath the uplifted zone of the rupture area.

Long-term variations also indicate that preseismic subsidence and postseismic uplift took place before and after the mainshock. These pressure fluctuations are now recognized as a preseismic seafloor subsidence and postseismic seafloor uplift before and after the earthquake, respectively. The cabled observatory was installed in 1999, i.e., four years before the mainshock, and pressure gauge data recorded about 0.03 m/yr of the continuous subsidence of the seafloor before the mainshock (Mikada et al., 2006). After the mainshock, a continuous uplift was observed at a rate of 4 mm/day at the beginning with the rate of uplift asymptotically diminishing in the long term. These long-term variations are, however, not confirmed in their reliability, since the long-term response changes of the pressure sensors have not been well analyzed in the past. As Takahashi and Kasahara (2003) pointed out preseismic changes such as the formation of a seismic gap before

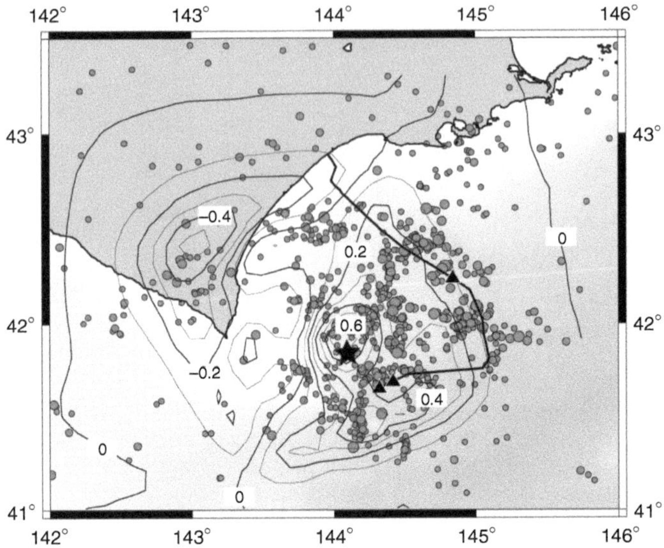

Figure 7 Estimated coseismic uplift at the time of the 2003 Tokachi-oki earthquake (Mikada et al., 2006). Because of displacement distribution for the plate boundary, a large area of vertical displacement is formed, mainly offshore, and relatively minor subsidence takes place along the coastline. The numbers on the contours are in meters and the solid circles are for post-seismic events whose estimated magnitudes are more than three. The units of the vertical displacement values are meters.

the 2003 Tokachi-oki earthquake, there should be certain processes ongoing in the focal region of the earthquake. However, the sensors of the above long-term passive monitoring system could not detect any change other than that of seismicity or long-term seafloor subsidence. This fact implies three major future directions: (i) observations should be made as close as possible to the hypocenter of plate boundary earthquakes to detect signals with a good signal-to-noise ratio, (ii) improvement of our understanding of seismogenic processes enough to model the nucleation of earthquakes, and (iii) introduction of active schemes to monitor seismogenic processes at depth.

4. Advances in Asperity Study

It has been hypothesized that the source regions for plate boundary earthquakes have stronger localized mechanical coupling, i.e., asperities (Kanamori and Stewart, 1978), along plate boundaries. The 2003 Tokachi-oki earthquake which took place 51 years after the 1952 Tokachi-oki earthquake has shown that not only the hypocenter location but the asperity

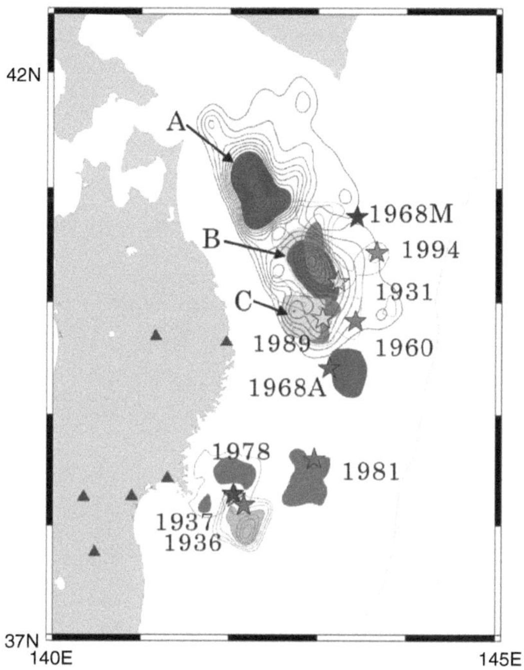

Figure 8 Asperity map for major plate boundary earthquakes, Off Tohoku, northeastern Japan, derived from teleseismic waveform modeling by Yamanaka and Kikuchi (2004). The stars indicate the locations of the initial ruptures. The contour lines show the moment release distribution for every 0.5 meters. Areas within the value of half the maximum slip for each seismic event are shaded as asperities.

distribution are very similar to each other (Yamanaka and Kikuchi, 2003). Recently, the study of earthquake source mechanisms has revealed that the location of asperities seems fixed in space away from the hypocenter, i.e., the location of initial breaks, and is a region of low aftershock activity after the asperities which have ruptured for seismogenic zones in northeastern Japan for the past 70 years (Yamanaka and Kikuchi, 2004; Figure 8). These findings indicate that the spatial distribution of asperities, is closely related to seismic radiation of possible future plate boundary earthquakes. If we assume that the asperities are space invariant for at least several decades, we might be able to map asperities, which are strongly related to mechanical coupling, along the plate interfaces.

Fujie et al. (2002) found that there are strong reflections from a plate interface, east of Tohoku along the Japan Trench subduction zone, in their seismic reflection profiling and that the strong reflections could be mapped onto a plate interface of low seismicity of micro-earthquakes, i.e., aseismic regions. They suggested an existence of materials with low compressional wave velocities along the plate interface to explain the observed amplitudes

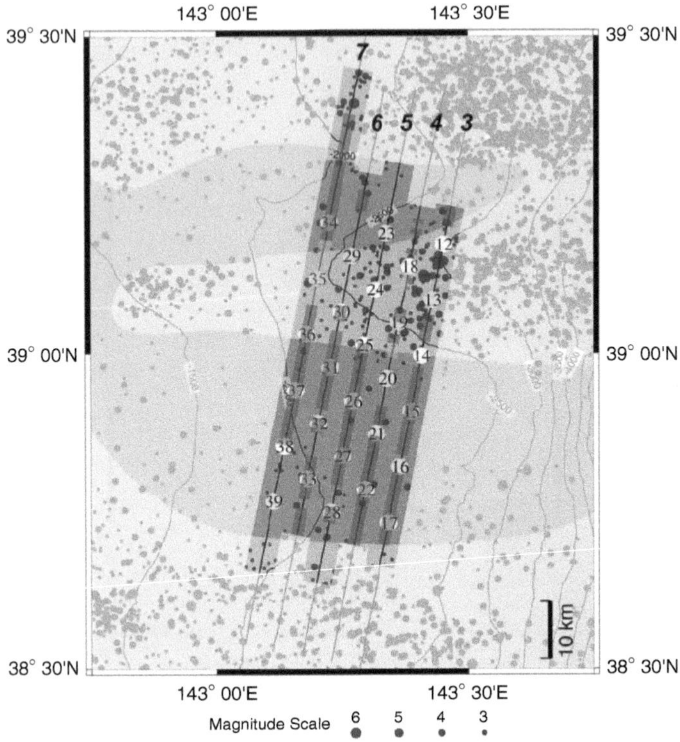

Figure 9 Map showing the difference in reflection coefficients for the plate interface (Mochizuki et al., 2005). The numbered circles are locations of OBSs used in this study. Non-numbered solid circles are epicenters of earthquakes with magnitudes greater than 2.5 observed by the on-land seismic network of Tohoku University during a period from 1975 to 2002. Two distinctive aseismic regions are seen for an area from 38° 42'N to 39° 00'N and from 39° 06'N to 39° 18'N. The seismically active area on the top-right of the figure is a part of the asperities shown in Figure 8. This figure indicates that reflection amplitudes in seismic reflection profiling could map aseismic regions.

of seismic signals. Mochizuki et al. (2005) conducted reflection surveys for a similar analysis as Fujie et al. (2002) and concluded that the aseismic regions could be caused by fluids and clays produced from sediments, which are subducting with the plate. They also proposed that aseismic slips would be dominant along the plate interfaces of observed strong seismic reflections (Figure 9). In this regard, we may assume that what Mochizuki et al. (2005) proposed can imply mapping of non-asperity regions. If we compare the results from the above studies (Figures 8 and 9) which were obtained for the same areas with different spatial scales, then we see that there is less overlap between the asperities (Figure 8) and the strong reflections (Figure 9) in an area near 39°N and 143°E. The strong reflections found by Fujie et al. (2002), seem to surround the asperity of the 1968A earthquake (Figure 8).

Since the spatial scale of these two studies are so different, this interpretation is not deterministic, but we could guess that at least non-asperities could be mapped with the help of active surveys, such as seismic reflection profiling.

5. Exertion of Earthquake Monitoring Capability

It would be right to say that any earthquake-related phenomena are observable if and only if they are associated with dynamic changes in environmental physical parameters, i.e., water pressure change, oscillation, etc., in the system of inertia. Earthquake models generally use static values of pore pressure to explain that interstitial fluids support the load between two fault planes (Brodsky and Kanamori, 2001). Seismologists think that pore fluids play an important role in the generation of earthquakes both to lower friction at the interface of two materials facing each other in fault zones, and to reduce the load normal to the fault plane (Hubbert and Rubey, 1959). The importance of hydrological observations, or the detection of changes in physical properties in fault zones, is, therefore, claimed. Unfortunately, there are not many observations that have been done monitoring fluid pressures except for off-line systems such as Circulation Obviation Retrofit Kits (called as CORK; Davis et al., 2002 for example). Since passive monitoring of such properties at depth require drilling or any other schemes to enable direct access for installing sensors or observation systems.

The physical properties of materials in fault zones may be manifested in earthquake behavior along the subduction zone interface (Bilek and Lay, 1998). Tsuru et al. (2005) conducted a time-lapse seismic reflection survey before and after the 2003 Tokachi-oki earthquake and found that the reflection coefficients of some reflectors near the source area exhibited a spatiotemporal evolution. They used the evolution pattern and the aftershock observations of Shinohara et al. (2004), and proposed that the change is caused by gradual fluid flow. The magnitude of the reflectivity change implies a 1.6% increase in porosity, a 2.2-5.6 MPa reduction in shear strength under the condition of the subsequent permeability in the plate boundary shear zone being 10^{-15} m^2. Tsuru et al. (2005) have shown that a time-lapse seismic survey can be applied to detect changes in physical properties for inter-plate earthquakes.

Ma et al. (2003) analyzed ground motion data of the 1999 Chi-Chi earthquake in Taiwan and found that a fault lubrication mechanism can be applied to the explanation of the difference in the frequency contents of seismographs between the two regions near the earthquake fault. They also proposed that fluid pressure in the fault zone with respect to the order of asperity defines the slips in the fault zone and the frequency contents of the radiations.

If we interpret the above findings, the following necessary conditions could be summarized towards monitoring of earthquake faults for future rupture:

(1) Mapping of the order of asperity over the fault plane is necessary to know where possible large slips are likely to take place,
(2) The fluid pressure distribution at the fault interface needs to be estimated with respect to the distribution of asperity, and
(3) Time-lapse reflection profiling would contribute to estimate large changes in physical properties in seismogenic zones before and after an inter-plate earthquake.

Therefore, it would be fair to say that future observation and monitoring require both space and time axes to understand earthquake phenomena.

6. Discussions

6.1. Passive Monitoring of Seismogenic Processes

In the previous chapters, we listed the scientific results from passive monitoring. Since cabled observatories are deployed near or right above the seismogenic zones, the acquired data from the cabled observatories could bring new findings in geodesy, seismology, tsunami disaster mitigation, etc. Tsunami observation has proven that it is important to have additional data to constrain the fault parameters or to monitor geodetic seafloor movement as a function of time (Mikada et al., 2004). Offshore tsunami observation detected tsunamis with enough leading time to the arrival to the shore (Matsumoto and Mikada, 2005). Seismic records have shown that the spatial coverage by sensors over a wide focal plane of earthquakes enhances the resolution in the hypocenter determination (Shinohara et al., 2004). Multidisciplinary seafloor environment sensors have worked to detect underwater debris flow associated with an inter-plate boundary earthquake (Mikada et al., 2006). In these past studies, it has been proven that the data obtained using cabled observatories are indispensable to understand the nature of earthquakes in terms of disaster mitigation, understanding their generation mechanisms, etc.

Since the deployment of cabled observatories is always associated with difficulties in funding, the number of observatories that has been installed is limited. However, we need to exert our efforts to improve the situation by means of increasing the number of sensors to enhance the spatial coverage near plate boundaries for further understanding of generation of earthquakes at plate boundaries. In terms of disaster mitigation, real-time observations should be realized and the installation of cabled systems is the most practical solution.

We are now working towards the next step in understanding of earthquakes, and should consider the observation systems reflecting our past experience. We have learnt that the tsunami generation process is related to the velocity of geodetic seafloor movement and that the turbidity current is a phenomenon associated with an earthquake. Multi-disciplinary observation would be an example that would necessitate other means as part of cabled observations so that earthquake related phenomena, including those unknown in the past, could be detected and analyzed together with seismogenic processes.

6.2. Active Monitoring

Time-lapse reflection surveys have been proven to be an important scheme to acquire any changes in physical properties near the fault interface for which the asperity is mapped as a function of space through the reflection profiling technique (Fujie et al., 2002; Sato et al., 2005; for example). As Tsuru et al. (2005) attempted, the method should lead to monitoring of variations of fault zone properties after steady efforts of signal-to-noise enhancement for reflections in the acquired data.

Tsuru et al. (2005) assumed a continuous fluid flow for 3.4 years to explain the difference in reflection coefficients of a reflector before and after the 2003 Tokachi-oki inter-plate earthquake, as alluded to in the above. The reflection coefficient for the subsequent survey has risen to 2.7 times that for the initial survey. Since it is not possible to see if any expulsion of fluid took place at the inter-plate earthquake, a continuous fluid flow is the only possible assumption to explain the changes in reflection coefficient. However, as already reported by many authors, the combination of both steady and intermittent fluid flow seems to be a solution explaining the observed permeability and geochemical state of sediments and décollement at least at three subduction zones, i.e., Nankai, Costa Rica, and Barbados (Moore and Silver, 2002). Seismological advances now require continuous monitoring of such properties to understand how these changes take place in time. For these objectives of acquiring changes physical properties, active monitoring schemes including not only seismic reflection profiling but also the development of high-precision methodologies to detect such changes at depth are indispensable.

6.3. The inclusion of Space-Time Axes in Observations

Most of plate boundary inter-plate boundary earthquakes take place offshore. Major crustal deformations take place, accordingly, offshore. Although high density seismological and geodetic observations are now possible on land, we would like to emphasize that long-term and high precision measurements or observations offshore would be one of the most important directions for detecting any changes related to generation of subduction-related earthquakes, i.e., seismogenic and inter-seismic processes, at much closer locations to the seismogenic zone. Also, the inclusion of observations to detect any time-dependent changes of physical properties in the vicinity of seismogenic zones is the most indispensable scheme for a breakthrough in current seismology. Any interstitial fluid flow near an earthquake fault would change the pore pressure in time and space. Changes in physical properties might, accordingly, be not only time-dependent but space-dependent as well. We believe that the current technology of cabled observatories or any offshore observational schemes should be enhanced to accommodate future extension of the observations in terms of precision and detectability for changes of physical properties near earthquake fault both in space and time. Moreover, methodologies need to be developed to bring information on physical properties from the deep earthquake fault in the form of detectable signals. We think that active monitoring is one of the feasible directions towards the ultimate objectives of understanding earthquake phenomena. Figure 10 depicts a proposed direction called the Exploration of Asperities-Reflectors System (EARS; Tsuruga et al., 2006) for integrating the-above-mentioned mapping, monitoring, and real-time continuous monitoring of the seismic activity in seismogenic zones.

7. Conclusions

We have summarized the scientific achievement using data from the Japanese cabled observatories. Introducing recent advances in earthquake seismology such as earthquake asperities, time-lapse reflection profiling, etc., we would like to mention that it is very important to have the observatories offshore. Our conclusions are the following:

(1) A real-time cabled observatory has shown the ability:
    (i) to improve the accuracy of hypocenter determination and, hence, our understanding of tectonics through the spatial variation of aftershock seismicity,
    (ii) to highlight the importance of temporal geodetic changes in the generation of tsunami, and
    (iii) to detect tsunami with a long lead-time to their arrival at the shore,
(2) Mapping of physical properties along an earthquake fault plane is necessary for future understanding of earthquake generation, and
(3) Development of active monitoring schemes is of importance to detect spatiotemporal changes in physical properties along the earthquake fault.

We are convinced that one of the best routes to take would be the deployment of cabled observatories for detection and the development

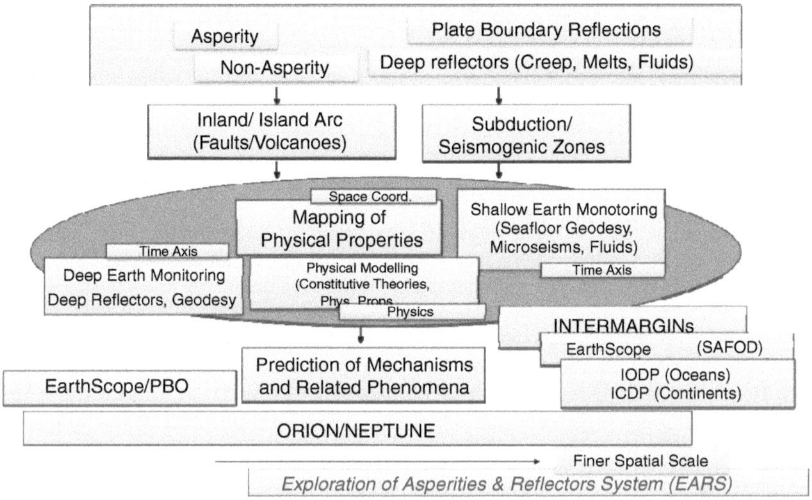

Figure 10 Schematic conceptual diagram of the Exploration of Asperities-Reflectors System (EARS; Tsuruga et al., 2006). As discussed in the text, we think that it is fundamental to understand earthquake generation as a function of time and space. Since any research may be aliased either in space or in time, allied efforts to integrate all available products are essential to cover the variation of wide spectra. In particular, long-term monitoring by cabled observatories, mapping of physical properties by active source surveys, and monitoring of time-varying physical properties are indispensable tools.

of active sources which have the ability to radiate signals to bring any information on changes of physical properties at depth back to the cabled observatory.

REFERENCES

Asakawa, K., Mikada, H., Kawaguchi, K., Iwase, R., Hirata, K., Goto, T., Mitsuzawa, K., Matsumoto, H., Watanabe, T., Suyehiro, K., 2002. Envisioned network system for future underwater observations. In: Proc. of Techno-Ocean '02. [CD-ROM].

Bilek, S.L., Lay, T., 1998. Variation of interplate fault zone properties with depth in the japan subduction zone. Science 281, 1175–1178.

Brodsky, E.E., Kanamori, H., 2001. Elastohydrodynamic lubrication of faults. J. Geophys. Res. 106 (B8), 16,357–16,374.

Davis, E.E., Becker, K., Dziak, R., Cassidy, J., Wang, K., Lilley, M., 2002. Hydrological response to a seafloor spreading episode on the Juan de Fuca ridge. Nature 430, 335–338.

Fujie, G., Kasahara, J., Hino, R., Sato, T., Shinohara, M., Suyehiro, K., 2002. A significant relation between seismic activities and reflection intensities in the Japan Trench region. Geophys. Res. Lett. 29. doi:10.1029/2001GL013764.

Hino, R., Kanazawa, T., Hasegawa, A., 1996. Interplate seismic activity near the northern Japan Trench deduced from ocean bottom and land-based seismic observations. Phys. Earth Planet. Inter. 93 (1–2), 37–52.

Hirata, K., Aoyagi, M., Mikada, H., Kawaguchi, K., Kaiho, Y., Iwase, R., Morita, S., Fujisawa, I., Sugioka, H., Mitsuzawa, K., Suyehiro, K., Kinoshita, H., 2002. Real-time geophysical measurements on the deep seafloor using submarine cable in the southern kurile subduction zone. IEEE J. Oceanic Eng. 27, 170–181.

Hirata, K., Takahashi, H., Geist, E.L., Satake, K., Tanioka, Y., Sugioka, H., Mikada, H., 2003. Source depth dependence of micro-tsunamis recorded with ocean-bottom pressure gauges: The January 28, 2000 Mw 6.8 earthquake off Nemuro Peninsula, Japan. Earth Planet. Sci. Lett 208, 305–318.

Hirose, H., Hirahara, K., 2002. A model for complex slip behavior on a large asperity at subduction zones. Geophys. Res. Lett. 29 (22). 2068. doi:10.1029/2002GL015825.

Hubbert, M.K., Rubey, W.W., 1959. Role of fluid pressure in mechanics of overthrust faulting. Geol. Soc. Am. Bull 70, 115–166.

Ichiyanagi, M., Takahashi, H., Motoya, Y., Kasahara, M., Hirata, K., Mikada, H., 2004. Hypocenter determination of Off Tokachi earthquakes for 2001 to 2002. Bull. Seismol. Soc. Jpn. 56 (4), 529–536 (in Japanese with English abstract).

Kanamori, H., Stewart, G.S., 1978. Seismological aspects of the Guatemala earthquake of February 4, 1976. J. Geophys. Res. 83, 3427–3434.

Ma, K.-F., Brodsky, E.E., Mori, J., Ji, C., Song, T.-R.A., Kanamori, H., 2003. Evidence for fault lubrication during the 1999 Chi-Chi, Taiwan, earthquake (Mw7.6). Geophys. Res. Lett 30 (5). 1244. doi:10.1029/2002GL015380.

Matsumoto, H., Mikada, H., 2005. Fault geometry of the 2004 off Kii-peninsula earthquake inferred from offshore pressure waveforms. Earth Planet. Space 57 (3), 161–166.

Mikada, H., Matsumoto, H., Watanabe, T., 2004. Wanted long-term observations? Experience from a recent plate boundary earthquake. In: Proc. 1st International Workshop on Active Monitoring in the Solid Earth Geophysics, Task Group for Active Monitoring, International Union of Geodesy and Geophysics, Mizunami, Japan, Paper S2-04.

Mikada, H., Mitsuzawa, K., Matsumoto, H., Watanabe, T., Morita, S., Otsuka, R., Sugioka, H., Baba, T., Araki, E., Suyehiro, K., 2006. New discoveries in dynamics of an M8 earthquake—Phenomena and their implications from the 2003 Tokachi-oki Earthquake using a long term monitoring cabled observatory. Tectonophysics 426, 95–105.

Mochizuki, K., Nakamura, M., Kasahara, J., Hino, R., Nishino, M., Kuwano, A., Nakamura, Y., Yamada, T., Shinohara, M., Sato, T., Moghaddam, P.P., Kanazawa, T., 2005. Intense PP reflection beneath the aseismic forearc slope of the Japan Trench subduction zone and its implication of aseismic slip subduction. J. Geophys. Res. 110 (B01302). doi:10.1029/2003JB002892.

Moore, J.C., Silver, E., 2002. Fluid flow in accreting and eroding convergent margins. JOIDES Jour. 28 (1), 91–96.

Nadeau, R.M., McEvilly, T.V., 1999. Fault slip rates at depth from recurrence intervals of repeating micro-earthquakes. Science 285, 718–721.

Nagai, R., Kikuchi, M., Yamanaka, Y., 2001. Comparative study on the source process of recurrent large earthquakes in Sanriku-Oki region: The 1968 Tokachi-Oki earthquake and the 1994 Sanriku-oki earthquake. J. Seismol. Soc. Jpn. 52, 267–289 (in Japanese with English abstract).

Obara, K., 2002. Nonvolcanic deep tremor associated with subduction in southwest Japan. Science 296, 1679–1681.

Sato, H., Hirata, N., Koketsu, K., Okaya, D., Abe, S., Kobayashi, R., Matsubara, M., Iwasaki, T., Ito, T., Ikawa, T., Kawanaka, T., Kasahara, K., Harder, S., 2005. Earthquake source fault beneath Tokyo. Science 309, 462–464.

Shinohara, M., Yamada, T., Kanazawa, T., Hirata, N., Kaneda, Y., Takanami, T., Mikada, H., Suyehiro, K., Sakai, S., Watanabe, T., Uehira, K., Takahashi, N., Nishino, M., Mochizuki, K., Sato, T., Araki, E., Hino, R., Uhira, K., Shiobara, H., Shimizu, H., 2004. Aftershock observation of the 2003 Tokachi-oki earthquake by using dense ocean bottom seismometer network. Earth Planet. Space 56, 295–300.

Takahashi, H., Kasahara, M., 2003. Seismic activity off Tokachi region, Hokkaido, Japan. J. Seismol. Soc. Jpn. 56, 393–403 (in Japanese with English abstract).

Tsuru, T., Park, J.-O., Kido, Y., Ito, A., Kaneda, Y., Yamada, T., Shinohara, M., Kanazawa, T., 2005. Did expanded porous patches guide rupture propagation in 2003 Tokachi-oki earthquake? Geophys. Res. Lett. 32 (L20310). doi:10.1029/2005GL023753.

Tsuruga, K., Kasahara, J., Mikada, H., Yamaoka, K., Fujii, N., 2006. Mapping of non-asperities and continuous-active monitoring of interior of the earth's crust. J. Geography 115 (1), 51–71 (in Japanese with English abstract).

Watanabe, T., Takahashi, H., Ichiyanagi, M., Okayama, M., Takada, M., Otsuka, R., Hirata, K., Morita, S., Kasahara, M., Mikada, H., 2006a. Seismological monitoring on the 2003 Tokachi-oki Earthquake, derived from Off-Kushiro permanent cabled OBSs and land-based observations. Tectonophysics 426, 107–118.

Watanabe, T., Kasahara, M., Mikada, H., 2006b. Background seismicity, a megathrust earthquake and following aftershock activity at the southern end of the Kurile subduction zone, recorded by JAMSTEC off-Kushiro cabled geophysical observatory and land-based observation network. In: Proc. 4th International Workshop on Scientific Use of Submarine Cable and Related Technologies (SSC'06), Paper 049. [CD-ROM].

Watanabe, T., Otsuka, R., Mikada, H., Hirata, K., Takahashi, H., Kasahara, M., Suyehiro, K., 2003. Double seismic zone in the Hokkaido island, southern Kurile arc, derived from Off-Kushiro permanent OBS and land-based observations. In: Proc. IEEE, 3rd Inter. Workshop Sci. Use Submarine Cables Related Tech. pp. 133–136.

Yagi, Y., 2004. Source rupture process of the 2003 Tokachi-oki earthquake determined by joint inversion of teleseismic body wave and strong ground motion data. Earth Planet. Space 56, 311–316.

Yamanaka, Y., 2004. Offshore Southeast of the Kii Peninsula Earthquakes. EIC Seismological Note 152/153 (in Japanese).

Yamanaka, Y., Kikuchi, M., 2004. Asperity map along the subduction zone in northeastern Japan inferred from regional seismic data. J. Geophys. Res. 109 (B07307). doi:10.1029/2003JB002683.

Yamanaka, Y., Kikuchi, M., 2003. Source processes of the recurrent Tokachi-oki earthquake on September 26, 2003, inferred from teleseismic body waves. Earth Planet. Space 55, e21-24.

CHAPTER 27

Seasonal Changes in Vibroseismic Monitoring Experiments in Siberia

Aleksandr F. Emanov[a], Victor S. Seleznev[a], Innokentiy S. Chichinin[b], Victor M. Soloviev[a], Irina V. Zhemchugova[a], *and* Vladimir N. Kashun[a]

Contents

1. Introduction

The idea of a geophysical medium (Sadovsky et al., 1987) as something constantly varying in time has attracted experts in seismic methods for decades. Seismological research on changes in seismic wave velocities over time (Zubkov and Solovieva, 1986) has been conducted repeatedly in large-earthquake zones. Moreover, seismic monitoring of the medium using explosive sources (Myachkin, 1978; Myachkin et al., 1982; Gamburtsev et al., 1994; Gamburtsev, 1992; Seismic Studying of Focal Zones, 1983) has been conducted in many regions of the world. Further studies of small variations in the medium will undoubtedly be carried out in the future.

The challenges of vibroseismic monitoring are considered in a number of papers (Nickolaev et al., 1974; Alekseev et al., 1982). The attractiveness of this research direction is based on the fact that monitoring stress-strain states of seismically active zones in the Earth's crust can be done using vibroseismic signals, under constant conditions of excitation and recording. The high repeatability of seismic signal radiation was shown in a number of experiments (Yushin et al., 1981; Emanov et al., 1986, 1989) and inspires

[a] Geophysical Survey SB RAS, Pr. Akad. Koptyuga, 3, Novosibirsk, 630090, Russia
[b] Institute of Geophysics SB RAS, Pr. Akad. Koptyuga, 3, Novosibirsk, 630090, Russia

E-mail address: sel@gs.nsc.ru (V.S. Seleznev).

Handbook of Geophysical Exploration: Seismic Exploration, Volume 40
ISSN 0950-1401, DOI: 10.1016/S0950-1401(10)04032-2

confidence in such monitoring. Some conclusions about the usability of large vibrators (Alekseev et al., 2005) for monitoring purposes can be made from the following estimate: The vibrator described below reliably generated seismograms with good signal/noise ratios at distances of 300-400 km, using 30-45-minute-long vibration sessions. (The TNT equivalent of such a source would be 3-5 tons, if exploded in a group of shallow holes or in a water reservoir.) In a monochromatic radiation mode, the vibroseismic oscillations are clearly detected at distances up to 1500 km.

When performing experiments with this vibrator, investigators noticed that winter seismograms are essentially different from summer seismograms, and that understanding this difference could be the key to maintaining the high repeatability of seismograms. In this paper, we consider the experimental results from studying wave-field changes in time, gathered from vibroseismic monitoring. The main point of the experiments described below is as follows: The frequency characteristics of vibrating source radiation, located outdoors, evidently vary as the ground freezes or thaws. At the Bystrovka vibroseismic polygon, experiments were performed with two vibrators. In one part of the experiment, the vibrators were placed outdoors alongside one another, with one of them located in a covered underground pavilion. One of the vibrators, a 100 ton stationary vibrator used to monitor the Earth' crust, has a frequency range overlapping its resonance frequency. A 50 ton vibrator was operated above its resonance frequency and located outdoors for one year, and then located in an underground pavilion the following year. Seismic receivers were isolated to avoid temperature-induced variations.

The experiments were conducted in two regions: the Baikal rift zone and the Altai-Sayan fold area. We monitored changes in the emitted seismic field over time and also studied peculiarities in seismic wave attenuation within deep fault zones. Both these activities can be carried out through vibroseismic monitoring—the first measurement in time and the second in space—with the constant and known radiation characteristic of seismic waves.

2. Experiment Using Vibroseismic Monitoring

A stationary 100 ton vibrator was located at the Bystrovka vibroseismic polygon, situated on the right shore of the Ob' sea, 50 km from Novosibirsk (Alekseev et al., 2005). The Novosibirsk seismological station (at Klyuchi) was used as a recording point 49 km from the source. At this station, a three-component receiver is positioned on a concrete base at the CK-1P thermostat pavilion, rigidly in contact with bedrock. An east-west horizontal receiver was directed to a source-receiver line at a 40° angle, while a south-north horizontal receiver was directed at a 50° angle. The azimuth of the vibrator (from the point of observation) has a 230° angle. The orientation of the "source-receiver" line corresponded to the bearing of the Kolyvan' fold zone from the southwest to the northeast. Recording of seismic waves was carried out using Alpha-Geon seismological equipment (Gluchov et al., 1994), which recorded the displacement velocity.

The systematic recording of vibroseismic oscillations at the Novosibirsk station was begun in the summer of 1997. Each "single observation" (the experiment was conducted overnight) consisted of several vibrosessions (up to four sessions following one another), with the interval between individual observations varying from one day to two weeks. The observations (which are still ongoing) were conducted frequently in the autumn and spring, and seldom in the summer and winter. Each vibrosession was also recorded by a reference sensor located at a 3 m depth under a vibrator, using the CB-5 detector, located in a waterproof container. In the spring of 1998, seismic observations were supplemented by regular measurements of ground temperature down to a depth of 3 m.

Note that using digital seismological equipment allowed us to conduct uniform seismological observations and vibroseismic monitoring of that medium. The possibility of using the Alpha-Geon equipment to obtain seismograms from a vibrator was demonstrated in several experiments (Solodilov et al., 1995), in which (at a substantial distance from a vibrator) seismic noise, recorded in time intervals at a specified point some distance away from the vibrator, could be coordinated with vibrator activity. Using a computer, we could calculate the cross-correlation function between a record and a programmed sweep-signal from the vibrator. As a result, seismograms of good quality were obtained.

During the seismic-monitoring experiments, we maintained the same frequency range of 6.25-9.5 Hz. For a 100 ton vibrator, this frequency range spans the entire spectrum of source resonance. It is established that vibrator radiation at resonant frequencies is most sensitive to changes in ground conditions. In winter 1998, the experiment was supplemented with the inclusion of a 40 ton vibrator, so that a single observation consisted of two sessions with a 100 ton vibrator and two with a 40 ton vibrator. For the latter, a frequency band above its resonance of 8.5-11 Hz was selected. Excluding the source resonance from the working frequency band reduces the energy of emitted oscillations, but at the same time should also reduce the influence of seasonal changes in ground conditions on radiation. In the event that seasonal changes in ground conditions under a source exclusively affect the spectrum of excited oscillations, the transition to above-resonance frequencies should make seasonally independent seismograms possible.

In the next experimental step—started on November 23, 1999, and completed on April 25, 2001—a 40 ton vibrator was placed in the underground pavilion (basically an earth-house). Soil at the bottom of the pavilion was thawed all year around. The pavilion dimensions are 10 × 12 m, with a depth relative to the ground surface of 4 m.

In what follows, we present an overview of the obtained data, then consider the data in more detail and explain the phenomena observed.

Figure 1 shows a summary seismogram for P-waves, while Figure 2 illustrates the same for S-waves, obtained from the 100 ton vibrator during the period January 1997-March 1999. These seismograms are to be considered in tandem with Figure 3. The plots of Figure 3(a) were obtained by recording, for each vibrating session, the oscillations of a point at a depth of 3 m under a vibrator. On these vibrograms, we recovered impulse seismograms through calculating the cross-correlation function between the vibrogram and the sweep-signal (identical to how seismograms were accumulated in Figures 1 and 2). An amplitude spectrum for impulse seismograms at a depth of 3 m under a vibrator is shown in Figure 3(a).

This sensor has the same temperature (+1 °C) all year round. On an impulse seismogram, one impulse is recorded, the amplitude spectrum of which could be considered the spectrum of oscillations emitted by the vibrator. In Figure 4, the seismogram amplitude spectrum (Figure 1) is given for the time interval of 7.5-9.5 seconds (for P-wave arrival times), and in Figure 5(c) for the time interval of 14-17 seconds (Figure 2), corresponding to the time interval for recording S-waves together with Rayleigh waves. The seismic traces (Figures 1 and 2) correspond with spectra (Figure 4) on the recording dates, marked in the margins of these figures.

Figure 4(a) shows that in the winter (January-March 1997), there is rather good repeatability from one vibrosession to another; only at the end of March were there minor changes in radiation. Then, in April, a sharp change in the radiation spectrum occurred. The weather conditions during this period were characterized by a wide increase in temperature and abundant thawing of snow. As this took place, the ground thawed approximately 1 m over one week's time, and vibrator operation changed from winter radiation conditions to summer radiation conditions. In May, the frequency characteristic was smoothly transformed, having turned to a practically rectangular shape towards June 1997. The slow transformation process of the radiation spectrum can be connected to further thawing of a frozen ground bed under the vibrator.

The rectangular form of the radiation spectrum was maintained up to the middle of November 1997. At this time, the stable negative temperatures ran to −30 °C, and the process of ground freezing under a vibrator progressed rapidly. In spite of this, however, the vibrator radiation corresponded to summer characteristics. In January 1998, the vibrator radiation characteristics differ radically from the previous observations, specifically taking a typical winter form. A wave picture (Figures 1 and 2) and the observations in the near field show that the vibrator has changed its radiation from the summer mode to winter mode at the turn of December and January. Further winter types of a radiation remained to the middle of April. In 1998, spring developed very slowly: stable above-freezing temperatures only become settled in May. Collected data show

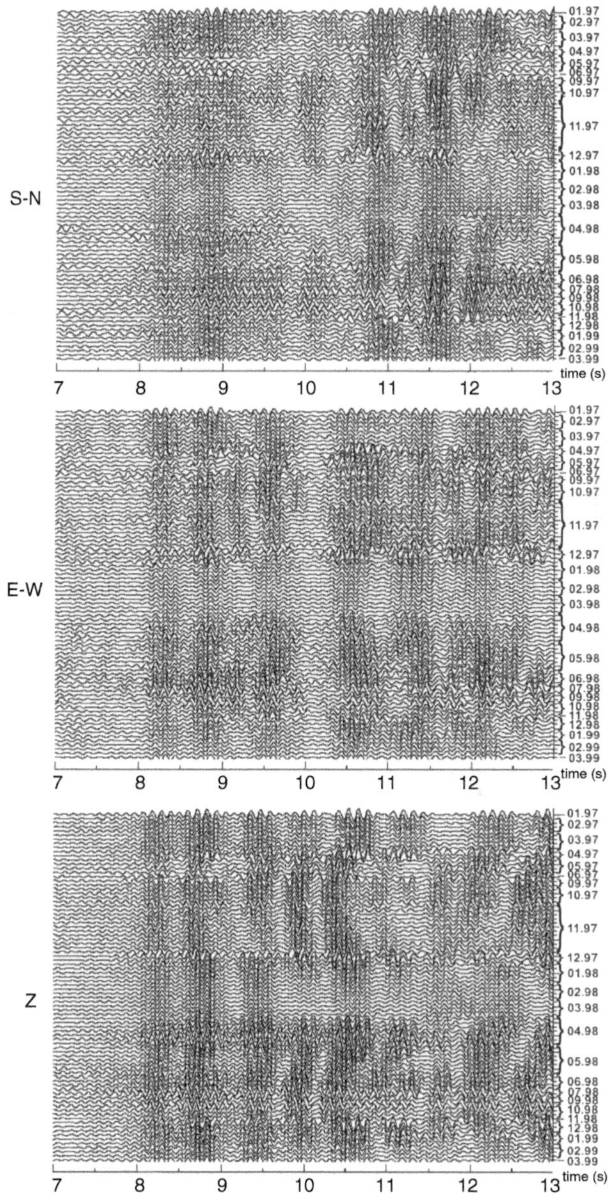

Figure 1 Seismograms recorded by a ternary sensor at the Novosibirsk seismic station from January 1997 through March 1999, with a time interval corresponding to the approximate arrivals of P-waves. The wave source was a fixed 100 ton vibrator. Recording was conducted at the Novosibirsk seismic station's north-south and east-west components; together with the Z (vertical) component. The vibrator azimuth from the recording point is 230°. Each trace is the result of stacked correlations of recorded vibrosessions with a 0.25-9.5 Hz frequency range.

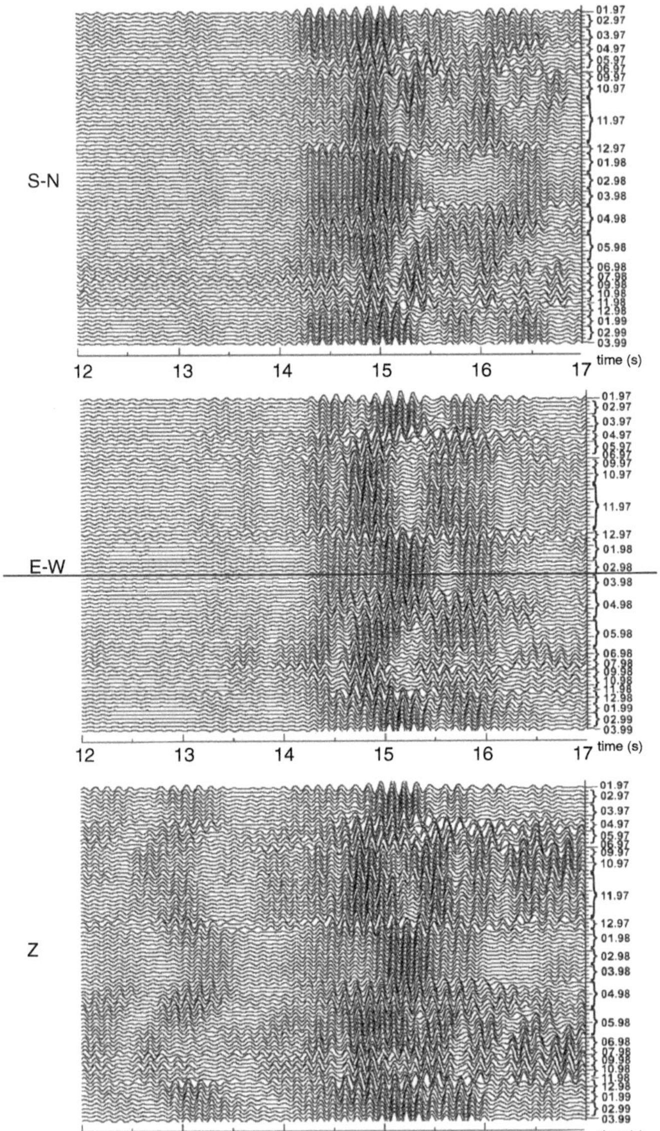

Figure 2 Seismograms recorded by a ternary sensor at the Novosibirsk seismic station from January 1997 through March 1999, with a time interval corresponding to the S-wave arrival time. Otherwise, all parameters are identical to Figure 1.

that the vibrator changed over to the summer radiation mode beginning in early June 1998. The transition period from winter to summer during that year was more extended compared to the previous year, and the measurements were conducted at a two-fold rate. Consequently, in 1998,

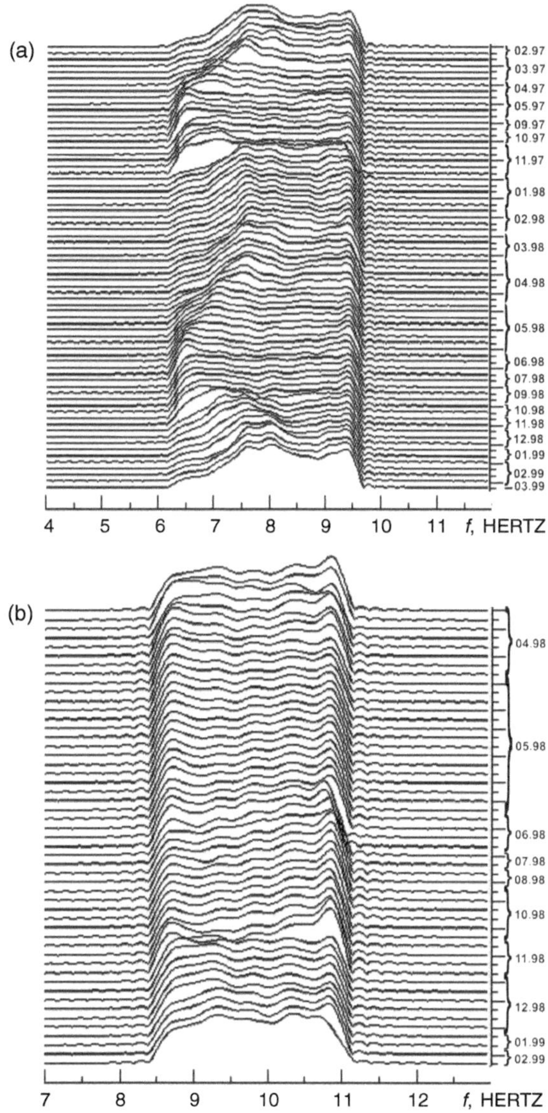

Figure 3 Seasonal changes in amplitude spectra for seismic signals from the sensors at 3 m depth under the (a) 100 ton vibrator and (b) 40 ton vibrator. Braces are on the right stacked spectra of radiated signals, obtained over a one-month period. Labels indicate the month and year.

more detailed information was obtained with respect to wave-field changes as the ground thawed. Further, up to 1999, radiation spectra repeated their patterns from summer to winter and vice versa. We therefore observed two types of stable vibrator radiation, that for summer and that for winter,

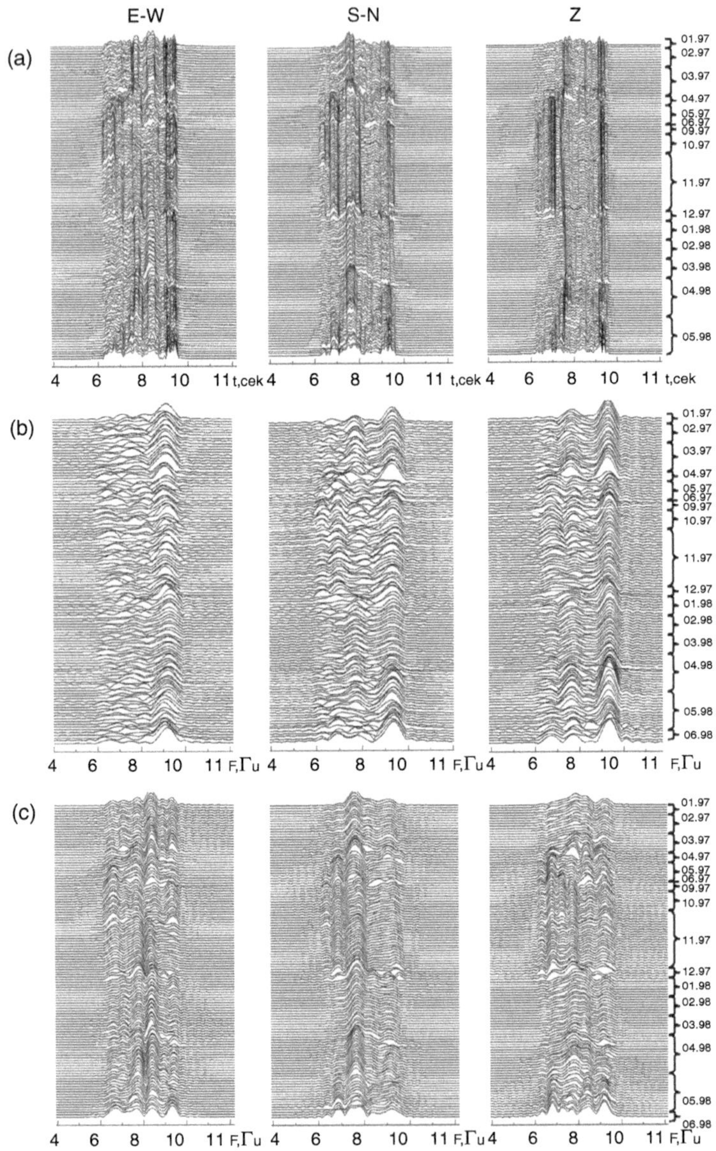

Figure 4 Seasonal changes in the amplitude spectra of seismograms recorded by a ternary sensor at the Novosobirsk seismic station. The wave source is the 100 ton vibrator. Seismograms corresponding to the given spectra are presented in Figures 1 and 2: (a) whole spectrum of seismograms; (b) spectrum for time intervals of P-wave arrival times; (c) spectrum for time intervals of S-wave arrival times.

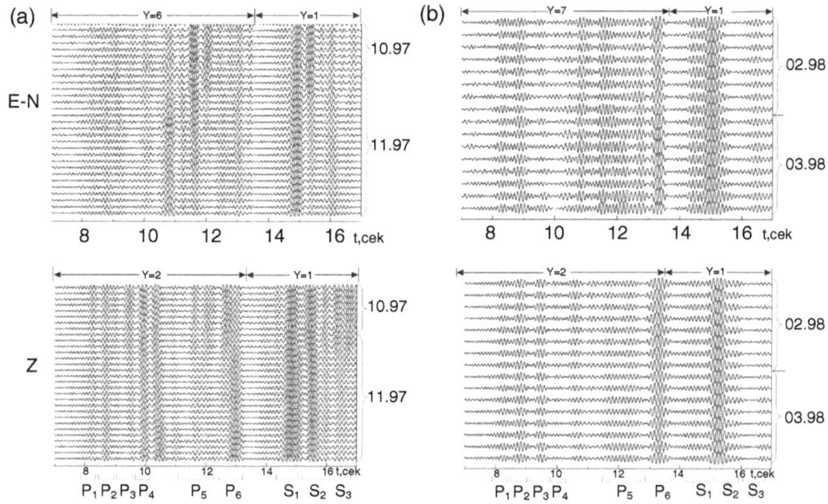

Figure 5 Data fragments from the study of wavefield recurrence over time with an invariable wave source and receiver. The source-receiver distance is 49.1 km: (a) seismograms recorded during the period of ground freezing, October-November 1997; (b) seismograms recorded during the period of stable-state ground, February-March 1997. Labels on the right indicate the month and year.

along with transition periods, the durations of which varied depending on weather conditions in the spring and autumn. The correlation between the temperature measurements in the ground and the radiation characteristics shows that in spring, the main changes to the spectrum take place when the ground thaws to ~1 m depth. At thicknesses of frozen ground less than 1 m, the vibrator operates in a summer-mode radiation spectrum, but the spectrum changes rapidly at greater than 1 m depth of frozen ground.

Returning to Figures 1 and 2, we see that the wave fields at a distance also have summer and winter modes with transition periods between them. In Figures 1 and 2, different gains are used, owing to the essentially weaker intensity of P-wave fields than that of S-wave fields. Detailed observations from vibroseismic monitoring show that the wave arrival times are not influenced by seasonal changes in source radiation. This fact allows us to use the kinematics of seismic waves for monitoring purposes without applying corrections. The dynamic characteristics of waves undergo large changes in time connected to transitions from winter to summer and vice versa.

At the beginning of the experimental setup, we were not sure whether the wave fields in the near and far zones would exhibit the same form they had one year before, because of possible irreversible changes in experimental conditions and the characteristics of the equipment used. The obtained data demonstrate that this concern was groundless. The changes in wave field dynamics have a cyclic character, although the calendar dates for the changes in wave dynamics do not strictly coincide. To update the seasonal changes in wave dynamics, we conducted a series of observations over several years, maintaining temperature control over the frozen ground underneath the vibrator.

Because of the large data volume, in Figure 1 we present just selected data. We noted previously that when the ground was frozen, the change in the spectrum begins when that freezing condition extends to about 1 m depth—until then, the radiation mode is still that of summer. Figure 5 presents data consisting of the fragments from Figure 1. For detailed consideration of the changes in wave dynamics, we selected two full-size fragments: the first fragment corresponds to an initial period of frozen ground that is less than 1 m thick ("summer"), the second one represents a stable period of greater than 1 m of frozen ground, following a "winter" source-radiation mode. For the first fragment, the radiation spectrum has not yet undergone noticeable changes, and in seismograms, a smooth modification in wave dynamics is observed. The second fragment of the series (Figure 5(b)) is characterized by high wave-dynamic stability.

Let us consider the waves recorded on seismograms in greater detail. The arrival time of the first longitudinal wave is 8.27 s (P_1), and the first transverse wave S_1 is 14.28 s. The used gain factor is shown in the Figure 5 seismograms. As can be seen, the transverse waves have high amplitudes—they are recorded at a minimum amplification. The longitudinal waves recorded by a horizontal receiver are weak. For their visualization, amplifications six and seven times larger were used. Compare these to the winter seismograms obtained from February-March 1998. The radiation spectra (Figure 3(a)) indicate an essential attenuation of oscillations for frequencies below 7 Hz. On seismograms (Figure 5(b)), this attenuation is not obvious, but on amplitude spectra of these records (Figure 4) it is very clear.

From the seismograms in Figure 5(b), and from other record samplings obtained during "stable" radiation periods, we note the high repeatability of the seismograms for different vibrosessions. The seismograms are identical, both for the times of wave arrivals and in the characteristics of wave dynamics. Fluctuations in amplitudes arise if the wave amplitude is commensurate with the noise level (the horizontal component of arrival times for longitudinal waves).

Returning to the seismograms for a period of ground freezing under the vibrator (Figure 5(a)) recorded from October 5 to November, 22—it is possible, within this monitoring-series fragment, to dynamically distinguish stable waves from waves whose dynamics have changed over time. The P_1, P_2, P_4 waves are stable, but the P_3, P_5, and P_6 waves are unstable. The arrival times of unstable waves are practically constant, but their amplitudes are rather surprising. If the P_3 and P_5 waves attenuate for any reason, the P_6 wave's energy changes from one phase to another. The freezing of the ground for the considered series fragment has not reached the point at which the radiation spectrum changes, but the rock freezing process affects the dynamics of some waves at a long distance from the vibrator. In comparing the seismograms in Figures 5(a) and 5(b), we note that prominent and stable dynamics vanish, starting with the P_4 wave in Figure 5(b). This is because the change in dynamics at the transition from summer to winter is not identical for each wave. We can distinguish dynamically stable waves from the wave field and perform monitoring analysis using these waves only. Note that the kinematics of any waves used in monitoring does not depend on the freezing or thawing of the ground. What is the reason for such differences in the dynamics of different seismic waves? Spectral analyses of seismic records show (Figure 4) that not only do the spectra of longitudinal waves differ from those of transverse waves, but also that the spectra of different longitudinal waves differ from each other. It turns out that the radiation spectrum is much wider than the spectra of waves recorded at a distance, and each wave with different arrival times has its individual spectrum. The distinctions in the wave spectrum and radiation spectrum can partly explain the changes in seismogram dynamics, but fail to explain *all* changes in dynamics. This fact is confirmed once again by an experiment with a 40 ton vibrator operated at frequencies above resonance with a stable radiation spectrum of oscillations in both summer and winter (Figure 3(b)). Figure 6 shows the seismograms obtained from this source at the Novosibirsk station.

In rather stable radiation spectra over time, we have a change in wave spectra, with their dynamics in transition from the winter to summer radiation mode. This finding once again confirms that, over a full year, freezing and thawing of the ground under a vibrator affects not only the radiation spectrum, but also very likely the directional diagram of the source as well. The first arrivals of longitudinal waves have stable time waveforms, with the only difference being that during the winter radiation period, such waves have somewhat smaller amplitude than during the summer radiation period. The dynamics of longitudinal waves for the latter arrival is essentially variable. Consider a wave with arrival time 9.8 s. During winter radiation, such a wave is hardly visible on a seismogram (beginning of April 1998 and January-February 1999). During the transition period from the winter to summer radiation mode (April-May), it is clearly distinguishable, while during summer, this wave has an intermediate intensity between the summer transition mode and the winter mode. In transition from the summer radiation mode to the winter mode, this wave increases its intensity. (Unfortunately, this period is represented only in a few traces.) Turning to the intense S-N-component first arrivals of transverse waves (14.28 s), we find that, using reliable phase correlation of a wave over time, we encounter deterioration in wave compactness during the summer. This leads us to conclude that for any given wave, the spectrum width varies from summer to winter.

This experiment showed that, for seismograms obtained at a long distance from the source, the transition in frequency domain (which

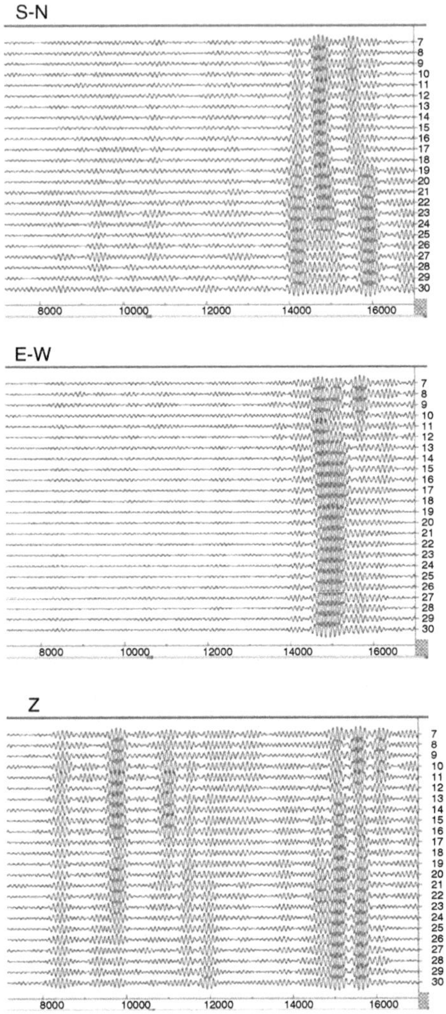

Figure 6 Vibroseismic monitoring of the changes from winter to summer in 1998. The 40 ton vibrator was placed on the surface. Recordings were made at the Novosibirsk seismic station. The source-receiver distance is 49.1 km; S-N = south-north component; EW = east-west-component; Z = vertical component. Day, month, and year are indicated on the right.

does not in practice depend on a season) is influenced by freezing and thawing processes of the ground under a source. The most likely reason for seismograms showing seasonal changes is the dependence of the directional diagram on the freezing and thawing of ground under the source. In terms of significant features, the dynamic characteristics of waves should be used for waves that are stable over time; for the remaining waves, an adjustment in

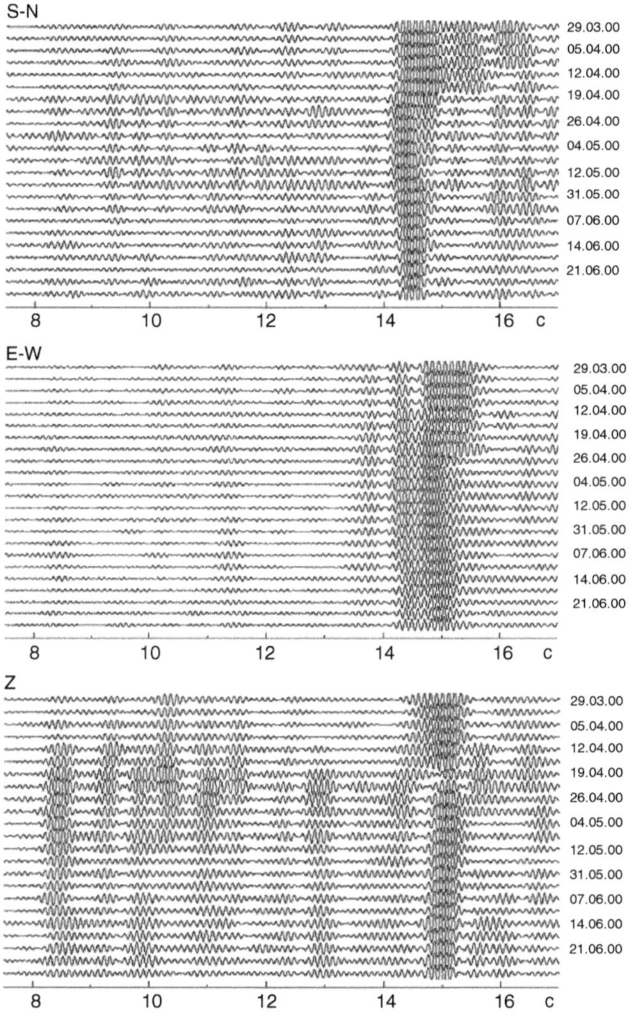

Figure 7 Vibroseismic monitoring of changes from winter to summer in 2000. The 40 ton vibrator was placed in an underground pavilion. Recordings were made at the Novosibirsk seismic station. Source-receiver interval was 49.1 km. S-N = north-south component; E-W = east-west-component; and Z = vertical component. Day, month, and year are indicated on the right.

dynamic characteristics, eliminating periodic seasonal changes, is required. The presented data allow us to study how wave dynamics change over time and to account for these variations in our interpretations.

Figure 7 presents correlograms of the transitional period from winter to summer for the source located in the underground pavilion. The source works on thawed ground in the summer and winter. However, we notice the influence of both frost penetration processes and ground thawing on this source. The differences between winter and summer seismograms are still apparent.

Notice that, except for the basic harmonic, the vibrating source radiated over the entire set of multiple harmonics. These harmonics have one spectral structure; semi-multiple harmonics have another. For surface vibrators, semi-multiple harmonics are narrow-band and are equally susceptible to frost penetration into, and thawing of, the ground.

By placing the vibrator in an underground pavilion, we stabilized its radiation on semi-multiple harmonics, and this radiation becomes the same both in summer and in winter. The basic and multiple harmonics continue to depend on ground frost penetration and thawing processes.

3. Areal Vibroseismic Monitoring

Areal (regional) monitoring was conducted in the Baikal rift zone. Whereas in the above-described experiment, we studied modifications in wave fields over time using fixed source-receiver positions, at Baikal we recorded stable source radiation simultaneously for different receiver locations. It was noted during the recording of vibrational signals that in different (approximately equidistant from a source) locations, different durations for vibrosessions were necessary to obtain seismograms of the same signal/noise ratio. This research was conducted to compare wave characteristics at different distances from the vibrator, using the same level of radiation, to determine the relationship between the dynamic parameters of waves and the fault tectonics of the region.

The observation scheme is shown in Figure 8. The vibrator was located on the outskirts of Babushkin, a village 1 km from the Lake Baikal shore. The first profile of observations, with about a 20 km pitch between recording points, crosses zones of deep faults in the Khamar-Daban Ridge and Gusinoozyersk Valley. The last recording point on this profile is in the area of Khar'yasta, 125 km from the vibrator. In addition, we also recorded vibroseismic signals near the village of Gurulba, 93 km from the vibrator. As opposed to the Babushkin-Khar'yasto profile, which goes across a fault line, the Babushkin-Gurulba profile covers a stable block of the Earth's crust, bypassing fault zones.

For recording, we used the Bereza digital correlation seismic station (Geza and Yushin, 1981), which, after completion of a vibrosession, gave real-time output of the impulse seismograms. A number of findings along and across tectonic structures resulted from the use of the Bereza seismic station, mounted in a cross-country vehicle. At all recording locations, we obtained seismograms (with good signal/noise ratios) in the vibrator frequency band of 6.25-11 Hz. For all seismograms derived from the

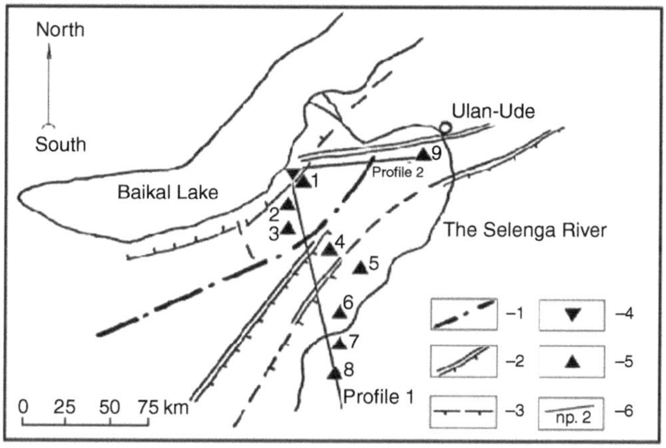

Figure 8 Location map and neotectonic framework: (1) boundaries of areas with different ages of neotectonic activity; (2) major faults; (3) normal faults; (4) vibrators; (5) seismic stations (note that digits near triangles are distances from vibrator, in kilometers); (6) = profiles: Pr. 1 is the Babushkin-Kharyasta line, Pr. 2 is the Babushkin-Gurul'ba line.

experiment, the first arrivals were longitudinal refracted P-waves with recorded crustal penetration depths of up to 10 km. Figure 9 presents the amplitude spectra of the P-wave at different observation *p*.

The principal feature of the recorded data is that the spectrum width of the longitudinal refracted P-wave is different at various observation points for the same source signals. At a 7 km distance from the vibrator, the intensive components exist on all radiation frequencies. At a 20 km distance, only the low-frequency components of the spectrum remain in the spectra. This point is located immediately behind a deep fault oriented along the Khamar-Daban ridge. At the subsequent recording location (38 km), the spectrum of the refracted P-wave extends again. Moreover, the spectral analysis shows that the waves reflected from the crustal horizon, and recorded at a distance of 20 km, have a wide band spectrum after arrival of the head P-wave. Under such conditions, propagation of waves in the Earth's crust can occur if there is a local zone within a deep fault at depths up to 5 km, with an anomalous absorption of high frequencies. Such conditions ensure an anomalous low-frequency spectrum in refracted waves 20 km away, with no effect on the spectra of refracted waves at greater distances because of the propagation of seismic rays deeper than the fault's depth. Further displacement of the refracted-wave spectrum to low frequencies is noted at the recording station in the Gusinozyorsk Valley, beyond the zones of deep faults. (See the spectrum of P-wave at points located at distances of 85, 98, and 125 km.) But for all that, it is remarkable that the loss of high frequencies occurs not only in the P-wave, but in waves arriving afterwards. This fact indicates that these deep faults have an increased absorption of seismic waves

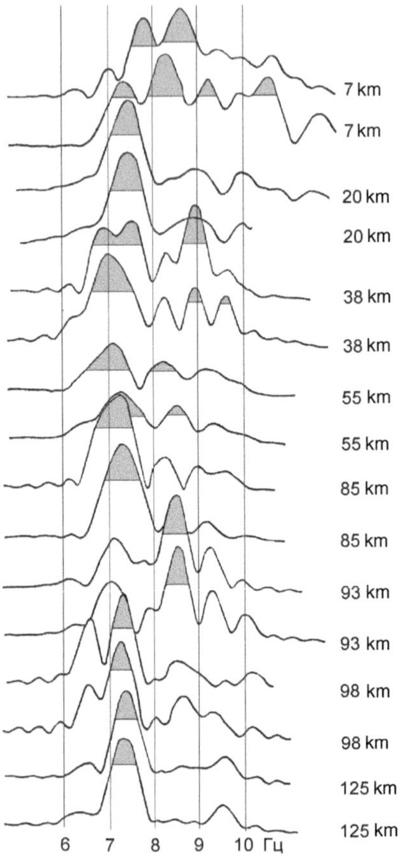

Figure 9 Amplitude spectra of a wave refracted in the crust, at different offset distances.

over the entire range of depths. The spectrum of the refracted P-wave 93 km away (Gurulba) appears to be essentially wide band. In this case, the ray goes through a stable block of the Earth's crust, passing through zones of faults with an increased absorption of seismic waves.

Using the data from these experiments, we made a quantitative estimation of differences in seismic-wave absorption across the fault zones and inside the stable block, applying an established method (Berzon et al., 1962). For the Babushkino-Khar'yasta profile, the average absorption decrement for longitudinal waves was evaluated as 0.039, and the value of Q-factor as $Q_p = 82$. For the Babushkin-Gurulba direction, the average value of absorption decrement was 0.02; accordingly, the value of the Q-factor was $Q_p = 157$.

According to reference data, the absorption decrement for different rocks, conditions of formation, and depths of occurrence lie within rather broad limits. From laboratory measurements (at $P = 400$ kg/cm^2), the decrement of absorption for granites, and gneissose granite, is in the 0.03-0.07 range (Gamburtsev, 1992). From other sources, the decrement of absorption for the rock fundaments under natural bedding conditions is 0.03-0.4 for alluvial soils and 0.001-0.05 for consolidated materials (Myachkin, 1978). The obtained value of the absorption decrement is close to the conditions of the changed foundations, possessing a reduced Q-factor in comparison to consolidated rocks. The anomalous attenuation could possibly be produced by a faulting zone of crystalline rocks within the survey section. As a whole, the obtained value of Q-factor, $Q_P = 157$, is close to Q_P values obtained by investigators for the top part of the Earth's crust in various regions of Siberia and the Far East.

4. Discussion

Resonating vibrators generate powerful seismic waves in rock, with the ability to generate seismograms over distances as much as 300-400 km, enabling experiments involving vibroseismic monitoring of the Earth's crust and the study of its velocity structure.

Because of the nonlinear effects from vibration excitation, except for the first harmonic, a train of multiple and semi-multiple harmonics (with factors 1/2, 3/2; 2; 5/2; 3; etc.) radiates into the environment. Use of multiple harmonics extends the capabilities of scientific investigation.

The high repeatability of seismograms generated under the same weather conditions makes possible the use of high-power vibrators for vibromonitoring, with the goal of studying small changes in the Earth's crust caused by tectonic processes. However, ground freezing has a profound effect on the dynamic characteristics of recorded wavefields. Placing a vibrator in a warm underground pavilion during the winter does not eliminate these seasonal changes in wavefields. On the other hand, semi-multiple harmonics radiated from such sources become insensitive to the ground's freezing or thawing. The fundamental and integer-multiple harmonics are stable across the spectrum, but changes in wavefield dynamics occur, owing to seasonal changes in the directional diagram of the source.

An experimentally determined strict seasonal recurrence of wavefields, year after year, may serve as the physical foundation for removing the seasonal effect of freezing/thawing processes, as a result of seismic monitoring in regions with severe climates.

The high repeatability of vibrators gives scientists the ability to study changes in seismic wave attenuation in space. Revealed differences in crust attenuation along the length and breadth of a fault zone show that active monitoring using powerful vibrators is a promising technology for studying the attenuation characteristics of the Earth's crust.

5. Conclusions

It has been found that the processes of freezing and thawing of the ground near a seismic vibrator changes the frequency characteristics of radiated oscillations and the directional diagram of the source. A cyclic seasonal change in the dynamics of seismic waves is observed, compared with its ideal repetition for the same state of ground conditions in different years.

In various experiments, powerful seismic vibrators excited a series of multiple and semi-multiple harmonics. The semi-multiple harmonics change their characteristics with the freezing and thawing of the ground they travel through, and are found to be more sensitive to seasonal changes in the ground than multiple harmonics. Fully multiple harmonics change according to their degree of multiplicity—the higher the multiplicity, the greater the sensitivity to ground freezing and thawing.

A vibrator placed in an underground pavilion (10 × 12 m) where the ground does not freeze produces first and fully multiple harmonics that become stable only by the source spectrum, while the directional diagram of the source continues to undergo seasonal changes. On the other hand, semi-multiple harmonic components cease to be affected by freezing and thawing of the ground.

From the data processing of areal recordings, the existence of anomalies (up to a factor of 2) in the attenuation of seismic waves in Khamar-Daban ridge (Baikal rift zone) has been detected.

REFERENCES

Alekseev, A.S., Chichinin, I.S., Korneev, V.A., 2005. Powerful low-frequency vibrators for active seismology. Bulletin of the Seismological Society of America 95, 1–17.

Alekseev, A.S., Ryashentsev, N.P., Chichinin, I.S., 1982. How to look deep into the planet. Science in the USSR 3, 30–37 (in Russian).

Berzon, I.S., Epinatieva, A.M., Pariyschaya, G.N., Starodubrovskaya, S.P., 1962. Dynamic Performance of Seismic Waves in Real Mediums. Publishing House of AS USSR, Moscow, 490 pp. (in Russian).

Emanov, A.F., Kuzmenko, A.P., Mokshanov, M.A., Popov, U.P., Seleznev, V.S., Yushin, V.I., Danilov, I.A., Bakh, A.A., 1989. Experiments on vibration studying of the Earth's crust. In: Working Out and Research of Nonexplosive Sources of Seismic Signals. VNIIOENG, Moscow, pp. 48–53 (in Russian).

Emanov, A.F., Kuzmenko, A.P., Seleznev, V.S., 1986. Outcomes of studying of wave field from powerful centrifugal vibrator. In: Radiation and Recording of Vibroseismic Signals. IGG SB AS USSR, Novosibirsk, pp. 105–120 (in Russian).

Gamburtsev, A.G., 1992. Seismic Monitoring of Lithosphere. Nauka Publishing House, Moscow, 200 pp. (in Russian).

Gamburtsev, A.G., Alexandrov, S.I., Belyakov, A.S., 1994. Atlas of time variations of natural processes, V. 1. In: Harmony and Chaos in Lithosphere and Other Spheres. UIPhE RAS, Moscow, 176 pp. (in Russian).

Geza, N.I., Yushin, V.I., 1981. About some features of correlation accumulation of weak vibroseismic signals. In: The Problems of Vibration Seismic Sounding. IGG SB AS USSR, Novosibirsk, pp. 99–111 (in Russian).

Gluchov, Yu.I., Solodilov, L.N., Yasyulevich, N.N., Semeikin, N.P., 1994. Digital recorder of seismic signals RSS Alpha-Geon: Prospecting and protection of the Earth's interior, N. 10, pp. 38–40 (in Russian).

Myachkin, V.I., 1978. Processes of Earthquake Development. Nauka Publishing House, Moscow, 232 pp. (in Russian).

Myachkin, V.I., Dolbilkina, N.A., Solovieva, R.P. (Eds.), 1982. Seismic Studying of Focal Zones. Nauka Publishing House, Moscow, Moscow, IPhE AS USSR, 194 pp. (in Russian).

Nickolaev, A.V., Artyushkov, E.V., Chichinin, I.S., Galkin, I.N., 1974. Vibration studying of the Earth, Moscow, VINITY, dep. N. 2549-74 (in Russian).

Sadovsky, M.A., Bolhovitinov, M.G., Pisarenko, V.F., 1987. Deformability of Geophysical Medium and Seismic Process. Nauka Publishing House, Moscow, 100 pp. (in Russian).

Seismic Studying of Focal Zones, 1983, Moscow, Nauka Publishing House, 275 pp. (in Russian).

Solodilov, L.N., Seleznev, V.S., Emanov, A.F., Yasyulevich, N.N., Semeikin, N.P., 1995. Apparatus-technical system for DSS and joint network of seismological observations and vibroseismic monitoring: Prospecting and protection of the Earth's interior, N. 5, pp. 37–39 (in Russian).

Yushin, V.I., Geza, N.I., En Din, Yun, 1981. About experimental appraisal of possibility of correlation accumulation of vibroseismic signals for DSS purposes. Geology and Geophysics 8, 123–126 (in Russian).

Zubkov, S.I., Solovieva, R.P., 1986. Catalogue of Seismic Precursors of Earthquakes. Nauka Publishing House, Moscow, 267 pp. (in Russian).

CHAPTER 28

Estimates of Crustal Stress Sensitivity in the Lake Baikal Region, from Active Vibroseismic Monitoring and Solid Earth Tide Data

Vyacheslav I. Yushin[a] *and* Nikolay I. Geza[a]

Contents

1. Introduction

The problem of monitoring variations in seismic wave propagation has arisen in seismology in relation to earthquake prediction, which, according to Stacey (1969), means "a search for appropriate methods of elastic stresses identification". The dependence of wave velocities on stresses inside a rock massif is established, but the subtlety of the expected effects requires special approaches to monitoring and data interpretation. There are two similar parameters proposed to estimate the sensitivity of seismic velocities

[a] Trofimuk Institute of Petroleum Geology and Geophysics SB RAS, prosp. Akad. Koptyuga, 3, Novosibirsk, 630090, Russia

Handbook of Geophysical Exploration: Seismic Exploration, Volume 40
ISSN 0950-1401, DOI: 10.1016/S0950-1401(10)04033-4

to internal stresses within a rock *in situ* (De Fazio et al., 1973; Reasenberg and Aki, 1974; Bungum et al., 1981)

$$\nu = \frac{\Delta V}{V \cdot \Delta P}, \quad \text{or} \quad \nu = \frac{\Delta V}{V \cdot \Delta P} \cdot 100\ [\%/\text{bar}], \tag{1}$$

where V is the velocity as a function of lithostatic pressure (stress) P. Whereas Western scientists usually use the parameter ν, in the Russian scientific literature, a different nonlinear parameter K is used, determined by the nondimensional equation (Gushchin and Shalashov, 1981; Nikolaev, 1987; Nazarov, 1991; Vasiliev et al., 1987; Gamburtsev, 1992)

$$K = \rho \cdot V \cdot \frac{\Delta V}{\Delta P}, \tag{2}$$

where ρ is density. These two parameters are related through the equation

$$K = \nu \cdot \rho \cdot V^2. \tag{3}$$

Inasmuch as ν is often estimated in bars, and relative velocity variations $\Delta V/V$ are sometimes expressed in percentages, the relationship (3) can involve a coefficient 10^{-7}:

$$K = 10^{-7} \cdot \nu[\%/\text{bar}] \cdot \rho \cdot V^2. \tag{4}$$

The parameters υ and K express the nonlinearity of the geophysical medium, which does not cause a loss of wave energy. At the same time, absorption can also be nonlinear depending on wave amplitude. Nazarov (1991) proposed to distinguish the two types of nonlinearity as reactive and dissipative, respectively.

The velocity-stress relationships for different rocks, for pressures up to 30 kbar (corresponding to 100 km depth), have been clearly established from laboratory experiments (Clark, 1966; Christensen, 1974) and have been in general confirmed by deep seismic sounding. However, the issue of absolute critical stress remains open to discussion, because the proposed estimates differ by orders of magnitude, from 10 to 1000 bar (Stacey, 1969; Press and Brace, 1968). Hence, there is another important quantitative relationship between observed velocity variation and the related *in situ* stress change. In this respect, studies of seismic velocity variations related to solid Earth tides can be a helpful tool, with the latter serving as a natural calibrator of tectonic stress.

The known theoretical feasibility of detecting tidal effects by seismic methods involves the following (De Fazio et al., 1973; Reasenberg and Aki, 1974; Bungum et al., 1981). Gravity variations $\Delta g/g$ of 2×10^{-7} cause a crustal strain $\Delta l/l$ on the order of 2×10^{-8}. The empirical mean rigidity μ of the consolidated crustis about $5 \cdot 10^5$ bar. Thus, the peak-to-peak amplitude of tidal stress changes in the crust $\Delta P = \mu \Delta l/l$ is about 0.01 bar. (Stacey (1972) assumes 0.05 bar.) To estimate the tidal component of velocity changes, $\Delta V/V$, we must know the stress sensitivity of the seismic velocity (Eq. (1)), which is known to vary functionally over a broad range of rock lithology, fracturing, and depth-dependent lithostatic pressure (300 bar/km). The maximum ν shown in laboratory tests is 0.1%/bar, attributed to samples containing cracks. These conditions are restricted *in situ* to depths as shallow as a few tens of meters. Laboratory tests of rock-sample triaxial compression under pressures from 0.1 to 12 kbar (corresponding to depths from 0.3 to 40 km) showed ν varying from 0.01 to 0.001%/bar, and even smaller.

An integral estimate for velocity stress sensitivity of the crust can also be obtained independently of laboratory experiments, solely from seismic data if seismic velocities at different depths are known. If Eq. (1) is rewritten as

$$\frac{\Delta V}{V(P)} = \nu \cdot \Delta P \tag{5}$$

and assuming $\nu = \tilde{\nu} = \text{const}$, where $\tilde{\nu}$ is the mean value, obtained by integrating this equation (in the sense of Stieltjes), we derive the mean stress sensitivity of velocity for the given crustal thickness:

$$\tilde{\nu} = \frac{\ln V_2 - \ln V_1}{P_2 - P_1}, \tag{6}$$

where V_1, V_2, P_1, P_2 are velocities and pressures for the surface and deep reflector, respectively.

As an example (which we will need below), let us use this method to estimate the integral velocity-stress sensitivity of the crust for the wave reflected from the Moho boundary in the area of the San Andreas Fault (California). According to Feng and McEvilly (1983), the P-wave velocity V_1 in the near-surface layer is about 3 km/s, and the boundary velocity V_2 at the Moho boundary ($H = 25$ km, $P_2 = 7500$ bar) is 7.85 km/s. Substituting these values into (6) and assuming that $P_1 = 1$ bar, we obtain $\upsilon \approx 0.013\%/\text{bar}$ ($K = 130$).

For the Baikal region, where $V_1 = 3$ km/s, $V_2 = 8.1$ km/s, $H_2 = 40$ km, and $P_2 = 12{,}000$ bar (Puzyrev, 1981), the same value ν estimated by Eq. (6) is about 0.008%/bar, or $K = 75$. Hence, the tidal travel-time change $\Delta t/t = \Delta V/V$ of a wave reflected from the Moho must be, according to Eq. (5), about 1.3×10^{-6} and 0.8×10^{-6} for California and Baikal, respectively. Such subtle changes can be detected at a resolution of at least 0.2×10^{-6}. A lower resolution is acceptable for shallower depths, where ν can attain 0.1%/bar, but it should be above 10^{-5} in any case

(Reasenberg and Aki, 1974; Bungum et al., 1981). Given that the available resolution of seismic methods is no better than 10^{-4}; seismic detection of tidal effects would be thought impossible. However, widely published experimental results (De Fazio et al., 1973; Reasenberg and Aki, 1974; Rykunov et al., 1981; Bungum et al., 1981; Gamburtsev, 1992; Soloviev, 1993; Akhiyarov et al., 1980) provide evidence that such detection is possible. Here we consider three papers that (in our view) are the most reliable, and that base their findings on high-quality rigorous experiments.

The experiment described in De Fazio et al. (1973) was performed along a roughly vertical shaft of a mine cut in coarse-grained calcite (marble). Seismic signals were generated by a hydraulic shaker, operated at a quartz-normalized frequency of 500 Hz. Measurements were taken of the phase difference between two geophones—the remote geophone 30 m below the surface and the reference geophone separated by about 300 m along the mine shaft. The shaker was placed 37 m further down from the reference geophone. Waveform analyses showed that the principal seismic mode was represented by a Rayleigh wave-like tunnel wave with a phase velocity of 3000 m/s. The experiment lasted 37 h and recorded a periodic phase change corresponding, as calculated, to a velocity change of $\Delta V/V = 10^{-3}$, at a precision on the order of 10^{-4}. A comparison of the observed velocity changes with the actual earth-tidal strain in three horizontal directions showed good correlation in one azimuth. De Fazio et al. (1973) found this correlation satisfactory, although their plots show little reason for optimism because of the striking discord between the main periods of velocity and strain changes—noted by the authors themselves. Such a substantial velocity contrast, if it really were caused by a solid Earth tide, would correspond to $v = 10\%/$ or $K = 20{,}000$ according to Eqs (1) and (3). At the same time, inasmuch as the velocity stress sensitivity v from laboratory tests on similar rock is in the range of 0.01 to 0.1%/bar at pressures under 100 bar, $\Delta V/V$ must be 2 to 3 orders of magnitude lower than observed in that experiment. De Fazio et al. (1973) explain the discrepancy by the presence of much larger cracks in the rock *in situ*, which calls into doubt any direct extrapolation from laboratory tests using small samples.

In another experiment (Reasenberg and Aki, 1974) performed in a granite quarry, the signals were generated by an air gun suspended in a water-filled cylindrical hole and were received by an array of geophones at a distance of 200 m. Measurements were taken at the time interval between the onset of the pulse at a reference near-field accelerometer and the first arrival of the wave at the geophone. The phase velocity of the direct wave is 1.2 km/s; the travel time is 170 ms. An average of over 300 shots for each seismogram yielded a standard deviation of 3×10^{-4}. The experiment revealed a visible periodic travel-time variation with a double peak-to-peak amplitude of about 1 ms, which corresponds to a velocity change of $\Delta V/V \sim 5 \times 10^{-3}$ (much greater than in the previous experiment).

Taking into account the observed strain, which likewise turned out to be greater, Reasenberg and Aki estimated the *in situ* velocity stress sensitivity as 20%/bar.

The two experiments were restricted to the shallow crust, but were carried out in crystalline rock typical of deeper crust as well, and thus are of interest. As far as the true crustal variations are concerned, the majority of known publications (see for instance, Gamburtsev, 1992; Popandopulo, 1982) deal with processing of earthquake records based on inaccurate raw data, and the results differ strongly from those obtained by more precise active methods. A surprising reverse relationship between the amplitude of observed velocity changes and the precision was noted by Soloviev (1993), who proposed considering these estimates as strictly boundary values representing the detection limit of seismic methods.

In this respect, we should mention the paper by Clymer and McEvilly (1981) concerning long-term travel-time monitoring of reflected waves, including Moho reflections. The paper contains plots of two experiments, two days long each, with combined strain and travel-time records, showing no evident correlation between the two processes. Clymer and McEvilly (1981) mentioned that "an apparent response of the travel time to the solid earth tide was observed during the first three... experiments, which will be discussed in a separate paper." However, as Dr. McEvilly informed us, they cancelled the promised publication, after realizing that what in fact they observed were the shallow effects. Inasmuch as the experiments focused on diurnal travel-time trends for waves reflected from the Moho, we tried to draw boundary information from them. The first one coincided in time with a nearly maximum tide (0.22 mGal), so there was no reason to expect greater tide effects in other experiments. It is reasonable to assume that the hypothetical tide effect in those three other experiments may have been detected because of low noise levels. As a result, the upper limit of tidal effects can be determined from maximum travel-time differences recorded in those "failed" experiments (about 3 ms at a travel time of 8 seconds) and estimated at 0.4×10^{-3}. Assuming that the tidal stress change ΔP is 0.01 bar, the N obtained for California from a wave ray reflected from the Moho should (according to Eq. (1)) be within 4%/bar. This result contrasts sharply with the indirect estimate of 0.013%/bar. However, taking into consideration the private explanation by Clymer and McEvilly (1981), we have to evaluate their result as one in which no tidal effects were observed in the consolidated crust, and the estimate of 4%/bar should be considered as a boundary value for instrumental and methodological estimations.

In October 1991, we ran a two-week active monitoring test in a seismic area of the southern Baikal, to study tidal effects on seismic velocities in the consolidated crust. Some preliminary results were published in Yushin et al. (1994). Repeated vibroseismic soundings were carried out along a 125.5 km long profile from the southeastern side of Lake Baikal (town of Babushkin) northwestward to the village of Kharat (Figure 1), with the use of

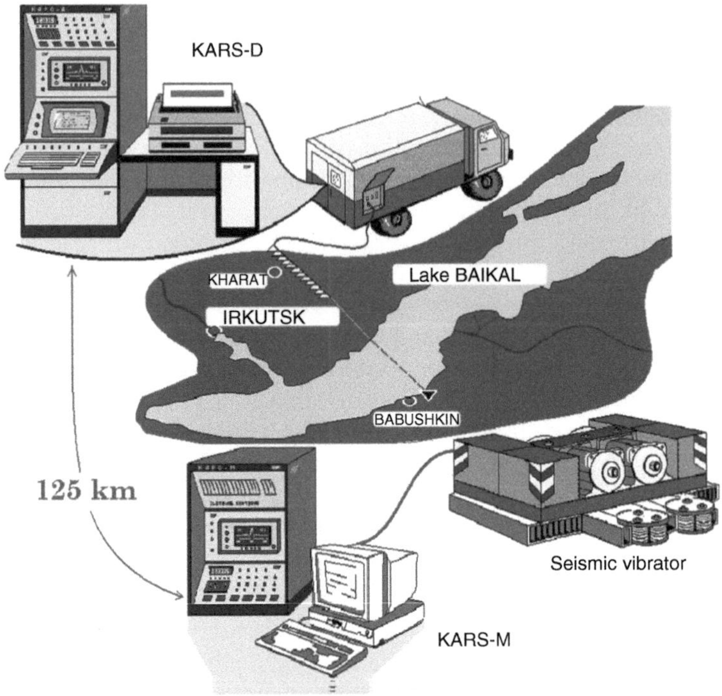

Figure 1 Active vibroseismic monitoring setup at Lake Baikal. KARS-M: precision control system of the powerful vibrator. KARS-D: field center of far registration and seismic information processing enables correlation transformation of previously optimized nonlinear frequency modulated signals. Track of examining Babushkin-Kharat: 125 km.

an eccentric 100 ton vibrator (Alekseev et al., 2005) operated at frequencies from 5 to 10 Hz. Surveys with stationary, superheavy vibrators were carried out using optimized nonlinear sweeps (Geza, 1998).

No tidal effects were detected, even with the high-precision equipment (exceeding the known capabilities of previous work). The present paper is a result of more detailed processing of a large data volume from that experiment, in search of more precise boundary estimates of nonlinear parameters and the stress sensitivity of seismic velocity in the crust of a highly active, hazardous seismic region.

2. Data Processing and Interpretation

Four typical seismograms obtained by repeated soundings are shown in Figure 2. Each seismogram is a result of a real-time direct correlation of vibroseismic signals during acquisition. The signals were received by a 12-channel inline geophone array oriented azimuthally toward the source.

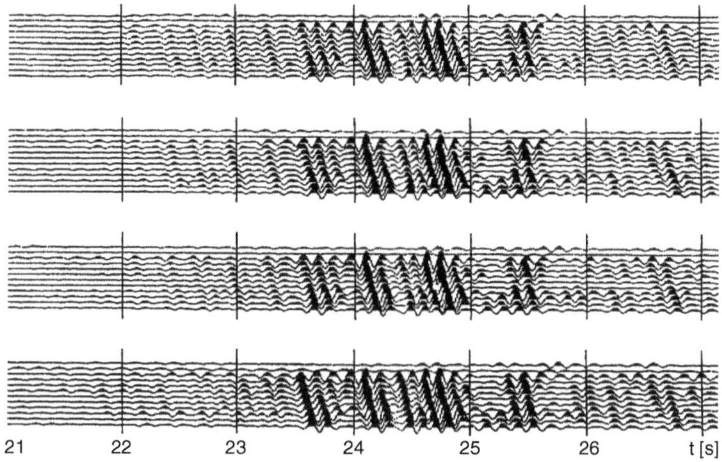

Figure 2 Four typical correlated records of active vibroseismic monitoring. Each seismogram results from a convolution of a 20 min optimized nonlinear sweep signal. The in-line seismic array points to the vibrator and includes 12 channels recording 11 cross-line geophone arrays each. The step between channels is 100 m. The upper two channels are X and Y components respectively, the other nine are Z components. These seismograms illustrate the typical quality of repeated and stacked records.

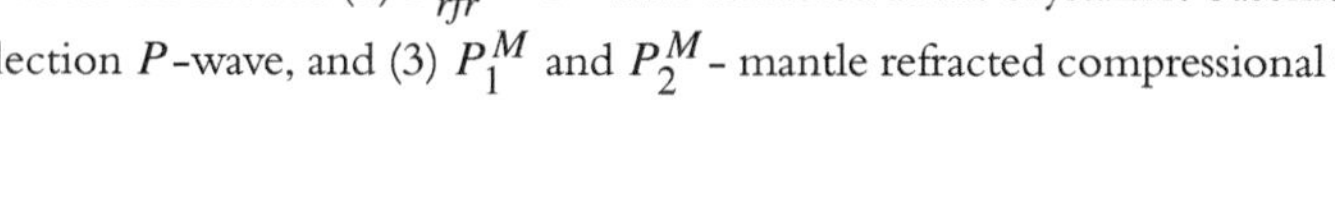

Figure 3 A seismogram cleared of additive noises by stacking 16 separate seismograms. The lowest line shows the graphic gain level, which decreases 10 times from 22 to 23 s. Three identified waves are shown: (1) P^{Φ}_{rfr} – P-wave refracted at the crystalline basement (2) P^{M}_{ref}- Moho reflection P-wave, and (3) P^{M}_{1} and P^{M}_{2} - mantle refracted compressional waves.

The upper three channels recorded respectively X, Y, and Z components of a three-component geophone. The other nine channels recorded Z components.

The geophones were arranged into clusters of 11 for each channel and spaced at 10 m within a group. The clusters were oriented across the profile and spaced at 100 m (distance between the centers of clusters). As can be seen in Figure 2, the differences in these seismograms can be observed only by very close inspection. After visual analyses, 60 seismograms were selected for statistical analysis. In spite of a fairly high signal-to-noise ratio for individual seismograms, the interpretation was performed using a seismogram specially filtered from any incoherent noise obtained as a stack sum of 16 records. This seismogram is shown in Figure 3.

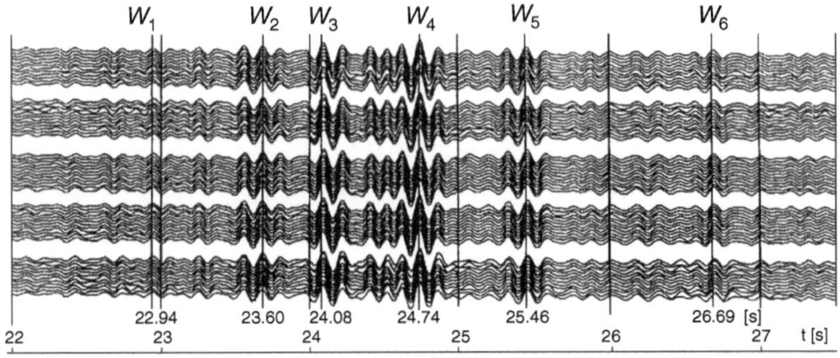

Figure 4 The calendar gather of seismograms from 10-day-long monitoring, illustrating the data quality. Each trace is the integrated result of one sounding session, derived from a seismogram like that shown in Figure 2 (above) by beam-forming the stack of nine Z-channels. W_1—wave refracted at the crystalline basement (P_{rfr}^{Φ}); W_2—Moho reflected P-wave (P_{ref}^{M}); W_3 and W_4—mantle refracted compressional waves (P_1^M and P_2^M); W_5 and W_6—refracted and repeated waves from Moho and upper mantle.

Interpretation involved comparison with DSS results for this region (Puzyrev, 1981). To enhance the signal-to-noise ratio, we transformed the obtained multichannel inline seismograms into a single-channel areal record by stacking over the 9 traces with identical transversally positioned Z-channel geophones. The channels were stacked with a 10 ms time shift, which corresponded to the orientation of the maximum radiation pattern with respect to the apparent velocity of 10 km/s (beam-forming array). Theoretically, this configuration enabled a three-fold increase in signal-to-noise ratio, and each stacked record resulted from the directed reception of a single sweep by a 800×100 m array of 99 geophones. A fragment from the seismogram gather of 60 stacked traces (Figure 4) covers six of the seven analyzed wavelets.

The true first arrival was from a low-amplitude P-wave refracted at the crystalline basement with a travel time of 21.6 s, which is visible only on the stacked seismogram in Figure 3. This wave is unusable for monitoring because of the poor signal-to-noise ratio (note, by the way, that the wave is barely detectable even if excited by explosions). The first arriving wave is followed by several refraction waves (one with a travel time of 22.94 s, labeled as W_1 in Figure 4 and included in the analyzed selection). The travel time of 23.6 s corresponds to the prominent arrival of a Moho reflection, labeled as P_{ref}^{M} (W_2), whose parameters can be analyzed using its maximum at the second dominant half-period.. The other five waves (W_3-W_7) included in the selection cannot be unambiguously interpreted, but are certainly of deep origin, as indicated by the high apparent velocities (8-9 km/s). The two most prominent ones (W_3 and W_4) are tentatively attributed to the mantle (P_1^M and P_2^M, respectively).

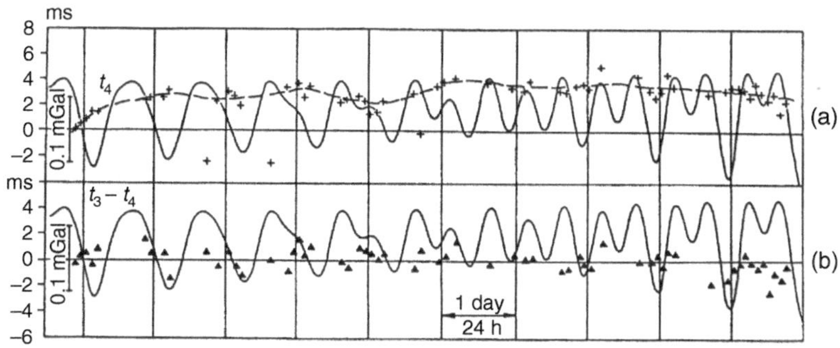

Figure 5 Deffered travel-time (markers) and tidal gravity variations (continuous line): reference wavelet W_4 (a); variations in deffered travel-time for wavelets W_3 and W_4 (b).

3. Measurements and Correction of Travel Time and Amplitude Variations

Plots of travel time and tidal strain variations obtained from a processed seismogram gather recorded during a 10-day monitoring period are shown in Figure 5, together with a theoretical plot of the tidal force. Below, we explain the plotting procedure.

Precise measurement of travel-time fluctuations recorded at different sessions was achieved by calculating the cross-phase spectra of the respective wavelet pulses. The wavelet amplitude was found as a maximum absolute discrete counting of a given pulse. Residual noise was estimated separately for each wavelet, as a standard deviation σ_n on the time interval before P-wave first arrivals. Since the source and receiver were automatically synchronized (by a special quartz timer of precision 5×10^{-9} s), their misfit is reflected on seismograms as an arrival shift common to all waves. The timer was repeatedly verified or adjusted against radio signals of the National Time Survey, and the respective corrections were introduced into the measured arrivals during data processing. Figure 5(a) shows travel-time variations (corrected for clock drifts at both the source and the receiver) for the reference wavelet W_4 and a theoretical plot of tidal force. Travel-time variations of the other six waves were studied as differences between their travel-times and the reference travel time t_4 of wave W_4 after subtraction of invariable components. Figure 5(b) shows most significant differential travel-time variations t_3-t_4. It is obvious that in this case, only wavelet W_4 bears time effects common to the entire seismogram.

These effects may be produced by synchronization errors remaining after data correction, by local rheological changes at the source and receiver areas (due, for instance, to weather effects on the ground) or, finally, by a tidal component shared by all wavelets. These components will not be evident

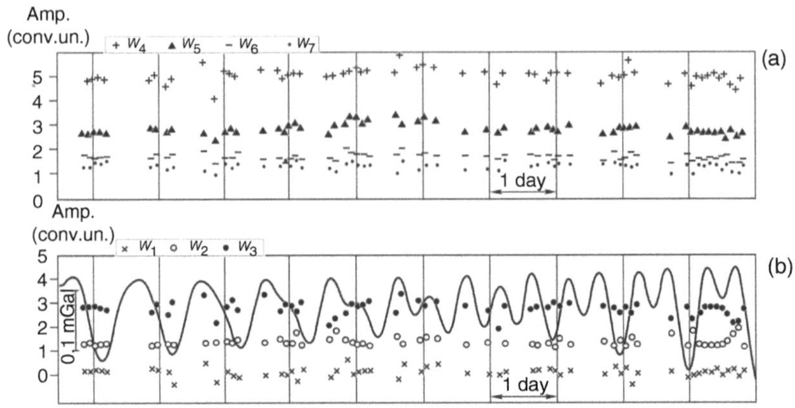

Figure 6 Amplitudes of seven different wavelets W_1-W_7 (markers) obtained during 10 days of measurements together with tidal gravity variations (continuous line). Amplitudes are in conventional units. The graph grid shows midnights.

in travel-time differences (t_i-t_4). However, since the rays of different waves do not follow the same paths, the tidal effects on them will be dissimilar, and thus the travel-time difference variations will bear individual features of these effects, in a clearer form than the variation t_4.

Apart from three anomalous arrivals of W_4 on October 15, 16, and 18, the time series t_4 involves a slow positive trend (dashed line in Figure 5(a)) against which random deviations of measured travel times are within 1 ms. Over the first day, the increment reached +3 ms and then stabilized at +4 ms with respect to the original state. This trend may in principle have tectonic causes, but is most likely produced by shallow rheological effects caused by weather and long-lasting compression of the soil under the vibrator. Isolated anomalies possibly produced by synchronization errors (since they have no correlation in adjacent records) are not evident in travel-time difference series, and residual noise at the respective seismograms was even lower than the average.

The amplitudes of the same seven wavelets are plotted in Figure 6. Note that unlike the travel-time plots, the amplitudes are shown as absolute rather than relative values.

4. Statistical Analysis of Wave Arrival and Amplitude Series: Boundary Estimates

4.1. Visual examination

As a rule, variations detected during continuous surveys require further analysis to reveal inherent relationships, latent periodicity, trends, etc. It is known from numerous psychophysical experiments that a harmonic signal with amplitude B can be visually detected against ambient noise of the same frequency with standard deviation σ, if the signal-to-noise ratio (on the basis of the mean square amplitude) is above 2:

$$\gamma = \frac{B}{\sqrt{2} \cdot \sigma} \geq 2. \tag{7}$$

In this limit case, the scattering σ_Y^2 of the observed values of series Y equals the total magnitude of the harmonic signal and the random noise

$$\sigma_Y^2 = \frac{B_2}{2} + \sigma^2. \tag{8}$$

Thus it follows, with regard to condition (7), that if a harmonic component cannot be distinguished by visual analysis, its amplitude does not exceed the value

$$B_{max} \leq \sqrt{\frac{8}{5}} \cdot \delta_Y \approx 1.26\sigma. \tag{9}$$

In particular, having subtracted the positive trend from the series $\{t_4\}$ (see Figure 5(a)), we obtain $\sigma_Y = \sigma_{t_4} = 0.5$ ms. Because no visible relationship with the tide effects is observed, we can state, according to (9), that the amplitude of the tidal component, if it exists, does not exceed

$$B_{max} \leq 1.26 \cdot 0.5 = 0.63 \text{ ms}, \tag{10}$$

which corresponds to a relative variation of W_4 phase velocity ($t = 24.74$ s)

$$\frac{\Delta V}{V_0} = \frac{B_{max}}{t_0} < 2.5 \cdot 10^{-5}; \tag{11}$$

smaller than $2.5 \cdot 10^{-3}\%$.

4.2. Analysis of random-scatter in travel times

Ambient noise in seismograms is the first physically evident cause of travel-time scatter. Its effect can be predicted using the known formula (Gonorovskii, 1986):

$$\sigma_t = \frac{\sigma_N}{2\sqrt{2}\pi f_0 A}, \tag{12}$$

where σ_t is the standard deviation of travel times for a given wavelet with amplitude A, σ_N is the standard deviation of ambient noise, and f_0 is the predominant signal frequency.

If a seismogram results from the correlation of a vibrosignal, as in the given case, then f_0 is, as a rule, the central frequency of the noise spectrum. Expression (12) is valid for signal-to-noise ratios over 10—the usual case, otherwise monitoring does not make sense.

Noise in a travel-time difference series $\{t_i\text{-}t_j\}$ is estimated by the formula

$$\sigma_{t'} = \frac{\sigma_N}{2 \cdot \sqrt{2} \cdot \pi \cdot f_0} \cdot \sqrt{\frac{1}{A_i^2} + \frac{1}{A_j^2} - \frac{2 \cdot r_N(t_i - t_j)}{A_i \cdot A_j}}, \tag{13}$$

where $r_N(t_i - t_j)$ is the background-normalized noise component. Given that $r_N < 1$, this expression is always valid. Comparing the obtained $\sigma_{t'}$ with σ_t calculated immediately from the series $\{t_i - t_j\}$, we can conclude that if $\sigma_t \approx \sigma_{t'}$, the scatter is chiefly caused by ambient noise; if $\sigma_t \gg \sigma_{t'}$, some other significant cause or causes should be investigated. The case when $\sigma_t < \sigma_{t'}$ is physically impossible and attests to a serious error in processing.

4.3. Revealing latent relationships using the correlation method

A correlation analysis reveals the relationship between two series or between a series and a known function obscured by random measurement errors. It is better suited for a discontinuous time series than for spectral analysis. To estimate the hypothetical tidal component in the variations of some parameters, we must consider the following statistical problem. Let X and Y be two functions of two other independent random variables Φ and N:

$$X = G \sin \Phi \tag{14}$$

$$Y = B \sin \Phi + N, \tag{15}$$

where G and B are constants. G is assumed known (amplitude of gravity variations), N (random error in measurements of some parameter Y of a seismic wave) has a normal distribution with standard deviation σ_N, and Φ is distributed uniformly over the 0-2π interval. We wish to estimate a constant B in a selection of n independent measurements (number of measurements taken during the monitoring period, or the length of the series). Physically, value Φ models a tidal phase, which is certainly not a random component. However, if we consider that measurements are taken at arbitrary moments of time, such an approach is valid, because this assumption does not improve the final estimate with respect to the real situation. The solution of this problem leads to the following result. The correlation coefficient of arbitrary values

X and Y is determined by the following expression:

$$r_{XY} = \frac{\gamma}{\sqrt{1+\gamma^2}}, \tag{16}$$

where γ is the amplitude signal-to-noise ratio in an observation series

$$\gamma = \frac{B}{\sqrt{2} \cdot \sigma_N}. \tag{17}$$

The relationship (16) enables us to evaluate the original signal-to-noise ratio (17) using the correlation coefficient between series X and Y:

$$\gamma = \frac{r_{XY}}{\sqrt{1 - r_{XY}^2}}, \tag{18}$$

where the true value r_{XY} can be replaced by its empirical estimate. Thus, the algorithm of statistical evaluation for the tidal component B includes the following steps:

1. calculation of the standard deviation of series Y;
2. calculation of r_{XY};
3. calculation of the initial signal-to-noise ratio γ by formula (18);
4. evaluation of B by formula obtained from (17), which leads to

$$B = \sqrt{2} \cdot \gamma \cdot \sigma_N, \tag{19}$$

where, since $\gamma \ll 1$, measurement errors σ_N can be replaced with the standard deviation σ_Y of series Y. The accuracy and stability of this estimate can be verified via the standard deviation σ_r of the empirical correlation coefficient, which is known (Livshits and Pugachev, 1963) to depend on the volume of selection n as

$$\sigma_r = \frac{1 - r_{XY}^2}{\sqrt{n}}. \tag{20}$$

The value of the constant G is in this case of no significance.

Let us now estimate the tidal component in travel-time change for the reference wavelet W_4 using these equations and real data. The highest correlation between tidal effect and variation t_4 (excluding the trend) was at $r^* = 0.17$, associated with a 4 h shift in the travel-time series $\{t_4\}$ relative to the tide curve. The rms error in this estimate was $\sigma_r = 0.12$, according to Eq. (20)—i.e., of low statistical significance. Nevertheless, assuming the empirical value $r^* = 0.17$ to be the most reliable, the respective signal-to-noise ratio $\gamma \approx r = 0.17$ and the amplitude of the tidal component by Eqs (18) and (19) is:

$$B_{max} < \sqrt{2} \cdot 0.17 \cdot 0.5 = 0.12 \text{ ms.} \tag{21}$$

Therefore, using an additional correlation analysis, we arrived at a detectable limit of the tidal component more than five times lower than its visible magnitude (Eq. (10)), with a travel-time relative value of 0.5×10^{-5}.

The obtained result should be considered in terms of probability. It indicates that if a correlation between travel time and tide does exist, it can be estimated by the above value. This is in agreement with observations, but the experimental data and its volume are insufficient for a more detailed conclusion. Moreover, the rough statistical estimates of the relationship between the two series appear too optimistic, since they ignore the true probability of data distribution remaining unattainable.

To estimate the probable natural scatter of the empirical correlation coefficient in the case when a series lacks any invariable component, we performed a numerical experiment on the correlation of independent arbitrary-number selections at $n = 60$, as in nature with the real tidal function. Some selections contained a shifted series with a correlation coefficient of up to 0.15-0.2. Consequently, the detectable limit of tidal-velocity variation is overestimated, in spite of the above formal value, and in reality, it must be much lower than 0.5×10^{-5}.

Estimates of the correlation between the travel-time differences $\{t_i\text{-}t_j\}$ of the other six wavelets and the tidal component did not show absolute values above that for $\{t_4\}$, and hence do not contradict the obtained upper boundary value.

4.4. Boundary estimate of velocity sensitivity to stresses

Let us find, according to (1), "visual" υ_1 and "statistical" υ_2 boundary estimates of the velocity stress sensitivity for the consolidated crust in a given region, using the obtained boundary estimates B_{max} (10) and (21) for the amplitude of the tidal component in travel-time variations, assuming $\Delta P = 0.01$ bar and $t_0 = 25$ s:

$$\nu_1 = \frac{\Delta V}{\Delta P \cdot V_0} < \frac{B_{max}}{\Delta P \cdot t_0} = 0.25 \cdot 10^{-2} \text{ bar}^{-1} = 0.25\%/\text{bar}, \tag{22}$$

$$\nu_2 < 0.05 \cdot 10^{-2} \text{ bar}^{-1} = 0.05\%/\text{bar}. \tag{23}$$

A comparison of the latter estimate with the implicit value of 0.008%/bar shows that positive detection of the tidal effect in variations of P-wave velocities requires about an order of magnitude higher precision. In addition to improving synchronization and extending the time series, the idea of investigating travel-time differences in P- and S-waves that are free from synchronization errors appears promising, but has not yet been achieved in our experiments.

4.5. Analysis of amplitude variations

Observed variations in the amplitudes of seven wavelets are shown in Figure 6. Mean and mean square amplitudes for each wavelet in the selection are given in Columns 4 and 5 of Table 1. Additionally, we determined noise scatter on wave-free segments (before first arrivals) for each seismogram $\sigma^2_{N_q}$ and the mean over all seismograms σ^2_N:

$$\sigma^2_N = \frac{1}{n} \cdot \sum_{q=1}^{n} \sigma^2_{N_q} = 10 \text{ c.u. (conventional units).} \tag{24}$$

It is seen from Table 1 that the amplitude scatter is everywhere above the mean noise (Eq. (24)) and obviously depends on the respective amplitudes. It indicates that the ambient noise is not the only cause of scatter, and that there is a component acting as a random magnifying coefficient. This component might be related to the tide, but with no obvious correlation between the amplitude variations and the tidal strain (Figure 6(b)), it is reasonable to check a simpler hypothesis first, that of vibrator instability. In this case, the statistical model of fluctuations can appear as the sum of the random magnifying component of amplitude change and ambient noise.

Consider the following: let amplitudes A_i and A_j of two wavelets fit the following statistical equations:

$$A_i = (1 + \delta) \cdot \overline{A}_i + N_i, \tag{25}$$

$$A_j = (1 + \delta) \cdot \overline{A}_j + N_j, \tag{26}$$

where $\overline{A}_i$ and $\overline{A}_j$ are mathematical expectations for amplitudes of W_i and W_j, and N_i and N_j represent random ambient noise at the moment of measuring amplitudes A_i and A_j, respectively. N_i and N_j have zero mean values, identical scatter σ^2_N, and correlation coefficients $r_N(i, j)$; δ is a nondimensional random variable with a zero mean value and a standard deviation $\Delta \ll 1$, which characterizes mutually dependent multiplicative fluctuation of amplitudes A_i and A_j. It is assumed that the variation δ is the same for all wavelets. The physical value $(1 + \delta)$ describes the component of amplitude changes common to all wavelets produced by random fluctuations of the sounding source amplitude. We wish to evaluate the standard deviation Δ (multiplicative instability of the amplitude).

Table 1 Statistical analysis of wavelet amplitude variations.

Wavelet characteristics				Wavelet amplitude		Coefficient of amplitude correlation between a pair of wavelets and its confidence interval[b]						
Symbol	Nature	Travel time, s	Mean amplitude	Amplitude scatter[a] rms	%	W_1	W_2	W_3	W_4	W_5	W_6	W_7
W_1	P^{Φ}_{rfr}	22.94	117	18	15	1	-	+[c]	+[c]	+[c]	+[c]	+[c]
W_2	P^{M}_{ref}	23.68	242	19	8		1	−0.56 ±0.27	-	-	-	−0.45 ±0.31
W_3	P^{M}_{1}	24.08	375	29	8			1	0.66 ±0.22	0.58 ±0.26	0.47 ±0.30	0.39 ±0.33
W_4	P^{M}_{2}	24.74	504	28	5.5				1	+[c]	0.71 ±0.49	+[c]
W_5	Refracted and repeated waves from Moho and upper mantle	25.46	281	21	7.5					1	0.43 ±0.31	0.40 ±0.32
W_6		26.69	167	13	8						1	+[c]
W_7		27.74	126	15	12							1

[a] Standard deviation of amplitudes is in absolute conventional units and in percents of mean amplitude of a wavelet.

[b] Confidence interval $\mathcal{E} = \pm 3(1 - r_{Wi,Wj})/\sqrt{60}$.

[c] + Marks only the sign of the correlation coefficient if its value is lower than confidence interval.

The correlation moment $R_A(i, j)$ between measured amplitudes of a wavelet pair represented by (25) and (26), with regard to these assumptions, will, after transformation, appear as follows:

$$R_A(i, j) = M \cdot [(A_i - \overline{A}_i) \cdot (A_j - \overline{A}_j)] = \Delta^2 \cdot \overline{A}_i \cdot \overline{A}_j + \sigma_N^2 \cdot r_N(i, j) \tag{27}$$

where $r_N(i, j)$ is the correlation coefficient of noise in one and the same seismogram, apparently equal to the background-normalized noise at a time shift equal to the time difference between arrivals of W_i and W_j. This function can be easily estimated from seismograms, but to a first approximation it can be replaced with the theoretical expression (Gonorovskii, 1986)

$$r_N(i, j) = \frac{\sin \pi F \cdot (t_i - t_j)}{\pi F \cdot (t_i - t_j)} \cdot \cos 2\pi \cdot f_0 \cdot (t_i - t_j), \tag{28}$$

where F and f_0 are sweep bandpass and mean frequency, respectively. In this case, $F = 4$ Hz and $f_0 = 7.5$ Hz. Substituting t_i and t_j into Eq. (28) for different wavelet pairs (Table 1, Column 3), we can estimate the respective noise correlation coefficients. If we express the standard deviation Δ from Eq. (27) and correlation moment $R_A(i, j)$ as $\sigma_i \cdot \sigma_j \cdot r_A(i, j)$, where σ_i, σ_j and $r_A(i, j)$ are the standard deviations and correlation coefficient of the measured amplitudes of two wavelets (A_i and A_j) respectively, we obtain:

$$\Delta = \sqrt{\frac{\sigma_i \cdot \sigma_j \cdot R_A(i, j) - \sigma_N^2 \cdot r_A(i, j)}{\overline{A}_i \cdot \overline{A}_j}}. \tag{29}$$

This expression enables us to estimate (within the limits of the assumed physical model) the sweep amplitude instability, by separating it from the ambient noise. The inequality deltas estimated from different wavelet pairs (i, j), and/or obtaining imaginary values will attest to the inadequacy of the chosen model.

The right half of Table 1 contains a matrix of amplitude pair-correlation coefficients for all seven wavelets with statistically significant coefficients (above the double standard deviation, which, according to Eq. (20), is 0.26). Some wavelets show fairly strong amplitude correlation, which appears to confirm the hypothetic model of their relationship through variations in vibratory amplitude. The level of wave relationship Δ, calculated by formula (29) is 5-6% for most wavelets. However, the statistically significant negative correlation between the W_2 amplitude and some wavelets is plaguing, and cannot be accounted for by the simple model presented by Eqs (25) and

(26). At present, we can explain it by the exotic mechanism of changing radiation patterns, which in fact can cause different effects on seismic rays. Actually, there are no physical reasons for amplitude variations in the unbalanced vibrator as a machine; the actual force is strictly related to sweep frequency, which is reproduced with an accuracy of 10^{-8}. All truly observable variations of sweep amplitude are undoubtedly related to external effects, either near field or remote. If we assume variability in the medium (caused by weather effects or by vibration itself), we may hypothesize that this influence affects both the intensity and pattern of seismic radiation.

5. Boundary Estimate of Wave Amplitude Sensitivity to Crust Stresses

We can obtain the upper limit of wave-amplitude stress sensitivity through a statistical analysis of observation series, assuming the probable existence of a relationship between changes in wavelet amplitudes and tidal strain. This approach is similar to that for wave velocity, and likewise involves two types of estimates: visual examination (a_1) and correlation analysis (a_2). For the visual estimate, we must know the standard deviations σ_A of both the wavelet amplitude change and mean wavelet amplitude. (Both values are given in Table 1.) Applying the same derivation as in Eq. (9), we find that if there is no visible relationship with the tidal strain G, the boundary estimate of the tidal effect on wavelet amplitude cannot be larger than

$$\Pi_1 < 2 \cdot \sqrt{\frac{2}{5}} \cdot \sigma_A = 1.26 \cdot \sigma_A, \tag{30}$$

where Π is the hypothetical tidal effect on amplitude A, and $\sigma_A = 28$ units for wavelet W_4 (see Table 1), waveform $\Pi_1 < 35$ units. A better constrained Π_2 can be obtained by analogy with Eq. (19) after calculating the correlation coefficient r_{GA} for a series of observed amplitudes A and tidal component G:

$$\Pi_2 < |r_{GA}| \cdot \sqrt{2} \cdot \sigma_A = |r_{GA}| \cdot \frac{\sqrt{5}}{2} \cdot \Pi_1. \tag{31}$$

The maximum value of r_{GA}, 0.2, was observed at a time shift of three hours (0.09 without the shift). According to (31), $\Pi_2 < 8$ units. By analogy with velocity stress sensitivity (Eq. (1)), the amplitude sensitivity can be estimated as

$$a = \frac{\Pi}{A \cdot \Delta P}. \tag{32}$$

Assuming that $\Delta P = 0.01$ bar and substituting Π in Eq. (32) with the obtained boundary estimates Π_1 and Π_2 and mean amplitude of the wavelet W_4 ($\overline{A}_4 = 504$), we find that $a_1 < 7$ bar^{-1} (700%/bar) according to visual estimate, and $a_2 < 1.6$ bar^{-1} (160%/bar) according to statistical estimate.

Unfortunately, we do not know the theoretical predictions for dissipative sensitivity (related to attenuation) to match with the obtained estimates. Thus, they should be considered as the attainable limit of instrumental detection for solid Earth tide effects on variations in seismic wavelet amplitudes, provided that the active monitoring is carried out by an eccentric vibrator, which is at present the most stable source.

6. Conclusions

1. Active vibroseismic monitoring was performed in the Lake Baikal region with the aim of testing a method for detecting stress and strain variations in the crust of hazardous seismic regions. The specific problem was determining the sensitivity of P-wave velocities in the consolidated crust to stress changes caused by the solid Earth tide.
2. A series of travel times and wave amplitudes reflected from the Moho boundary and uppermost mantle generated the basis for statistical and visual analysis of their relationship to gravity variations.
3. Velocity variations were measured to a precision of 2×10^{-5}, or nearly two orders of magnitude better than in earlier experiments by De Fazio, Reasenberg, Aki, etc.
4. The revealed variations in wave velocities and amplitudes are related to external surface effects and microseismic noise.
5. The absence of an obvious correlation with Earth tides permitted us to determine the upper limit of possible velocity stress sensitivity in the Earth crust, which proved to be 1.5 orders of magnitude lower than obtained by other investigators. It is 0.15%/bar, much closer to the results of laboratory tests on rock samples than other *in situ* studies have found.
6. Besides the usual additive (microseismic) noise and multiplicative (source force) fluctuations, the statistical analysis of wavelet amplitudes revealed an unexpected negative wavelet correlation, which can be physically explained by variability in the vibrator's radiation pattern.

REFERENCES

Alekseev, A.S., Chichinin, I.S., Korneev, V.A., 2005. Powerful low-frequency vibrators for active seismology. Bulletin of the Seismological Society of America 95, 1–17.

Akhiyarov, V.Kh., Petrosyan, L.G., Shimelevich, Yu.G., 1980. On changing of the geophysical features of temporary variations of high-frequency microseism in upper Earth lays under the action of tidal phenomena. Doklady Akademii Nauk SSSR 252 (3), 577–581 (in Russian).

Bungum, H., Yortenberg, E., Rizbo, T., 1981. Application of vibroseismic vibrations generated by hydroelectric dam for the study of variations of seismic velocities. In: Studies of the Earth by Non-Explosion Seismic Sources. Nauka, Moscow, pp. 248–259 (in Russian).

Christensen, N.I., 1974. Compressional wave velocities in possible mantle rocks to pressures 30 kilobars. Journal of Geophysical Research 79 (2), 407–412.

S. Clark (Ed.) 1966. Handbook of Physical Constants. Yale University, New Haven, CT.

Clymer, R.W., McEvilly, T.V., 1981. Travel-time monitoring with VIBROSEIS. Bulletin of the Seismological Society of America 71 (6), 1903–1907.

De Fazio, T.L., Aki, K., Alba, J., 1973. Solid Earth tide and observed change *in situ* seismic velocity. Journal of Geophysical Research 78 (8), 1319–1322.

Feng, R., McEvilly, T.V., 1983. Interpretation of seismic reflection profiling data for the structure of San Andreas fault zone. Bulletin of the Seismological Society of America 73, 1701–1720.

Gamburtsev, A.G., 1992. Seismic Monitoring of the Lithosphere. Nauka, Moscow. (in Russian).

Geza, N.I., 1998. Vibratory signal spectrum management with use of nonlinear sweeps (as example is Baikal rift zone monitoring). Distributed information processing. In: Proceedings of International Seminar, June 23–25, SB RAS, Novosibirsk, pp. 446–456.

Gonorovskii, I.S., 1986. Radiotechnical Circuits and Signals. Radio and Communication, Moscow. (in Russian).

Gushchin, V.V., Shalashov, G.M., 1981. On possibility of using the nonlinear seismic effects in problems of vibratory illumination of the earth. In: Studies of the Earth by Non-Explosion Seismic Sources. Nauka, Moscow, pp. 144–155 (in Russian).

Livshits, N.A., Pugachev, V.N., 1963. Probabilistic Analysis of Automatic Control Systems. Sovetskoe Radio, Moscow, 1. (in Russian).

Nazarov, V.Ye., 1991. Nonlinear damping a "sound on sound" in metals. Akusticheskii zhurnal 37 (6), 1177–1182.

Nikolaev, A.V., 1987. Problems of nonlinear seismics. In: Problems of Nonlinear Seismics. Nauka, Moscow, pp. 5–20 (in Russian).

Popandopulo, G.A., 1982. Actions of earth tides on velocities of seismic waves. Doklady Akademii Nauk SSSR 262 (3), 580–586.

Press, F., Brace, U.F., 1968. Development of earthquake prediction problem. In: Earthquake Prediction. Mir, Moscow, pp. 32–61 (Russian translation).

Puzyrev, N.N. (Ed.), 1981. The Baikal Interior (from seismic data). Nauka, Novosibirsk. (in Russian).

Reasenberg, P., Aki, K., 1974. A precise continuous measurements of seismic velocity for monitoring *in situ* stress. Journal of Geophysical Research 79 (2), 399–406.

Rykunov, L.N., Khavroshkin, O.B., Tsyplakov, V.V., 1981. Moon-solar tidal periodicity in spectrum-lines. In: Problems of Nonlinear Geophysics. VNIIYG, Moscow, pp. 109–112 (in Russian).

Soloviev, V.S., 1993. Temporary changing of the seismic wave velocity. Doklady Earth Sciences 329 (2), 166–168 (in Russian).

Stacey, F.D., 1969. Physics of the Earth, 2nd ed. John Wiley, New York.

Vasiliev, Yu.I., Vidmont, N.A., Gvozdev, A.A., et al., 1987. The direct measurements of seismo-radiation stress at the ground. In: Problems of Nonlinear Seismic. Nauka, Moscow, pp. 149–151 (in Russian).

Yushin, V.I., Geza, N.I., Velinsky, V.V., et al., 1994. Vibroseismic monitoring at the Baikal region. Journal of Earthquake Prediction Research 3, 119–134.

Author Index

SUBJECT INDEX

FSC
www.fsc.org
MIX
Papier | Fördert
gute Waldnutzung
FSC® C083411

Kolibri 360 GmbH
Ferdinand-Jühlke-Straße 7
99095 Erfurt, Deutschland
produktsicherheit@kolibri360.de